THE PRESENTATION AND SETTLEMENT OF CONTRACTORS' CLAIMS

THE PRESENTATION AND SETTLEMENT OF CONTRACTORS' CLAIMS

Second Edition

Geoffrey Trickey

FRICS, ACIArb

and

Mark Hackett

MA, MSc, FRICS, ACIArb

London and New York

First published 1983 by E & N Spon

This edition published 2001 by Spon Press
11 New Fetter Lane, London EC4P 4EE

Simultaneously published in the USA and Canada
by Spon Press
29 West 35th Street, New York, NY 10001

Spon Press is an imprint of the Taylor & Francis Group

© 2001 Geoffrey Trickey and Mark Hackett

Typeset in Sabon and Helvetica by Steven Gardiner Ltd
Printed and bound in Great Britain by TJ International Ltd, Padstow, Cornwall

British Library Cataloguing in Publication Data
A catalogue record for this book is available from the British Library

Library of Congress Cataloging in Publication Data
Trickey, Geoffrey.
Presentation and settlement of contractors' claims / Geoffrey Trickey and Mark Hackett. – 2nd ed.
p. cm.
Includes index.
1. Construction contracts – Great Britain. I. Hackett, Mark. II. Title.
KD1641.T74 2000
343.41′078624 – dc21   00-027912

ISBN 0 419 20500 4

CONTENTS

CONTENTS

Each chapter is sub-divided into sections: for example, Section 2.7 is Section 7 in Chapter 2. Each section is further sub-divided into numbered paragraphs (some of which may be grouped under a suitable heading): paragraph 2.7.13 is paragraph 13 within section 2.7.

PREFACE TO FIRST EDITION

'If a thing's worth doing its worth doing badly' so G. K. Chesterton wrote in *What's Wrong with the World*. I recognize that this quotation will give my critics an easy target. But it has become increasingly clear to me that one aspect of the design team's work in the construction industry which undermines the rest is the detailed evaluation of Contractors claims, a subject on which there are many questions and few answers. I therefore fall back on the above quotation for the necessary impetus to undertake this task.

I soon discovered, however, that the two greatest difficulties in writing a book on this subject are knowing how to start and where to stop. When preparing an outline synopsis of my proposals I was asked whether it would be written from the Employers' or the Contractors' standpoint or both, and therein lies the first problem, not to avoid the question – for the answer is neither – but to overcome the widely held view that the settlement of claims is a giant game of poker where nothing will beat a full house of fading memories.

Moreover, the word 'claim' in connection with building contracts is an emotive one. Purists will point out that the word does not exist in the Standard Forms issued by the Joint Contracts Tribunal dealt with in this book: pragmatists will reply that it is a word often used and widely understood.

Broadly this book is concerned with those occurrences during the progress of a building contract which potentially give rise to differences of opinion as to the Contractors', and to a lesser extent the Employers', monetary entitlement. It covers not only claims for loss due to disruption and delay, but also calculation of payments for fluctuations – changes in the price of labour or materials – and the effects of determination of the employment of the Contractor by either party to the Contract.

Lectures and seminars on the niceties of claims in building contracts abound but they mostly fall short of the final practical solution. My purpose then, is to examine those areas in the Contract which have a claim potential; to determine, so far as one is able, the rights and obligations of the parties in those circumstances; and to show by examples how the monetary entitlement can be ascertained. It is intended to be a practical book, culminating in a detailed example related to one fictional building contract. It should therefore assist both the student and the practitioner. The broad pattern is established by considering the following questions in sequence:

(a) What clauses of the Contract apply and how are they to be interpreted?
(b) What elements of cost are involved?
(c) What is the monetary entitlement?

In some instances the whole of this pattern can be traced within one chapter – for example Fluctuations – in others it has been necessary to review the contractual and practical background in greater depth prior to considering the rules for evaluation. For this reason claims for delay and disruption and the like span more than one chapter and deal in some detail with such related subjects as the calculation of extensions of time.

The text is not directed to any one profession. It is hoped that it will have an appeal to the Architect or Supervising Officer, who primarily has the duty to determine extensions of time and to ascertain loss and expense under the JCT Forms; to the Quantity Surveyor (named in the Contract or acting for the Contractor) to whom the task of ascertainment and/or settlement is often delegated; and to the Contracts Manager and Clerk of Works whose records can do more than anything else to pave the way for a timely and, therefore, an equitable settlement.

The book deals with the *Joint Contracts Tribunal (JCT) Standard Forms of Contract* in both the 1963 and 1980 *Private* Editions and in particular the With Quantities variants. (Where extracts from the forms are reproduced these are taken from the *Local Authorities* edition.)

Part 1 contains an introductory chapter which considers building contracts generally and compares the provisions of the 1963 Edition with the 1980 Edition of the Standard Form in relation to claims.

Part 2 deals with the 1963 Edition (July 1977 revision). Contracts are still being let on a variant of the 1963 Edition and claims and disputes arising from it are likely to continue for many years. Moreover parts of the text of the 1980 Edition have simply repeated the comparable parts of the 1963 Edition.

Part 3 deals with the 1980 Edition. In some instances it has been possible simply to include a short reference to Part 2; in others it has been necessary to prepare virtually a completely new text.

Part 4 contains the closing chapter which deals with the control of claims, suggests ways in which claims might be contained and, when they do occur, how the resultant costs can be minimized.

I am grateful to RIBA Publications Ltd for permitting me to reproduce parts of the text of the Standard Forms, and the National Federation of Building Trades Employers (now the Building Employers Confederation) for permission to reproduce extracts from the "Green" Form of Sub-Contract; to Michael Merritt RIBA for assistance in preparing the charts and in the preparation of the worked examples.

Finally my thanks are due both to Cyril Smith of my office for reading through the draft and to Cliff Cowen FRICS for his comments on some of my earlier thoughts; in fairness, however, I should say that whilst I have always taken note of errors of fact to which they have drawn my attention they are in no way responsible for the many opinions expressed herein.

PREFACE

It is now over 17 years since the first edition of this book was published. The first edition dealt exclusively with claims under the 1963 and 1980 Editions of the Joint Contracts Tribunal's Standard Forms of Building Contract.

Since that time, whilst the basic form of the 1980 Edition of that contract has not changed dramatically, a number of other contracts have come into the public domain albeit that the use of the JCT forms, in one version or another, still dominates the market. Thus the major change in this the second edition is the introduction of further forms of contract.

In addition to basing the first five chapters of this latest edition on the 1998 version of JCT's Standard Form of Building Contract, we have included chapters on five additional JCT Forms. There are also chapters on the 1998 version of the Government Form GC/Works/1 and on the Engineering and Construction Contract (formerly the NEC). The chapters from the first edition which dealt specifically with Fluctuations and Determination have been dropped.

However, the original aim of the book remains, namely: *"to examine those areas in the Contract which have a claim potential; to determine, so far as one is able, the rights and obligations of the parties in those circumstances; and to show by examples how the monetary entitlement can be ascertained. It is intended to be a practical book, culminating in a detailed example related to one fictional building contract. It should therefore assist both the student and the practitioner."*

The text is not directed to any one group. It will have an appeal to the Contractor, who will be concerned to ensure that he is properly reimbursed; to the Architect or others involved in determining the entitlement to extensions of time and ascertaining the amount of any loss and/or expense; to the Quantity Surveyor (either named in the Contract or acting for the Contractor) to whom the task of ascertainment or settlement is often delegated; and to the Contracts Manager and Clerk of Works whose records can do more than anything else to pave the way for a timely and just settlement.

Part 1 contains an introductory chapter that considers building contracts generally.

Part 2 deals with the Joint Contracts Tribunal Standard Form of Building Contract 1998 Edition Private with Quantities *(JCT 98 PWQ)*. This forms the basis for the book against which other forms of contract are contrasted and compared in later chapters.

Part 3 covers forms of building contract other than JCT 98 PWQ, namely:

- JCT Nominated Sub-Contract Conditions 1998 Edition

- JCT Agreement for Minor Building Works 1998 Edition
- JCT Standard Form of Building Contract With Contractor's Design 1998 Edition
- JCT Intermediate Form of Building Contract for Works of Simple Content 1998 Edition
- JCT Standard Form of Management Contract 1998 Edition
- General Conditions of Contract for Building & Civil Engineering Major Works GC/Works/1 With Quantities (1998)
- Engineering and Construction Contract (being the Second Edition of the New Engineering Contract).

Part 4 contains a fully worked example which demonstrates in detail how the principles put forward in the text of the book would be applied in practice to one theoretical project based on a JCT 98 PWQ contract.

We are grateful to the following organisations for permitting us to reproduce parts of their forms of Contract:

- The Joint Contracts Tribunal Limited for extracts from their suite of building contracts
- The Controller of Her Majesty's Stationery Office for extracts from GC/Works/1 With Quantities (1998) produced by the Property Advisers to the Civil Estate and published by The Stationery Office Limited with Crown Copyright
- Thomas Telford Publishing Limited (wholly owned by the Institution of Civil Engineers) of Thomas Telford House, 1 Heron Quay, London E14 4JD, Tel 020 7987 6999, e-mail ttbooks@ice.org.uk for extracts from the Engineering and Construction Contract.

We are also grateful to Michael Merritt RIBA for assistance in preparing charts and diagrams for the first edition which are repeated in this second edition.

A debt of gratitude is also owed to the following individuals who provided us with an invaluable critique of the book during the drafting stages:

- Pierce Hackett ARICS
- Simon Tolson BA(Hons), FCIArb, FFB, a Partner in the practice of specialist construction lawyers Fenwick Elliott
- John Quier FRICS, a chartered quantity surveyor in private practice.

Special mention, and thanks, must go to Ian Robinson BSc(Hons), LLB(Hons), ARICS, ACIArb, a Partner in the legal support services group of Davis Langdon & Everest. Ian's research and tenacity throughout the drafting process made this book possible.

Whilst we have been fortunate enough for experienced practitioners to have given up scarce time to let us have their comments, it is we and not they who bear the responsibility for the many opinions expressed throughout the book.

GGT and MEH
September 2000

TABLE OF CASE LAW

TABLE OF STATUTES

LIST OF EXAMPLES

Part 1

GENERAL

1

INTRODUCTION

1.1 INTRODUCTION

1.1.1 For most clients, to have a building designed and erected is only a means to an end, not an end in itself. It is a means of expanding a manufacturing process, of housing staff, of solving housing problems, of operating an hotel or the like. Having finally made the decision to build, probably after months if not years of debate, clients are impatient for a result.

1.1.2 Generally, clients believe that their financial commitment is precisely established at the outset. They observe that, from an early stage of the design process, detailed estimates of time and money are given; changes in cost are monitored as the design is developed; tenders are usually called for on the basis of very detailed measurement and the rates for valuing any variations are established as part of the tendering process in advance of letting the contract. Against this background of care, inevitably purchased at the client's expense, it must be bewildering to discover, as sometimes is the case, that progress on site has lapsed and there appears to be little that can be done about it; that the final cost and completion date are impossible to predict; and that the final settlement is arrived at out of exasperation rather than by evaluation. The cost plan may be based upon countless data from previous schemes, the Contract Bills measured against a precise set of rules, but the contractor's claim is too often settled by going to war.

1.1.3 It is, therefore, the aim of this book to encourage a more systematic approach to the matter of claims. The first edition of this book was prepared based on the 1963 Edition of the JCT Standard Form of Building Contract; the 1980 Edition appeared shortly before the book went to print and was thus dealt with by comparing or contrasting it with the provisions of the 1963 Edition. In the present edition of this book, the Joint Contracts Tribunal Standard Form of Contract 1998 Edition Private With Quantities (incorporating Amendments 1 and 2) is the base; and is referred to as JCT 98 PWQ throughout. Later chapters contain comparisons with other forms of building contract so far as the subject of claims is concerned. Accordingly this book covers the forms of Building Contract set out in the following table.

Forms of Building Contract referred to in this book

Contract Form	Reference used in this book	Chapter reference
JCT Standard Form of Building Contract 1998 Edition Private With Quantities	JCT 98 PWQ	2, 3, 4 and 5
JCT Nominated Sub-Contract Conditions (NSC/C) 1998 Edition	NSC/C	6
JCT Agreement for Minor Building Works 1998 Edition	JCT 98 MW	7
JCT Standard Form of Building Contract With Contractor's Design 1998 Edition	JCT 98 WCD	8

JCT Intermediate Form of Building Contract for works of simple content 1998 Edition	JCT 98 IFC	9
JCT Standard Form of Management Contract 1998 Edition	JCT 98 MC	10
General Conditions of Contract for Building & Civil Engineering Major Works GC/Works/1 With Quantities (1998)	GC/Works 98	11
The Engineering and Construction Contract (being the Second Edition of the New Engineering Contract)	NEC/2	12

1.1.4 The JCT 98 PWQ contract that is assumed is generally the *With Quantities* variant. Where the *Without Quantities* variant is referred to, the Specification or the Schedule of Rates will take the place of references in this book to the Contract Bills. However, before launching into the detail of these contracts it is necessary to make some general comments.

1.2   BUILDING CONTRACTS GENERALLY

1.2.1 One must look to four places to determine the rights and obligations of the parties to a Contract:

a)   the terms of the Contract be they express and/or implied;
b)   the findings of the Courts on relevant issues – this is referred to as Common Law;
c)   statute, i.e. Acts of Parliament and the like; and
d)   to a lesser extent, custom and trade usage.

1.2.2 It is beyond the scope of this book, and certainly beyond our competence, to become too involved in details of b) and c). However, as this book is planned to be a practical one in which solutions are offered, it is inevitable that we make judgments on what we believe to be the relevance of the rulings of the Courts where appropriate. We recognise that we run the risk of criticism from lawyers who understandably complain about the Quantity Surveyor's predisposition to become amateurs in their field. However, it generally falls to the Quantity Surveyor to settle monetary matters under building contracts and this means that he must make a judgment on the interpretation of the terms in the Contract; he cannot put up every item on which there is a difference of view for a decision by an Adjudicator, Arbitrator or by the Courts. Thus there will be many views expressed in this book that can be challenged and there will almost certainly be areas that will be affected by future decisions of the Courts.

1.2.3 Being conscious of this potential for criticism, we have stated how we have arrived at our various conclusions, not because we believe the logic to be unchallengeable, but so that the reader may follow our reasoning and be better able, should he so choose, to disagree.

1.2.4 Before considering the wording in any form of contract in detail, it is necessary to determine the background of Common Law against which it operates and it is with some misgivings that we step into this minefield – particularly as it would seem that the Court of Appeal or the House of Lords keeps moving those mines whose location we thought had been determined. It may be useful, however, to consider what Common Law provisions might have applied, and in some instances still do apply, had not the Contract, signed by the parties, purported to have dealt with the matter.

1.2.5 We would add that the views expressed apply to the law of England and Wales; we have not ventured to consider the niceties of the different legal considerations that apply in Scotland.

1.2.6 As a general rule, the parties who have entered into a Contract will not lose their Common Law rights in any particular instance unless either:

a) the terms of the Contract clearly purport to cover fully such rights and Common Law rights are not specifically reserved; or

b) a term in a standard Contract which would have covered such rights has been deleted by agreement (*Mottram Consultants Ltd -v- Bernard Sunley & Sons Ltd (1974) 2 BLR 28*) (see para. 2.8.3).

As to a) above, it is to be noted that a party's Common Law rights will only be excluded or limited if very clear words are used (see *Billyack -v- Leyland Construction Ltd [1968] 1 All ER 783*) or if the words are expressed to be exhaustive of all rights (as was the case with clause 44.4 of the Model Form MF/1 Conditions of Contract as decided in *Strachan & Henshaw -v- Stein Industrie (UK) Ltd*).

1.2.7 Where the Contract lays down clear rules of remedy any Common Law entitlement would generally be superseded, this being on the premiss that implied terms from Common Law will yield to express terms contained in the Contract. However, where the contract terms leave some ground uncovered, a Common Law remedy may still exist. Thus there may well be circumstances in which the Architect – or the Project Manager, as the case may be – cannot respond to an apparently legitimate claim from the Contractor in the absence of a specific provision in the Contract dealing with the matter, but the Contractor may still retain his Common Law entitlement.

1.2.8 Occasionally, a written contract will make clear the extent to which Common Law rights will prevail notwithstanding the written terms. For example, in JCT 98 PWQ, clause 26 lays down the procedures to be followed in the event of a fairly comprehensive list of causes of disruption due to the Employer's act or default and it is suggested that the operation of this clause, if unqualified, might well have restricted the Common Law entitlement. However, clause 26.6 goes on to state that *The provisions of clause 26 are without prejudice to any other rights and remedies which the Contractor may possess*. This clearly gives the Contractor the opportunity of falling back on to his Common Law rights.

1.2.9 Conversely, the written terms of a contract can also be drafted so as to give one party a greater degree of protection than would be available to him at Common Law. Even so, the Architect, Quantity Surveyor or Project Manager, when exercising the powers conferred on them by the Contract, must act strictly in accordance with the terms of the relevant clauses. Simply because they are required to exercise a role often reserved for the courts, they cannot assume a general mantle of judicial responsibility.

1.2.10 Examples of this principle can be seen in clause 25 of JCT 98 PWQ which deals with extensions of time for completion, which in this form of contract, requires the Architect to *fix* a new *Completion Date*. The grounds for extension of time broadly cover not only the default of the Employer (for which a remedy exists at Common Law) but also cover events outside the control of both parties, e.g. exceptionally adverse weather conditions or delay by Statutory Authorities, neither of which would normally be grounds entertained by the Courts for an extension of time in the absence of an express provision. On the other hand, delay by the Employer in granting possession of the site on the date named in the Contract is a clear breach by him, but in JCT 98 PWQ (unless clause 23.1.2 is stated in the Contract Appendix to apply) it is not a ground for extension of time, so cannot be dealt with by the Architect.

1.2.11 The Articles of Agreement and Conditions of Contract do not contain all the obligations to be imposed upon the contractor. Many additional obligations and the specification of workmanship and materials will probably appear in the Contract Bills or Specification – and, beware, in Collateral Warranties. Thus, in considering the obligations of the parties to the contract, one must look both to the Conditions of Contract and beyond to the Contract Bills, Specification or any Collateral Warranty. However, the role of the Contract Bills is severely limited in JCT 98 PWQ by the inclusion of clause 2.2.1; this states:

> *Nothing contained in the Contract Bills shall override or modify the application or interpretation of that which is contained in the Articles of Agreement, the Conditions or the Appendix.*

1.2.12 The above express term overrides the Common Law principle derived from *Robertson -v- French (1803) 4 East 130* which would otherwise pertain namely that bespoke conditions prevail over printed conditions. The above quoted express term, as it appeared in JCT 63, was considered by the House of Lords in the case of *English Industrial Estates Corporation -v- Wimpey (George) & Co (1972) 7 BLR 122* in which the Contract Bills included provisions that were at variance with those in the Contract Conditions. Lord Justice Denning considered, in this instance, that the requirements of the Contract Bills so formed the basis of the Contractor's calculations that this alone overrode the restriction that the clause in the contract sought to impose which "... *should have taken second place.*" However, he was alone in coming to this view. Lord Justice Stevenson summed up his views on the effect of provision in the Contract Bills (which was the view of the majority) in the following words:

> *Insofar as they (the Contract Bills) repeat or copy printed conditions or amendments which have been added to the printed conditions ... they add nothing to those conditions. Insofar as they introduce further contractual obligations ... they may add obligations which are consistent with the obligations imposed by the conditions, but they do not affect them by overriding or modifying them or in any other way whatsoever.*

1.2.13 This is the interpretation that is relied upon in this book, albeit that some may find it a little odd that a document that has been drafted to suit a particular Employer's requirements may end up yielding instead to requirements which were not drafted for anybody in particular. Thus the Contract Bills may, for example, require a Method Statement to be produced and although JCT 98 PWQ contains no such requirement this will become a binding obligation upon the Contractor; but the Contract Bills may not seek to define the meaning of words in the contract conditions, e.g. qualify or define *exceptionally adverse weather conditions* as used in clause 25.4.2 of JCT 98 PWQ.

1.2.14 It is a common misconception of many Contractors, Architects and Quantity Surveyors that where there is delay and/or disruption to a building project which could not have been envisaged at tender stage, the Contractor is entitled either to reimbursement of loss and/or expense or to extra time or both. This is not so (see *Fairweather & Co. Ltd -v- London Borough of Wandsworth (1988) 12 BLR 40*). The Employer would only be liable if the problems were within his power to control or were otherwise stated to be at his risk. It is the Contractor who takes all the remaining risks unless the terms of the Contract say otherwise. Thus under a lump sum contract with a fixed Date for Completion the make up of the Contractor's entitlement to time and money is as shown in the following table.

Contractor's entitlement to time and money under a Lump Sum Contract

Time	Money
The contract period	The Contract Sum
Extensions of time specifically provided for in the Contract	Monetary adjustments specifically provided for in the Contract
Any Common Law entitlement still prevailing	Any Common Law entitlement still prevailing

All other eventualities are at the Contractor's risk and are deemed to have been covered in the Contractor's price.

1.2.15 The Common Law position on the apportionment of risk, in the absence of express terms to the contrary, was stated in the *Moorcock (1889) 14 PD 64* to operate so as *"not to impose on one side all the perils of the transaction, or to emancipate one side from all chances of failure, but to make each party promise in law as much, at all events, as it must have been in the contemplation of both parties that he should be responsible for in respect of those perils or chances."* This is to suggest a fair and reasonable apportionment of risk between the parties in the absence of express terms to the contrary – where express terms do exist, the apportionment of risk can of course be more precisely ascribed by the draftsmen.

1.2.16 In the context of the contractual entitlement to time and money under a lump sum contract, it is worth noting what rules the Courts might apply in establishing the amount of damages payable if, for example, the regular progress of the works was disrupted by the Employer particularly as in the case of *F. G. Minter -v- Welsh Health Technical Services Organisation* referred to at para. 4.9.1. It was stated that the provisions (of JCT 63) in relation to the reimbursement of direct loss and/or expense were the same as those relating to the recovery of damages at Common Law. The rules for establishing damages were systematically laid down in two cases; *Hadley -v- Baxendale (1854) 9 Ex 341* and *Victoria Laundry (Windsor) -v- Newman Industries (1949) 2 KB 528.* The important part of those rules in this context is that the damages will be the loss which was reasonably foreseeable by the parties (at least by the party who committed the breach) at the time that they made the Contract as being the natural result of the breach. The emphasis on those rules when applied to building contracts must be on the reference to what was in the minds of the parties *"... at the time they made the contract."* Contractors often overlook this limitation to their Common Law remedy when suggesting that an ascertainment of loss and/or expense is inequitable, in that it may not fully cover their costs. They tend to overlook that there is another party to the Contract, the Employer, who could not reasonably be expected to have entered into an Agreement in which his slightest stumble might bring an unheralded claim for reimbursement.

1.3 BACKGROUND TO THE CONTENTS OF THIS BOOK

1.3.1 The title of the book refers to "Claims". No apology is offered for this although the word does not appear in JCT 98 PWQ nor in most forms of building contract. It is used here to describe the demands by the Contractor to the effect that he is not receiving his proper monetary entitlement.

1.3.2 Earlier reference was made to the apparent lack of precision in the matter of settling claims. This is due to one of three reasons:

a) the terms used in the Contract are not as clear as they might be – it is, of course, easier to criticise contracts than to draft them;

b) settlements are all too often made months, if not years, after the events to which they relate when memories have faded and records are found to be inadequate;

c) an element of conjecture is always required; for example, where there is a justifiable claim for disruption to the regular progress of the Works, the monetary entitlement will be established by comparing the actual costs that were properly incurred in undertaking the disrupted work with the estimated costs of **what would properly have been incurred** had the problem under investigation not taken place. No amount of factual information will prove what the estimated costs might have been. (See also para. 4.2.7 and Example 4.4 referred to at para. 4.4.8.)

1.3.3 Chapters 2 to 4 are concerned with the scope usually attributed to the word "Claims" (namely extensions of time and the ascertainment of direct loss and/or expense) under JCT 98 PWQ. Although we have not dealt with direct loss and/or damage arising out of determination in this book there follows, for those who are interested, a schedule of relevant clauses.

Schedule of clauses relating to direct loss and/or damage arising out of determination

CONTRACT	CLAUSE	DETERMINATION BY
JCT 98 PWQ	27.7.1.2	Employer
	28.4.3.4	Contractor
	28A.5.5	Employer or Contractor
NSC/C	7.5.4	Contractor
	7.8.2.3	Sub-Contractor
	7.10.5	Employer of Contractor's employment
	7.11.1.5	Employer of Contractor's employment
JCT 98 MW	7.2.3	Employer
	7.3.3	Contractor
JCT 98 WCD	27.6.6.1	Employer
	27.7.1.2	Employer
	28.4.4.4	Contractor
	28A.6.5	Employer or Contractor
JCT 98 IFC	7.6(g)	Employer
	7.7.1(b)	Employer
	7.11.3(d)	Contractor
	7.18.5	Employer or Contractor
NAM/SC	27.3.4.1	Contractor
	28.3.2.6	Sub-Contractor
JCT 98 MC	7.6.5.1	Employer
	7.7.1.3	Employer
	7.11.3.4	Management Contractor
	7.17.5	Employer or Management Contractor
	7.23	At will by Employer
WC/2	7.5.4	Management Contractor
	7.9.2.3	Works Contractor
	7.11.3.5	Employer of Management Contractor's employment

1.3.4 Chapter 5 deals with the stance often taken by either side to a dispute and how to react to it, and considers how claims can be prevented or their extent limited.

1.3.5 In Chapters 6 to 12, the position established in regard to JCT 98 PWQ is contrasted and compared with similar provisions in other forms of building contract.

1.3.6 Then follows a complete worked example where a theoretical claim, submitted by a Contractor is analysed, responded to in detail, and a final ascertainment made.

1.3.7 The Preface declared the intention that this would be a practical book. To this end examples have, whenever possible, been introduced to demonstrate the points being made with, following the conclusion, a detailed example of a claim relating to a fictional Contract. The Contract in the example, in common with Part 2 of this book, is deemed to be based upon the JCT Standard Form of Building Contract 1998 Edition Private With Quantities, incorporating Amendments 1 and 2.

1.3.8 The term "the Architect" has been used throughout the text of Part 2 although such references are deemed to include references to "the Architect/the Contract Administrator".

Part 2

JOINT CONTRACTS TRIBUNAL STANDARD FORM OF BUILDING CONTRACT 1998 EDITION PRIVATE WITH QUANTITIES

2

EXTENSIONS OF TIME

2.8   THE RELEVANT EVENTS   2.8.1–2.8.75

Clause 25.4.1 force majeure   2.8.2–2.8.3

Clause 25.4.2 exceptionally adverse weather conditions   2.8.4

Clause 25.4.3 loss or damage occasioned by any one or more of the Specified Perils.   2.8.5

Clause 25.4.4 civil commotion, local combination of workmen, ...   2.8.6–2.8.7

Clause 25.4.5 1 compliance with the Architect's instructions under clauses 2.3, 2.4.1, 13.2, 13.3, 13A.4.1, 23.2, 34, 35 or 36;   2.8.8

     Generally   2.8.9–2.8.14

     clauses 2.3 and 2.4.1 discrepancies   2.8.15–2.8.17

     clauses 13.2 and 13.3 Variations   2.8.18–2.8.22

     clause 13A.4.1 Contractor's 13A Quotation not accepted   2.8.23

     clause 23.2 Architect's instructions-postponement   2.8.24

     clause 34 Antiquities   2.8.25

     clauses 35 and 36 Nominations   2.8.26

Clause 25.4.5.2 compliance with the Architect's instructions in regard to the opening up for inspection of any work ...   2.8.28

Clause 25.4.6.1 where an Information Release Schedule has been provided, failure of the Architect to comply with clause 5.4.1   2.8.29–2.8.32

Clause 25.4.6.2 failure of the Architect to comply with clause 5.4.2.   2.8.33–2.8.44

Clause 25.4.7 delay on the part of Nominated Sub-Contractors or Nominated Suppliers ...   2.8.45–2.8.50

Clause 25.4.8.1 the execution of work not forming part of this Contract by the Employer ...   2.8.51–2.8.54

Clause 25.4.8.2 the supply by the Employer of materials and goods ...   2.8.55

Clause 25.4.9 the exercise after the Base Date by the United Kingdom Government of any statutory power which directly affects the execution of the Works ...   2.8.56–2.8.57

Clause 25.4.10 the Contractor's inability ... to secure such labour (or such goods or materials) ...   2.8.58–2.8.60

Clause 25.4.11 the carrying out by a local authority or statutory undertaker of work ...   2.8.61

2.1 INTRODUCTION

2.1.1 It is necessary in a book dealing with claims to consider extensions of time primarily for the following reasons:

a) extensions of time will affect the extent of the Employer's right to claim damages for delayed completion;

b) many of the events which are grounds for an extension of time are also grounds for claiming reimbursement of loss and/or expense;

c) an analysis of the extensions of time will reveal when the disrupted or varied work ought to have been carried out; this is often not the same as the time at which it was carried out and thus the circumstances (e.g. weather conditions) under which the work should have been executed can and should be taken into account;

d) so far as JCT 98 PWQ is concerned, the Architect is required to state in writing to the Contractor what extension of time, if any, he has made for those events which also appear listed under clause 26 as matters that may also give rise to an entitlement to extra reimbursement by way of recovery of loss and/or expense (clauses 26.3 and 26.4.2).

2.1.2 For those who are familiar with construction contracts, the extension of time provisions contained therein are often taken for granted and it is therefore worth considering the position were they not included.

2.1.3 All the standard forms of construction contract contain a provision for the insertion of a date for completion which, if not achieved, renders the Contractor liable to the Employer for damages which almost invariably are in the form of liquidated and ascertained damages; the method of calculating liquidated and ascertained damages is dealt with at Sections 2.13 and 2.14. The Employer must not do anything lawfully or unlawfully to prevent the Contractor from completing the Works by the due date or he will forfeit his right to liquidated and ascertained damages unless time can be, and is, properly extended under express contractual provisions. This follows the "prevention principle" – followed in a trio of cases, namely *Holme -v- Guppy (1838) 3 M.&W. 387*, *Wells -v- Army & Navy Co-operative Society (1902) 86 LT 764* and *Dodd -v- Churton [1897] 1 QB 562* – that a party loses his right to insist on the performance of an obligation where it is through that party's own fault that the obligation cannot be performed, e.g. the late issue of instructions.

2.1.4 In the absence of express contractual provisions enabling the Employer to allow additional time to the Contractor in recognition of default, the completion date and the right to levy liquidated and ascertained damages are displaced and time is "at large" as soon as the Employer, through his default, prevents the Contractor achieving that completion date.

2.1.5 In his work "Liquidated Damages and Extension of Time" (Blackwell Scientific Publications), Brian Eggleston excellently summarises the meaning of "time at large" in the following way;

> *The phrase time at large is much loved by Contractors. It has about it the ring of plenty; the suggestion that the Contractor has as much time as he wants to finish the Works. This is not what it means. Time becomes at large when the*

obligation to complete within the specified time for completion of a contract is lost. The definition then becomes to complete within a reasonable time. The question of what is a reasonable time ... is most certainly not as and when the Contractor sees fit.

2.1.6 From the above, it may be seen that the extension of time provisions within the standard forms of construction contract override the position which would otherwise pertain at Common Law and provide the Employer with a means of extending the date for completion in a measure commensurate with his own default, thus setting a new date from which liquidated and ascertained damages may run.

2.1.7 In determining the extent to which delays and claims should be allowed within the terms of a contract (as opposed to being left to be resolved in the courts), the draughtsmen of the JCT Standard Forms clearly had in mind two criteria, namely:

a) events which are in the control of, or are caused by, the act or default of the Employer or his agents are to give the Contractor an entitlement to an extension of time for completion and also an entitlement to be reimbursed the loss and/or expense arising;

b) events which are outside the control of either party, often termed "neutral events" to give the Contractor an entitlement to an extension of time for completion but not to recover any loss and/or expense.

2.1.8 Broadly speaking, the first criterion is in line with the provisions that apply in Common Law; any default by the Employer or his agents will be legitimate grounds for a claim by the Contractor. But the second criterion goes much further. In general, unless the terms of the Contract provide otherwise, a Completion Date would only be set aside or subject to review where an act or default of the Employer or his agents prevented its achievement. But JCT 98 PWQ, and many other standard forms of building contract, extend this facility to causes of delay outside the control of either party, e.g. *exceptionally adverse weather conditions*. Clearly, the second criterion benefits the Contractor since it provides him with grounds for an extension of time (and thereby relief from liquidated and ascertained damages) which would not exist at Common Law. Another view of the second criterion is that it is a fair apportionment of risk in that the Employer is denied the right to levy liquidated and ascertained damages during the period of delay caused by a "neutral event" whilst the Contractor bears the costs of prolongation without the right to recover these costs from the Employer.

2.1.9 The table below illustrates the comparison in JCT 98 PWQ between clause 25 (which deals with the Contractor's entitlement to extension of time for completion) and clause 26 (which deals with the grounds for which he is able to claim reimbursement of his loss and/or expense); the dangers of making such a comparison are referred to at paras 2.5.2 and 4.4.13.

Comparison of the causes listed in clause 25 with those in clause 26

Extracts from clause 25.4 – grounds for extension of time for completion (termed Relevant Events)	Clause	Comparable extracts from clause 26.2 – grounds for reimbursement for loss and/or expense (termed list of matters)	Clause
Force majeure	25.4.1	None	
Exceptionally adverse weather conditions	25.4.2	None	
Specified perils	25.4.3	None	
Civil commotion and strikes	25.4.4	None	
Architect's instructions under clauses:	25.4.5.1		
2.3 and 2.4.1 Discrepancies		Discrepancies (see note 1)	26.2.3
13.2 Variations (including variations instructed under clause 13A.4.1)		Variations (including variations instructed under clause 13A.4.1) (see note 2)	26.2.7
13.3 Provisional sums		Provisional sums (see note 3)	26.2.7
23.2 Postponement		Postponement	26.2.5
34 Discovery of antiquities		Discovery of antiquities	34.3.1
35 Nominated Sub-Contractors		None	
36 Nominated Suppliers		None	
Architect's instructions requiring opening up for inspection	25.4.5.2	Architect's instructions requiring opening up for inspection (see note 4)	26.2.2
Where an Information Release Schedule has been provided failure of the Architect to comply with clause 5.4.1	25.4.6.1	Where an Information Release Schedule has been provided failure of the Architect to comply with clause 5.4.1	26.2.1.1
Failure of the Architect to comply with clause 5.4.2	25.4.6.2	Failure of the Architect to comply with clause 5.4.2	26.2.1.2
Nominated Sub-Contractors or Nominated Suppliers	25.4.7	None	
Work by the Employer or those for whom he is responsible	25.4.8.1	Work by the Employer or those for whom he is responsible	26.2.4.1
Supply of materials by the Employer	25.4.8.2	Supply of materials by the Employer	26.2.4.2
Restriction in the availability of men, materials, fuel or energy due to Government exercising its statutory powers	25.4.9	None	
Contractor's inability to secure:			
Such labour as is essential for the Works	24.4.10.1	None	
Such materials as are essential for the Works	25.4.10.2	None	

Extracts from clause 25.4 – grounds for extension of time for completion (termed Relevant Events)	Clause	Comparable extracts from clause 26.2 – grounds for reimbursement for loss and/or expense (termed list of matters)	Clause
Work by statutory authorities	25.4.11	None	
Failure to give ingress to or egress from the site	25.4.12	Failure to give ingress to or egress from the site	26.2.6
Deferment of giving possession of the site (where this option has been selected)	25.4.13	Deferment of giving possession of the site (where this option has been selected (see note 5)	26.1
Approximate quantities proving not to be an accurate forecast of the work actually required	25.4.14	Approximate quantities proving not to be an accurate forecast of the work actually required	26.2.8
Change in Statutory require-ments requiring changes to Performance Specified Work	25.4.15	None	
Terrorism or threat of same and/or activity of authorities in dealing with this	25.4.16	None	
Compliance or non compliance by Employer with his obligations arising out of the Construction (Design and Management) Regulations 1994	25.4.17	Compliance or non compliance by Employer with his obligations arising out of the Construction (Design and Management) Regulations 1994	26.2.9
Suspension by the Contractor of the performance of his obligations under the Contract to the Employer	25.4.18	Suspension by the Contractor of the performance of his obligations under the Contract to the Employer (see note 6)	26.2.10

The foregoing descriptions are extracts only and the detailed clauses and their effects will be reviewed elsewhere in this book.

Notes:

1. The documents, in or between which discrepancies or divergencies might give rise to an extension of time under clause 25.4.5.1, are those listed in clause 2.3 namely:

    a) the Contract Drawings;

    b) the Contract Bills;

    c) Architect's instructions (except those requiring a variation);

    d) Architect's drawings or documents issued under clauses 5.3.1.1 (descriptive schedules and the like), 5.4 (information referred to in the Information Release Schedule and further drawings or details) or 7 (setting out drawings);

    e) the Numbered Documents (documents annexed to the Articles of Nominated Sub-Contract Agreement).

(continued)

21

Comparison of the causes listed in clause 25 with those in clause 26

Notes *(continued)*

(The documents, in or between which discrepancies or divergences might give rise to an extension of time under clause 25.4.5.1, are also referred to in clause 2.4 namely the Contractor's Statement in respect of Performance Specified Work and any instruction of the Architect issued by him after receipt by him of the Contractor's Statement.) Only discrepancies in or between a), b) and e) above (the Contract Drawings, the Contract Bills and the Numbered Documents) give rise to an entitlement to claim reimbursement of loss and/or expense under clause 26.2.3.

2. Excluding a variation to work for which there has been a confirmed acceptance of a Quotation given under clause 13A.

3. Excluding the expenditure of provisional sums through the Contractor as a result of a tender by him under clause 35.2.1.

4. There is some restriction on the Contractor's entitlement to claim reimbursement of any direct loss and/or expense (para. 3.5.8).

5. This is not one of the Matters listed in clause 26.2 but is included in the introductory clause 26.1 as a ground for claiming reimbursement of direct loss and/or expense (para. 3.5.4).

6. The suspension must not be frivolous or vexatious.

2.1.10 Note that in the table at para. 2.1.9 all items which appear under the first column but not in the second column represent causes beyond the control of either party (i.e. they are "neutral events"); whereas items which appear under both (with the exception of the discovery of antiquities) represent causes due to the act or default of the Employer. At Common Law the former would not be grounds either for an extension of time for completion or for claiming reimbursement of loss and/or expense, unless the terms of the contract so provided.

2.1.11 It is this apparent leaning towards the Contractor by way of introducing "neutral events" as grounds for granting an extension of time that has resulted in the mounting criticism, particularly from the private sector, of the JCT Forms of Contract. It can, of course, be argued that to make "neutral events" a Contractor's risk (e.g. *exceptionally adverse weather conditions*) will result in a disproportionate increase in the Contractor's tender to cover those damages for the delay that might ensue. On the other hand, the pressures of competition will control the tenderer's temptation to overload this and the party in control of the works on site (the Contractor) is the better placed to mitigate the effects on progress of any event particularly if he has the incentive to do so. Contractors will point out that they already have such an incentive in that they get no recompense for the extra costs to them, such as extended use of their site establishment due to any such delay, only relief from damages for delay which they will say is the much lesser amount.

2.1.12 Whatever the merits or otherwise of the present strategy relating to clause 25, it has currently been drawn so widely as to persuade some Contractors and their advisors that the onus is now on the Employer to prove that any delay is due to the Contractors' default before the Employer can levy liquidated and ascertained damages – or at least to claim that any cause of delay outside the Contractor's control is a proper ground for extension of time. This is not so. The Contractor undertakes to complete the Works by the Completion Date which will only be extended for the reasons given in the Contract and then only to the extent that such recognised events cause an inevitable delay to the Completion Date. As long as clauses 24 and 25 have been properly implemented, then if the date of Practical Completion occurs after the Completion Date the Employer will be entitled to liquidated and ascertained damages for that period of delay.

2.2   THE PURPOSE OF AN EXTENSION OF TIME CLAUSE

2.2.1 As previously observed at paras 2.1.4 and 2.1.5, at Common Law if an Employer or his agent prevents a Contractor from completing his work by the contracted Completion Date then the Contractor's obligation is transformed from one of having to complete by that date to one of completing within a reasonable time, unless there is an express power in the contract for the Employer to set a new date for completion in the new circumstances and that this power is properly implemented; clause 25 contains that express power in JCT 98 PWQ.

2.2.2 It will be seen, therefore, that the existence and proper operation of the clause is, insofar as a breach by the Employer or his agent has caused the delay, for the benefit of the Employer in order to preserve his rights. Thus the Employer who seeks to hide behind a technical failure of the Contractor fully to comply with the provisions of clause 25 as a means of denying an extension of time might find his euphoria short-lived; he may well lose all entitlement to liquidated and ascertained damages for delay. Architects also should be aware that failure properly to implement the provisions of clause 25 may, rather than set the scene for Adjudication, Arbitration or Legal Proceedings on the amount of an extension of time, set the seal on the Employer's prospect of claiming any liquidated and ascertained damages at all. It is for this reason that Architects, when operating under JCT 98 PWQ, must remember that when giving their final decision on extensions of time they are required to take into account all Relevant Events whether they have been previously notified or not (see paras 2.6.8, 2.7.34 and 2.7.35).

2.3   DELAY IN GRANTING AN EXTENSION OF TIME

2.3.1 It is often argued that if the Architect does not grant an extension of time fairly soon after the event is brought to his notice or at least by Practical Completion, time is "at large" and the Contractor simply has to demonstrate that he has completed in a reasonable time in order to obtain shelter from a claim from the Employer for liquidated and ascertained damages for delay. Unfortunately, it would seem that the law is far from clear on this albeit that in *Temloc Ltd -v- Errill Properties Ltd (1987) 39 BLR 39*, Lord Justice Croom Johnson did indicate that the 12 week period referred to in clause 25.3 (i.e. the period post practical completion when the Architect is to declare the final extension of time due to the Contractor) is "directory only as to time" and is not something which, if not complied with, would invalidate the Employer's right to liquidated and ascertained damages.

2.3.2 Paraphrased extracts from the case of *Miller -v- L.C.C. (1934) 151 LT. 425* are often appended to Contractor's claims to support the argument that the Architect has no right to grant extensions of time retrospectively after Practical Completion; certainly it would seem that this was the judgement in that case. But in the later case of *Amalgamated Building Contractors Ltd -v- Waltham Holy Cross UDC (1952) ALL ER 452*, Lord Denning made it clear that in his view the decision in *Miller -v- L.C.C.* turned on the particular wording of the relevant clause in that Contract which prevented a retrospective extension of time. He went on to state that under the wording in the *Amalgamated Building Contractors* case (which wording was similar to that in JCT 98 PWQ) retrospective extensions of time were clearly permitted. He pointed out that if a Contractor passed the Completion Date due to his own inefficiency and then the Employer or his agent created a delay, the Employer must be able to extend the Completion Date even though it had passed. However, having clarified matters to that extent, his Lordship added that the position might have been different if the earlier delay had been created by the Employer.

2.3.3 Thus one can only suggest that generally the Architect is entitled to grant extensions of time after the event and probably entitled to do so after Practical Completion, of the works, provided that he makes his award as soon as he is able having regard to the circumstances. But one must not overlook the fact that normally a Contractor is only entitled to an extension of time when there is a physical delay to progress not a theoretical view of what might have been. That is to say it is upon there being an actual delay to progress that the Contractor will register the matter with the Architect and it is an estimate of the practical effect of that delay upon the Completion Date with which the Architect must then be concerned.

2.3.4 If the works are not completed by the Completion Date due to an earlier default of the Contractor it is clear from the *Amalgamated Building Contractors* case that an extension of the Completion Date can be considered retrospectively due to a later delay. However, if the earlier delay is due to a default of the Employer, the Architect is generally on much less firm ground because the law on this situation has not been defined and, as Examples 2.1 and 2.2 indicate, it may be difficult for the Architect to defend his position by arguing that he was unable to estimate the effect of the earlier delay by the Completion Date.

Example 2.1 Granting timely extensions of time – I

A Contract has a Completion Date at the end of week 40. At week 35 there is a delay on a critical element that comes within the scope of clause 25 and the delay lasts for 3 weeks (see *Figure 2.1*). By the end of week 40 the Architect should be aware not only of the fact of the delay but he should also have formed an estimate of the effect of it by at least week 38. In a practical sense, therefore, he could and should grant an extension of time of 3 weeks giving a revised Completion Date before the existing one expired.

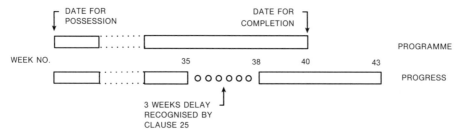

Figure 2.1 Delay on a critical element.

Example 2.2 Granting timely extensions of time – II

Taking the same example but with the delay of more than 3 weeks (say 8 weeks) gives the result in *Figure 2.2*. Again by the end of week 40 the Architect should be aware not only of the fact of the delay but also should have formed the view that an estimate of the delay would be at least 5 weeks. He is thus in position before the existing date expires, to grant an extension of at least 5 weeks.

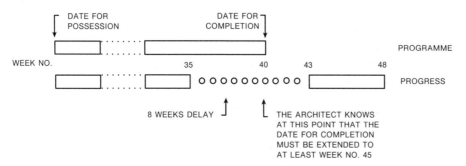

Figure 2.2 Further delay on a critical element.

2.3.5 Whatever the strict legal position, at a practical level, the Architect would be wise to repair the completion date as early as possible. Not only would this avoid the possibility of a legal battle, the result of which might be difficult to predict, but it would also help the Contractor who has the task of planning the residue of his work notwithstanding the legal arguments; and it would clarify for the Employer the date upon which he could expect the completion of his building.

2.3.6 Figure 2.3 indicates in visual form the position in the *Amalgamated Building Contractors* case given similar facts to those postulated for *Figures 2.1* and *2.2*.

2.3.7 Without doubt the Contractor, who has the duty to plan his work to meet the Completion Date, may find this difficult if there is an acknowledged delay to progress but no quick response by the Architect in granting an extension of time. In such circumstances, the Contractor may argue that as no extension has been granted he must complete by the Completion Date currently in being and that this may entail

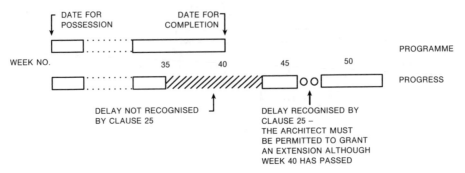

Figure 2.3 Delays resulting from different causes.

additional resources (shift and overtime working for example) the cost of which he will pass on to the Employer by way of what are sometimes referred to as "constructive acceleration claims". Alternatively, he may argue that as the fact of a delay is indisputable but no extension of time is granted then time is "at large" and he need only complete within a reasonable time. It is suggested that both of these views are incorrect.

2.3.8 As soon as it is reasonably apparent that the progress of the Works is, or will be, delayed the Contractor will, so far as the facts justify it, have an entitlement in due course to an extension of time provided that the Contract contains provisions for extending the Completion Date. The Contractor cannot pretend that the ultimate extension of time will be at the whim or fancy of the Architect and that until this is known he is in a planning vacuum. The Contractor, who will be in as good a position as the Architect to know the real effect of any such delay, must proceed regularly and diligently with the works relying upon the Architect ultimately granting an extension of time which is fair and which properly and reasonably reflects the facts (such a decision can in any event, be challenged at Adjudication, Arbitration or in substantive Legal Proceedings).

2.3.9 Notwithstanding the above general remarks, the position relating to the timing of granting extensions of time under JCT 98 PWQ is clear; namely that extensions of time can be granted after the Completion Date and the date of Practical Completion have passed (clause 25.3.3). There is a possible complication however; this concerns a provision in JCT 98 PWQ that the Contractor must give an estimate of the expected delay in the completion of the Works as part of the data submitted to the Architect in support of a notice of delay. The Contractor might be tempted (at least until the Architect responds) to claim to have carried out the Works as if his estimate of the extension of time was going to be accepted by the Architect. In such circumstances, the Architect would be wise to notify the Contractor of any reservation to the Contractor's estimate of the delay as soon as it is apparent to the Architect.

2.3.10 One further pressure to grant extensions of time in accordance with the programme laid down in clause 25 is that unless this is done it is not possible to "freeze" the adjustments for fluctuations in prices under the provisions of clauses 38.4.8, 39.5.8 or 40.7.1 whichever is applicable. Under these clauses, the rate of payment is frozen at that prevailing at the Completion Date provided that the Architect has ... *in respect of every written notification by the Contractor under clause 25, fixed or confirmed in* *writing such Completion Date as he considers to be in accordance with clause 25. (See section 2.12.)*

2.4 CLAUSE 25 CONSIDERED IN DETAIL

2.4.1 Before the Liquidated and Ascertained Damages clause (clause 24) can be activated, the Architect must have granted such extensions of time as he thinks fit under clause 25. Thus clause 25 is considered in detail before considering clause 24.

2.4.2 Clause 25 reads as follows:

25 Extension of time

25.1   In clause 25 any reference to delay, notice or extension of time includes further delay, further notice or further extension of time.

25.2   .1   .1   If and whenever it becomes reasonably apparent that the progress of the Works is being or is likely to be delayed the Contractor shall forthwith give written notice to the Architect of the material circumstances including the cause or causes of the delay and identify in such notice any event which in his opinion is a Relevant Event.

.1   .2   Where the material circumstances of which written notice has been given under clause 25.2.1.1 include reference to a Nominated Sub-Contractor, the Contractor shall forthwith send a copy of such written notice to the Nominated Sub-Contractor concerned.

25.2   .2   In respect of each and every Relevant Event identified in the notice given in accordance with clause 25.2.1.1 the Contractor shall, if practicable in such notice, or otherwise in writing as soon as possible after such notice:

.2   .1   give particulars of the expected effects thereof; and

.2   .2   estimate the extent, if any, of the expected delay in the completion of the Works beyond the Completion Date resulting therefrom whether or not concurrently with delay resulting from any other Relevant Event

and shall give such particulars and estimate to any Nominated Sub-Contractor to whom a copy of any written notice has been given under clause 25.2.1.2.

25.2   .3   The Contractor shall give such further written notices to the Architect, and send a copy to any Nominated Sub-Contractor to whom a copy of any written notice has been given under clause 25.2.1.2, as may be reasonably necessary or as the Architect may reasonably require for keeping up-to-date the particulars and estimate referred to in clauses 25.2.2.1 and 25.2.2.2 including any material change in such particulars or estimate.

25.3   .1   If, in the opinion of the Architect, upon receipt of any notice, particulars and estimate under clauses 25.21.1, 25.2.2 and 25.2.3:

.1   .1   any of the events which are stated by the Contractor to be the cause of the delay is a Relevant Event and

27

.1   .2   the completion of the Works is likely to be delayed thereby beyond the Completion Date

the Architect shall in writing to the Contractor give an extension of time by fixing such later date as the Completion Date as he then estimates to be fair and reasonable. The Architect shall, in fixing such new Completion Date, state:

.1   .3   which of the Relevant Events he has taken into account and

.1   .4   the extent, if any, to which he has had regard to any instructions issued under clause 13.2 which require as a Variation the omission of any work or obligation and/or under clause 13.3 in regard to the expenditure of a provisional sum for defined work or for Performance Specified Work which results in the omission of any such work,

and shall, if reasonably practicable having regard to the sufficiency of the aforesaid notice, particulars and estimates, fix such new Completion Date not later than 12 weeks from receipt of the notice and of reasonably sufficient particulars and estimate, or, where the period between receipt thereof and the Completion Date is less than 12 weeks, not later than the Completion Date.

If, in the opinion of the Architect, upon receipt of any such notice, particulars and estimate, it is not fair and reasonable to fix a later date as a new Completion Date, the Architect shall if reasonably practicable having regard to the sufficiency of the aforesaid notice, particulars and estimate so notify the Contractor in writing not later than 12 weeks from receipt of the notice, particulars and estimate, or, where the period between receipt thereof and the Completion Date is less than 12 weeks, not later than the Completion Date.

25.3   .2   After the first exercise by the Architect of his duty under clause 25.3.1 or after any revision to the Completion Date stated by the Architect in a confirmed acceptance of a 13A Quotation in respect of a Variation the Architect may in writing fix a Completion Date earlier than that previously fixed under clause 25 or than that stated by the Architect in a confirmed acceptance of a 13A Quotation if in his opinion the fixing of such earlier Completion Date is fair and reasonable having regard to any instructions issued after the last occasion on which the Architect fixed a new Completion Date

–   under clause 13.2 which require or sanction as a Variation the omission of any work or obligation; and/or

–   under clause 13.3 in regard to the expenditure of a provisional sum for defined work or for Performance Specified Work which result in the omission of any such work.

Provided that no decision under clause 25.3.2 shall alter the length of any adjustment to the time required by the Contractor for the completion of the Works in respect of a Variation for which a 13A Quotation has been given and which has been stated in a confirmed acceptance of a 13A Quotation or in respect of a Variation or work for which an adjustment to the time for completion of the Works has been accepted pursuant to clause 13.4.1.2 paragraph A7.

25.3   .3   After the Completion Date, if this occurs before the date of Practical Completion, the Architect may, and not later than the expiry of 12 weeks after the date of Practical Completion shall, in writing to the Contractor either

.3   .1   fix a Completion Date later than that previously fixed if in his opinion the fixing of such later Completion Date is fair and reasonable having regard to *any of the Relevant Events, whether upon reviewing a previous decision or otherwise and whether or not the Relevant Event has been specifically notified by the Contractor under clause 25.2.1.1; or*

25.3 .3 .2 *fix a Completion Date earlier than that previously fixed under clause 25 or stated in a confirmed acceptance of a 13A Quotation if in his opinion the fixing of such earlier Completion Date is fair and reasonable having regard to any instructions issued after the last occasion on which the Architect fixed a new Completion Date*

— *under clause 13.2 which require or sanction as a Variation the omission of any work or obligation; and/or*

— *under clause 13.3 in regard to the expenditure of a provisional sum for defined work or for Performance Specified Work which result in the omission of any such work; or*

.3 .3 *confirm to the Contractor the Completion Date previously fixed or stated in a confirmed acceptance of a 13A Quotation.*

Provided that no decision under clause 25.3.3.1 or clause 25.3.3.2 shall alter the length of any adjustment to the time required by the Contractor for the completion of the Works in respect of a Variation for which a 13A Quotation has been given and which has been stated in a confirmed acceptance of a 13A Quotation.

25.3 .4 *Provided always that:*

.4 .1 *the Contractor shall use constantly his best endeavours to prevent delay in the progress of the Works, howsoever caused, and to prevent the completion of the Works being delayed or further delayed beyond the Completion Date;*

.4 .2 *the Contractor shall do all that may reasonably be required to the satisfaction of the Architect to proceed with the Works.*

25.3 .5 *The Architect shall notify in writing to every Nominated Sub-Contractor each decision of the Architect under clause 25.3 fixing a Completion Date and each revised Completion Date stated in the confirmed acceptance of a 13A Quotation together with, where relevant, any revised period or periods for the completion of the work of each Nominated Sub-Contractor stated in such confirmed acceptance.*

25.3 .6 *No decision of the Architect under clause 25.3.2 or clause 25.3.3.2 shall fix a Completion Date earlier than the Date for Completion stated in the Appendix.*

25.4 *The following are the Relevant Events referred to in clause 25:*

25.4 .1 *force majeure;*

25.4 .2 *exceptionally adverse weather conditions;*

25.4 .3 *loss or damage occasioned by any one or more of the Specified Perils;*

25.4 .4 *civil commotion, local combination of workmen, strike or lock-out affecting any of the trades employed upon the Works or any of the trades engaged in the preparation, manufacture or transportation of any of the goods or materials required for the Works;*

25.4 .5 *compliance with the Architect's instructions*

.5 .1 *under clauses 2.3, 2.4.1, 13.2 (except for a confirmed acceptance of a 13A Quotation), 13.3 (except compliance with an Architect's instruction for the* expenditure of a provisional sum for defined work or of a provisional sum for Performance Specified Work), 13A.4.1, 23.2, 34, 35 or 36; or

.5 .2 in regard to the opening up for inspection of any work covered up or the testing of any of the work, materials or goods in accordance with clause 8.3 (including making good in consequence of such opening up or testing) unless the inspection or test showed that the work, materials or goods were not in accordance with this Contract;

25.4 .6 .1 where an Information Release Schedule has been provided, failure of the Architect to comply with clause 5.4.1;

25.4 .6 .2 failure of the Architect to comply with clause 5.4.2;

25.4 .7 delay on the part of Nominated Sub-Contractors or Nominated Suppliers which the Contractor has taken all practicable steps to avoid or reduce;

25.4 .8 .1 the execution of work not forming part of this Contract by the Employer himself or by persons employed or otherwise engaged by the Employer as referred to in clause 29 or the failure to execute such work;

.8 .2 the supply by the Employer of materials and goods which the Employer has agreed to provide for the Works or the failure so to supply;

25.4 .9 the exercise after the Base Date by the United Kingdom Government of any statutory power which directly affects the execution of the Works by restricting the availability or use of labour which is essential to the proper carrying out of the Works or preventing the Contractor from, or delaying the Contractor in, securing such goods or materials or such fuel or energy as are essential to the proper carrying out of the Works;

25.4 .10 .1 the Contractor's inability for reasons beyond his control and which he could not reasonably have foreseen at the Base Date to secure such labour as is essential to the proper carrying out of the Works; or

.10 .2 the Contractor's inability for reasons beyond his control and which he could not reasonably have foreseen at the Base Date to secure such goods or materials as are essential to the proper carrying out of the Works;

25.4 .11 the carrying out by a local authority or statutory undertaker of work in pursuance of its statutory obligations in relation to the Works, or the failure to carry out such work;

25.4 .12 failure of the Employer to give in due time ingress to or egress from the site of the Works or any part thereof through or over any land, buildings, way or passage adjoining or connected with the site and in the possession and control of the Employer, in accordance with the Contract Bills and/or the Contract Drawings, after receipt by the Architect of such notice, if any, as the Contractor is required to give, or failure of the Employer to give such ingress or egress as otherwise agreed between the Architect and the Contractor;

25.4 .13 where clause 23.1.2 is stated in the Appendix to apply, the deferment by the Employer of giving possession of the site under clause 23.1.2;

25.4 .14 by reason of the execution of work for which an Approximate Quantity is included in the Contract Bills which is not a reasonably accurate forecast of the quantity of work required;

25.4 .15 delay which the Contractor has taken all practicable steps to avoid or reduce consequent upon a change in the Statutory Requirements after the Base Date *which necessitates some alteration or modification to any Performance Specified Work;*

25.4 .16 *the use or threat of terrorism and/or the activity of the relevant authorities in dealing with such use or threat;*

25.4 .17 *compliance or non compliance by the Employer with clause 6A.1;*

25.4 .18 *delay arising from a suspension by the Contractor of the performance of his obligations under the Contract to the Employer pursuant to clause 30.1.4.*

2.5 COMPARISON OF CLAUSES 25 AND 26

2.5.1 Clause 25 cannot be read in isolation. In particular, it must be read in conjunction with clauses 23 and 24 which are complementary to it and the effects of which are considered later in this chapter at section 2.11. Moreover, it cannot be isolated from clause 26 either as this covers matters which are similar to some of the events in clause 25 although for quite different purposes.

2.5.2 There are dangers in considering clause 25 alongside clause 26; equally there are dangers in ignoring that consideration. The danger in considering them together is that it might be tempting to conclude that the one depends upon the other or at least that the amount of loss and/or expense is exactly *pro rata* to the extension of time or vice versa (see also para. 2.1.9). On the other hand, because many of the causes of delay under clause 25 are the same as the grounds for additional reimbursement under clause 26, it is equally dangerous to ignore the facts and judgments accepted in determining extensions of time when later considering loss and/or expense. What is required is a proper understanding of what the relevant clauses say and what their objectives are. (See also para. 4.4.13.)

2.5.3 In any event clauses 26.3 and 26.4.2 require that the Architect state in writing to the Contractor what extensions of time he has granted in respect of those *Relevant Events* which also qualify as *matters* under clause 26 giving rise in principle to a monetary claim. (See also para. 2.9.2.)

2.5.4 Clause 25 enables the Architect to amend the Completion Date in a variety of circumstances and in consequence alter the Contractor's liability for liquidated and ascertained damages for delay. As has been suggested earlier, this facility is there to protect the Employer's rights as much as to reduce the Contractor's liability (para. 2.2.2). Clause 26 details the rules for determining the Contractor's entitlement to reimbursement for loss and/or expense if variations are introduced or if the Contract is disrupted by the Employer or his agents.

2.5.5 There can be an entitlement to an extension of time without an entitlement to reimbursement of loss and/or expense (see table at para. 2.1.9). On the other hand, there can be an entitlement to reimbursement of loss and/or expense without an entitlement to an extension of time; for example, a delay or disruption on an element of the building which, although it has involved the Contractor in extra expense due to loss of output, has not and does not become critical to completion of the contract on time. This is illustrated in Example 2.3.

31

Example 2.3 Delay on a non-critical element

A Contractor plans to undertake part of the landscaping on a building contract from time to time whilst the building is under construction as and when labour becomes available. In practice, work on this landscaping is interrupted by variations. Lack of progress on this is not critical to the completion of the building as a whole so no extension of time is granted. However, output from the men working on the landscaping is reduced because of the interruptions by the variations, so direct loss and/or expense is payable which would equate to the cost of the loss of output.

2.5.6 Even where an event gives rise to a claim for both an extension of time and reimbursement of loss and/or expense, it by no means follows that the one is a measure of the other. One long delay, due say to the postponement of the whole of the works for three weeks, might cost the Contractor far less in terms of disruption (particularly if he had prior warning of such a postponement and assuming that he could redeploy his men to other sites) than a whole series of minor delays totalling a 3-week extension to the Completion Date (see also paras 4.4.14 and 4.4.17). Moreover, when granting an extension of time, under clause 25.3.1 the Architect must make a reasonable estimate of the delay which will necessarily entail a logical analysis of the delay rendered to completion, i.e. he must generally look forward; whereas when determining the entitlement to loss and/or expense he must ascertain it after it has been incurred. As always in the matter of claims, every event must be investigated and evaluated on its merits. It is no good, to use the words of Recorder Toulson QC, to make an "impressionistic assessment" of the situation (see *John Barker Construction Ltd -v- London Portman Hotel (1996) 12 Const. LJ 277*).

2.6 CLAUSE 25 – THE BROAD COMPONENTS

2.6.1 Clause 25 is best seen as comprising six parts:

a) the requirement for the Contractor to give timeous notice of delay, and to give further details to assist the Architect in establishing any extension of time (clause 25.2. refers);

b) the requirement for the Architect in certain circumstances to grant extensions of time during the construction period or to notify the Contractor that an extension of time is, in his view, not justified (clause 25.3.1 refers);

c) the requirement or permission for the Architect to review the Completion Date after completion (clause 25.3.3 refers);

d) the effect of omissions upon extensions of time (clauses 25.3.2 and 25.3.3.2 refer)

e) the list of events which entitle the Architect to consider granting an extension of time – these are termed Relevant Events in JCT 98 PWQ (clause 25.4 refers);

f) the general obligations imposed upon the Architect and Contractor in relation to their conduct (clauses 25.3.4, 25.3.5 and 25.3.6 refer).

2.6.2 This general process is illustrated in *Figure 2.4.*

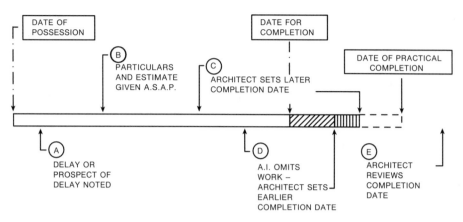

Figure 2.4 General process of granting extensions of time.

2.6.3 Event A   A delay, present or future, is reasonably apparent. The Contractor gives notice including:

a)   material circumstances;
b)   causes of delay;
c)   identifying any Relevant Event.

Any Nominated Sub-Contractor mentioned in the material circumstances, must be informed.

2.6.4 Event B   For each and every Relevant Event the Contractor must give, as soon as practicable:

a)   particulars of the expected effects;
b)   an estimate of the effect, if any, upon the Completion Date;
c)   to any Nominated Sub-Contractor notified under Event 'A' above, the details of the particulars and estimate.

2.6.5 Event C   The Architect asks himself:

a)   is the event referred to in the notice a 'Relevant Event'?;
b)   if so, will it delay the Completion Date?

If the answer is "yes" to both questions, the Architect fixes a new Completion Date and states which of the Relevant Events he has taken into account (as this is the first exercise of his duty under clause 25.3.1 the Architect will not have taken any omissions into account but on every subsequent occasion when fixing a new Completion Date he must state the extent to which he has taken omissions into account). If the Architect does not believe that an extension of time is justified he must so notify the Contractor. All this within 12 weeks of receipt of adequate particulars and estimate and before the Completion Date, whichever is the earlier. All Nominated Sub-Contractors must be informed.

2.6.6 Event D   The Architect issues an instruction omitting work. He can take this into account and fix an earlier Completion Date provided:

a) this is not earlier than the original Date for Completion and does not erode the period of time allowed for a variation for which there has been a confirmed acceptance of a clause 13A Quotation or in respect of a Variation or work for which an adjustment to the time for completion of the Works has been accepted pursuant to clause 13.4.1.2A7 (see paras 3.3.5 and 3.4.5);

b) the instruction of the omission took place after the last extension of time was granted or after a new Completion Date has been stated in a confirmed acceptance of a clause 13A quotation (strangely, acceptance of an adjustment to the time for completion of the Works pursuant to clause 13.4.1.2A7 is not part of this proviso).

2.6.7 Event E   If the Completion Date has passed the Architect may, and within 12 weeks of Practical Completion must, review the Completion Date and either

a) confirm it, or

b) extend it, or

c) reduce it – but only if there has been an instruction for the omission of work or obligations issued since the last fixing of a new Completion Date or if a new Completion Date has been stated in a confirmed acceptance of a clause 13A quotation (again, note that the acceptance of an adjustment to the time for completion of the Works pursuant to clause 13.4.1.2A7 is not part of the proviso).

2.6.8 In his final review, the Architect will take into account all Relevant Events whether notified or not. The reason for this provision is referred to at para. 2.2.2. This requirement, which is often overlooked in the early stages of a project, has very wide implications for the Architect. The effect of it is to require the Architect to note the incidence and the effect upon progress and completion of every Relevant Event even where the Contractor has not reported it. Thus the Architect must make his own arrangements to ensure that he is kept fully aware throughout the contract of the progress of the Works and of the causes and effect of all delays. (See also paras 2.7.34 and 2.7.35.)

2.7   CLAUSE 25 – THE DETAILED COMPONENTS

2.7.1 The following paragraphs elaborate upon the above process. Clause 25 commences by making it clear that reference to delay, notice, or extension of time includes further delay, further notice or further extension of time. Moreover, in the Definitions at clause 1.3, the term Completion Date refers to the Date for Completion as stated in the Appendix to the Contract or any other date fixed under clauses 25 or 13A whereas the term Date for Completion refers only to the date fixed and stated in the Appendix. Strangely, acceptance of an adjustment to the time for completion of the Works pursuant to clause 13.4.1.2A7 does not feature in the definition of Completion Date at clause 1.3 but it is suggested that this apparent oversight on the part of the JCT draughtsman does not undermine the logic of this thesis. That is, the term Completion Date includes all extensions of time granted from time to time whereas the term Date for Completion does not.

Adjustment to the Completion Date resulting from a confirmed acceptance of a clause 13A Quotation

2.7.2 In January 1994, the JCT introduced Amendment No 13 to JCT 80. It provided the facility for the cost and time effects of a proposed Variation to be agreed prior to its sanction. This is more fully considered in Chapter 3. This route provides for adjustment to be made to the time required for the completion of the Works (clause 13A.3.2.3). This in effect takes the place of any extension of time in respect of such a Variation and this agreed change to the contract period, once accepted and confirmed, cannot later be changed by the Architect under clause 25 (clauses 25.3.2 and 25.3.3).

2.7.3 Reference is made in clause 13A to the ... *adjustment to the time required by the Contractor for completion of the Works* ... in contrast with other opportunities for changing the Completion Date which are referred to as ... *fixing such later date as the Completion Date* ... Note that in making the *adjustment to the time required for completion of the Works* under clause 13A.2.2 the Date for Completion (i.e. that stated in the Appendix to the Contract Conditions) can be brought forward and thus the original contract period shortened. This is the only express provision in the contract for reducing the original contract period.

Adjustment to the time for completion of the Works resulting from the acceptance of a Contractor's Price Statement pursuant to clause 13.4.1.2A7

2.7.4 In April 1998, JCT issued Amendment 18 to JCT 80 which, inter alia, introduced into clause 13.4.1 an alternative method for the agreement of the cost and time effects of Variations, provisional sum work and work covered by Approximate Quantities. This is more fully considered in Chapter 3. This route like that under clause 13A, provides for an adjustment to the time for the completion of the Works (clause 13.4.1.2A7.2) although, unlike clause 13A, it does not lay down the procedure for the translation of the accepted adjustment of time into a revised Completion Date. Nonetheless it is suggested that such an acceptance takes the place of any extension of time in respect of such Variations, provisional sum work and work covered by Approximate Quantities and cannot later be changed by the Architect under clause 25 (clause 25.3.2).

2.7.5 Like clause 13A reference is made in clause 13.4.1.2A to the ... *adjustment to the time for the completion of the Works* ... in contrast with the other opportunities for changing the Completion Date which are referred to as ... *fixing such later date as the Completion Date* ... Unlike clause 13A, there is no express provision for fixing a date earlier than the Date for Completion given in the Appendix to the Contract but neither is this expressly prohibited.

Clause 25.2 – notices and details to be provided by the Contractor

2.7.6 With the exception of adjustments to the time required by the Contractor for completion of the Works accepted pursuant to clauses 13A and 13.4.1.2A, the event that sets the machinery for granting an extension of time in motion in JCT 98 PWQ is a delay, or the prospect of a delay in the progress of the Works, becoming apparent. The giving of such notices are therefore mandatory and must be given even if the Contractor is not entitled to an extension of time. The purpose of this requirement to give notice is to keep the Architect fully informed as to any delay in progress throughout the project duration. Failure so to keep the Architect informed may well hamper him in determining the proper extension of time in due course and a Contractor who is clearly not fulfilling this obligation to give notice every time that a delay occurs should be warned of this prospect.

2.7.7 A Contractor who receives an Architect's instruction requiring a substantial variation to be executed at some future date, which when executed will inevitably affect progress, must give notice of this prospect at the time he gets the instruction. The clause does not say to whom the delay should be apparent; the delay, or the prospect of a delay, simply has to be such as to be reasonably noticeable even if it is not noticed. Contractors must therefore scrutinise all instructions from the Architect upon their receipt. For as will be seen later, to fail in giving the proper notice will defer the Contractor's right to an extension of time until after Practical Completion is reached. The notice from the Contractor must, in accordance with clause 25.2.1.1:

a) be in writing;
b) be given ... *forthwith* ... upon the delay, or the prospect of a delay, becoming reasonably apparent;
c) give the material circumstances including the cause or causes of the delay; and
d) identify from those causes any event which in the opinion of the Contractor is a Relevant Event, i.e. one recognised by clause 25.4.

2.7.8 Thus the Contractor must include in his list every cause of delay whether or not they are categorised as Relevant Events under clause 25.4; for example, he may well include in his list failure of a domestic sub-contractor – even though this will not rank for an extension of time. But where, and only where, a Relevant Event is identified in the notice the Contractor must in accordance with clause 25.2.2 give, in respect of each and every such Relevant Event:

a) particulars of the expected effects; and
b) an estimate of the delay in the completion of the Works.

2.7.9 This further information must be given in the original notice if this is practicable, and if not practicable, as soon as possible after the original notice has been issued. The fact that the Contractor does not necessarily have to give these extra details at the time of the original notice should not be taken as implying that the Contractor can wait until the event has passed and worked its way through the programme; this extra data comprises the *expected* effects and an *estimate* of the expected delay in completion. So as soon as it is practicable to make such forecasts the Contractor must do so.

2.7.10 Such further information must be kept up to date as may be necessary, or as the Architect may reasonably require (clause 25.2.3). Thus, even if the Contractor does not consider it necessary to provide updated information, the Architect can insist upon its supply as long as he is acting reasonably. It would seem to be very much in the Contractor's interests to keep this information up to date. Delays to progress often have a continuing impact on the project and a Contractor who has given the initial data and who chooses not to keep this up to date may well be implying that no continuing delay exists; the requirement to supply further data does not depend upon the Architect asking for it since it has to be supplied ... *as may be reasonably necessary.*

Clause 25.3.1 – extension of time granted by the Architect during the period of construction

2.7.11 The Contract requires the Architect to consider extensions of time both during construction and again after Practical Completion. However, the requirement to consider granting an extension of time during construction is largely, even entirely, dependent upon the Contractor first having given the notice referred to above as laid down in clause 25.2.1.1; clause 25.3.1, which deals with these **interim** extensions of time, begins *If, in the opinion of the Architect, upon receipt of any notice* ... If no notice is given the Architect is not compelled to grant an extension of time. Indeed he may well be acting outside his powers if he were to do so. Moreover the notice, to comply with the Contract, must be issued *forthwith* upon it becoming reasonably apparent that the progress of the Works is being or is likely to be delayed. Thus although, in JCT 98 PWQ , the Architect is put under some pressure in that there is a time limit within which extensions of time must be granted, the Contractor, if he wants an extension of time to be considered before Practical Completion, appears to have an even more onerous task. Failure to comply with the strict terms of the Contract relating to the timing and content of the notice and particulars may prejudice the right to an extension of time until after Practical Completion is reached.

2.7.12 If the Architect does receive the notice, particulars or estimates referred to in the clause he must, under clause 25.3.1, form an opinion on:

a) whether any of the events notified by the Contractor is a Relevant Event; and
b) whether the completion of the Works is likely to be delayed thereby beyond the Completion Date.

2.7.13 If the answer to both questions is in the affirmative then the Architect must fix such later Completion Date ... *as he then estimates to be fair and reasonable.* The word *then* in this context envisages that he may later form a different view, but should nevertheless grant an extension at the time. The Architect would be well advised only to grant, at this interim stage, that part of an extension of time which he can estimate with confidence; because the review required of him after Practical Completion only enables him to confirm or increase any extension given in a previous decision; he cannot reduce it unless omissions have been instructed since the last extension of time (para. 2.7.36). This is not to say that the Architect can give a paltry extension of time and appear to comply; he must give such extension as he estimates at the time to be fair and reasonable. However, to be fair to both parties he should be aware that whilst he can increase this extension after Practical Completion he cannot reduce it.

2.7.14 The Architect, when considering granting an extension of time, is not concerned with the current delay to progress but with the likely effect of that delay upon the Completion Date. Thus in forming his view he will need to consider whether the delay is on a critical element and the effect of float in the Contractor's programme.

2.7.15 A simple delay in progress during the contract is not sufficient grounds for an extension of time; it must be of such a size or on such an element of the works as will have an inevitable effect upon the Completion Date. Reference to Example 2.4 will illustrate this proviso.

Example 2.4 The effect of naturally occurring float upon delay

Assume a Contract in which the roof to a particular building cannot be started until the external walls and the internal structural partitions are completed, the latter taking less time to erect than the former (see *Figure 2.5*). A delay to the internal partitions at any point in the period marked X will not affect the Date for Completion until the float which occurs naturally at Y, has been fully eroded.

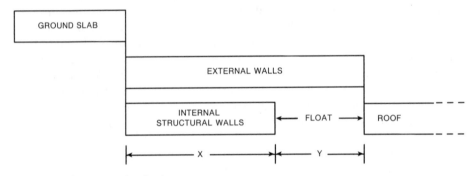

Figure 2.5 Float occurring in the programme.

2.7.16 A distinction must be drawn between naturally occurring float such as has been referred to above and float specifically introduced by the Contractor as a contingency against matters which are at his risk. Such matters may be:

a) unexceptional, albeit, adverse weather conditions;
b) lack of output or management by his workforce or that of his domestic Sub-Contractors;
c) rectification of errors in the works including the opening up and testing thereof;
d) failure to give proper notice of outstanding information;
e) failure of work by Nominated Sub-Contractors after Practical Completion of their works;
f) vandalism;
g) where any other sub-clause of clause 25.4 has been deleted, the events that would have been covered by that sub-clause.

2.7.17 The prudent Contractor will make a reasonable allowance for these in his programme either by an overall allocation at the end of the contract period or by marginal increases in each element of it as shown in *Figures 2.6* and *2.7*.

2.7.18 It is also to be recommended that this contingency allowance is referred to in the Contractor's programme because, it is suggested, the Architect would be bound to reflect this in any extension of time he grants, as shown in Example 2.5.

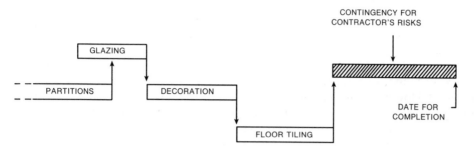

Figure 2.6 Float allocated at the end of a programme.

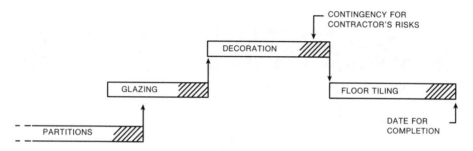

Figure 2.7 Float allocated to each element of construction.

Example 2.5   The effect of contingent float upon extensions of time

A Contractor prepares a programme for the works which allows 3 weeks in a 60-week programme to cover his risks (i.e. 5%). A delay occurs due to extra work being introduced which, at face value, the Architect estimates will result in a 5-week delay to the completion date. The Architect should grant an extension of time of $5\frac{1}{4}$ weeks for this item to give the Contractor comparable cover for the risks he shoulders; even the $\frac{1}{4}$ week would need to be assessed having regard to the circumstances – if the delay occurred during the winter months, for example, he may feel it proper to make more than an average allowance for unexceptional, albeit adverse, weather conditions.

2.7.19 Such a contingent float in the Contractor's programme should not be offset by the Architect against any delays which otherwise would rank for an extension of time – these contingent allowances are there to enable the Contractor to achieve his obligation to finish on time. On the other hand the Contractor has the right, but cannot demand the facilities in terms of the supply of outstanding details and the like, to work to a significantly shorter programme than that laid down in the Contract Conditions (paras 2.8.43 and 2.8.44). (For a consideration of the role of the Contractor's programme, see section 2.10.)

2.7.20 If the Architect, upon receipt of the relevant notice, particulars and estimate believes that an extension of time is not justified he must notify the Contractor accordingly (clause 25.3.1).

2.7.21 The matter of extensions of time is made more complicated by the introduction of clauses 13A and 13.4.1.2A. These provisions are referred to in greater detail in Chapter 3. However, it is to be noted that clause 13A allows a Completion Date to be agreed between the parties which is in advance of the original Date for Completion i.e. it permits a reduction in the original contract period and whilst not expressly providing for this, clause 13.4.1.2A does not prohibit it (see para. 2.7.5). This said, the Architect is barred from fixing a Completion Date in advance of the Date for Completion when operating the provisions of clause 25. This means that there can be circumstances in which the Architect may be debarred from granting an extension of time even though the circumstances warrant such a course as Example 2.6 shows.

Example 2.6   An anomaly resulting from the implementation of clause 13A

A contract has a Date for Completion at the end of week 50. Work is omitted by a variation for which there has been a confirmed acceptance of a clause 13A Quotation which provides for the Completion Date to be fixed at the end of week 45 (i.e. a reduction in the original contract period of 5 weeks). Subsequently there is an addition of work which is to be valued under clause 13.4.1.2B and which justifies an extension of time of one week which would have resulted in a Completion Date of the end of week 46. However, this would mean fixing a Completion Date four weeks earlier than the Date for Completion stated in the Appendix which is barred by clause 25.3.6. The Architect cannot in all conscience reinstate the original Date for Completion as the new Completion Date as being ... *fair and reasonable* ... (i.e. in effect giving an extension of time of 5 weeks) for his conclusion is that an extension of only one week is justified.

The timing of extensions of time

2.7.22 JCT 98 PWQ imposes a time limit upon the granting of these interim extensions of time and on notifying the Contractor that in the Architect's opinion no extension of time is justified. This time limit is 12 weeks from the receipt of such data ... *if reasonably practicable having regard to the sufficiency of the aforesaid notice, particulars and estimates* ... or by the Completion Date whichever is the sooner (clause 25.3.1).

2.7.23 If the data contained in the Contractor's notice, particulars and estimate are not sufficient to make it reasonably practicable to achieve this time limit the Architect is not relieved of his obligation to make the interim award; he is simply relieved of the need to make it within the 12 weeks. Thus the shortage of data may be grounds for the Architect to take longer in forming his view but the contract does envisage the Architect ultimately granting an extension of time at this interim stage even if the inadequacy of the data from the Contractor makes a decision within the 12 weeks impracticable.

2.7.24 The requirement in clause 25.3.1 that the final extension must be given within 12 weeks of receipt of the notices and the like is qualified as follows; ... *or, where the period between receipt* (of sufficient data) *and the Completion Date is less than 12 weeks, not later than the Completion Date.* Thus if the necessary data is only supplied, say, one week before the Completion Date the Architect, it might appear, is bound to respond in that one week as the Contract requires these interim extensions to be granted by the Completion Date at the latest. This provision might in practice be said to prevent the Architect from granting any interim extension in the context of Example 2.7.

Example 2.7 Delay after the expiry of the Completion Date

The Completion Date of a Contract is the 1st September but the Contractor fails to meet this date and after this date is passed there is a further delay caused to progress by delay on the part of a Nominated Sub-Contractor. The Architect would appear to be under no obligation to grant an extension of time because the Completion Date has already passed; although under clause 25.3.3 he **may** do so.

2.7.25 The wording of the whole of clauses 25.3.1 and 25.3.3 needs very careful consideration. The Architect's overriding obligation is contained in the wording between clause 25.3.1.2 and clause 25.3.1.4, namely to fix ... *such later date as the Completion Date as he then estimates to be fair and reasonable.* Note that the requirement as to what is fair and reasonable does not relate to the extension of time but, rather, to the Architect's opinion in respect thereof. The requirement to grant an extension of time – fix a new Completion Date – and to do so within a time limit of 12 weeks is dependent upon the sufficiency of the notice particulars and estimate.

2.7.26 Thus, in the example described above, it can be argued that as the data did not exist by the Completion Date it was clearly not sufficient to make it practicable to grant the extension of time by the Completion Date.

2.7.27 Clause 25.3.3 states, inter alia, *After the Completion Date, if this occurs before the date of Practical Completion, the Architect may* ... review and revise the fixing of any previous Completion Date in accordance with the somewhat limited provisions of that clause (paras 2.7.32 to 2.7.35).

2.7.28 What the clause seems to be saying is:

a) the Architect must make a fair and reasonable extension of time during the currency of the work on site if he is satisfied that the Completion Date will be delayed and that the Contract conditions have been complied with (e.g. proper notice and particulars are given);

b) once the Contractor has given all the necessary data the Architect must act within 12 weeks;

c) but if the existing Completion Date occurs during this 12-week period then the Architect must give his award before that date provided that the data from the Contractor makes this practicable;

d) where this is not practicable due for example to the inadequacy of the data from the Contractor, or where a delay occurs after the Completion Date, the Architect

41

may review and revise his earlier decision but is not bound to do so until the overall review he is required to undertake by not later than 12 weeks after the date of Practical Completion.

2.7.29 As a rule, the shorter the period from the supply of the necessary data to the existing Completion Date the more explicit and incontrovertible that data must be. If, for example, there remain only two days between the date upon which the Contractor supplies the data and the existing Completion Date, that data must be very clear and the arguments wholly acceptable to the Architect. For if the Architect has any questions to ask before he can accept the Contractor's submission, that data will not have been sufficient to make it reasonably practicable to grant the extension of time in the two days. Moreover, it may well be that the Architect considers that there are concurrent causes of delay and this may raise the question of whether payment may ensue (para. 3.6.1).

2.7.30 It is quite possible that the Architect having given an interim extension of time will increase this when making his final review. In the meantime liquidated and ascertained damages may have been charged and the payment of fluctuations may have been frozen in accordance with clauses 38.4.7, 39.5.7 or 40.7.1, whichever is applicable. However, there is provision for repaying any such liquidated and ascertained damages that transpire to have been overcharged (para. 2.11.8).

2.7.31 If the Architect does give an interim extension of time under clause 25.3.1 that transpires to have been inadequate, it would appear that he must await the review following the Completion Date before he can revise this judgment; when acting under clause 25.3.1, to grant an interim extension of time the Architect may only have one bite of the cherry up until the Completion Date has passed. Except when adjusting the time for completion of the Works pursuant to clauses 13A and 13.4.1.2A, the Contract does not require the Architect and Contractor to agree on extensions of, or reductions in, time; the Architect must fix such new Completion Dates as he judges to be fair and reasonable. Nevertheless, such a decision can be challenged at Adjudication, Arbitration or in Substantive Legal Proceedings.

Clause 25.3.3 – extensions of time granted by the Architect after Practical Completion

2.7.32 Within 12 weeks of the date of Practical Completion, the Architect must finalise his views on extensions of time. In doing so he must, in accordance with clause 25.3.3:

a) fix a Completion Date later than that previously fixed; or
b) fix a Completion Date earlier than that previously fixed if the omissions, which justify it, have been instructed since the last extension of time was granted (para. 2.7.38); or
c) confirm the Completion Date previously fixed.

2.7.33 But the Architect cannot do nothing; he must respond to one of these three options. Moreover, he has 12 weeks within which to make up his mind and this requirement is not qualified in any way even if there is insufficient data from the Contractor or no notification at all from him. The Architect should, therefore, canvass the Contractor at or just before the date of Practical Completion for any data that would assist the

Architect in forming his view on this matter. The Contractor would be well advised to respond fully as the Architect could not be blamed if he granted inadequate extensions of time due to insufficient information from the Contractor.

2.7.34 In forming this final view on extensions of time, the Architect is required not to limit his review to his earlier decisions on the Contractor's applications for extensions of time nor to the Relevant Events notified by the Contractor. He must take into account all Relevant Events whether notified by the Contractor or not (paras 2.2.2 and 2.6.8). In this context, he is required to fix a Completion Date that ... *is fair and reasonable having regard to any of the Relevant Events, whether upon reviewing a previous decision or otherwise and whether or not the Relevant Event has been specifically notified* ... (clause 25.3.3.1).

2.7.35 The requirement for the Contractor to provide notice of delay, particulars of its effect and an estimate of the expected delay in the completion of the Works is only a condition precedent to the granting of an interim extension of time, i.e. one to be granted before the Completion Date. The Contract clearly envisages that the Architect will, for the purposes of finalising the extensions of time, keep himself fully in the picture throughout the contract period as to the progress of work, the reasons for and the impact of delays that occur and their effect upon the Completion Date notwithstanding any failure by the Contractor so to notify him. Nevertheless, the Contractor should bear in mind that the need to give notice of a prospective delay is an unqualified one; that is to say it is required of him whether or not he is seeking an extension of time and whether or not the cause of the delay is a Relevant Event (clause 25.2.1.1).

Clauses 25.3.2 and 25.3.3.2 – the effect of omissions upon extensions of time

2.7.36 JCT 98 PWQ permits the Architect to take omissions into account when assessing extensions of time. However, this right is subject to very specific conditions given in clauses 25.3.2 and 25.3.3.2 which may be summarised as follows:

a) under no circumstances can the Architect fix a Completion Date in advance of that named in the Appendix to the Contract, i.e. he can never reduce the original contract period (clause 25.3.6). Nor can he alter the period of time included in respect of a Variation for which there has been a confirmed acceptance of a clause 13A Quotation or in respect of a Variation or work for which an adjustment to the time for completion of the Works has been accepted pursuant to clause 13.4.1.2A7;

b) before the right accrues to the Architect to consider the effect of omissions he must first have considered at least one extension of time and have either fixed a new Completion Date or have confirmed the existing one; or there must have been a change to the Completion Date as a result of a confirmed acceptance of a clause 13A Quotation (clause 25.3.2). The maximum scope he has for reducing the period for completion due to any such omissions is the aggregate of the extension of time to be granted by him up to that time otherwise he will have offended against the restriction referred to at a) above;

c) every time an extension of time is granted, a guillotine drops on the Architect's
    ability to consider the effect of any omission that has been instructed up to that
    date. Thus every time the Architect is disposed to fix a new, later, Completion
    Date or to confirm the previous Completion Date he must carefully review all
    omissions of work instructed since he last fixed a new Completion Date.

2.7.37 Example 2.8 illustrates these points.

Example 2.8   The effect of omissions upon extensions of time

Figure 2.8 illustrates events on a building Contract of 52 weeks' duration. At week 4,
before any delays have been notified to the Architect, he instructs an omission of work.
He judges that this omission is such that 2 weeks will be saved in the construction
period. At week 6 the Contractor notifies him of delays for causes (Relevant Events)
which in principle entitle him to an extension of time and he claims a 3-week extension
and submits adequate particulars thereof. At week 17 (i.e. within the 12-week period laid
down), the Architect grants an extension of three weeks by fixing a new Completion Date
at week 55; he cannot take the 2 weeks saving into account as this is the ... *first exercise
by the Architect of his duty under clause 25.3.1* ... (clause 25.3.2). At week 20, the
Architect issues an instruction omitting further work and in his judgment a further 5
weeks construction time will be saved. He decides to do nothing at this stage although
he is entitled under clause 25.3.2 to consider fixing an earlier Completion Date. At week
24 the Contractor notifies the Architect of further delays, submits adequate particulars
and claims a further 1-week extension. The Architect agrees to grant the 1-week extension, but at this stage he can take into account the saving of 5 weeks due to the second
omission of work as that instruction was issued after the last occasion upon which he
made an extension of time. He cannot, however, take into account the first omission of
work as it took place before that date. He is thus disposed to fix a Completion Date 4
weeks earlier than previously fixed (1-week extension less 5 weeks' reduction).
However, this would mean reducing the original contract period as the previous extension was only 3 weeks and this is not permitted (clause 25.3.6). At week 36 he therefore
fixes an earlier Completion Date at the original Date for Completion – week 52. On the
next occasion he considers an extension of time he can review only the omissions that
took place since week 36 for the guillotine referred to earlier (i.e. the limitation of his
right to go back in time to review omissions that have been instructed) drops on the ...
last occasion on which the Architect fixed a new Completion Date.

However, the position was different under the JCT 80 form prior to the Amendments
introduced in May 1993. Under the provisions of that earlier form the Architect could
review all the omissions that took place back to week 17 in this example because the
guillotine was ... *the last occasion on which the Architect ... made an extension of time.*
And at week 36 the Architect fixed a new Completion Date by making a reduction in time.

2.7.38 When reviewing the Completion Date after that date has passed or after the date of
Practical Completion, the Architect can again fix such an earlier Completion Date as
he deems appropriate as long as it is in respect of omissions instructed after the last

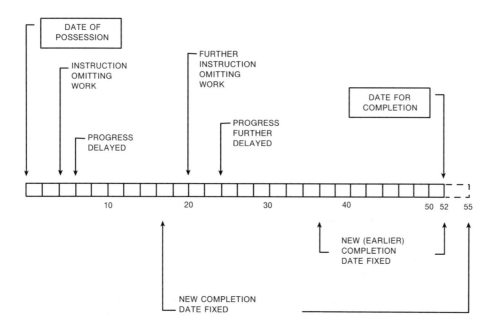

Figure 2.8 Series of events in a 52-week programme.

occasion on which he fixed a new Completion Date and as long as he does not fix a Completion Date earlier than that in the Appendix to the Contract nor erode any period established in respect of a confirmed acceptance of a clause 13A Quotation or in respect of a Variation or work for which an adjustment to the time for completion of the Works has been accepted pursuant to clause 13.4.1.2A7. It is therefore perfectly possible for the Contractor to be working to an extended Completion Date, to achieve completion on the day named only to find that, after the Architect has reviewed matters, the finishing post has been brought forward due to omissions and that liquidated and ascertained damages for delay become chargeable. However, this should not come as a complete surprise to the Contractor who is familiar with the provisions of the Contract. It is only omissions instructed since the date of the last extension of time that count, so that each time an instruction is received which omits work he will be aware of the possibility that the prevailing Completion Date will be brought forward. If he is wise he will, upon receipt of such an instruction to omit work, write to the Architect asking him to exercise his powers under clause 25.3.2 so that the Contractor may know where he stands.

2.7.39 The Contractor who gives no notice of delay during the currency of the work is not entitled to any extension of time until the Architect undertakes his review of the Completion Date under clause 25.3.3, i.e. after that Completion Date has passed. As no extension of time will therefore have been given up to that time – the Contractor not having asked for one – the Architect will have no opportunity to take account of the effect of any omissions at all for even the ability to consider omissions in the final review is conditioned by the fact that the omissions must have been instructed ... *after the last occasion on which the Architect fixed a new Completion Date* (clause 25.3.3.2).

2.7.40 It would appear that the unduly sophisticated provisions in relation to taking omissions into account when considering extensions of time have been designed to ensure that the Contractor is not caught unawares by working to a Completion Date which is subsequently brought forward in time as a result of earlier omissions. In practice, however, it may tempt the Architect to delay issuing instructions requiring an omission until he has granted at least his first extension of time. Equally it may tempt the Contractor to delay applying for an extension of time until after Practical Completion as a means of denying the Architect the opportunity of taking the effect of omissions into account. Such a manipulation of the administration of the Contract should be resisted.

2.7.41 The ability to take account of the effect of omissions on the Completion Date is a welcome feature; whether the tortuous conditions adopted make it worthwhile is questionable.

The advantages of promptly fixing a new Completion Date

2.7.42 It is important to note that all that is required of the Architect at the interim stage of considering extensions of time is that he determines the length of a delay ... *as he then estimates to be fair and reasonable* ...; it is a matter for the Architect's judgment not an ex post facto calculation of the delay. The Contract does not say "when the full effects of the cause have become manifest"; it requires the Architect to estimate. If this seems a difficult task, bear in mind that in the tender period the Contractor has to estimate, in a period of four to six weeks, the time necessary for the completion of the whole of the Works (or determine a management strategy necessary to meet the period laid down in the tender documents) and he will have been looking at the project for the first time.

2.7.43 On practical grounds, the earliest possible resolution of extensions of time will be of inestimable benefit to both parties. The Contractor's agent on site will have a new target at which to aim and his director, if he has only limited management available, will place it on the site where the Completion Date has been repaired. A project which has many proper grounds for extensions of time but none granted may find itself beset by indifferent management on the basis that if there is no target, there is no point in aiming.

2.7.44 From the Employer's viewpoint there are also benefits, quite apart from the benefit he will enjoy from the more certain management referred to above – after all, most clients would prefer to have their building finished on time rather than damages in lieu. He will benefit from the certainty of a new Completion Date rather than a vague feeling that he can only rely on possession of the building when it is completed whenever this might be; with the formalities of the granting of extensions of time following after like some irrelevant paper exercise. The Architect is the person charged with the responsibility of making the extension of time, and he can be forgiven for feeling at the time that it is apparent that a delay has occurred or will occur, that the matter is too confused to permit a clear judgment to be made. He can be tempted to conclude that after the passage of time all will become clearer, and that until that occurs he is not able to make a fair and reasonable extension of time. However, experience shows that the passage of time does little to clarify such issues: more likely it will confuse and obscure. The Architect may delay granting an extension of time due to say extra works discovered in the foundations on the understandable grounds that the real effect of this delay upon the whole of the works will only become clear when it has worked its way through the building. But in practice he will almost certainly discover by that time that there is a series of other delays which compound the problem rather than isolate it; for example, variations, exceptionally adverse weather conditions or late information. His reluctance is sometimes aided and abetted by a misunderstanding of the terms of clause 25: that the granting of a timely extension to the Completion Date will deny the Employer the possibility that the Contractor might make up some of the delay. But, it should be noted that, in fixing a new Completion Date the Architect must in any event take account of the proviso in clause 25.3.4 which requires the Contractor to ... *use constantly his best endeavours to prevent delay ... and to prevent the completion of the Works being delayed or further delayed ...* It is with this obligation on the Contractor in mind that the Architect must fix the new Completion Date. It is important therefore that when the Architect estimates any extension of time due to the Contractor, he makes proper allowance for the effect upon progress that the Contractor has, or should have, achieved in using his best endeavours.

2.7.45 This proviso, in clause 25.3.4 including the reference to ... *best endeavours* ..., is a very stringent one. It would seem to impose upon the Contractor an obligation to do all that it is in his power to do to reclaim the earlier completion date. In the case of *Sheffield District Railway Company -v- Great Central Railway Company (1911) 27 TLR 4512* "best endeavours" was said to require the party under such an obligation to "*leave no stone unturned ...*" and in the case of *IBM (UK) Limited -v- Rockware Glass Limited (1980) 2 Lloyds Rep 608* such a party was said to be required to "*... take all the steps in their power which are capable of producing the desired result being steps which a prudent, determined and reasonable owner, **acting in his own interest** and desiring to achieve that result, would take*". In *Keating on Building Contracts* (Sixth Edition, page 642), it is suggested that the obligation imposed by clause 25.3.4 does not contemplate the expenditure of substantial sums of money. The financial implications of such an obligation are considered at para. 4.2.11 to 4.2.13.

2.8   THE RELEVANT EVENTS

2.8.1 These are the events which in principle entitle the Contractor to claim an extension of time. Any event which occurs and delays the Completion Date not appearing in this list is at the Contractor's risk.

Clause 25.4.1 – force majeure

2.8.2 The term *force majeure* comes from French law (Code Napoléon) – the statement of its meaning was approved in *Lebeaupin -v- Crispin [1920] 2 KB 714*, thus,

> *This term is used with reference to all circumstances independent of the will of man, and which is not in his power to control ... thus war, inundations and epidemics are cases of force majeure, it has even been decided that a strike of workmen constitutes a case of force majeure.*

2.8.3 In Keating on Building Contracts (Sixth Edition, page 643), it is observed by reference to the Lebeaupin case that a *force majeure* clause *"... should be construed in each case with a close attention to the words which precede or follow it and with due regard to the nature and general terms of the contract."* This being so, the term clearly cannot cover items which appear expressly in their own right in clause 25.4. The term is also likely to exclude any of the causes struck out of clause 25.4; Lord Cross, in *Mottram Consultants Ltd -v- Bernard Sunley & Sons Ltd (1974) 2 BLR 28* stated: *"When the parties use a printed form and delete parts of it one can, in my opinion, pay regard to what has been deleted as part of the surrounding circumstances in the light of which one must construe what they have chosen to leave in"*, e.g. a strike of workmen under JCT 98 PWQ will not constitute *force majeure* since this eventuality is specifically catered for by virtue of clause 25.4.4 and even if clause 25.4.4 were to be deleted then such a strike would still not count as *force majeure* since one would pay regard to what the parties had deleted. In the same vein, one can comment with authority, that *force majeure* will not cover the usual incidents of life which must be taken into account, e.g. bad weather, football matches or a state funeral (per Mr Justice Bailhache in *Matsoukis -v- Priestman [1915] 1 KB 681* in a case concerning the general strike of 1912). So whilst little **specific** help can be offered on what *force majeure* does cover, one can decide what it does not cover: it does not extend to any of the other causes referred to in the remainder of clause 25, nor on the basis of the above extract, to any deleted therefrom. Perhaps the most apt comment on *force majeure* comes from the late Vincent Powell-Smith, the well-respected legal commentator on building contract matters. He once said *"The term is, I suspect, widely used simply because it is so imprecise"* [1].

Clause 25.4.2 – exceptionally adverse weather conditions

2.8.4 What constitutes exceptionally adverse weather conditions is not defined, nor is it possible to seek to introduce a definition or yardstick in the Contract Bills owing to the general embargo contained in clause 2.2.1 (paras 1.2.11 to 1.2.13). It is worth noting that JCT 98 PWQ refers to *"adverse"*, as opposed to *"inclement"* in JCT 63, hence weather conditions other than that which are inclement are to be taken into account, e.g. a prolonged drought. This said, the operative words "exceptionally adverse" do point to conditions which are seriously bad and rare.

Clause 25.4.3 – loss or damage occasioned by any one or more of the Specified Perils

2.8.5 This sub-clause requires little comment. The Specified Perils are defined in clause 1.3 as being ... *fire, lightning, explosion, storm, tempest, flood, bursting or overflowing of water tanks, apparatus or pipes, earthquake, aircraft and other aerial devices or articles dropped therefrom, riot and civil commotion, but excluding Excepted Risks.* On the surface, in that some of these occurrences such as bursting pipes may be due to the Contractor's negligence, the Contractor might appear in such circumstances to be entitled to an extension of time. However, it is suggested that the proviso at clause 25.3.4.1 ... *the Contractor shall use constantly his best endeavours to prevent delay* ... would eliminate this possibility.

Clause 25.4.4 – civil commotion, local combination of workmen, strike or lockout affecting any of the trades employed upon the Works or any of the trades engaged in the preparation, manufacture or transportation of any of the goods or materials required for the Works

2.8.6 In *Levy -v- Assicurazioni [1940] 3 All ER 427* it was decided that the words "civil commotion" amounted to a stage between a riot and a civil war. For the definition of a riot (the state to be exceeded for civil commotion to have been established), attention is drawn to the Public Order Act 1986 which provides that where twelve or more persons who are present together use or threaten unlawful violence for a common purpose and the conduct of them (taken together) is such as would cause a person of reasonable firmness present at the scene to fear for his personal safety, each of the persons using unlawful violence for the common purpose is guilty of riot.

2.8.7 As to the matter of strike or lockout, it should be noted that this extends to the manufacture of materials; thus even a strike in a foreign country affecting the manufacture of a piece of machinery destined for the work would be grounds for considering an extension of time. It would not extend, for example, to strikes in timber mills resulting in a general shortage of timber; one must be able to demonstrate that the materials in question were specifically destined for the works.

Clause 25.4.5.1 – compliance with the Architect's instructions under clauses 2.3, 2.4.1, 13.2 (except for a confirmed acceptance of a 13A Quotation), 13.3 (except compliance with an Architect's instruction for the expenditure of a provisional sum for defined work or of a provisional sum for Performance Specified Work), 13A.4.1, 23.2, 34, 35 or 36

2.8.8 This sub-clause does not deal with delay caused by the failure of the Architect to provide information timeously in accordance with clause 5.4; this is dealt with under clause 25.4.6. It should also be noted that this sub-clause 25.4.5.1 does not refer to every instruction issued by the Architect, only to those issued under the specific clauses mentioned, namely:

a) clause 2.3 discrepancies;
b) clause 2.4.1 discrepancies relating to Performance Specified Work;
c) clause 13.2 instructions requiring a variation (with exceptions);
d) clause 13.3 instructions on provisional sums (with exceptions);
   clause 13A.4.1 instructions requiring that work for which a Quotation under clause 13A has been submitted but not accepted is to be valued under clause 13.4.1 and not by reference to that Quotation;
e) clause 23.2 postponement of any work;
f) clause 34 antiquities;
g) clause 35 Nominated Sub-Contractors;
h) clause 36 Nominated Suppliers.

Generally

2.8.9 These clauses are considered in detail below but first it should be noted that the above list does not cover further information issued under clause 5.4.2, or drawings issued by a Nominated Sub-Contractor.

2.8.10 For the issue of further information under clause 5.4.2 or the issue of information by a Nominated Sub-Contractor to be a ground for an extension of time it must either have been the subject of a failure by the Architect to provide the information timeously in accordance with clause 5.4 or constitute a variation under clause 13.

2.8.11 Clause 5.4.2 states inter alia:

> *the Architect as and when from time to time may be necessary without charge to the Contractor shall provide him with 2 copies of such further drawings or details which are reasonably necessary to explain and amplify the Contract Drawings and shall issue such instructions (including those for or in regard to the expenditure of provisional sums) to enable the Contractor to carry out and complete the Works in accordance with the Conditions.*

2.8.12 This, it is often said, means the Contractor cannot be required to undertake any design work. But it is suggested that this is not entirely correct. It would seem perfectly possible under the Contract to specify in the Contract Bills that the Contractor is to design, say, the roof trusses having been given the spans, the rise, the superimposed loads and the like. This requirement to design will become part of ... *the Works in compliance with the Contract Documents* (clause 2.1). All that the Contractor can expect under clause 5.4, where the Contract Bills require the Contractor to design the roof trusses, is that the Architect provides the drawings necessary to enable the Contractor to carry out the Works including the preparation of the design drawings for those roof trusses, not that the Architect provides those design drawings himself. Certainly the Contractor's responsibility for the adequacy of that design is not clearly spelt out, but it would seem that it is possible in principle to require him to undertake such work. Without doubt, however, the better course where work is to be defined by performance or where the Contractor is to undertake some element of design is either to use the facilities in clause 42 (Performance Specified Work) of JCT 98 PWQ or to use the JCT Contractor's Designed Portion Supplement, as appropriate.

2.8.13 It is very common for specialist Nominated Sub-Contractors to be required to produce drawings or details for their works, and the responsibility for so doing is contained in the Warranty between the sub-contractor and the Employer (para. 6.1.6). So far as the main Contractor is concerned, he will still look to the Architect for the supply of this information. Any failure by a Nominated Sub-Contractor to supply this in time, or any discrepancies in this information, may be grounds for an extension of time under clauses 25.4.5 or 25.4.6 as if the failure were that of the Architect; moreover such failures may also be grounds for reimbursement to the main Contractor of any loss and/or expense under clause 26.

2.8.14 The Employer's remedy in these instances is to proceed against the Nominated Sub-Contractor for breach of warranty (para. 6.1.6).

Clauses 2.3 and 2.4.1 – discrepancies

2.8.15 These clauses deal with the discovery of discrepancies; they both contain the words:

> If the Contractor shall find any discrepancy in or divergence between ... he shall immediately give to the Architect a written notice specifying the discrepancy or divergence and the Architect shall issue instructions in regard thereto.

2.8.16 The Contract does not specifically charge the Contractor with a duty to look for any such discrepancies and the like; it simply records the machinery to be followed if they are discovered. However, Keating suggests that the Contractor's obligations should go further:

> It is an implied term of the contract that the Contractor in his workmanship will use the proper skill to be expected of a Contractor. It is suggested that a Contractor who fails to observe errors which ought to have been obvious to him as the Contractor in the circumstances and exercising such proper skill cannot take advantage of such failure [2].

2.8.17 Thus in determining the Contractor's entitlement under these clauses the Architect should consider the time at which a prudent Contractor would have noticed the discrepancy and to judge the necessary extension of time accordingly. On the basis of Keating's view, a Contractor who blunders on without noticing a fairly obvious discrepancy only to exacerbate the resultant delay (and cost) may well find himself left with the responsibility for some of it. In any event, the Contractor has a duty to search for the information available in good time in order to determine whether he requires any further data and this search alone, if properly carried out, should highlight obvious discrepancies. However, in exercising his judgment on this matter, the Architect should remember that it is his duty under the Contract to supply and be responsible for all necessary information. This is illustrated by an example in Chapter 4 – Example 4.12 (para. 4.6.5).

Clauses 13.2 and 13.3 – Variations

2.8.18 Instructions given under clauses 13.2 and 13.3 are those requiring a variation, but clause 25.4.5.1 excludes references to:

a) work in respect of which a 13A Quotation has been accepted by the Employer and confirmed by the Architect;
b) instructions relating to provisional sums for defined work;
c) instructions relating to provisional sums for Performance Specified Work.

2.8.19 The execution of work included as a provisional sum in the Contract Bills will not rank as a ground for extensions of time where it is in respect of defined work: that is to say, where the following information is given: (see footnote in JCT 98 PWQ to clause 1.3):

a) the nature and construction of the work;
b) how and where the work is fixed to the building and what other work is to be fixed thereto;

    c)   quantities indicating its scope and extent;

    d)   specific limitations relating to the work.

2.8.20 However, even then the execution of such work may qualify for an extension of time if in the event the work that is required differs from that described in the Contract Bills and thereby changes the conditions under which other work is executed. In this case this other work will be considered a variation under clause 13.2 (clause 13.5.5).

2.8.21 The practical effect of a variation upon the Contractor's progress may be quite different to that envisaged by the Architect. In particular, some materials may require a substantial lead time for ordering and the substitution of a new material during the progress of the work can have a disproportionate effect upon progress. Thus a close liaison between the Architect and Contractor on prospective variations is necessary.

2.8.22 All too often the prospect of variations is kept from the Contractor until the last minute, perhaps on the mistaken grounds that to reveal the intention too soon will be to give the Contractor a greater opportunity to exploit this to his own end. But even if the Contractor were so minded, his entitlement has to be supported by facts and he has a continuing obligation to proceed regularly and diligently with the Works. Indeed, if the Architect warns the Contractor in good time of prospective variations and the Contractor is thus made aware of areas of possible change, far from giving him an opportunity to exploit this potential, it gives far more bite to the proviso in clause 25.3.4.1 that ... *the Contractor shall use constantly his best endeavours to prevent delay...*; he is thus expected to take steps to mitigate the effect of such a change.

Clause 13A.4.1 – Contractor's 13A Quotation not accepted

2.8.23 This clause applies where the Contractor has submitted a Quotation under clause 13A which has not been accepted but where the work is nevertheless instructed. This will then be valued as if it were an ordinary variation plus a reasonable sum to cover the cost of preparing the Quotation. (See para. 3.4.8.)

Clause 23.2 – Architect's instructions – postponement

2.8.24 This refers to ... *the postponement of any work to be executed under the provisions of this Contract.* This does not refer to the delayed possession of the site or any part of it. Such a failure would be a breach of contract by the Employer and therefore the remedy should not be by means of an extension of time under clause 25 unless such a delay is covered by a relevant entry in the Appendix (para. 1.2.10); clause 23.1 states clearly that ... *possession of the site shall be given to the Contractor* ...

Clause 34 – Antiquities

2.8.25 Clause 34 deals with the discovery of antiquities and as well as giving the Architect fairly wide powers on what is to be done it also imposes upon the Contractor severe restrictions (including ceasing work if necessary). All this can be grounds for an extension of time.

Clauses 35 and 36 – Nominations

2.8.26 Under clauses 35 and 36 the Architect is empowered to issue instructions in the instances listed in the table below and any one of these instances may be considered as grounds for an extension of time provided that it is considered they will delay progress and ultimately the Completion Date. Such instructions are, however, not grounds for claiming reimbursement of loss and/or expense and this would seem to be a consistent policy, namely that only acts, omissions or defaults of the Employer, or those he controls, are grounds for claiming reimbursement of loss and/or expense and for extensions of time; whereas delays caused by events outside the control of both parties to the Contract are grounds for considering an extension of time only.

List of Architect's instructions possible under clauses 35 and 36 – Nominated Sub-Contractors and Nominated Suppliers

Clause	Purpose of the instruction
35.5.2	to remove any reasonable objection raised by the Contractor to a nomination instruction or cancelling the nomination instruction and then either omitting the relevant work or nominating another Sub-Contractor;
35.6	to nominate a Sub-Contractor;
35.9	to remove the grounds for non compliance with a nomination instruction or cancelling the nomination instruction and then either omitting the relevant work or nominating another Sub-Contractor;
35.18.1.1	to nominate a "substituted Sub-Contractor" to rectify defects which the Nominated Sub-Contractor has failed to rectify;
35.24.6	to instruct the Contractor to give a notice to the Nominated Sub-Contractor specifying a default;
35.24.7	to complete and/or remedy the work of a Nominated Sub-Contractor whose employment has been determined;
36.2	to nominate a Supplier.

2.8.27 The above list of instructions which qualify as Relevant Events under clause 25.4.5.1 appears to be an extensive one. However, the fact that an instruction is given under any one of the above headings will not of itself justify an extension of time; it has, in common with all the other Relevant Events, to have created a delay to the progress of the Works to the extent, or on such a critical element, that the Completion Date will also be delayed. In addition, the Contractor has an overriding duty to prevent delay and to proceed with the Works (paras 2.7.45 and 2.9.8).

Clause 25.4.5.2 – compliance with the Architect's instructions in regard to the opening up for inspection of any work ...

2.8.28 There is little one need say on this sub-clause except to note that unless the tests show the work or materials, etc., to be unsatisfactory then an entitlement to extensions of time will flow, even though such tests were required because faults were revealed from earlier tests in similar circumstances which revealed errors. The same does not apply when it comes to considering the reimbursement of direct loss and/or expense (para. 3.5.8).

Clause 25.4.6.1 – where an Information Release Schedule has been provided, failure of the Architect to comply with clause 5.4.1

2.8.29 In April 1998, JCT issued Amendment 18 to JCT 80 which, inter alia, introduced a wholly new option, described in the Sixth Recital, which provides ... *the Employer has provided the Contractor with a schedule ("Information Release Schedule") which states what information the Architect will release and the time of that release.* It also introduced clause 5.4.1 which requires that ... *the Architect shall ensure that 2 copies of the information referred to in the Information Release Schedule are released at the time stated in the Schedule ...*

2.8.30 In those instances where the Employer chooses to delete the Sixth Recital, clauses 5.4.1, 25.4.6.1 and 26.2.1.1 will be of no effect. But, where not deleted clause 5.4.1 places an obligation on the Architect to ensure that information on the Information Release Schedule is released at the time stated in the Schedule. Failure to do so, subject to an exception and a proviso, will result in entitlement to an extension of time and, by virtue of clause 26.2.1.1, reimbursement of any loss and/or expense incurred provided, of course, that a delay to the Completion Date and/or actual loss and/or expense is thereby caused. It is therefore incumbent on the Architect to ensure that the Information Release Schedule is realistic both in terms of when it is possible for the Architect to provide the information and, perhaps more importantly, when the information will be required by the Contractor to enable completion of the Works in accordance with the Conditions. This is to be commended as introducing some parity in the relationship between the Architect and the Contractor – it serves also to place the Architect in the role of being proactive in the production of information rather than, as some Contractors have been known to suggest, simply reactive.

2.8.31 The architect is exempted from his duty under clause 5.4.1 only if and to the extent that he is prevented from performing it ... *by the act or default of the Contractor or any person for whom the Contractor is responsible ...*

2.8.32 The operation of clause 5.4.1 is also subject to an important proviso which recognises that as the Works progress the timing of the release of information may need to be changed and expressly provides for such change to be made by further agreement between the Employer and the Contractor ... *which agreement shall not be unreasonably withheld or delayed ...*

Clause 25.4.6.2 – failure of the Architect to comply with clause 5.4.2

2.8.33 JCT also introduced a new clause 5.4.2, as part of Amendment 18. This new clause effectively combined the wording of pre-Amendment 18 clauses 5.4 and 25.4.6 and eliminated some but not all of the difficulties which previously existed because of them. It also dramatically switched the responsibility for ensuring the timeous provision of information from the Contractor to the Architect, effectively giving contractual form to the key principle in *Neodox -v- Swinton and Pendlebury BC (1958) BLR 34* where Mr Justice Diplock (as he then was) stated that information must be released to the Contractor in a reasonable time. Clause 25.4.6 in JCT 80 operated only when the Contractor had not ... *received in due time necessary instructions ... from the Architect ... for which he specifically applied in writing ...* whereas clause 25.4.6.2 of JCT 98 PWQ becomes operational upon the ... *failure of the Architect ...* to provide such data as may be necessary but it is not conditional upon prior receipt of a request from the Contractor for that information except as noted in the next paragraph. One has to ask if the Architect's assessment of the need for information, and the timing of its release, will coincide with that of the Contractor!

2.8.34 The only occasion when clause 5.4.2 requires action by the Contractor in relation to the timeous provision of information is ... *Where the Contractor is aware and has reasonable grounds for believing that the Architect is not so aware of the time when it is necessary for the Contractor to receive such further drawings, details or instructions ...* When and if such a situation arises then the Contractor is required to ... *advise the Architect of the time sufficiently in advance of when the Contractor needs such further drawings or details or instructions, to enable the Architect to fulfil his obligations ...* but only ... *if and to the extent that it is reasonably practicable to do so ...*

2.8.35 In JCT 80, clause 25.4.6 was long and tortuous and created many difficulties not least of which were those revolving around the perennial questions relating to the meaning of *due time* and whether or not the Architect could take advantage of the Contractor's delay when issuing further information. The meaning of *due time* is no longer an issue because that phrase is no longer used but the other question remains.

2.8.36 Clearly the Architect must provide any outstanding information including drawings issued by Nominated Sub-Contractors (para. 2.8.13) in sufficient time so as not to delay the Works.

2.8.37 Moreover, the Architect in forming his judgment as to when to provide the further information must work on the assumption that the Contractor is on programme to complete by the Completion Date notwithstanding that he may be in delay due to his own default or be in technical delay in that extensions of time are due but not granted to date.

2.8.38 Except to the extent included in the Information Release Schedule Clause 5.4.2 requires the Architect to provide the Contractor with:

a) Such further drawings or details which are reasonably necessary to explain and amplify the Contract Drawings;

b) Such instructions (including those for or in regard to the expenditure of provisional sums) to enable the Contractor to carry out and complete the Works in accordance with the Contract Conditions.

2.8.39 These are to be supplied ... *as and when from time to time may be necessary and ... at a time when, having regard to the progress of the Works, or where, in the opinion of the Architect, Practical Completion of the Works is likely to be achieved before the Completion Date, having regard to such Completion Date ...* It would seem that the Contractor would have no claim against the Employer for the late supply of information provided it was supplied in time not to delay or disrupt in any way the Contractor's progress. Thus the Architect can, in effect, in programming to supply outstanding information, take advantage of any delay in progress by the Contractor. Given that there is a clear provision in the contract (clause 25.3.3) to grant extensions of time after the Completion Date has passed, it would seem that the Architect can issue instructions after the Completion Date and, unless he physically disrupts and delays the progress of the Contractor, he would not be required to grant an extension of time (nor to ascertain loss and/or expense). Thus an instruction given, say, six weeks after the Completion Date which delays the Contractor's progress by one week would give rise to an extension of time of one week not seven weeks and a similar approach should be applied to the ascertainment of direct loss and/or expense. This interpretation is confirmed in the Judgment of Colman J in the case of *Balfour Beatty -v- Chestermount Properties Ltd (1993) 62 BLR 1.*

2.8.40 However, there are a number of dangers in taking advantage of a Contractor's delay when issuing instructions, quite apart from the tension this can create with those whose duty it is to plan the progress of the Works. First, the Architect may find at the last minute that some unforeseen hitch occurs in his drawing office which delays the issue of the outstanding information and puts the Employer in default. Second, it may not be sufficient simply to supply drawings just before the Contractor puts the work in hand. The Contractor may have other activities which will depend upon the supply of information, for example:

a) to plan the work;
b) to order the material;
c) to arrange the hire and/or delivery of plant;
d) to reorganise his work force;
e) to hire labour or to negotiate terms with his men.

2.8.41 These are matters which are within the control of the Contractor and of which the Architect will have little experience, so this again is an area for close liaison between the Architect and Contractor. In general, all information should be supplied at the earliest possible moment to give the Contractor ample opportunity to plan his work. Also, the Contractor may wish to catch up on the delay caused by his own default – indeed it is his duty to do so; the Architect must be in a position to respond to such changing circumstances. Nonetheless for the Architect to be in default he must be shown to have actually delayed the progress of the work, not offended against a theoretical target. Nevertheless, he treads a tight-rope blindfold if he relies too heavily on a Contractor's delay as grounds for not issuing outstanding instructions.

2.8.42 The Contractor cannot properly argue that, because the outstanding information was not forthcoming, he slowed down progress or failed to bring forward other work whilst awaiting this information; clause 23 requires the Contractor to proceed regularly and diligently with the works; and clause 25 requires him to ... *do all that may reasonably be required ... to proceed with the Works ...* On the other hand, he cannot speed up or concentrate unduly on the one part of the Works where he suspects the Architect may have difficulty in meeting his obligations to supply all the information in good time in order to exacerbate a potential delay: this is not proceeding *regularly* with the Works. It must be admitted that, in practice, neither of these events (slowing down or speeding up in the anticipation of a delay in instructions) is easy to prove. So the Architect is again strongly advised to produce all necessary data at the earliest opportunity to avoid these arguments. Where this is not practicable, for whatever reason, he should inform the Contractor as soon as possible. This gains the maximum potential from the requirement in clause 25 that the Contractor ... *use constantly his best endeavours to prevent delay...*; he can only do this most effectively if he knows what is going on.

2.8.43 The Contractor may of course seek to programme his work to finish markedly in advance of the Completion Date; more so than simply to provide a reasonable management contingency in his programme. Clause 23 recognises the Contractor's duty to complete ... *on or before the Completion Date.* Thus a Contractor who is falling behind programme must be allowed every reasonable facility to catch up, but, having regard to clause 5.4.2, cannot expect to receive data at such a time that will enable him to finish in advance of the Completion date.

2.8.44 This interpretation is confirmed by the Courts in *Glenlion Construction Ltd -v- The Guinness Trust (1987) 39 BLR 89* where it was decided that whilst the Contractor was free to complete before the Completion Date the Employer had no duty to enable him to do so.

Clause 25.4.7 – delay on the part of Nominated Sub-Contractors or Nominated Suppliers which the Contractor has taken all practicable steps to avoid or reduce

2.8.45 Whilst much has been written on the subject of Nominated Sub-Contractors, as a result of two cases which were heard in the House of Lords (*North West Metropolitan Regional Hospital Board -v- T.A. Bickerton & Son (1970) 1 ALL ER 1039*, and *Westminster City Corporation -v- Jarvis (1970) 7 BLR 64*), nomination is far less popular than when the first edition of this book was published albeit that a word or two is necessary now lest the tide should turn again.

2.8.46 In the *Bickerton* case, the House of Lords concluded that, as the work which was the subject of a prime cost sum could, under the terms of JCT 63, only be carried out by a firm nominated by the Architect, then the Architect had a duty to renominate if a Nominated Sub-Contractor failed. However, whilst it is convenient to consider the *Bickerton* case when looking at clause 25.4.7, it is very doubtful if this sub-clause is wholly relevant. First, it is at least questionable whether delay following failure by a Nominated Sub-Contractor resulting in the determination of his employment con- stitutes delay *on the part of Nominated Sub-Contractors* and second, the Contractor would be wise to look elsewhere for his remedy because clause 25.4.7 is not matched by a comparable sub-clause at clause 26; that is to say, it is grounds for an extension of time but not reimbursement of loss and/or expense. And a re-nomination does not come within the scope of clause 26.

2.8.47 The *Jarvis* case, on the other hand, was specifically concerned with clause 23 of JCT 63 which in this context is identical to clause 25 of JCT 98 PWQ. The House of Lords decided that delay to the main Contract caused by remedying defects in the Nominated Sub-Contractor's work which came to light only after the sub-contractor had finished his work and left the site, did not constitute *delay on the part of Nominated Sub-Contractors*. Had the clause read "delay on the part of or caused by" the outcome might have been different.

2.8.48 During the case, the clause suffered much criticism in that it was suggested to be unfair to the Employer that delay on the part of Nominated Sub-Contractors should reduce the Employer's right to damages – although their Lordships held that this did not apply in the *Jarvis* case because of the particular events. It must be emphasised that the Contractor will only be entitled to an extension of time (not to reimbursement of loss and/or expense) in the event of delay on the part of the Nominated Sub-Contractor; he would have to proceed against the Sub-Contractor to recover any additional costs. Moreover, the Contractor is required to take ... *all practicable steps to avoid or reduce* ... any delay on the part of Nominated Sub-Contractors and Suppliers (clause 25.4.7). What is envisaged is presumably that as the Contractor is in contract with the Nominated Sub-Contractor it is he who has a measure of control and sanction over the Nominated Sub-Contractor and is therefore in the best position to bring pressure to bear upon any lack of performance.

2.8.49 Following the controversy over clause 23(g) in JCT 63, some Employers have attempted to overcome the objections aimed at the clause. There are, however, practical difficulties in achieving this. One can seek to re-establish the rights of the Employer to obtain redress for delay by either making the Contractor liable for any failure by Nominated Sub-Contractors, by deleting the clause altogether, or by making the Sub-Contractor liable direct to the Employer – for example by use of the Warranty (para. 6.1.6). The deletion of clause 25.4.7 will simply remove *Delay on the part of Nominated Sub-Contractors* or Nominated Suppliers as grounds for an extension of time. It will not deprive the Contractor of grounds for extension where the Nominated Sub-Contractor is delayed by *force majeure, exceptionally adverse weather conditions* or any other of the remaining causes listed in clause 25. The removal of clause 25.4.7 leaves the Contractor liable for delay caused by a defaulting Sub-Contractor or Supplier nominated to him by the Architect. It is possible that the extra burden created by the removal of clause 25.4.7 could encourage the Contractor to increase his tender to cover this prospect; and it is likely that the removal of this clause would extend the Contractor's scope for objection to a nomination afforded to him by clause 35.5.1 which states *No person against whom the Contractor makes a reasonable objection shall be a Nominated Sub-Contractor*. The use of the Warranty Agreement – Form Agreement NSC/W – creates a direct relationship between the Employer and the Nominated Sub-Contractor in which the Sub-Contractor warrants, inter alia, that he will not so delay the Works as to entitle the Contractor to an extension of time under clause 25.4.7 nor so act that the Architect has to consider issuing an instruction to determine the Employment of that Sub-Contractor. (This Warranty also deals with the failure of the Nominated Sub-Contractor to supply information in due time also but this is considered later, para. 6.1.6.) If he does so fail he is in breach of his undertaking and the Employer can sue him for damages. In that the Nominated Sub-Contractor's failure would give the Contractor an extension of time (provided the other conditions of clause 25 have been met) and thus deny the Employer liquidated and ascertained damages for the relevant period, these lost damages are likely to be part measure of the

Employer's claim. Nominated Sub-Contractors are often reluctant (to say the least) to enter into an agreement where although their work is a very small part of the whole, they are nevertheless responsible for the financial consequences of delaying the whole contract – see Example 2.9.

Example 2.9   Possible extreme damages due from Nominated Sub-Contractors

A Nominated Sub-Contractor who was required to execute £2,500 worth of floor finishes in a £500,000 contract would be reluctant to shoulder the potential burden of damages for delay which could amount to £1,000 per week.

2.8.50 In calculating the extension of time to which the Contractor may be entitled as a result of a delay on the part of a Nominated Sub-Contractor, the Architect must have regard not only to the general matters already referred to in this chapter but also the programme to which the Sub-Contractor is working under the terms of his Sub-Contract; an incompatible Sub-Contract programme would almost certainly result in the Contractor giving a notice of non-compliance under clause 35.8.2 justified by the Architect under clause 35.9.2. In a well-ordered world the nomination would be made at the outset of the Contract, the Sub-Contract would include a period for execution which was properly co-ordinated with the main Contract programme (or vice versa) and the Architect would simply have to measure the Sub-Contractor's performance against this yardstick.

Clause 25.4.8.1 – the execution of work not forming part of this Contract by the Employer himself or by persons employed or otherwise engaged by the Employer as referred to in clause 29 or the failure to execute such work

2.8.51 Clause 29 which qualifies what is described here, envisages that any work which will be carried out by the Employer (or by persons engaged or employed by him) will fall under one of two headings: clause 29.1 refers to such work where it is sufficiently well described in the Contract Bills in which event the Contractor must permit the execution of that work; and clause 29.2 refers to such work which is not adequately described in the Contract Bills in which case the Employer first requires the consent of the Contractor to have that work put in hand. Moreover, it is the execution of such work, as well as the failure so to execute it, that can be a ground for an extension of time.

2.8.52 This wording is likely to cause difficulty. What if the work undertaken by such persons is executed exactly as was envisaged in the Contract Documents, is the Contractor entitled to claim that the progress of the Works is nevertheless delayed thereby? This would seem wholly illogical and may not succeed as an argument. It can be argued that the Contractor is bound to permit the execution of that work and thus should have made adequate provision in his programme for it. Thus its execution – unless the same is itself late and/or disrupted – should neither delay the progress nor the completion of the Contractor's work. But what if the Contractor is first delayed by say his own mismanagement and thus the work executed by those directly employed Contractors is undertaken out of sequence so far as the Contract Works are concerned, is the Contractor then entitled to claim delay? Or if the Contractor wishes to speed up the works, and this work by direct Contractors prevents this, is he entitled to claim that he is delayed thereby?

2.8.53 This clause is not qualified in the same way as clause 25.4.10 (inability to obtain men or materials) where the Contractor must be able to demonstrate that he could not reasonably have foreseen the eventuality at the Base Date (usually a date at or just before the date for return of tenders). It is interesting to look at clause 28.2.2.3 where the words ... *delay in* ... have been added to the words ... *execution of work* ... (and) ... *supply of materials* ... This seems to indicate that clause 25.4.8.1 in contrast to clause 28 is operable even if there is no default on the part of these persons. From this one can deduce that clause 25.4.8.1 is also operable even if there is no default by the Employer or those acting for him. There is, of course, the safeguard that in any event there must be actual delay to progress of the Works. Thus it would seem that provided that the Contractor can demonstrate that such delay was caused, he might be entitled to an extension of time even though the work of the direct Contractors was carried out as outlined in the Contract Documents. As a safeguard against this possibility one should clearly lay down in the Contract Bills what this work comprises, when it is to be executed and how it is to be integrated in with other work, and require the Contractor to make due allowance for this in his master programme. In such cases, notwithstanding the niceties of the drafting, it could clearly be demonstrated that the execution of such work in that manner could not have had any effect upon the Completion Date and thus would not warrant fixing a new Completion Date.

2.8.54 Such difficulties are also likely to be encountered with clauses 25.4.8.2 (supply of materials by the Employer) and 25.4.11 (work by local authorities or statutory undertakers) where the wording is similar.

Clause 25.4.8.2 – the supply by the Employer of materials and goods which the Employer has agreed to provide for the Works or the failure so to supply

2.8.55 Note that it is the supply as well as the failure so to do which is a ground for extension of time; to this extent, therefore, it is similar to, and suffers the same difficulties as, clause 25.4.8.1 (see above). But again an actual delay to progress must have been caused.

Clause 25.4.9 – the exercise after the Base Date by the United Kingdom Government of any statutory power which directly affects the execution of the Works ...

2.8.56 This clause appears to have been very widely drawn; *the exercise ... of any statutory power* is a very far reaching expression with a scope that it is not possible to begin to catalogue. Nevertheless, it should be noted that where it can be shown that the Government has exercised one of its statutory powers, before it can be considered as a ground for an extension of time, it must have:

a) been exercised after the Base Date (generally a date just before the date for return of tenders). Whether "exercise" in this context means "putting into operation" or "keeping in operation" is not clear; reference to the Base Date implies that it means the former;

b) restricted the availability or use of labour which is essential to the proper carrying out of the Works; or

c) prevented the Contractor from, or delayed him in, securing goods, materials, fuel or energy which are essential to the proper carrying out of the Works.

2.8.57 It is the exercise of statutory powers, not the failure to do so, that is referred to in this clause; moreover, it is the exercise of the statutory power, not an announcement to do so on a named future date, that is the criterion.

Clause 25.4.10 – the Contractor's inability for reasons beyond his control and which he could not reasonably have foreseen at the Base Date to secure such labour (or such goods or materials) as are essential to the proper carrying out of the Works

2.8.58 This is a thoroughly unsatisfactory provision from the Employer's viewpoint who could be expected to ask what the role of the Contractor is if it is not to assemble the necessary resources for the project. However, those who are disposed to delete it from the Form of Contract should bear in mind that clauses 38.4.8, 39.5.8, or 40.7.2 would also require amendment. These clauses state that one can only "freeze" fluctuations payments where the Contractor is in delay (see section 2.12) if clause 25 is included in the Contract unamended.

2.8.59 This clause should be contrasted with the previous clause. As clause 25.4.10 refers to matters beyond the control of the Contractor, it presumably embraces the effects of Government exercising its statutory powers. However, clause 25.4.9 goes further than clause 25.4.10, as the table below shows.

Comparison of the potential problems covered by clauses 25.4.9 and 25.4.10

Item	Clause 25.4.9	Clause 25.4.10
Labour essential to the proper carrying out of the Works	Restricting the availability or, restricting the use	Inability to secure
Materials, etc. essential to the proper carrying out of the Works	Preventing the Contractor from or delaying him in securing	Inability to secure:
	Goods	Goods
	Materials	Materials
	Fuel	
	Energy	

2.8.60 There are three points that need to be noted in respect of each of the sub-clauses of 25.4.10:

a) the reasons creating the inability must be beyond the Contractor's control and not reasonably foreseeable at the Base Date (usually a date just before the date for submitting tenders). What does … *inability for reasons beyond his* (the Contractor's) *control … to secure …* mean? While labour and materials exist, they can be secured for the Works – at a price. What price should the Contractor offer in order to seek to secure resources to meet the requirements of this clause – the rates used in the tender; the rates being paid in the market at the time the work is executed; or at any price? The only proper criterion is the rate for labour and materials currently being paid in the market at the time the work is executed. It is impracticable to apply the first criterion (the rate used in the tender) as it will be difficult to determine because the Contractor's price is a lump sum tender for the Works and, moreover, such a view would leave the Employer vulnerable on a matter which is properly the Contractor's risk. It is a nonsense to apply the last criterion because, at any price, one can always secure resources. It is suggested that the Contractor's obligation, so far as this clause is concerned, is to pay the market price for the resource current at the time it is executed – albeit that this may be well in excess of that envisaged at tender stage. But nevertheless even this interpretation leaves one with considerable difficulties. The market place is continually changing; if the Contractor finds that he cannot secure the necessary resources almost certainly he is offering yesterday's prices in today's market. In short this is, as suggested earlier, a thoroughly unsatisfactory provision;

b) the Contractor's inability to obtain labour or materials must have arisen since the Base Date, i.e. there must have been a deterioration in the supply situation since that date;

c) the labour or material in question must be **essential** for the proper carrying out of the works, not simply desirable or preferable.

Clause 25.4.11 – the carrying out by a local authority or statutory undertaker of work in pursuance of its statutory obligation in relation to the Works, or the failure to carry out such work

2.8.61 The following particular points should be noted:

a) the authority or undertaker must be exercising its statutory obligation and not be executing other work for which it has been engaged by the Employer.

b) the clause refers to … *in carrying out work …* as well as to … *the failure to carry out work*. To this extent it is very similar to clause 25.4.8.1 (q.v.). On the surface, therefore, it would appear that as soon as an authority or undertaker does any work in pursuance of its statutory obligation there is a ground for an extension of time under clause 25. However, this work must first create a delay to the progress of the Works and ultimately the Completion Date and if the work of the authority or undertaker had been fully described in the Contract Bills the Contractor would have a responsibility to make reasonable provision in his programme for such work. The work causing the delay referred to in this clause is that which is being undertaken by any person or organisation carried out *in pursuance of its statutory obligations*; it does not therefore cover work undertaken in accordance with a contract with the Employer. In this latter event any delay in such work would come within the terms of clause 25.4.8.1. The significance of this distinction is that there is no comparable provision in clause 26 (recovery of loss and/or expense) for delay by those working in pursuance of their statutory obligations whereas there is a comparable provision for disturbance caused by those engaged directly by the Employer under clause 29. Thus delay and disturbance by the former would entitle the Contractor to claim an extension of time, but any associated costs would be at the Contractor's risk. Whereas delay and disturbance by the latter would entitle the Contractor to claim both an extension of time and reimbursement of direct loss and/or expense. This distinction was emphasised in the case of *Henry Boot Construction Ltd -v- Central Lancashire New Town Development Corporation (1980) 15 BLR 1*, where Judge Fay QC said:

> *If the Employers contract with the statutory undertakers, they can contract to provide for what is to happen if the undertakers are guilty of delay, just as they can so provide if they employ an artist or a tradesman, and it is just that they should bear this risk, which they had the opportunity of safeguarding themselves against. If, however, without having a Contract the undertakers, using their statutory powers to fulfil their statutory obligation came on the scene and hindered the works and caused delay, then the consequential loss would be one like force majeure which can be laid at the door neither of the Employers nor of the Contractors, and so ... the loss would lie where it falls.*

Clause 25.4.12 – failure of the Employer to give in due time ingress to or egress from the site of the Works ...

2.8.62 The clause envisages that the Contract Drawings or Contract Bills lay down the rights and obligations of the Contractor (including an obligation to give notice) with regard to:

> Ingress (right of entry) to the site of the Works and egress (right of exit) from the site of the Works including the right to pass through or over:

| Land
Buildings
Way
Passage | which are adjoining or connected with the site and which are in the possession and control of the Employer. |

2.8.63 The Contractor is not, under this clause, entitled to wholly unrestricted access, only access to the extent laid down in the Contract Drawings and/or Contract Bills. Moreover, failure by the Contractor to secure passage through land, buildings and the like not in the control and possession of the Employer would not be grounds for an extension of time. The Employer must himself have defaulted in honouring the commitment given in the Contract Drawings or Contract Bills, to the extent that he has control over that commitment. However, it is open to the Architect and Contractor to agree alternative means of ingress or egress, if what was offered in the Contract Drawings or Contract Bills cannot be met, and if such an agreement is reached the Employer would not be in default under this clause.

Clause 25.4.13 – where clause 23.1.2 is stated in the Appendix to apply, the deferment by the Employer of giving possession of the site under clause 23.1.2

2.8.64 The right of the Employer to defer giving possession of the site is not an automatic one; the parties must have opted for this facility by making the appropriate entry in the Appendix. Moreover, the maximum deferment recognised by the Contract is 6 weeks; the parties can however opt for a lesser period than 6 weeks when making that entry in the Appendix.

Clause 25.4.14 – by reason of ... an Approximate Quantity ... which is not a reasonably accurate forecast of the quantity of work required

2.8.65 An approximate quantity is described in the Standard Method of Measurement 7th Edition [4] as a quantity of work required that cannot be accurately determined where such work can nevertheless be described and given in items in accordance with the rules contained therein (General Rules item 10.1 refers).

2.8.66 It is not possible to lay down a rule of thumb as to what is a reasonably accurate forecast of the quantities. It must be borne in mind, however, that the discrepancy in quantities must always be such as to create a delay in the Completion Date for it to generate an entitlement to an extension of time. In the event much will depend on the criticality of the work, which is the subject of the Approximate Quantity, to achieving the Completion Date.

2.8.67 It is suggested that where the forecast of an approximate quantity transpires to have been too high, the reduction that results from substituting the lower, accurate, quantity would not entitle the Architect to reduce an extension of time under clause 25.3.2. This clause requires the omission to have been ... *instructed or sanctioned by the Architect* ... The correction of approximate quantities does not require an instruction from the Architect. Work is not being omitted; approximate quantities are being corrected.

Clause 25.4.15 – delay which the Contractor has taken all practicable steps to avoid or reduce consequent upon a change in the Statutory Requirements after the Base Date which necessitates some alteration or modification to any Performance Specified Work

2.8.68 The Contractor is not liable, so far as time is concerned, for the effect of changes in Statutory Requirements (e.g. in Building Regulations or Planning Permission) if this has an impact upon Performance Specified Work. Note, however, that, as with delays by Nominated Sub-Contractors the Contractor must take *all practicable steps to avoid or reduce* any such delay.

Clause 25.4.16 – the use or threat of terrorism and/or the activity of the relevant authorities in dealing with such use or threat

2.8.69 Note that threats of terrorism, as well as acts of terrorism, qualify under this clause. There is no definition of what an "authority" is under this clause. It is suggested that it is restricted to such organisations as the Police, the Fire Brigade and the Local Authority and would not extend to a private individual or organisation unless it is acting in compliance with powers given to it by statute.

Clause 25.4.17 – compliance or non-compliance by the Employer with clause 6A.1

2.8.70 Clause 6A.1 requires the Employer to ensure that the Planning Supervisor (appointed under the Construction (Design and Management) Regulations 1994) carries out his duties and that there is someone to act as Principal Contractor for the purposes of those regulations if this is not the Contractor for the Works. Again, note that it is compliance as well as non-compliance which qualifies for consideration for extension of time. This seems to leave the Employer with all the risks of implementing these regulations given that this is also a matter referred to in clause 26.2 as being a ground for claiming reimbursement of direct loss and/or expense.

Clause 25.4.18 – delay arising from a suspension by the Contractor of the performance of his obligations under the Contract to the Employer pursuant to clause 30.1.4

2.8.71 This clause was first introduced by JCT in April 1998 via Amendment 18 to JCT 80 to give effect to section 112(4) of the Housing Grants, Construction and Regeneration Act 1996.

2.8.72 A new clause 30.1.4 was also introduced as a part of Amendment 18 to comply with sections 112(1) to (3) of the same Act. This clause permits the Contractor to ... *suspend the performance of his obligations under* ... (the) ... *Contract to the Employer until* ... *payment in full occurs* ... It is to be noted that the right to suspend extends to all of the Contractor's obligations under his contract with the Employer and not just the obligation under clause 2.1 to ... *carry out and complete the Works* ... It therefore extends to such matters as protection of the Works, site security and the Contractor's obligations under the various insurance and indemnity provisions. In fact JCT advise, in their Guidance Notes relating to this clause, that if the ... *suspension right is exercised by the Contractor both the Contractor and the Employer should seek expert insurance advice* ...

2.8.73 The right to suspend is ... *Without prejudice to any other rights and remedies which the Contractor may possess* ... i.e. it does not impinge upon any of his common law rights. However, the Contractor may only suspend his obligations if the Employer shall ... *fail to pay the Contractor in full by the final date for payment* ... and then only if ... *such failure shall continue for 7 days after the Contractor has given to the Employer, with a copy to the Architect, written notice of his intention to suspend* ... The written notice must also set down ... *the ground or grounds on which it is intended to suspend* ... It is worth noting that even when the optional, Supplementary

65

Provisions for EDI (first introduced in JCT 98 PWQ) apply, such notice must be given in writing and cannot validly be given by EDI (Annex 2 clause 1.4.2 applies).

2.8.74 For the avoidance of doubt clause 30.1.4 expressly provides that a suspension pursuant to it is not one to which clause 27.2.1.1 refers nor is it a failure ... to proceed regularly and diligently with the Works ... contemplated by clause 27.2.1.2 – either of which would entitle the Employer to pursue Determination of his contract with the Contractor.

2.8.75 Exercise by the Contractor of his right to suspend under clause 30.1.4 will result in an entitlement to an extension of time and, by virtue of clause 26.2.10, reimbursement of any loss and/or expense incurred. It should be noted, however, that the Contractor may only ... *suspend such performance of his obligations ... until such payment in full occurs* ... This seems to imply that immediately upon *payment in full* the Contractor must recommence the performance of all his obligations to the Employer. This may be so but it is suggested that the date on which payment in full occurs does not define the limit of the extension of time to which the Contractor is entitled or the loss and/or expense he may recover. As it will take some time for the Contractor to gear up after suspension it is further suggested that any reasonable time taken to do so should be included in the extension of time granted under clause 25.4.18 and that the reasonable cost consequences of doing so should be reimbursed as loss and/or expense under clause 26.2.10.

2.9   GENERAL MATTERS ARISING FROM CLAUSE 25

2.9.1 Before moving on, it is worth reviewing a number of matters often raised in connection with clause 25 and these are:

a) delays which result from two parallel causes – concurrent delays;
b) phased or sectional completion;
c) the overriding qualifications to clause 25.

Delays which result from two parallel causes

2.9.2 Often a delay can be said to be due to more than one of the causes listed in clause 25; for example, there can be a delay due to *exceptionally adverse weather conditions* and yet there may also be inadequate information for work to have proceeded due to a delay in its issue. This provokes the question "Under which sub-clause is the extension to be given?" The purist's answer is that it doesn't matter; only one of the delays which occur due to parallel causes will rank for an extension of time and, as the purpose of clause 25 is simply to extend the completion date, which Relevant Event is to be relied upon is immaterial. This ignores the question lurking in the background, namely "Is the Contractor entitled to reimbursement of his direct loss and/or expense under clause 26?" Costs arising from disruption due to exceptionally adverse weather conditions do not rank for reimbursement under clause 26 whereas those due to late information do – clause 26.2.1. The answer to the question, it is suggested, is that generally such costs do not rank for reimbursement (para. 3.6.1). Thus whilst it is irrelevant which part of clause 25.4 is used as a ground for granting an extension of time the Architect needs to exercise care when stating, as he is required to do under clause 26.3 or clause 26.4.2, which sub-clause of clause 25.4 in his view prevails. The Architect needs also to be aware of the judgment in *Henry Boot Construction (UK) Ltd -v- Malmaison Hotel (Manchester) Ltd [1999] All ER (D) 1118* which provides that if there are two concurrent causes of delay, one of which is a Relevant Event (e.g. exceptionally adverse weather conditions) and one of which is not (e.g. because the contractor has a shortage of labour), then the contractor is not necessarily disentitled from getting an extension of time simply because of the concurrent effect of the other event.

Phased or sectional completion

2.9.3 JCT issued a supplement entitled "Sectional Completion Supplement and Practice Note 1" 1998 Edition which lists all the amendments which it is necessary to make to JCT 98 PWQ to provide for sectional completion (i.e. one in which each phase or section must, as a matter of contractual obligation, be finished by a given date failing which damages will flow).

2.9.4 For this to be effective there must be:

a) identification of the relevant sections with an adequate description of those works (including reference to such matters as the completion and operation of an associated boiler house or the completion of mains services if these are necessary to permit the section to operate as planned);

b) separate values for each section;

c) separate defects liability periods for each section;

d) separate Dates of Possession for each section;

e) separate Dates for Completion for each section;

f) separate rates for liquidated and ascertained damages for each section.

2.9.5 Phased or sectional completion, as described above, must be distinguished from "Partial possession by Employer" referred to in clause 18 of the Contract and from an intention appearing only on the Contractor's programmes that certain buildings or parts of the building will be completed in advance of the whole Contract.

2.9.6 Clause 18 is simply there to provide the administrative and contractual machinery for dealing with the situation in which the Employer and Contractor agree as a matter of convenience, rather than as a contractual right or duty, that part of the Works will be completed and handed over in advance of completing the whole.

2.9.7 Employers often mistakenly rely upon achieving the handover of part of the Works in advance of completing the whole simply because the Contractor has planned his work in that way. If this intention comes to fruition clause 18 will apply but the Employer cannot rely on such advanced dates; if the Employer wishes to rely upon such earlier dates then these must be written into the Contract as an obligation by use of the Supplement referred to above. Example 2.10 illustrates this.

Example 2.10 The effect of a single Date for Completion where there is more than one building

A construction project comprises three separate buildings each of which is planned to be completed by the same Date for Completion. This is fixed in the Appendix to the contract at the end of week 40. Extra work is instructed on one of the buildings only. The Architect judges that this entitles the Contractor to an extension of time of 5 weeks. However, the Employer wishes to take possession of the remaining two buildings at the original Date for Completion. He cannot demand this. There is only one Completion Date and this has been extended by 5 weeks. The Contractor will therefore have met his obligations as long as the whole project is completed by that date – i.e. the end of week 45.

The overriding qualifications to clause 25

2.9.8 Save in respect of a *Boot -v- Malmaison* type situation (para. 2.9.2), it is worth re-emphasising two overriding matters which qualify the Contractor's entitlement to an extension of time:

a) there must be a delay to the progress of the Works. Simply illustrating that a delay might have ensued in other circumstances is insufficient. For this reason many of the long debates that have and no doubt will take place on the interpretation of clause 25.4.6 (late information) might quickly become academic if one first asks the question "Did this cause an actual delay to the progress of the Works which would result in the Completion Date being delayed?"; if not it is irrelevant;

b) the Contractor is obliged to ... *use constantly his best endeavours to prevent delay in the progress of the Works, howsoever caused, ... and shall do all that may reasonably be required to the satisfaction of the Architect to proceed with the Works.* If he sees the prospect of delay he must take steps to prevent the delay; he is also required to use his **best** endeavours which is an onerous requirement (paras 2.7.45 and 4.2.11 to 4.2.13).

2.9.9 The Contractor is required by clause 25.2 to send to any Nominated Sub-Contractor any notice in which he is referred to together with copies of any further particulars or estimates that follow such a notice. This will enable the relevant Nominated Sub-Contractor to be aware of any criticism levelled at him and, if he feels so disposed, to make representations.

2.9.10 The Architect also has a number of general obligations:

a) not to fix a Completion Date earlier than that stated in the Appendix to the Contract, i.e. he cannot reduce the original Contract Period (clause 25.3.6). (The Completion Date can be set earlier than that stated in the Appendix if this is established in a confirmed acceptance of a clause 13A Quotation (para. 3.4.5). But this process is known in the Contract as **stating a Completion Date** (by the Contractor) and not **fixing a Completion Date** (by the Architect));

b) not to erode the period set for a variation to which a confirmed acceptance of a clause 13A Quotation applies or in respect of a Variation or work for which an adjustment to the time for completion of the Works has been accepted pursuant to clause 13.4.1.2A7 (clauses 25.3.2 and 25.3.3.3);

c) to notify every Nominated Sub-Contractor in writing of his decision every time he fixes a new Completion Date (clause 25.3.5);

d) give certain particulars in relation to his extensions of time when it comes to ascertaining loss and/or expense (clauses 26.3 and 26.4.2);

e) in any review of the Completion Date made after the Completion Date has passed (including the mandatory review to be made within 12 weeks after the date of Practical Completion) to take account of the effect upon the Completion Date of all Relevant Events **whether notified or not.**

2.10   THE STATUS OF THE CONTRACTOR'S PROGRAMME

2.10.1 JCT 98 PWQ contains a provision in clause 5.3.1.2 whereby the Contractor is required to provide two copies of his master programme for the execution of the Works. He is also to amend this every time the Architect makes a decision under clause 25.3.1, i.e. every time the Architect decides to fix a new Completion Date or to confirm the existing one; and every time that there is a confirmed acceptance of a clause 13A Quotation. The Contractor apparently is not required to amend the programme when the Architect gives a reduction in the time for completion under clause 25.3.2 or when an adjustment to the time for completion of the Works in respect of a Variation of work has been accepted pursuant to clause 13.4.1.2A7, which is wholly illogical. The master programme is not a Contract Document and it is made clear that it cannot impose obligations beyond those contained in the Contract Documents. Moreover, the master programme does not have to have the consent or approval of the Architect and no reference is even made to the fact that the master programme should properly reflect the Dates of Possession and for Completion.

2.10.2 It is difficult to see why the master programme should be a Contract Document, a provision often pressed for by some commentators. Nothing would be achieved if it were except perhaps that it would be easier to identify it in that it would be signed by the parties to the Contract. However, a failure to meet any intermediate date except the Completion Date (e.g. a failure to meet the date for completing the foundations) would not give rise to any damages being payable unless, of course, a series of dates for sectional completion were incorporated into the Contract, which is quite a different matter (paras 2.9.3 to 2.9.7).

2.10.3 The master programme supplied by the Contractor under clause 5.3.1.2 is not binding on either party but is an almost indispensable yardstick for measuring performance and the inter-dependence of activities. The Contractor would be entitled to change the strategy of his programme if, for example, work fell behind due to his default or if in the event he concluded that his initial thoughts on programme were subsequently shown to have been misplaced or inconvenient. A Contractor showing reasonable cause for such a change should expect no resistance from the Architect; it is the Contractor's responsibility to proceed with the works and to plan or replan his method of so doing. The programme should be used as a tool to assist both the Contractor and the Architect to progress and administer the contract efficiently and not as a weapon with which to fight each other.

2.10.4 A master programme is, therefore, a statement made in good faith by the Contractor as to how he proposes at that point in time to execute the work and could, if unamended, form the basis upon which the Architect might reasonably make his estimation of extensions of time.

2.10.5 The provisions in the Contract give little power to the Architect in relation to his right to make representations to the Contractor over the details in or revisions to the master programme.

2.10.6 The Architect, the other members of the Design Team and not least the Employer need to be assured that the programme is a realistic one. Thus it would seem wise to ensure that provisions are made in the Contract Bills to add further obligations upon the Contractor in this respect – provided that they do not seek to *"override or modify the application or interpretation"* of the Contract conditions as this would offend against clause 2.2.1.

2.10.7 From the point of view of good management, a programme ought to be required to be revised on at least the following additional occasions:

a) when there is a delay to the progress of the Works and the Contractor has fallen significantly behind programme whether or not a new Completion Date is to be fixed by the Architect;

b) upon the Architect consenting to an extension of time to a Nominated Sub-Contractor;

c) when the Architect fixes an earlier Completion Date under clause 25.3.2;

d) when an adjustment to the time for completion of the Works in respect of a Variation or work has been accepted pursuant to clause 13.4.1.2A7;

e) when reasonably required by the Architect.

2.10.8 The Contract Bills should also contain a provision setting out the detailed requirements of the master programme and stating that it is to be to the approval of the Architect and in such a form as to act as a yardstick when it comes to assessing the effect of a delay upon the Completion Date, i.e. that it must indicate which activities are critical and which may float in the programme and to what extent.

2.10.9 It is not unknown for a Contractor to produce a programme that is apparently at variance with the requirement of the Contract Conditions, i.e. the Date for Completion shown on the programme is much earlier than the Date for Completion shown in the Appendix to the Contract Conditions and using for his authority the fact that clause 23 requires him to complete the Works ... *on or before the Completion Date* ... To accept that the provisions of clause 23 enables the Contractor to finish say 6 months early in all circumstances is to make a nonsense of clause 5.4.2 (paras 2.8.43 and 2.8.44) and this view was borne out by the case of *Glenlion Construction Ltd -v- The Guinness Trust (1987) 39 BLR 89* where it was decided that whilst the Contractor was free to complete before the Completion Date the Employer had no duty to enable him to do so.

2.10.10 When reviewing the master programme put forward by the Contractor, the Architect should note particularly whether the Contractor has properly reflected the Information Release Schedule (if there is one) and the Completion Date and, where he has not done so, bring this to the Contractor's attention forthwith. If the Contractor persists in his proposal he should be warned that the master programme does not meet with the Architect's approval (assuming that the Contract Bills required this to be so) and that the Contractor therefore runs the risk of being delayed without remedy. Moreover, if the master programme (or any subsequent revision thereto) is not a reasonable reflection of the Completion Date and the programmes (or prospective programmes for the work of Nominated Sub-Contractors and Suppliers) the Architect may well be justified in concluding that it does not represent a proper basis for considering extensions of time. This may mean that making an award of an extension of time will take much longer than would be the case if a representative programme existed.

2.10.11 The Architect's obligation to grant an extension of time during the progress of the Works is qualified by his right to receive sufficient details from the Contractor and the lack of an adequately detailed and updated programme may well hinder achieving this state.

2.10.12 However, the above is not to be confused with the prudent Contractor allowing a contingency in his master programme against delays which are his responsibility, e.g. lack of output by his men, less than exceptionally adverse weather conditions (paras 2.7.16 and 2.7.17). Some such contingency ought to be introduced, either in each element of the work or at the end of the programme and it is suggested that an Architect, when estimating an extension of time ought similarly to allow a comparable contingency.

2.11   CLAUSES 23 AND 24 CONSIDERED IN DETAIL

2.11.1 These clauses read as follows:

---

23   Date of Possession, completion and postponement

23.1 .1   *On the Date of Possession possession of the site shall be given to the Contractor who shall thereupon begin the Works and regularly and diligently proceed with the same and shall complete the same on or before the Completion Date.*

23.1 .2   *Where clause 23.1.2 is stated in the Appendix to apply the Employer may defer the giving of possession for a period not exceeding six weeks or such lesser period stated in the Appendix calculated from the Date of Possession.*

23.2   *The Architect may issue instructions in regard to the postponement of any work to be executed under the provisions of this Contract.*

23.3 .1   *For the purposes of the Works insurances the Contractor shall retain possession of the site and the Works up to and including the date of issue of the certificate of Practical Completion, and, subject to clause 18, the Employer shall not be entitled to take possession of any part or parts of the Works until that date.*

23.3 .2   *Notwithstanding the provisions of clause 23.3.1 the Employer may, with the consent in writing of the Contractor, use or occupy the site or the Works or part thereof whether for the purposes of storage of his goods or otherwise before the date of issue of the certificate of Practical Completion by the Architect. Before the Contractor shall give his consent to such use or occupation the Contractor or the Employer shall notify the insurers under clause 22A or clause 22B or clause*

22C.2 to .4 whichever may be applicable and obtain confirmation that such use or occupation will not prejudice the insurance. Subject to such confirmation the consent of the Contractor shall not be unreasonably withheld.

23.3 .3 Where clause 22A.2 or clause 22A.3 applies and the insurers in giving the confirmation referred to in clause 23.3.2 have made it a condition of such confirmation that an additional premium is required the Contractor shall notify the Employer of the amount of the additional premium. If the Employer continues to require use or occupation under clause 23.3.2 the additional premium required shall be added to the Contract Sum and the Contractor shall provide the Employer, if so requested, with the additional premium receipt therefor.

24 Damages for non-completion

24.1 If the Contractor fails to complete the Works by the Completion Date then the Architect shall issue a certificate to that effect. In the event of a new Completion Date being fixed after the issue of such a certificate such fixing shall cancel that certificate and the Architect shall issue such further certificate under clause 24.1 as may be necessary.

24.2 .1 Provided:
- the Architect has issued a certificate under clause 24.1; and
- the Employer has informed the Contractor in writing before the date of the Final Certificate that he may require payment of, or may withhold or deduct, liquidated and ascertained damages,

the Employer may, not later than 5 days before the final date for payment of the debt due under the Final Certificate:
either:

.1 .1 require in writing the Contractor to pay to the Employer liquidated and ascertained damages at the rate stated in the Appendix (or at such lesser rate as may be specified in writing by the Employer) for the period between the Completion Date and the date of Practical Completion and the Employer may recover the same as a debt:

or

.1 .2 give a notice pursuant to clause 30.1.1.4 or clause 30.8.3 to the Contractor that he will deduct from monies due to the Contractor liquidated and ascertained damages at the rate stated in the Appendix (or at such lesser rate as may be specified in the notice) for the period between the Completion Date and the date of Practical Completion.

24.2 .2 If, under clause 25.3.3, the Architect fixes a later Completion Date or a later Completion Date is stated in a confirmed acceptance of a 13A Quotation the Employer shall pay or repay to the Contractor any amounts recovered, allowed or paid under clause 24.2.1 for the period up to such later Completion Date.

24.2 .3 Not withstanding the issue of any further certificate of the Architect under clause 24.1 any requirement of the Employer which has been previously stated in writing in accordance with clause 24.2.1 shall remain effective unless withdrawn by the Employer.

2.11.2 The major points to note on clause 23 are as follows:

a) the Contractor is entitled to be given full possession of the site on the date stated and, if he is not, he may escape liability for liquidated and ascertained, as distinct from general, damages at the end of the day (see *Rapid Building Company Ltd -v- Ealing Family Housing (1984) 29 BLR 5*). Partial possession is not provided for in the Standard Form as drafted. The Contract, however, specifically contains an option (at clause 23.1.2) to include deferring possession of the site (up to a maximum of 6 weeks) to be admitted as grounds for claiming an extension of time and reimbursement of loss and/or expense; but the parties must make a positive decision to opt for this facility before entering into the Contract. When they do not so opt and where the Employer fails to give possession of the site, or part of it, he is in breach and the Contractor's remedy would be to proceed against the Employer at Common Law as there would be no remedy laid down in the Contract Conditions. However, the prudent Contractor is advised to register his objection at the time; for to accept delayed possession without protest may well wholly undermine any rights he may have at Common Law;

b) the starting date referred to in the Contract is the **Date of Possession** of the site not the date for commencement of the Works. The Contractor must of course have a reasonable time to gear himself up to undertake the Contract, and if this period is not given to the Contractor he will be expected to accommodate it within the contract period. If this is going to be difficult he should register his objection at the time; it will be too late to object after signing the Contract or after accepting the instruction to proceed;

c) the Contractor must proceed *regularly and diligently* (for an analysis of which see *West Faulkner Associates -v- Newham London Borough (1993) Const LJ 232* as per the Court of Appeal); it is the disruption to this process that may be grounds for an extension of time and/or reimbursement of direct loss and/or expense;

d) the Contractor must finish *on or before the Completion Date*. But the Contractor should be very wary of aiming to finish significantly in advance of the Completion Date as he cannot rely upon the provision of outstanding information to achieve this (paras 2.8.44 and 2.10.9);

e) the Date for Completion is qualified by the provisions relating to extensions of time. It is suggested that the Contractor, who has been subjected to a delay due to a Relevant Event but has yet to receive an extension of time in respect of that delay cannot slow down, nor speed up progress at the Employer's expense; he must proceed regularly and diligently on the basis that an extension of time will be forthcoming and that the extension will properly reflect the facts, facts of which he (the Contractor) ought to be aware; bearing in mind that any extension of time can be challenged at Adjudication, Arbitration or in Legal Proceedings;

f) clause 23.2 refers to ... *postponement of any work to be executed* ... This does not apply to failure to give possession of the site or any part of it.

2.11.3 Clause 24 lays down the procedure to be followed for the Employer to be able to deduct liquidated and ascertained damages.

2.11.4 No liquidated and ascertained damages can be imposed until the Architect has issued the certificate required by clause 24.1 to certify the fact that the Contractor has failed to complete the Works by the Completion Date. (See para. 2.11.8.)

2.11.5 All certificates are, unless otherwise specifically so provided in the Contract, to be sent to the Employer with a copy to the Contractor. In addition, the Employer must claim his entitlement to liquidated and ascertained damages; it is not an automatic right. Such a claim by the Employer must, to be in accordance with clause 24.2.1:

 a) be in writing;
 b) be made before the date of the Final Certificate.

2.11.6 Once the Employer has properly claimed his entitlement to liquidated and ascertained damages he may, by written notice to the Contractor … *not later than 5 days before the final date for payment of the debt due under the Final Certificate* … either require … *the Contractor to pay* … or advise … *that he will deduct from monies due to the Contractor* … *liquidated and ascertained damages at the rate stated in the Appendix (or such lesser amount as may be specified in writing by the Employer) for the period between the Completion Date and the date of Practical Completion* …

2.11.7 Unless these provisions are fully complied with the Employer is in danger of losing all his entitlement to liquidated and ascertained damages because properly issued certificates and claims under clause 24 are conditions precedent to that entitlement.

2.11.8 It is perfectly possible for the Employer to recover damages the moment the Completion Date has passed provided that the Architect issues a certificate in accordance with clause 24.1 and provided that the Employer serves the appropriate notices. Nevertheless, the Architect may subsequently issue a further extension of time (for example during the review period of 12 weeks after Practical Completion or if the Contractor gives notice that the Works are further delayed after the Completion Date is passed). Where he does so however, this will cancel the earlier notice and if the Contractor is still in culpable delay even after the Completion Date has been extended then a further certificate in accordance with clause 24.1 must be issued thereby re-entitling the Employer to deduct liquidated and ascertained damages if he so wishes. Clause 24.2.2 provides that the liquidated and ascertained damages previously deducted for the period for which any such further extension of time is granted be paid or repaid by the Employer to the Contractor.

2.11.9 Once the Employer has properly notified the Contractor in writing of his entitlement to liquidated and ascertained damages, he may then decide whether to deduct the appropriate amount from any monies due or which become due to the Contractor under the Contract; or to recover the amount as a debt. From the Employer's viewpoint, the former is less onerous and given that further extensions can be granted after the payment of damages has begun, and the relative simplicity of the Architect's certificate under clause 24.1, there appears no reason why the Employer should not commence deduction for damages just as soon as the Completion Date is passed albeit that the total period of the Contractor's default will still be unfolding provided of course that he serves all appropriate notices.

2.11.10 Liquidated and ascertained damages cannot be deducted by the Architect from amounts due on Interim or Final Certificates; it is for the Employer to make the deduction before making payment on such certificates. Whilst the Architect should remind the Employer of his right to make deduction from amounts due, once such deductions have been made there is no need to continue to make the deduction from all subsequent payments as the provisions in clause 30.2 relating to interim payments refer to deducting the amount for liquidated and ascertained damages from the amount ... *stated as due in Interim Certificates previously issued ...*

2.11.11 It is in the Employer's interest that the Architect issues a clause 24 certificate as soon as the Completion Date has passed; until this is done the Employer has no right to impose liquidated and ascertained damages for delay. If it is left until the end of the Contract there may be insufficient funds left fully to claim these damages by deduction. The Employer may, therefore, have to institute proceedings for recovery which may be a time consuming and costly process and fruitless should the Contractor have become insolvent.

2.11.12 There can be no claim for liquidated and ascertained damages by the Employer where the Contractor has failed to achieve some interim date (unless it is written into the Contract Appendix as a Date for Sectional Completion with a specific rate for liquidated and ascertained damages applying – see paras 2.9.3 to 2.9.7). It should be noted that clause 24 contains the machinery for implementing the claiming of liquidated and ascertained damages only – it would not apply if the parties had elected to deal with delay by means of general, unliquidated, damages.

2.11.13 With regard to the Employer's written notice of an intention to deduct or recover liquidated and ascertained damages, this matter was addressed in *J F Finnegan Ltd -v- Community Housing Association Ltd 8 Const LJ 311* from which the following principles are to be noted:

a) the requirement for notice in writing from the Employer under clause 24.2.1 is a condition precedent to the deduction or recovery of liquidated and ascertained damages by the Employer;

b) there are two matters which must be addressed in the written notice namely, i) whether the Employer is claiming a payment or deduction of liquidated and ascertained damages and ii) whether the requirement relates to the whole or a part (and, if so, what part) of the sum for liquidated and ascertained damages;

c) the Employer's written notice can be given contemporaneously with the deduction of liquidated and ascertained damages. This has now been superseded by the Housing Grants, Construction and Regeneration Act, 1996 and the consequential wording of clause 24.2.1 in JCT 98 PWQ.

2.11.14 A final point to note in relation to the Employer's written notice is that whilst a certificate under clause 24.1 will be invalidated on the occasion of a further extension of time being awarded, the Employer's written notice still stands because clause 24.2.3 provides that it *shall remain effective unless withdrawn by the Employer.* Notwithstanding this provision, some legal commentators consider that the Employer may still be required to serve fresh notice under clause 24.2.1 and until such time as clause 24.2.3 is tested it would be wise for the Employer to issue a fresh notice on each occasion of a Certificate of Non-Completion having been issued.

2.12   THE EFFECT OF CLAUSES 38, 39 AND 40 ON EXTENSIONS OF TIME

2.12.1 On the surface it may seem strange that a clause dealing with the reimbursement of fluctuations must be taken into account in considering the application of clause 25.

However clauses 38, 39 and 40 – the clauses dealing with fluctuations – have provisions whereby increases in prices which occur after the Completion Date has passed are not reimbursable or allowable, i.e. the level of payments for fluctuations is "frozen" at the level prevailing at that date (see in particular clauses 38.4.7, 39.5.7 and 40.7.1). However, the level of adjustments prevailing up to that Completion Date would continue to apply to outstanding work.

2.12.2 Under such "freezing" provisions the Contractor would face the disallowance of further increases in labour and material if for any reason the Architect has not granted proper extensions of time, resulting in a premature Completion Date continuing to apply. Therefore, clauses 38 to 40 provide for certain safeguards as a condition of "freezing" the index, these are that:

a) clause 25 (Extension of time) is incorporated in the Contract and its printed text is unamended. For those who are disposed to deleting certain Relevant Events from the Contract, on the grounds that a greater portion of the risk should be placed on the shoulders of Contractors (see paras 2.1.10 et seq.), consequential amendments to clauses 38.4.8, 39.5.8 or 40.7.2 (depending on the fluctuations option in force) will be necessary if the prospect of paying for fluctuations during a period of culpable delay is to be avoided.

b) the Architect has ... *in respect of every written notification by the Contractor under clause 25, fixed or confirmed in writing such Completion Date as he considers to be in accordance with clause 25.*

2.12.3 The Architect must therefore have responded to every such notice from the Contractor either by fixing a new Completion Date (i.e. by granting an extension of time) or by confirming the existing Completion Date (i.e. by deciding not to grant an extension of time). Thus the Contractor who gives written notice of delay under clause 25.2 but thereafter fails to provide the further data required by that clause, thus preventing the Architect from forming a view on extensions of time, could effectively undermine the freezing provisions at least until after Practical Completion. It might have been different had the "freezing" provisions of clauses 38 to 40 simply required the Architect to have complied with clause 25.3.1. In that event where the Architect was prevented from granting an extension of time or of confirming the existing Completion Date because the Contractor had failed to provide the data, the Architect could properly argue that he had complied with the clause. However, clauses 38 to 40 refer to the Architect having ... *fixed or confirmed in writing such Completion Date as he considers to be in accordance with clause 25.* This requires the Architect to undertake a positive act which the Contractor's failure has prevented.

2.12.4 All too often clauses dealing with fluctuations are framed against the background of inflation rather than a reduction in prices. In the provision being considered in these paragraphs, if prices fall after the extended completion date the Contractor receives a bonus for being late in achieving completion!

2.12.5 At the time of writing the second edition of this book, inflation is at a very low level indeed which takes much of the sting out of the above criticisms. However, this may not always be the case; and in any event it is unsatisfactory in principle to have provisions in a standard form of contract which are so clearly inequitable and/or capable of misuse.

2.13 THE CALCULATION OF THE RATE FOR LIQUIDATED AND ASCERTAINED DAMAGES

2.13.1 In a contract where completion to time is a requirement, the entitlement to damages in the event of delay can take one of two forms. Either it can be left to be determined, if necessary by the Courts, if and when the delay occurs; this is referred to in this book as general damages.

2.13.2 Alternatively, it can be resolved in advance by estimating the damage the Employer would suffer due to delay; these are known as liquidated (i.e. calculated, agreed) damages; or in the language of the JCT Standard Form "liquidated and ascertained damages".

2.13.3 Once the parties have decided on the use of liquidated and ascertained damages and the amount has been determined, the Courts will generally not set these aside. For them to be set aside they must be shown to have been pitched at such a level as to be a penalty on the Contractor rather than being a reasonable attempt at a pre-estimate of the damage that will be suffered by the Employer if there is delayed completion. The fact that it might be difficult to determine such loss in advance is no ground for avoiding this facility of liquidated damages. On the contrary this is one good reason for introducing it; namely that both parties agree at the outset of a contract that element which it might be difficult to calculate at a later date so as to reduce an area of uncertainty. Consequently, the fact that the damages actually suffered by delay differs from the amount of liquidated and ascertained damages is no ground for setting them aside.

2.13.4 The view is often expressed that if a building is destined never to create a tangible profit for its owner then no liquidated damages can be claimed in the event of delayed completion.

2.13.5 However, this is to take a very narrow view of the benefits of a building. Consider the building of a Church for example, the benefits of which may be difficult to evaluate, at least in this world. Nevertheless, those who commission such a project have taken the view that investing money in this way is more worthwhile than any alternative, e.g. a commercial building which would give a more tangible return. Delay in completion in any event will increase the cost of financing the work. Moreover, on those buildings which require a subsidy to operate, one cannot argue that the cost of the subsidy is saved by any delay: the life of the building (and therefore the total burden of the subsidy) is not reduced by late completion; it is only postponed. Thus there are good reasons for calculating liquidated and ascertained damages on an apparently loss-making project based on the average commercial venture.

2.13.6 As to distinguishing between liquidated damages (which are enforceable) and a penalty (which is not enforceable), the principles relating to each were summarised by Lord Dunedin in *Dunlop Ltd -v- New Garage Co Ltd [1915] AC 79 (HL)* in the following way:

> 1. *"Though the parties to a contract who use the words "penalty" or "liquidated damages" may prima facie be supposed to mean what they say, yet the expression used is not conclusive. The court must find out whether the payment stipulated is in truth a penalty or liquidated damages ...*

2. *The essence of a penalty is a payment of money stipulated as in terrorem of the offending party; the essence of liquidated damages is a genuine covenanted pre-estimate of damage.*

3. *The question whether a sum stipulated is a penalty or liquidated damages is a question of construction to be decided upon the terms and inherent circumstances of each particular contract, judged as at the time of the making of the contract, not as at the time of the breach.*

4. *To assist this task of construction various tests have been suggested, which if applicable to the case under consideration may prove helpful, or even conclusive. Such are:*

   (a) *It will be held to be a penalty if the sum stipulated for is extravagant and unconscionable in amount in comparison with the greatest loss that could conceivably be proved to have followed from the breach.*

   (b) *It will be held to be a penalty if the breach consists only in not paying a sum of money, and the sum stipulated is a greater sum than the sum which ought to have been paid ...*

   (c) *There is presumption (but no more) that it is a penalty when "a single lump sum is payable by way of compensation, on the occurrence of one or more or all of several events, some of which may occasion serious and others but trifling damage" (referring to Lord Elphinstone -v- Monkland Iron and Coal Co (1886) 11 App Cas 332 (HL).*

   (d) *It is no obstacle to the sum stipulated being a genuine pre-estimate of damage that the consequences of the breach are such as to make precise pre-estimation almost an impossibility. On the contrary, that is just the situation when it is probable that pre-estimated damage was the true bargain between the parties."*

2.13.7 Against the backdrop of the principles advocated by Lord Dunedin, it is not practicable to give examples which wholly exhaust the range of calculations for liquidated and ascertained damages but the following check list of items to be considered may be helpful:

a) it should be a genuine pre-estimate of damage (it does not matter that a precise pre-estimate is not possible as the Court will not take an overly technical and/or pedantic approach in the construction of a liquidated and ascertained damages charge, see *Philips Hong Kong Ltd -v- AG of Hong Kong (1993) 61 BLR 41*);

b) it should be the damage naturally arising from the delay and what the parties might reasonably have supposed, at the time the Contract was signed, was likely to be the outcome of the delay;

c) it must not be a penalty;

d) it should comprise the loss that the Employer may suffer in the event of delay and therefore the following should be considered in the calculation:

   (i) interest payable during the period of delay on cash paid out, i.e. on land costs (including purchase of old properties on the site), on interim payments on the building contract, professional fees and direct Contracts carried out by the Employer (e.g. demolition, diversions, soil investigation);

   (ii) loss of profit or return on the new building;

   (iii) loss of rent;
   (iv) rent of alternative premises and the cost of decanting;
   (v)   extra payments under the fluctuations clauses (the amount is dependent upon which clauses of the Contract are used);
   (vi) inflation or postponement costs of later direct contracts, e.g. installation of machinery, fitting out contracts, removal;
   (vii) alternative costs incurred in failing to take possession on the agreed date, e.g. staff costs.

2.13.8 It is sometimes argued that to put anything approaching a realistic rate in for liquidated and ascertained damages will encourage the Contractor to turn his energies to undermining the Employer's right to claim damages by pressing heavily for extensions of time, rather than trying to meet his obligations to complete on time. The fallacy of this argument is that it implies that extensions of time are there for the asking; they are not. There has to be delay, the delay has to be such as to be likely to affect the Completion Date and has to have been caused by one or more of the Relevant Events envisaged by the Contract. These will be matters of fact. Although constant demands for extensions of time from a Contractor can be time consuming for the design team to deal with, as long as their house is in order and the demands are responded to resolutely but reasonably, the Contractor will almost certainly return his attention to meeting his contractual obligations rather than trying to obscure them.

2.13.9 All too often the client's needs in this matter of timely completion are overlooked or at least undervalued in either the formulation of the Contract Documents or in the management of the Contract and the Contractor's difficulties are given pride of place. This is a short sighted view; the decision to build, upon which the building industry depends, must be largely influenced by the confidence that a prospective Employer may feel for the venture. Part of this confidence will come from a realisation that the industry will take seriously its obligation to complete on time or, failing this, be prepared to make adequate amends.

2.13.10 There may, however, be circumstances in which the real damage to the Employer due to delay will be out of proportion to the value of the building, e.g. when it is planned to install expensive machinery after completion of the building contract to a complex programme and the cost of that machinery can dwarf the cost of the building. The cost of such a delay can be real and very extensive and, moreover, the Contractor could be aware or made aware of this possibility before entering into the Contract and thus in theory be made liable for swingeing liquidated and ascertained damages. The cost of meeting this obligation, or of covering the risk of failing to do so, may itself result in a disproportionate amount being added to the Contract Sum; the Contractor is, after all, in the building business not the insurance business. If it is felt that this overpricing is a real prospect on any project, a satisfactory compromise may well be to reduce the liquidated and ascertained damages to a figure generally commensurate with the losses due to delay on an average building of comparable value and introduce a commitment on the Contractor regularly to monitor and report upon the progress of his work and to insist that he notify, in a structured way, all potential delays of which he ought to be aware. In this way the Employer will be forewarned of any delay and can take steps to mitigate his resultant losses.

2.13.11 A comprehensive paper on the calculation of Liquidated and Ascertained Damages has been published by the Society of Chief Quantity Surveyors in Local Government [5]. However, the effect upon the rate for Liquidated and Ascertained Damages by payments under the fluctuations clause needs further consideration. (See section 2.14.)

2.14 THE EFFECT OF FLUCTUATIONS PAYMENTS ON THE CALCULATION OF THE RATE FOR LIQUIDATED AND ASCERTAINED DAMAGES

2.14.1 In the case of *Peak Construction (Liverpool) Ltd -v- McKinney Foundations Ltd (1970) 1 BLR 111* it was held that the Contractor had an entitlement to extra payments under the fluctuations clauses up until Practical Completion even if this date was beyond the Completion Date; notwithstanding that any delay may have been due to his default. (The provisions in the Contract relating to the payment of fluctuations were similar to those of the JCT 63. JCT 98 PWQ ameliorates, but not necessarily eliminates, this position somewhat in that the fluctuations calculations are frozen, once the Completion Date has passed, at the level prevailing at that date (see section 2.12). The remedy for the Employer, if any, was said to be through the damages clause.) It follows, therefore, that in assessing the correct level for Liquidated and Ascertained Damages due allowance must be made for the additional fluctuations that might become payable during any period of overrun.

2.14.2 There are, however, difficulties in assessing the effect of fluctuations upon the rate for liquidated and ascertained damages; these are given below. As will be seen, this assessment involves an element of conjecture but perhaps no more than does the calculation of the other elements:

a) the Contract Sum may have to be estimated because the fluctuations payments are likely to bear a relationship to the value of the Contract;

b) the level of inflation prevailing during the period of delay must be predicted;

c) either the length of the prospective delay must be estimated or one must assume a uniform rate of inflation during the delay and no compounding effect;

d) most, if not all, of the hours worked and generally all of the materials used after the Completion Date and up to Practical Completion would have been required in the execution of the works had the project finished to time and thus would be likely to have qualified for fluctuations payments albeit at a reduced rate in any event. For a proper estimate of what would have been paid had the Contract not been in delay it is necessary to "concertina" the whole payment pattern and recalculate the fluctuations payable on this theoretical basis (para. 4.7.13);

e) the level of fluctuations payable will vary dependent upon which route for the calculation of reimbursable fluctuations (either clause 38, 39 or 40) is selected in the Contract.

2.14.3 The following is a formula for assessing the effect of anticipated fluctuations upon the rate for Liquidated and Ascertained Damages:

Contract Sum × inflation rate (1) × 60% (2) × fluctuations factor (3)

Notes

(1) The inflation rate must be for the same period as the rate for liquidated and ascertained damages (e.g. monthly).

(2) Interim payments throughout a building contract are usually heavier in the second half of the contract period than in the first (the pattern of cumulative expenditure follows an "S" curve) thus an average of 60% (rather than 50%) is adopted for the purpose of the above formula.

(3) The fluctuations incurred on a building contract are rarely fully reimbursed under the fluctuations clauses. From past experience the following factors, representing the recovery likely to be obtained by the Contractor of the costs of inflation, should therefore be used when assessing the likely impact of fluctuations on the rate for Liquidated and Ascertained Damages:

Clause 38: the reimbursement under this clause is negligible thus no allowance need be made.

Clause 39: 66% increased by the percentage stated at the end of that clause.

Clause 40: Theoretical full reimbursement less any Non-Adjustable Element.

2.14.4 Given that level of reimbursement of fluctuations can be "frozen" at the Completion Date (see section 2.12) then the amount paid for fluctuations after that date is passed (and thus the amount to be included in the rate for Liquidated and Ascertained Damages) is reduced; however, this reduction is unlikely to be more than 15% in practice.

REFERENCES

[1] Vincent Powell-Smith (1995) *Contracts Journal*, 6 July.

[2] *Keating on Building Contracts,* 6th Edition (1995) Sweet and Maxwell, London, pp. 534–535.

[3] R. E. Millard, (1963) The New RIBA Form of Contract. *Journal of the Royal Institute of British Architects,* February, pp. 47 et seq.

[4] *Standard Method of Measurement of Building Works Seventh Edition Revised 1998*, The Royal Institution of Chartered Surveyors and the Construction Confederation.

[5] Society of Chief Quantity Surveyors in Local Government (1981) *Assessment of Liquidated Damages on Building Contracts, 1976* (Revised 1981).

3

VARIATIONS AND DISRUPTION

3.1 INTRODUCTION

3.1.1 There are three possible components of the valuation of a variation:

a) the valuation of the variation itself;

b) the valuation of the effect of that variation upon other work; and

c) the ascertainment of any loss and/or expense directly arising from the regular progress of the work having been materially affected by that variation the reimbursement for which is not covered by any other provision in the contract.

3.1.2 In JCT 98 PWQ there are three primary routes available for valuing variations and any work executed by the Contractor for which there were provisional sums or approximate quantities in the Contract Bills:

a) at the instigation of the Architect under clause 13.2.3 a quotation may be offered by the Contractor for the work which is the subject of a proposed variation which quotation is subsequently accepted by the Employer and confirmed by the Architect before the work involved in that variation goes ahead. This procedure is governed by clause 13A and covers all three of the components mentioned above. This route is considered in more detail later in this chapter in section 3.4;

b) the Employer and Contractor agree between them a different route for valuing varied work. This is inferred from clause 13.4.1.1 where in setting out the general rules for valuing varied work it says ... *unless otherwise agreed by the Employer and the Contractor* ...; and this option is not restricted to the provisions of clause 13A. It is always open to the Employer and Contractor to vary any of the terms of a Contract between them and so this somewhat gratuitous provision will not be referred to in this book again save to note that any such agreement should be in writing and should deal comprehensively with the varied work and its cost effects namely to cover fully each of the three components referred to above.

c) the varied work and its effect upon other works is valued by reference to rates set out in the Contract Documents, namely in the Schedule of Rates or in the Contract Bills as the case may be. The rules for such valuations are given in clause 13.5. In JCT 98 PWQ, these rules may be implemented by one of two alternative routes. Either the Contractor may choose to submit a statement of the price he requires for executing the varied work to the Quantity Surveyor who may accept it or reject it in whole or in part; this procedure is governed by clause 13.4.1.2 Alternative A and is considered in more detail later in this chapter at paras. 3.3.2 to 3.3.6. Or, if the Contractor does not so choose, the varied work has to be valued by the Quantity Surveyor; this procedure is governed by clause 13.4.1.2 Alternative B.

3.1.3 Clause 13 provides that where varied work is executed it is valued by reference to the rules contained in that clause which primarily envisages such a valuation being by reference to the rates and prices in the Contract Bills. The clause also sets out the rules for valuing the effect of any variation upon other work. But where the execution of a variation has a material effect upon the regular progress of the Works generally the Contractor is entitled to be reimbursed any loss and/or expense which arises directly therefrom. Except insofar as the Contractor may separately attach to a Price Statement provided under clause 13.4.1.2A the amount he requires in lieu of any ascertainment of direct loss and/or expense under clause 26.1 this loss and/or expense is not dealt with in clause 13 but in clause 26 of the Contract since clause 13.5 closes with the words ... *Provided that no allowance shall be made under clause 13.5 for any effect upon the regular progress of the Works or for any other direct loss and/or expense for which the Contractor would be reimbursed by payment under any other provision in the Conditions* ... The dividing line between the straightforward valuation of variations at Contract rates adjusted as required by the Contract on the one hand, and reimbursement of such extra costs as loss and/or expense on the other, is an important one – not least because the methods of valuation and the administrative safeguards for each vary (see section 3.7). Accordingly this chapter is concerned with the detailed wording of the relevant clauses, the basic valuation of variations and the dividing line between the valuation of variations and the ascertainment of loss and/or expense. Detailed calculations of the quantum of loss and/or expense appear in Chapter 4.

3.1.4 There is a clear distinction between what is covered by clause 13.5 and what is covered by clause 26. The former envisages an evaluation firmly rooted in the Contract Bill rates and prices; the latter moves out of this pre-determined territory into the uncharted seas of the loss and/or expense directly incurred. The evaluation of the former is carried out by the Quantity Surveyor; whilst the approval in principle for the latter, the ascertainment of the amount due, and the implementation of the strict controls relating to the reimbursement of loss and/or expense, is undertaken by the Architect. Moreover, the latter contains a number of safeguards; e.g. the need for the Contractor to give prior notice and to provide data upon request (para. 3.4.10). The Architect may instruct the Quantity Surveyor to undertake the ascertainment but he cannot delegate the responsibility for forming an opinion on whether a claim for reimbursement of loss and/or expense is justified.

3.1.5 This clear distinction between the authority of the Architect and the authority of the Quantity Surveyor is abandoned, for no apparently good reason, where the variation is to work which has been the subject of an accepted quotation given under clause 13A and where the variation or any work executed by the Contractor for which there were provisional sums or approximate quantities in the Contract Bills is the subject of an accepted Price Statement under clause 13.4.1.2A. In such instances both the valuation of the varied work and the ascertainment of any related direct loss and/or expense is put in the hands of the Quantity Surveyor. Moreover, the necessary and strict management rules which circumscribe the reimbursement of loss and/or expense in clause 26 (see para. 3.4.9) are jettisoned.

3.1.6 The wording of parts of clause 13 is not as precise as it might have been and disputes as to what can be valued by the Quantity Surveyor at Contract rates and what costs must be ascertained by the Architect will inevitably arise and this will be referred to later in this chapter in section 3.7.

3.1.7 It is not sufficient to conclude that, for example, clause 13 applies to a particular circumstance: which part of clause 13 applies is an equally important distinction. The rules for evaluating variations are contained in clause 13.5 and envisage the use of Contract Bill rates as the foundation of the evaluation of all variations. Even the operation of ... *fair rates and prices* ... to which one must resort where clause

13.5.1.3 or ... *a fair valuation* ... where clause 13.5.6 applies – cannot, it is suggested, be undertaken without reference back to the basic components of the Contract rates, i.e. the general estimate of productivity, levels of profit and overheads, labour rates and the like.

3.1.8 Clause 13.5 is very widely drawn. It refers to the valuation of work similar or dissimilar in character to that included in the Contract Bills; it refers to work executed under similar or dissimilar conditions to that included in the Contract Bills; it refers to the fact that the prices in the Contract Bills must include the prices in the preliminaries (e.g. plant, supervision, overheads, profit). Where work which is not the subject of a variation is nevertheless executed under different conditions as a result of that variation (e.g. ceiling finishes being executed in smaller areas than anticipated at tender stage due to a variation introducing more partitioning) the extra cost of so executing that work would fall to be valued under clause 13.5.5.

3.1.9 Most of the cost effects of a variation can be dealt with under clause 13.5 and must, in accordance with clause 13.5.3.3, include any extended or increased use of supervision or plant. Clause 26 on the other hand would cover the extra cost of any disruption to the Works as a whole caused by executing a variation; it would not cover the correction of errors in, or shortfall in reimbursement arising from the use of, the prices in the Contract Bills (paras 3.3.17 to 3.3.20).

3.1.10 Claims arising under clause 26 for reimbursement of direct loss and/or expense are payable only where they are not reimbursed by ... *a payment under any other provision* ... in the Contract. The main provision under which payment might have been made is clause 13 – variations and provisional sums. It is therefore necessary to consider evaluation under clause 13 if one is to do justice to an analysis of ascertainment under clause 26.

3.1.11 The distinction between a valuation made under clause 13, on the one hand, and an ascertainment of direct loss and/or expense under clause 26, on the other, is important as the former must generally be related to rates in the Contract Bills whereas the latter must be related to the Contractor's reasonable losses or expenses directly incurred. Further reference is made to this distinction in section 3.7.

3.2   THE DEFINITION OF VARIATIONS

3.2.1 All variations instructed by the Architect and all work executed by the Contractor in respect of the expenditure of provisional sums in the Contract Bills must be valued in accordance with clause 13. Clause 13 provides for the valuation of work (but not prime cost sums) arising from the expenditure of *all* provisional sums. However, clause 13.5.3.3 precludes the allowance of preliminary items in the valuation of work resulting from the expenditure of a defined provisional sum.

3.2.2 A variation under the terms of the Contract is defined in clause 13.1; and clauses 13.2, 13.3 and 13.4.1.1 contain further provisions relating to variations and their valuation. These clauses read as follows:

13 Variations and provisional sums

13.1   The term "Variation" as used in the Conditions means:

13.1   .1   the alteration or modification of the design, quality or quantity of the Works including

.1   .1   the addition, omission or substitution of any work,

.1   .2   the alteration of the kind or standard of any of the materials or goods to be used in the Works,

.1   .3   the removal from the site of any work executed or materials or goods brought thereon by the Contractor for the purposes of the Works other than work materials or goods which are not in accordance with this Contract;

13.1   .2   the imposition by the Employer of any obligations or restrictions in regard to the matters set out in clauses 13.1.2.1 to 13.1.2.4 or the addition to or alteration or omission of any such obligations or restrictions so imposed or imposed by the Employer in the Contract Bills in regard to:

.2   .1   access to the site or use of any specific parts of the site;

.2   .2   limitations of working space;

.2   .3   limitations of working hours;

.2   .4   the execution or completion of the work in any specific order;

but excludes

13.1   .3   nomination of a sub-contractor to supply and fix materials or goods or to execute work of which the measured quantities have been set out and priced by the Contractor in the Contract Bills for supply and fixing or execution by the Contractor.

13.2   .1   The Architect may issue instructions requiring a Variation.

13.2   .2   Any instruction under clause 13.2.1 shall be subject to the Contractor's right of reasonable objection set out in clause 4.1.1.

13.2   .3   The valuation of a Variation instructed under clause 13.2.1 shall be in accordance with clause 13.4.1.1 unless the instruction states that the treatment and valuation of the Variation are to be in accordance with clause 13A or unless the Variation is one to which clause 13A.8 applies. Where the instruction so states, clause 13A shall apply unless the Contractor within 7 days (or such other period as may be agreed) of receipt of the instruction states in writing that he disagrees with the application of clause 13A to such instruction. If the Contractor so disagrees, clause 13A shall not apply to such instruction and the Variation shall not be carried out unless and until the Architect instructs that the Variation is to be carried out and is to be valued pursuant to clause 13.4.1.

13.2   .4   The Architect may sanction in writing any Variation made by the Contractor otherwise than pursuant to an instruction of the Architect.

13.2   .5   No Variation required by the Architect or subsequently sanctioned by him shall vitiate this Contract.

13.3   The Architect shall issue instructions in regard to:

13.3   .1   the expenditure of provisional sums included in the Contract Bills; and

13.3   .2   the expenditure of provisional sums included in a Nominated Sub-Contract.

13.4   .1   .1   Subject to clause 13.4.1.3

(continued)

13 Variations and provisional sums *(continued)*

- all Variations required by an instruction of the Architect or subsequently sanctioned by him in writing, and
- all work which under the Conditions is to be treated as if it were a Variation required by an instruction of the Architect under clause 13.2, and
- all work executed by the Contractor in accordance with the instructions by the Architect as to the expenditure of provisional sums which are included in the Contract Bills, and
- all work executed by the Contractor for which an Approximate Quantity has been included in the Contract Bills

shall, unless otherwise agreed by the Employer and the Contractor, be valued (in the Conditions called "the Valuation"), under Alternative A in clause 13.4.1.2 or, to the extent that Alternative A is not implemented by the Contractor or, if implemented, to the extent that the Price Statement or amended Price Statement is not accepted, under Alternative B in clause 13.4.1.2. Clause 13.4.1.1 shall not apply in respect of a Variation for which the Architect has issued a confirmed acceptance of a 13A Quotation or is a Variation to which clause 13A.8 applies.

13.4 .1 .2 Alternative A: Contractor's Price Statement

Paragraph:

A1 Without prejudice to his obligation to comply with any instruction or to execute any work to which clause 13.4.1.1 refers, the Contractor may within 21 days from receipt of the instruction or from commencement of work for which an Approximate Quantity is included in the Contract documents or, if later, from receipt of sufficient information to enable the Contractor to prepare his Price Statement, submit to the Quantity Surveyor his price ("Price Statement") for such compliance or for such work. The Price Statement shall state the Contractor's price for the work which shall be based on the provisions of clause 13.5 (valuation rules) and may also separately attach the Contractor's requirements for:

.1 any amount to be paid in lieu of any ascertainment under clause 26.1 of direct loss and/or expense not included in any accepted 13A Quotation or in any previous ascertainment under clause 26;

.2 any adjustment to the time for the completion of the Works to the extent that such adjustment is not included in any revision of the Completion Date that has been made by the Architect under clause 25.3 or in his confirmed acceptance of any 13A Quotation. (See paragraph A7.)

A2 Within 21 days of receipt of a Price Statement the Quantity Surveyor, after consultation with the Architect, shall notify the Contractor in writing either

.1 that the Price Statement is accepted

or

.2 that the Price Statement, or a part thereof, is not accepted.

A3 Where the Price Statement or a part thereof has been accepted the price in that accepted Price Statement or in that part which has been accepted shall in accordance with clause 13.7 be added to or deducted from the Contract Sum.

A4   Where the Price Statement or a part thereof has not been accepted:

.1   the Quantity Surveyor shall include in his notification to the Contractor the reasons for not having accepted the Price Statement or a part thereof and set out those reasons in similar detail to that given by the Contractor in his Price Statement and supply an amended Price Statement which is acceptable to the Quantity Surveyor after consultation with the Architect;

.2   within 14 days from receipt of the amended Price Statement the Contractor shall state whether or not he accepts the amended Price Statement or part thereof and if accepted paragraph A3 shall apply to that amended Price Statement or part thereof; if no statement within the 14 day period is made the Contractor shall be deemed not to have accepted, in whole or in part, the amended Price Statement;

.3   to the extent that the amended Price Statement is not accepted by the Contractor, the Contractor's Price Statement and the amended Price Statement may be referred either by the Employer or by the Contractor as a dispute or difference to the Adjudicator in accordance with the provisions of clause 41A.

A5   Where no notification has been given pursuant to paragraph A2 the Price Statement is deemed not to have been accepted, and the Contractor may, on or after the expiry of the 21 day period to which paragraph A2 refers, refer his Price Statement as a dispute or difference to the Adjudicator in accordance with the provisions of clause 41A.

A6   Where a Price Statement is not accepted by the Quantity Surveyor after consultation with the Architect or an amended Price Statement has not been accepted by the Contractor and no reference to the Adjudicator under paragraph A4.3 or paragraph A5 has been made, Alternative B shall apply.

A7 .1   Where the Contractor pursuant to paragraph A1 has attached his requirements to his Price Statement the Quantity Surveyor after consultation with the Architect shall within 21 days of receipt thereof notify the Contractor

.1 .1   either that the requirement in paragraph A1.1 in respect of the amount to be paid in lieu of any ascertainment under clause 26.1 is accepted or that the requirement is not accepted and clause 26.1 shall apply in respect of the ascertainment of any direct loss and/or expense; and

.1 .2   either that the requirement in paragraph A1.2 in respect of an adjustment to the time for the completion of the Works is accepted or that the requirement is not accepted and clause 25 shall apply in respect of any such adjustment.

A7 .2   If the Quantity Surveyor has not notified the Contractor within the 21 days specified in paragraph A7.1, clause 25 and clause 26

(continued)

13   Variations and provisional sums (continued)

           shall apply as if no requirements had been attached to the Price Statement.

.1   .2   **Alternative B**

           The Valuation shall be made by the Quantity Surveyor in accordance with the provisions of clauses 13.5.1 to 13.5.7.

.1   .3   *The valuation of Variations to the sub-contract works executed by a Nominated Sub-Contractor in accordance with instructions of the Architect and of all instructions of the Architect and of all instructions issued under clause 13.3.2 and all work executed by a Nominated Sub-Contractor for which an Approximate Quantity is included in any bills of quantities included in the Numbered Documents shall (unless otherwise agreed by the Contractor and the Nominated Sub-Contractor concerned with the approval of the Employer) be made in accordance with the relevant provisions of Conditions NSC/C.*

13.4   .2   *Where under the instruction of the Architect as to the expenditure of a provisional sum a prime cost sum arises and the Contractor under clause 35.2 tenders for the work covered by that prime cost sum and that tender is accepted by or on behalf of the Employer, that work shall be valued in accordance with the accepted tender of the Contractor and shall not be included in the Valuation of the instruction of the Architect in regard to the expenditure of the provisional sum.*

3.2.3 In addition to defining a variation under the Contract, clause 13 makes it clear that a sub-contractor cannot be nominated to undertake work which has been measured in the Contract Bills for execution by the Contractor. Moreover, clause 4.1.1 makes it clear that the Contractor can make reasonable objection to the changes in the Contractor's obligations envisaged by clause 13.1.2. This latter clause recognises the possibility that the Employer may impose obligations or restrictions or change the requirements set out in the Contract Bills in regard to:

a)   access to or use of the site;
b)   limitations of working space or hours; or
c)   executing or completing the work in any specific order

by an instruction from the Architect. Such changes, if not objected to by the Contractor, become variations and are valued in accordance with the rules laid down in clause 13. If the Contractor does raise objection to such proposed changes, and if the Architect does not accept the objection as being reasonable, the dispute may be referred to Adjudication, Arbitration or Legal Proceedings.

3.2.4 Clause 13.4.1.1 also makes it clear that the following shall be treated as variations for the purposes of valuation:

a)	correction of departures from the method of measurement laid down, and correction of errors in description or in quantity or omissions of items from the Contract Bills	Clause 2.2.2.2
b)	variation to the works instructed by the Architect in relation to a divergence from Statutory Requirements	Clause 6.1.3
c)	work executed and materials supplied in emergency compliance with Statutory Requirements	Clause 6.1.4.3
d)	the alteration or modification to any Performance Specified Work following a change in Statutory requirements	Clause 6.1.7

e) work where the conditions under which it is executed are substantially changed by:
    (i) executing a variation
    (ii) other work executed pursuant to the expenditure of a provisional sum for undefined work
    (iii) other work executed pursuant to the expenditure of a provisional sum for defined work where that work differs from that described in the Contract Bills
    (iv) other work in respect of which an Approximate Quantity in the Contract Bills transpires not to have been an accurate quantity

Clause 13.5.5

f)	work where the conditions under which it is executed are substantially changed by the execution of Performance Specified Work	Clause 13.5.6.6
g)	the restoration, replacement or repair of loss or damage occasioned by risks covered by the Joint Names Policy	Clauses 22B.3.5 and 22C.4.4.2

3.2.5 Architect's instructions issued in relation to antiquities and the like discovered on site do not themselves constitute variations although clause 34.3.1 does give the Contractor the right to claim direct loss and/or expense arising therefrom.

3.3 THE VALUATION OF VARIATIONS UNDER CLAUSE 13

The requirements of clause 13.4.1.1

3.3.1 All variations as defined above and any work executed by the Contractor for which there were provisional sums on approximate quantities in the Contract Bills (with the exception of the items listed in the following table) must be valued in accordance with one of the two alternative procedures in clause 13.4.1.1 (see para. 3.1.2).

Exceptions to the rule that variations must be valued in accordance with clause 13.4.1.1

Item	Basis of Valuation
a) Variation to which clause 13A applies – a Quotation submitted by the Contractor and accepted by the Employer and confirmed by the Architect	The Quotation as accepted by the Employer and confirmed by the Architect – see also para. 3.4.7 (clause 13A.3.2)
b) Variations to work instructed in a) above	Fair and reasonable valuation by the Quantity Surveyor (to include reimbursement for any direct loss and/ or expense) having regard to the content of the accepted quotation – see also para. 3.4.9 (clause 13A.8)
c) Work for which tenders are accepted by or on behalf of the Employer for work covered by a prime cost sum arising from a provisional sum	The accepted tender for that work (clause 13.4.2)
d) Variations to Nominated Sub-Contract works	The form of Sub-Contract NSC/C unless otherwise agreed between the Contractor and Nominated Sub-Contractor with the approval of the Employer (clause 13.4.1.3)
e) Performance Specified Work or variations thereto	Broadly in line with the rules for valuing variations generally and so no separate reference is made to this (clauses 13.5.6 and 7)

3.3.2 Except when otherwise agreed by the Employer and the Contractor varied work is to be valued under the Alternative A procedure in clause 13.4.1.2 provided that the Contractor decides to implement that procedure and to the extent that the resultant Price Statement is accepted by the Quantity Surveyor. If the procedure is not implemented or to the extent that the Price Statement is not accepted the varied work is to be valued under the Alternative B procedure in clause 13.4.1.2, that is valuation by the Quantity Surveyor in accordance with the clause 13.5 valuation rules.

3.3.3 The outline of the Alternative A procedure laid down in clause 13.4.1.2 is set out in *Figure 3.1*.

3.3.4 The price for the varied work included in the Contractor's Price Statement (and in any amended Price Statement) has to be based on the clause 13.5 valuation rules. These are

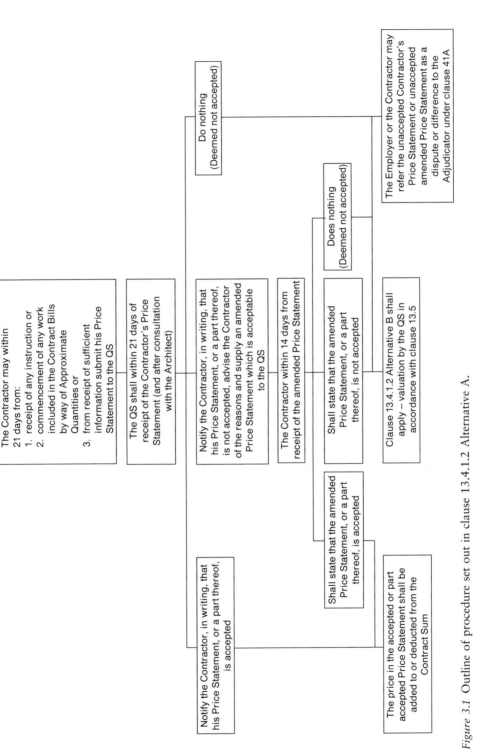

The Contractor may within 21 days from:
1. receipt of any instruction or
2. commencement of any work included in the Contract Bills by way of Approximate Quantities or
3. from receipt of sufficient information submit his Price Statement to the QS

The QS shall within 21 days of receipt of the Contractor's Price Statement (and after consultation with the Architect)

Notify the Contractor, in writing, that his Price Statement, or a part thereof, is not accepted, advise the Contractor of the reasons and supply an amended Price Statement which is acceptable to the QS

Notify the Contractor, in writing, that his Price Statement, or a part thereof, is accepted

Do nothing (Deemed not accepted)

The Contractor within 14 days from receipt of the amended Price Statement

Shall state that the amended Price Statement, or a part thereof, is accepted

Shall state that the amended Price Statement, or a part thereof, is not accepted

Does nothing (Deemed not accepted)

The price in the accepted or part accepted Price Statement shall be added to or deducted from the Contract Sum

Clause 13.4.1.2 Alternative B shall apply – valuation by the QS in accordance with clause 13.5

The Employer or the Contractor may refer the unaccepted Contractor's Price Statement or unaccepted amended Price Statement as a dispute or difference to the Adjudicator under clause 41A

Figure 3.1 Outline of procedure set out in clause 13.4.1.2 Alternative A.

considered in detail at paras 3.3.23 to 3.3.46. If the Price Statement is accepted the price is added to or deducted from the Contract Sum (clause 13.4.1.2A3).

3.3.5 The Contractor **may** (but not **shall**) separately attach to his Price Statement:

1. The amount that he requires to be paid in lieu of any ascertainment of loss and/or expense arising from the varied work provided that such loss and/or expense has not been included in any accepted 13A Quotation or in any previous ascertainment under clause 26.
2. The adjustment that he requires to the time for the completion of the Works as a result of the varied work provided that such adjustment of time has not been previously included in any revision of the Completion Date under clause 25.3 or in any accepted 13A Quotation.

3.3.6 It is to be noted that it is the **Quantity Surveyor** (albeit after consultation with the Architect) who, within 21 days of receipt of the Contractor's Price Statement and attachments, shall notify the Contractor that:

1. The amount required in lieu of any ascertainment of loss and/or expense is, or is not, accepted. If it is not accepted then the amount of any loss and/or expense shall be ascertained under clause 26.1 (see section 3.5). If it is accepted then the amount required is to be added to the Contract Sum (clause 13.7). However, as with clause 13A, it seems anomalous that the strict rules which apply in clause 26 to the ascertainment of direct loss and/or expense arising from variations generally, appear to have been abandoned when it comes to varied work valued pursuant to the Alternative A procedure. The abandoned rules are considered in paras 3.4.10 and 3.4.11.
2. The adjustment required to the time for completion of the Works is, or is not, accepted. If it is not accepted then clause 25 shall apply (see section 2.7). If it is accepted it is not clear what is to happen for, unlike clause 13A, clause 13.4.1.2A does not lay down any procedure for the translation of the accepted adjustment required to the time for completion of the Works into a revised Completion Date (see para. 2.7.4).

3.3.7 The requirements of clause 13.4 rarely presented any difficulty prior to JCT 98 PWQ and whilst the concept of the Contractor's Price Statement was an introduction in this latest edition the underlying clause 13.5 valuation rules remained unchanged. Consequently it is unlikely that the Price Statement will introduce any new difficulties in relation to the valuation of varied work. However, it is necessary to refer to three related matters:

a) the need to ensure that prices in the Contract Bills are consistent;
b) the pricing of preliminaries;
c) the correction of "errors".

Consistency of prices

3.3.8 Before accepting a tender the Quantity Surveyor should undertake a detailed inspection of the prices in the Contract Bills to ensure that they are consistent throughout, and where they are not, to come to an agreement with the Contractor, before signing the Contract, to make them so. Whether such amendments result in a change in the tender will depend on which of the alternative methods for correcting errors has been adopted (para. 3.3.16).

3.3.9 The prices should be consistent throughout the Bill, but they need not be the same. A price for excavation within the building proper need not be the same as an identically described item in the external works; but the difference should be capable of a logical explanation. Difficulty can be caused in valuing variations if inconsistent prices appear in the Contract Bills. It is in the Contractor's interest as well as the Employer's to resolve any discrepancies prior to signing the Contract. Once the Contract is signed those prices are binding for the valuation of variations.

The pricing of preliminaries

3.3.10 Such a scrutiny of the Bill prices must also extend to the pricing of the preliminaries section of the Bill. It is important that the Contractor is required, through the Contract Bills, to provide a full and consistent breakdown of this element of his work as these too form part of the valuation of variations referred to in clause 13. Such a breakdown should provide at least the following information in a consistent and logical pattern:

a) a distinction between items of plant and the like priced in the preliminaries and those priced in the measured work rates;

b) cost of setting up plant, huts, temporary works and services;

c) cost of clearing away last and making good;

d) hire period and rates assumed together with the start and finish dates for each item of plant, huts, and the like;

e) running and operating costs and attendance on last;

f) a detailed make up of the extent, timing and nature of all supervisory site staff not included in the prices of measured work;

g) cost of providing fuel, water and insurances, often expressed as a percentage of the value of work to be executed;

h) any standing time or waiting time to be allowed for in the pricing of plant and the like.

3.3.11 The worked example at the end of this book contains a build up of a preliminaries section which is based upon prices contained in *Spon's Architects' and Builders' Price Book* [1] . It is clear which items in this example are time or value based and which, therefore, may require to be taken into account in the valuation of a variation.

3.3.12 They are summarised as follows:

a) examples of items or persons involved during the whole or most of the Contract Period, and which are related to time, which would be adjusted if a variation has affected the length of time of their involvement:
General Foreman
Ganger
Telephones
Welfare facilities

b) examples of items or persons involved during part of the contract period and which are related to time; each to be treated on their merits in assessing whether the execution of a variation has affected the length of time of their involvement:

    Assistant Foreman
    Storeman
    Trades Foreman
    Huts
    Lighting, heating and attendance
    Plant
    Scaffolding
    Hoist

c) examples of items which are related to value and which would generally be adjusted in the valuation of a variation in relation to its value:

    Insurance
    Small plant
    Water for the Works
    Temporary lighting and power
    Daily travelling and fare allowances

d) examples of items of a non-recurring nature but delay in the execution of which may result in their being increased in cost due to inflation:

    Erection   huts, offices
    Dismantling   of   plant
    Haulage to the site   scaffolding
    Haulage from the site   hoist

3.3.13 When valuing a variation at the prices in the Contract Bills therefore, it is necessary to determine whether executing the variation entailed:

a) greater use of **time-based** resources, in which event the items listed at (a) and (b) should be considered on their merits;

b) increase in **value**, in which event the items listed at (c) should be included in the valuation;

c) delay in executing any of the **non-recurrent** items, in which event fluctuations and/or disruption of the items listed at (d) should be included in the valuation.

Correction of errors

3.3.14 There are two types of error with which one might be faced when considering valuing variations; those appearing in the documents sent to tender – ... *error in description or in quantity or omission of items* ... as clause 2.2.2.2 calls them – and those perpetrated by the tenderer ... *any error whether of arithmetic or not in the computation of the Contract Sum* ... (clause 14.2). The former are corrected and deemed to be variations, thus the Contractor is not responsible for the adequacy of the quantities or descriptions in the Contract Bills; whilst the latter are ... *deemed to have been accepted by the parties hereto* (clause 14.2).

3.3.15 It is, therefore, particularly important that both the Contractor's estimator and the Quantity Surveyor appointed under the Contract should closely peruse the rates in the Contract Bills and calculations so that any such error is suitably dealt with in the

Contract Documents prior to signing the contract and in a manner which does not amend the Contract Sum.

3.3.16 The *Code of Procedure for Single Stage Selective Tendering* [2] issued by the National Joint Consultative Committee for Building (NJCC) provides alternative methods of dealing with errors discovered in a tender prior to the signing of the contract. In the first, the lowest tenderer, having been given details of his error, has the option of confirming or withdrawing his offer (not of amending it). In the second, the lowest tenderer has the option of confirming his offer or amending it to remove the error and running the risk of no longer being the lowest tenderer. In either case, if the tenderer opts for confirming his tender the tender sum remains unchanged but the errors are corrected and the priced Contract Bills endorsed indicating that all rates or prices (excluding preliminary items, contingencies, prime costs and provisional sums) inserted therein by the tenderer are to be considered as reduced or increased in the same proportion as the corrected total of such items exceeds or falls short of the tendered price of such items. Once the Contract is signed, those amended prices in the Contract Bills form the basis for determining the value of variations. The *Code of Practice for the Selection of Main Contractors* [3] issued by the Construction Industry Board also provides alternative methods of dealing with errors discovered in a tender prior to the signing of the contract. Where the Construction Industry Board's Code is in use, the tender documents are required to state whether overall price or the rates are dominant. Where the overall price is dominant, the tenderer will be required either to stand by his tender or to withdraw from the process; this is equivalent to Alternative 1 under the NJCC Code. If the rates are stated to be dominant, the client *may* request an amended tender price to accord with the rates given by the tenderer; this is broadly equivalent to Alternative 2 under the NJCC Code save that, under the NJCC scheme of things, it is the tenderer who is given the opportunity of confirming his offer or amending it.

3.3.17 Nevertheless, Contractors sometimes discover "errors" in their Contract Bill prices after the contract is signed and demand the correction of such "errors" before they are used for valuing variations – "I don't mind sticking by my errors in pricing for the work in the Contract Sum but it is unreasonable to expect them to be used for any extra work that the Client decides to add later" is how the thesis is often put. On the surface this is a very reasonable and understandable view. However, it indicates a fundamental misunderstanding of both the intent and the wording of the Contract. The prices in the Contract Bills form part of the Contract solely for the purpose of valuing variations; that they were also used in the compilation of the Contract Sum is not contractually relevant. There is no right in a lump sum contract to remeasure the whole of the Works and simply apply the bill rates to the quantities – this much was established in *London Steam Stone Sawmills Co. -v- Lorden (1900) HBC (4th Ed)*. The Contractor offers to complete the whole of the work in the Contract Bills for the Contract Sum; in addition he offers to undertake variations at the prices entered into the Contract Bills. It is nonsense, therefore, to demand that these prices can be overturned and set aside on the one occasion for which they were designed. Useful judicial guidance on this thorny issue was given in *Dudley Corporation -v- Parsons and Morris (1967) Building Industry News 17.02.67* where it was decided that the Contractor was entitled only to the quoted rate notwithstanding that it was clearly erroneous, related to a significantly greater quantity and was bad for the Contractor.

More recent authority on this point was provided by HH Judge Humphrey Lloyd QC in the case of *Henry Boot Construction Ltd -v- Alstom Combined Cycles Ltd (TCC 1999)*. In that case, the Plaintiff sought to benefit from an exceptionally high rate on the occasion of a Variation being ordered in respect of a category of work which was the subject of that rate; the Defendant, faced with the prospect of paying somewhat over the odds for the varied work, sought to avoid that exceptionally high rate by arguing that the varied work should instead be valued by reference to a fair and reasonable level of pricing – it was also apparent that the rate was high by reason of an error in the completion of the Plaintiff's tender and the Defendant saw the valuation of the Variation as presenting an opportunity to correct this error. The Plaintiff, a Contractor, was in the relatively unusual position of seeking to be held by the pricing in the Contract Bills and received support for this stance from HH Judge Humphrey Lloyd QC whose judgment was that a variation could not be used as a pretext to unravel and correct mistakes made by a Contractor in his pricing since this would "be completely inconsistent with the wording of such a contract and the philosophy to be derived from it". As noted from the 11 April 2000 edition of the Times, HH Judge Humphrey Lloyd QC's judgment has since been confirmed by the Court of Appeal. Once the contract is entered into there can, by definition, be no such thing as "errors" in the Contract Bill rates or prices. This is perhaps easier to demonstrate in the *Without Quantities* Forms where the Schedule of Rates (which takes the place of the rates in the Contract Bills) is appended to the Contract solely for the purpose of valuing variations. No-one would argue that they should not apply to the valuing of variations – this is their sole purpose. The position should be no different under the *With Quantities Form* where the wording is, for this purpose, identical.

3.3.18 Alternatively, it is often argued that any shortfall in the valuation of variations by the use of "erroneous" rates can be remedied by resorting to clause 26 which provides for reimbursement of direct loss and/or expense arising from a variation. But clause 26 would not apply to costs which are, in principle, covered by clause 13 albeit that the reimbursement falls short of the Contractors' costs or expectations. Clause 26 applies only to items not coming within the scope of clause 13, thus any element of cost which is, in principle, to be dealt with in clause 13 cannot be resurrected under clause 26. Clause 26 does not exist as a safeguard against the Contractor recovering less than his costs when executing variations; he contracts to undertake variations at the rates included in the Contract Bills with the safeguard that if he is involved in disruption as a result of the variation he may recover the direct loss and/or expense involved in that disruption.

3.3.19 If clause 26 existed to enable the Contractor to make up his costs on each variation, clause 13 would be meaningless. Variations are envisaged by the Contract and are not breaches of it since clause 13.2.5 provides that ... *No variation required by the Architect or subsequently sanctioned by him shall vitiate (the) Contract* ... Clause 13 specifies in advance the rules that must apply if and when a variation occurs, in much the same way as liquidated and ascertained damages do in the event of delay in completion – even if the damages figure turns out to be wrong it cannot be changed (unless it transpires to be a penalty) (para. 2.13.3); nor can the rates in the Contract Bills be changed if they are exceeded by the Contractor's costs. He shoulders the same responsibilities and risks on the valuation of variations as he does on the original Contract Sum; by the same token he is entitled to the same opportunities to profit from his efficiency or an exceptionally high rate as discussed in relation to the *Henry Boot* case (see para. 3.3.17).

3.3.20 This is not to say that, in the event of the Contractor making a genuine mistake in pricing the rate, the Architect or Employer can take advantage of this by ordering substantial additions on the underpriced work or substantial omissions of overpriced work. A variation should be a change reasonably and naturally occurring in the course of developing the design and in response to changes reasonably required by the Employer; not artificially contrived in order to take advantage of an oversight.

The detailed valuation rules: clause 13.5

3.3.21 Apart from the exceptions referred to earlier (table at para. 3.3.1), all variations are to be valued in accordance with clause 13.5 together with the ancillary rules in clauses 13.6 and 13.7. These clauses read as follows:

13.5 .1 *To the extent that the Valuation relates to the execution of additional or substituted work which can properly be valued by measurement or to the execution of work for which an Approximate Quantity is included in the Contract Bills such work shall be measured and shall be valued in accordance with the following rules:*

.1 .1 *where the additional or substituted work is of similar character to, is executed under similar conditions as, and does not significantly change the quantity of, work set out in the Contract Bills the rates and prices for the work so set out shall determine the Valuation;*

.1 .2 *where the additional or substituted work is of similar character to work set out in the Contract Bills but is not executed under similar conditions thereto and/or significantly changes the quantity thereof, the rates and prices for the work so set out shall be the basis for determining the valuation and the valuation shall include a fair allowance for such difference in conditions and/or quantity;*

.1 .3 *where the additional or substituted work is not of similar character to work set out in the Contract Bills the work shall be valued at fair rates and prices;*

.1 .4 *where the Approximate Quantity is a reasonably accurate forecast of the quantity of work required the rate or price for the Approximate Quantity shall determine the Valuation;*

.1 .5 *where the Approximate Quantity is not a reasonably accurate forecast of the quantity of work required the rate or price for that Approximate Quantity shall be the basis for determining the Valuation and the Valuation shall include a fair allowance for such difference in quantity.*

Provided that clause 13.5.1.4 and clause 13.5.1.5 shall only apply to the extent that the work has not been altered or modified other than in quantity.

13.5 .2 *To the extent that the Valuation relates to the omission of work set out in the Contract Bills the rates and prices for such work therein set out shall determine the valuation of the work omitted.*

13.5 .3 *In any valuation of work under clauses 13.5.1 and 13.5.2:*

.3 .1 *measurement shall be in accordance with the same principles as those* governing the preparation of the Contract Bills as referred to in clause 2.2.2.1;

.3   .2   allowance shall be made for any percentage or lump sum adjustments in the Contract Bills; and

.3   .3   allowance, where appropriate, shall be made for any addition to or reduction of preliminary items of the type referred to in the Standard Method of Measurement, 7th Edition, Section A (Preliminaries/General Conditions); provided that no such allowance shall be made in respect of compliance with an Architect's instruction for the expenditure of a provisional sum for defined work.

13.5   .4   To the extent that the Valuation relates to the execution of additional or substituted work which cannot properly be valued by measurement the Valuation shall comprise:

.4   .1   the prime cost of such work (calculated in accordance with the "Definition of Prime Cost of Daywork carried out under a Building Contract" issued by the Royal Institution of Chartered Surveyors and the Building Employers Confederation (now Construction Confederation) which was current at the Base Date) together with percentage additions to each section of the prime cost at the rates set out by the Contractor in the Contract Bills; or

.4   .2   where the work is within the province of any specialist trade and the said Institution and the appropriate body representing the employers in that trade have agreed and issued a definition of prime cost of daywork, the prime cost of such work calculated in accordance with that definition which was current at the Base Date together with percentage additions on the prime cost at the rates set out by the Contractor in the Contract Bills.

Provided that in any case vouchers specifying the time daily spent upon the work, the workmen's names, the plant and the materials employed shall be delivered for verification to the Architect or his authorised representative not later than the end of the week following that in which the work has been executed.

13.5   .5   If:

compliance with any instruction requiring a Variation or

compliance with any instruction as to the expenditure of a provisional sum for undefined work or

compliance with any instruction as to the expenditure of a provisional sum for defined work to the extent that the instruction for that work differs from the description given for such work in the Contract Bills or

the execution of work for which an Approximate Quantity is included in the Contract Bills to such extent as the quantity is more or less than the quantity ascribed to that work in the Contract Bills

substantially changes the conditions under which any other work is executed, then such other work shall be treated as if it had been the subject of an instruction of the Architect requiring a Variation under clause 13.2 which shall be valued in accordance with the provisions of clause 13.

13.5   .6   .1   The Valuation of Performance Specified Work shall include allowance for the addition or omission of any relevant work involved in the preparation and production of drawings, schedules or other documents:

.6   .2   *the Valuation of additional or substituted work related to Performance Specified Work shall be consistent with the rates and prices of work of a similar character set out in the Contract Bills or the Analysis making due allowance for any changes in the conditions under which the work is carried out and/or any significant change in the quantity of the work set out in the Contract Bills or in the Contractor's Statement. Where there is no work of a similar character set out in the Contract Bills or the Contractor's Statement a fair valuation shall be made;*

.6   .3   *the Valuation of the omission of work relating to Performance Specified Work shall be in accordance with the rates and prices for such work set out in the Contract Bills or the Analysis;*

.6   .4   *any valuation of work under clauses 13.5.6.2 and 13.5.6.3 shall include allowance for any necessary addition to or reduction of preliminary items of the type referred to in the Standard Method of Measurement, 7th Edition, Section A (Preliminaries/General Conditions);*

.6   .5   *where an appropriate basis of a fair valuation of additional or substituted work relating to Performance Specified Work is daywork the Valuation shall be in accordance with clauses 13.5.4.1 or 13.5.4.2 and the proviso to clause 13.5.4 shall apply;*

.6   .6   *if:*

compliance with any instruction under clause 42.11 requiring a Variation to Performance Specified Work or

compliance with any instruction as to the expenditure of a provisional sum for Performance Specified Work to the extent that the instruction for that Work differs from the information provided in the Contract Bills pursuant to clause 42.7.2 and/or 42.7.3 for such Performance Specified Work

substantially changes the conditions under which any other work is executed (including any other Performance Specified Work) then such other work (including any other Performance Specified Work) shall be treated as if it had been the subject of an instruction of the Architect requiring a Variation under clause 13.2 or, if relevant, under clause 42.11 which shall be valued in accordance with the provisions of clause 13.5.

13.5   .7   *To the extent that the Valuation does not relate to the execution of additional or substituted work or the omission of work or to the extent that the valuation of any work or liabilities directly associated with a Variation cannot reasonably be effected in the Valuation by the application of clauses 13.5.1 to .6 a fair valuation thereof shall be made.*

Provided that no allowance shall be made under clause 13.5 for any effect upon the regular progress of the Works or for any other direct loss and/or expense for which the Contractor would be reimbursed by payment under any other provision in the Conditions.

13.6   *Where it is necessary to measure work for the purpose of the Valuation the Quantity Surveyor shall give to the Contractor an opportunity of being present at the time of such measurement and of taking such notes and measurements as the Contractor may require.*

13.7 *Effect shall be given to a Valuation under clause 13.4.1.1, to an agreement by the Employer and the Contractor to which clause 13.4.1.1 refers, to a 13A Quotation for which the Architect has issued a confirmed acceptance and to a valuation pursuant to clause 13A.8 by addition to or deduction from the Contract Sum.*

3.3.22 The working of this clause can best be summarised by means of the table below. Following this table each relevant sub-clause is considered in detail and examples of the relevant variations are given. The wording in the Contract has been summarised in this table.

Summary of the Valuation rules in clause 13.5

Type of variation			Valuation rule	Clause
1. Additional and substituted works that can properly be valued by measurement	a)	Work of similar character, executed under similar conditions to, and with quantities not changed significantly from, those in the Contract Bills	Contract Bill rates and prices	13.5.1.1
	b)	As above but with quantities significantly changed and/or work not executed under similar conditions	As above but with a fair allowance made for the change in quantity and/or conditions	13.5.1.2
	c)	Work not of a similar character to that in the Contract bills	Fair rates and prices	13.5.1.3
	d)	Work the quantities of which were reasonably accurately forecast in the Approximate Quantities in the Contract Bills	Contract Bill rates or prices for the Approximate Quantity	13.5.1.4
	e)	Work the quantities of which were not reasonably accurately forecast in the Approximate Quantities in the Contract Bills	Contract Bill rates or prices for the Approximate Quantity with a fair allowance for the difference in quantity.	

Type of variation		Valuation rule	Clause
2.	Omitted work	Contract Bill rates and prices	13.5.2
3.	Additional or substituted work which cannot properly be valued by measurement	Daywork	13.5.4
4.	Work for which the conditions of execution have been changed by a variation, by the execution of work covered by a provisional sum or by changes to the Approximate Quantities ascribed thereto in the Contract Bills	Treat as if that work were the subject of the variation, i.e. clause 13 applies	13.5.5
5.	Variations other than omitted, additional or substituted work (e.g. changes in restrictions imposed by the Employer)	Fair valuation	13.5.7
6.	Work which cannot reasonably be valued in accordance with clauses 13.5.1 to 5	Fair valuation	13.5.7

Summary of the Valuation rules in clause 13

3.3.23 The rules for valuation are further qualified by clause 13.5.3 from which it is clear that where measurements are taken for the purpose of valuation, such measurements shall be in accordance with the same rules as govern the preparation of the Contract Bills and that, under clause 13.6, the Quantity Surveyor must give the Contractor the opportunity of being present when measurements are taken. It is suggested that this refers to both measurements taken on site and those taken from drawings away from the site. Allowance must also be made for any percentage or lump sum adjustments in the Contract Bills and for the adjustment of preliminaries (clause 13.5.3).

3.3.24 Under clause 13.7, the Valuation of variations must be added to or deducted from the Contract Sum; moreover, under clause 3 ... *as soon as such an amount is ascertained in whole or in part such amount shall be taken into account in the computation of the next Interim Certificate* ...

3.3.25 The proviso to clause 13.5 makes it clear that reimbursement of loss and/or expense due to the regular progress of the work being affected must be dealt with not under clause 13 but under clause 26. (See section 3.7.)

Clause 13.5.1

3.3.26 Clause 13.5.1 lists a series of rules by which a variation is valued. Clauses 13.5.1.1 and 13.5.1.2 require the use of the rates and prices contained in the Contract Bills where the work comprising the variation is of a similar character to work priced in the Contract Bills. Where such work (i.e. work of a similar character) is executed under similar conditions to those envisaged by the Contract Bills, the relevant rates and prices in those Contract Bills apply to the varied work as they stand. Note that the word used is "similar" not identical. What is similar and dissimilar can only be a matter of judgment for the Quantity Surveyor in each case. Where the conditions are not similar to those envisaged in the Contract Bills (or where there is a significant change in the quantity stated in the Contract Bills), the relevant rates and prices in the Contract Bills apply to the varied work but shall ... *include a fair allowance for such difference in conditions and/or quantity.*

3.3.27 This principle also extends to the prices for items in the preliminaries section of the Contract Bills, e.g. plant, supervision and on-costs.

3.3.28 There is a view expressed from time to time that any work executed as a variation, simply by virtue of it being a variation, must have been executed under dissimilar conditions and therefore the rates and prices in the Contract Bills cannot apply. Clearly this is not what the clause says. To accept that view would render clause 13 wholly meaningless.

3.3.29 Changed character might mean brickwork in smaller areas, or more complicated setting out. Changed conditions might mean work on a different floor or in a restricted space. The list of examples of potential changes in character and conditions is almost limitless. Nevertheless, even where valuing work under clause 13 which requires one to depart from the unadulterated use of the rates and prices in the Contract Bills (by including allowances in those rates and prices or valuing by means of a fair valuation as referred to in clauses 13.5.1.2, 13.5.1.3, 13.5.1.5 and 13.5.7), one must still have regard to the basic components of the price assumed in the tender; that is:

a) the labour rate;
b) the labour output;
c) the cost of materials;
d) the waste factor on materials;
e) the use of plant;
f) the level of supervision and other preliminary items;
g) the provision of temporary works;
h) head office overheads and profit.

3.3.30 Although the range of variables in the character of work or in the conditions under which it is to be executed is almost limitless, it is suggested that this does not extend to different market conditions resulting, say, in labour rates being enhanced which could render an otherwise adequate rate unprofitable albeit that actual market conditions as encountered at the time the Variation is ordered need to be distinguished from the Contractor's forecast as to market conditions when he formulates his tender (this latter eventuality is addressed in para. b) of Example 3.1 set out below).

3.3.31 Given the data implicit in the Contract Bill rates and prices, when forming an evaluation under clause 13.5.1.2 or 13.5.1.3, one should ask the question "What rate or price would the tenderer have used for this item had he known at tender stage that the work (now the subject of a variation order) was required given the level of the rates and prices included in the Contract Bills?". Example 3.1 illustrates many of the changes in rate that can come about under such an evaluation.

Example 3.1 Valuing Variations

A price of £14.00/m^2 appears in the Contract Bills for block partitions. They occur repetitively on all four floors of the building. A variation requires similar blockwork to be constructed in the plant room on the roof. To the basic rate of £14.00 allowance must be made for the following:

a) reduced labour output. The tenderer may have assumed an average output of completed blockwork of 3.80 m^2 per hour based upon:

   Ground Floor   4.00 m^2/h
   First Floor   3.85 m^2/h
   Second Floor   3.75 m^2/h
   Third Floor   3.60 m^2/h

   The plant room being on the roof with difficult access may require a labour output of 3.40 m^2/h – an increase of 11.7% on the labour element of the rate. The fact that the Contractor may have failed to achieve the above pattern is not relevant; valuations under clause 13.5.1.2 must be related to Contract Bill prices and not to cost records;

b) similarly if the blockwork in the Contract Bills was to have been executed over a period, then, on a fixed price contract, the rate would be deemed to be the average prevailing over that period. Work in the plant room may well have been executed after the end of the period for erection of the general blockwork, in which event an additional allowance must be made for inflation;

c) the blockwork in the plant room is likely to have been executed in smaller quantities than those generally envisaged in the Contract Bills, and, although coming within the same description laid down by the *Standard Method of Measurement of Building Works* [4], an allowance must be made for any reduction in output. The waste factor allowed on materials may also have to be increased together with any additional cost of purchasing where work is executed in small quantities;

d) the hoist may have had to be moved or extended to serve the blocklaying in the plant room and this must be included in the evaluation under clause 13.5.3.3;

e) time based and value based preliminaries – referred to in para. 3.3.12 – must also be included as appropriate under clause 13.5.3.3;

f) the tender allowance percentage for head office overheads and profit must be added to the additional cost elements of the valuation if not already included in the rates used.

3.3.32 But in any event the basic valuation must be carried out at rates and prices included in the Contract Bills plus adjustment for each of the above factors.

3.3.33 It must be emphasised that similar scope may exist for reducing the price in the Contract Bills if the circumstances support it.

3.3.34 There are exceptions to the application of the rates and prices from the Contract Bills to the valuation of variations. Clause 13.5.1.2 refers to making a ... *fair allowance* ... to the Bill rates for work not of a similar character and clause 13.5.1.3 refers to ... *fair rates and prices.* The prices in the Contract Bills do not apply to variations when ... *any work or liabilities directly associated with a Variation cannot reasonably be effected in the valuation by the application of clauses 13.5.1 to 13.5.6; in such cases* a ... *fair valuation* ... must be made (clause 13.5.7). It is impossible to give a comprehensive list of the occasions on which it would be necessary to make such a ... *fair allowance* ... or resort to ... *fair rates and prices* ... but the important point to note is that a fair valuation must still have regard to the items listed at para. 3.3.29; that is, to the component parts of the Contract Bill prices and not by reference to the Contractor's costs.

Clause 13.5.4

3.3.35 This provides that work shall be valued on a daywork basis where it cannot properly be valued by measurement. Note the careful wording of this clause ... *cannot properly be valued by measurement.* There may be circumstances where it is possible to measure the work, but it would not be proper to do so. For example breaking out the wall of a brick manhole, built earlier as part of the contract, to accommodate the connection of a drain introduced as a variation, might properly be valued as daywork.

3.3.36 The daywork total, arrived at by applying the definitions referred to in the clause, are subject to the addition of the percentages set out by the Contractor in the Contract Bills. Provision must therefore be made for these percentage additions to be entered in the priced Bill. They should also be linked to provisional sums of money in respect of labour, materials and plant so that the resultant totals form part of the tender, otherwise the tenderer could quote uncompetitive percentages for daywork without affecting his tender figure.

3.3.37 The clause envisages two types of definition of prime cost of daywork:

a) *Definition of Prime Cost of Daywork carried out under a Building Contract* issued by the Royal Institution of Chartered Surveyors (RICS) and the Building Employers Confederation (now the Construction Confederation) [5].

b) Any other similar definition of prime cost of daywork agreed between the RICS and any specialist body.

3.3.38 The Definition of prime cost of daywork under a building contract generally presents little difficulty other than when valuing plant costs. The definition of labour and materials costs is clearly set out in the document and refers in effect to the prices payable being those current when the relevant work is executed with the percentage additions covering certain Incidental Costs, Overheads and Profit. The position with plant costs is less clear. The Definition, so far as plant is concerned, states *"The rates for plant shall be as provided in the building contract"*. Thus such a provision – the rates to apply to such plant – must be made in the Contract Bills. It is usual to achieve this by reference to the *Schedule of Basic Plant Charges* published by the RICS [6] which establishes rates for the various items of plant. Thus the Contractor when tendering must also allow in his percentage not only for the incidental costs and the like but also for increases in price from the date of the Schedule to the date of his tender and for any further increases during the Contract; however, there is not normally a substantial amount involved.

3.3.39 The *Schedule of Basic Plant Charges* referred to above only applies to plant already on site; plant specifically hired for daywork has to be the subject of a separate agreement and care should be taken to ensure that if the percentage additions to plant rates referred to above are to apply to plant brought specially on to site for daywork purposes, the rates for that plant should be suitably adjusted.

3.3.40 There are, at present, only two other Definitions of Prime Cost of Daywork agreed with the RICS for use in England and Wales. These are:

a) Definition of the Prime Cost of Daywork carried out under a Heating, Ventilating, Air Conditioning, Refrigeration, Pipework and/or Domestic Engineering Contract [7];

b) Definition of Prime Cost of Daywork carried out under an Electrical Contract [8].

Reference to these together with a provision for the percentage adjustments must be made where appropriate in the Bills of Quantities before being sent to tender.

3.3.41 Thereafter, calculation of payments on a daywork basis should be a fairly straightforward operation. The Contractor is required to submit vouchers to the Architect or his authorised representative (usually the Clerk of Works) not later than the end of the week following that in which the work was executed. The Architect's task thereafter is to verify, or check the correctness of, the data. Thus, provided that the data recorded on daywork sheets are a correct record of the resources properly expended on the particular item of work, the costs would generally be admitted into the account. Moreover, if the daywork sheets are not verified it does not deprive the Contractor of his right to payment; he has only to ensure that the sheets are ... *delivered for verification* ... It is therefore important that the Architect, or the Clerk of Works if this task is delegated to him, does respond at the time. It is possible that on some projects daywork sheets will be submitted in abundance, and the temptation is to put them on one side to attend to "more important matters". This is a dangerous course to follow, for as time passes the Architect will be less well equipped to challenge them with any authority and the more difficult it will become to do other than include them at their face value, if indeed the Quantity Surveyor judges that valuation by daywork is the appropriate course.

3.3.42 The fact that a variation is to be priced as daywork is recognition that it cannot properly be valued by measurement and this is a decision made by the Quantity Surveyor, subject of course to the Contractor's right to challenge this in Ajudication, Arbitration or Legal Proceedings. Neither the fact that the Contractor submits daywork sheets nor the fact that the Architect may verify the data on those sheets, creates a contractual right to have the work valued as daywork; this is a matter entirely for the Quantity Surveyor's judgment.

Clause 13.5.5

3.3.43 It was suggested earlier in this chapter that the issue of a variation may have an impact beyond the work directly involved, i.e. it could cause additional expense to be incurred on other elements not referred to in the variation. A variation omitting work can also have an impact on remaining work in that it may change the conditions under which that remaining work is executed, e.g. omitting a significant value of reinforcement could mean that the cost of the crane needed to move reinforcement around the site has to be recovered over a smaller quantity of work and thus an enhanced rate should be applied as in the following Example 3.2. (In practice it is likely that there would be greater flexibility in the use of the crane. But for the purpose of achieving greater clarity in demonstrating the principles involved this has been ignored in the following three examples.)

Example 3.2 Valuing Variations – the effect of omissions

A crane cost £5,000 to set up, test, dismantle and remove. It costs £600 per week to hire and maintain over a 52-week period and £550 per week to operate over two 20-week periods (it is to stand idle for the intervening 12 weeks). It is expected to fulfil the Contract Bill requirements of moving some 5,000 tonnes of material, predominantly reinforcement. 500 tonnes (i.e. 10%) of these materials are omitted by a variation order but the work is executed over the same period. The cost of this plant per tonne, incorporated as part of the rates in the Contract Bills, is:

	£
Set up and removal	5,000
Hiring and maintaining £600.00 × 52 weeks	31,200
Operating £550.00 × 40 weeks	22,000
	58,200

i.e. £11.64 per tonne for 5,000 tonnes. However, with 500 tonnes omitted the new costs will be:

	£
Set up and removal	5,000
Hire and maintaining £600.00 × 52 weeks	31,200
Operating £550 × 36 weeks (40 weeks less 10%)	19,800
	56,000

i.e. £12.44 per tonne – an extra of £0.80 per tonne – for the remaining 4,500 tonnes valued under clause 13.5.1.2 (as required by clause 13.5.5) making a fair allowance for the difference in conditions.

3.3.44 By the same token, a comparable increase in reinforcement, moved by the same plant with the same basic costs would result in a lowering of the price to be paid in accordance with clause 13.5.5, as shown in Example 3.3.

Example 3.3 Valuing Variations – reducing the rate following an increase in quantity

	£
Set up and removal	5,000
Hiring and maintaining £600 × 52 weeks	31,200
Operating £550 × 44 weeks (40 weeks plus10%, it being assumed that it is practicable to move the additional material during the 12-week idle time)	24,200
	60,400

This is £2,200 more than the £58,200 covered by the Contract Sum and moves an extra 500 tonnes; thus the cost of the plant attributable to this extra reinforcement is only £4.40 per tonne.

3.3.45 The rate could also be reduced if, in the Contract Bills, the quantity of a particular item was just too great to be supported by one piece of plant and a whole additional piece of plant was allowed in the tender, albeit that it would be working at well below maximum output. This would result in a higher rate being quoted overall than might otherwise have prevailed. If the quantity of work were reduced by a variation and this variation was instructed in good time, the Contractor might be able to do without the uneconomic plant and a reduced rate could be applied to the remaining work as illustrated in Example 3.4.

Example 3.4   Valuing Variations – reducing the rate following a reduction in quantity

Assume the same basic rates as given above but it is judged at tender stage that one crane is insufficient to cope and another, smaller, crane is allowed for part of the time; this secondary crane costs £3,500 to set up and remove, £300 per week to hire and maintain for 20 weeks, and £275 per week to operate for 15 weeks. The following calculation would then result if the omission of 500 tonnes of reinforcement meant that the supplementary plant was not required. The Contract Bill rate would be as follows:

	£
Primary plant as before	58,200
Secondary plant:	
Set up and remove	3,500
Hire and maintain £300 × 20 weeks	6,000
Operating £275 × 15 weeks	4,125
	71,825

i.e. £14.37 per tonne for 5,000 tonnes. The Contract Bills would have a rate which included this £14.37 per tonne. But the omission of 500 tonnes – if this meant that there was now no need for the secondary plant – would enable the remaining 4,500 tonnes to be revalued at £12.93 per tonne (£58,200 – the cost of the primary plant – divided by 4,500 tonnes); that is, a reduction of £1.44 per tonne.

3.3.46 However, two points must be stressed in connection with this last example. First, the reduced requirement for moving materials must be within the capacity of the primary plant and second, the Contractor must have been notified in good time of the variation omitting the reinforcement, so that he could avoid ordering the secondary plant. This again emphasises the need for the Architect to keep the Contractor in the picture.

3.4   CLAUSE 13A CONSIDERED IN DETAIL

The general procedure

3.4.1 In January 1994, the Joint Contracts Tribunal introduced Amendment No. 13 which added clause 13A to JCT 80. This clause, the text of which is set out below, provides for the time and cost effects of a proposed variation being established in advance of its sanction. The outline of the procedure laid down in clause 13A is set out in *Figure 3.2* following the reproduction of the clause:

13.A   Variation instruction – Contractor's quotation in compliance with the instruction

13A   Clause 13A shall only apply to an instruction where pursuant to clause 13.2.3 the Contractor has not disagreed with the application of clause 13A to such instruction.

13A.1 .1   The instruction to which clause 13A is to apply shall have provided sufficient information to enable the Contractor to provide a quotation, which shall comprise the matters set out in clause 13A.2 (a "13A Quotation"), in compliance with the instruction; and in respect of any part of the Variation which relates to the work of any Nominated Sub-Contractor sufficient information to enable the Contractor to obtain a 3.3A Quotation from the Nominated Sub-Contractor in accordance with clause 3.3A.1.2 of the Conditions NSC/C. If the Contractor reasonably considers that the information provided is not sufficient, then, not later than 7 days from the receipt of the instruction, he shall request the Architect to supply sufficient further information.

13A.1 .2   The Contractor shall submit to the Quantity Surveyor his 13A Quotation in compliance with the instruction and shall include therein 3.3A Quotations in respect of any parts of the Variation which relate to the work of Nominated Sub-Contractors not later than 21 days from
        the date of receipt of the instruction
or   if applicable, the date of receipt by the Contractor of the sufficient further information to which clause 13A.1.1 refers
whichever date is the later and the 13A Quotation shall remain open for acceptance by the Employer for 7 days from its receipt by the Quantity Surveyor.

13A.1 .3   The Variation for which the Contractor has submitted his 13A Quotation shall not be carried out by the Contractor or as relevant by any Nominated Sub-Contractor until receipt by the Contractor of the confirmed acceptance issued by the Architect pursuant to clause 13A.3.2.

13A.2 The 13A Quotation shall separately comprise:

13A.2 .1   the value of the adjustment to the Contract Sum (other than any amount to which clause 13A.2.3 refers) including therein the effect of the instruction on any other work including that of Nominated Sub-Contractors supported by all necessary calculations by reference, where relevant, to the rates and prices in the Contract Bills and including, where appropriate, allowances for any adjustment of preliminary items;

13A.2 .2   any adjustment to the time required for completion of the Works (including where relevant stating an earlier Completion Date than the Date for Completion given in the Appendix) to the extent that such adjustment is not included in any revision of the Completion Date that has been made by the Architect under clause 25.3 or in his confirmed acceptance of any other 13A Quotation;

13A.2 .3   the amount to be paid in lieu of any ascertainment under clause 26.1 of direct loss and/or expense not included in any other accepted 13A Quotation or in any previous ascertainment under clause 26;

13A.2 .4   a fair and reasonable amount in respect of the cost of preparing the 13A Quotation and, where specifically required by the instruction, shall provide indicative information in statements on

13A.2 .5   the additional resources (if any) required to carry out the Variation; and

13A.2 .6   the method of carrying out the Variation.

Each part of the 13A Quotation shall contain reasonably sufficient supporting information to enable that part to be evaluated by or on behalf of the Employer.

13A.3 .1   If the Employer wishes to accept a 13A Quotation the Employer shall so notify the Contractor in writing not later than the last day of the period for acceptance stated in clause 13A.1.2.

13A.3 .2   If the Employer accepts a 13A Quotation the Architect shall, immediately upon that acceptance, confirm such acceptance by stating in writing to the Contractor (in clause 13A and elsewhere in the Conditions called a "confirmed acceptance"):

.2   .1   that the Contractor is to carry out the Variation;

.2   .2   the adjustment of the Contract Sum, including therein any amounts to which clause 13A.2.3 and clause 13A.2.4 refer, to be made for complying with the instruction requiring the Variation;

.2   .3   any adjustment to the time required by the Contractor for completion of the Works and the revised Completion Date arising therefrom (which, where relevant, may be a date earlier than the Date for Completion given in the Appendix) and, where relevant, any revised period or periods for the completion of the Nominated Sub-Contract work of each Nominated Sub-Contractor; and

.2   .4   that the Contractor, pursuant to clause 3.3A.3 of the Conditions NSC/C, shall accept any 3.3A Quotation included in the 13A Quotation for which the confirmed acceptance has been issued.

13A.4 If the Employer does not accept the 13A Quotation by the expiry of the period for acceptance stated in clause 13A.1.2, the Architect shall, on the expiry of that period either

13A.4 .1   instruct that the Variation is to be carried out and is to be valued pursuant to clause 13.4.1;

or

13A.4 .2   instruct that the Variation is not to be carried out.

13A.5 If a 13A Quotation is not accepted a fair and reasonable amount shall be added to the Contract Sum in respect of the cost of preparation of the 13A Quotation provided that the 13A Quotation has been prepared on a fair and reasonable basis. The non-acceptance by the Employer of a 13A Quotation shall not of itself be evidence that the Quotation was not prepared on a fair and reasonable basis.

(continued)

13.A   Variation instruction – Contractor's quotation in compliance with the instruction (continued)

13A.6 If the Architect has not, under clause 13A.3.2, issued a confirmed acceptance of a 13A Quotation neither the Employer nor the Contractor may use that 13A Quotation for any purpose whatsoever.

13A.7 The Employer and the Contractor may agree to increase or reduce the number of days stated in clause 13A.1.1 and/or in clause 13A.1.2 and any such agreement shall be confirmed in writing by the Employer to the Contractor. Where relevant the Contractor shall notify each Nominated Sub-Contractor of any agreed increase or reduction pursuant to this clause 13A.7.

13A.8 If the Architect issues an instruction requiring a Variation to work for which a 13A Quotation has been given and in respect of which the Architect has issued a confirmed acceptance to the Contractor such Variation shall not be valued under clause 13.5; but the Quantity Surveyor shall make a valuation of such Variation on a fair and reasonable basis having regard to the content of such 13A Quotation and shall include in that valuation the direct loss and/or expense, if any, incurred by the Contractor because the regular progress of the Works or any part thereof has been materially affected by compliance with the instruction requiring the Variation.

3.4.2 For the procedure in clause 13A to apply the Architect must require this to be so in the original instruction. And, because the effect of such an instruction is to seek a tender for the proposed variation, that instruction should contain data comparable in content and detail to that in the original tender documents (e.g. Addendum Bills of Quantities, Drawings and the like).

3.4.3 Under clause 13.2.3 the Contractor can, within 7 days of receipt by him of that instruction (or within such other period as may be agreed), disagree with the application of clause 13A to the proposed work. The clause does not state with whom the Contractor must agree such a change in the 7-day period but it would seem prudent to assume that this has to be an agreement with the Employer given the fact that it is the Employer who accepts any quotation given under clause 13A and it is the Employer who agrees to changing other periods laid down in clause 13A (see clause 13A.7). If the Contractor disagrees with the application of clause 13A to a proposed variation that variation instruction will be of no effect unless the Architect then further instructs that the variation shall go ahead and shall be valued in accordance with the general rules for valuing variations in accordance with clause 13.4.1.1 (clause 13.2.3). The Contractor does not have to give any reason for his disagreement; it is a matter entirely for him.

3.4.4 If the Contractor does not disagree with the application of clause 13A to a proposed variation he may nevertheless conclude that the data included in the variation instruction is an insufficient basis for submitting his Quotation, which Quotation must include for all the cost and time effects of that work. In this event, within 7 days of receipt of the instruction (or such longer or shorter period as may be agreed between the Employer and the Contractor in writing (clause 13A.7)), the Contractor must ask the Architect to provide further data. Within 21 days of the receipt by the Contractor of adequate data (or such longer or shorter period as may be agreed between the Employer and the Contractor in writing (clause 13A.7)) he must submit

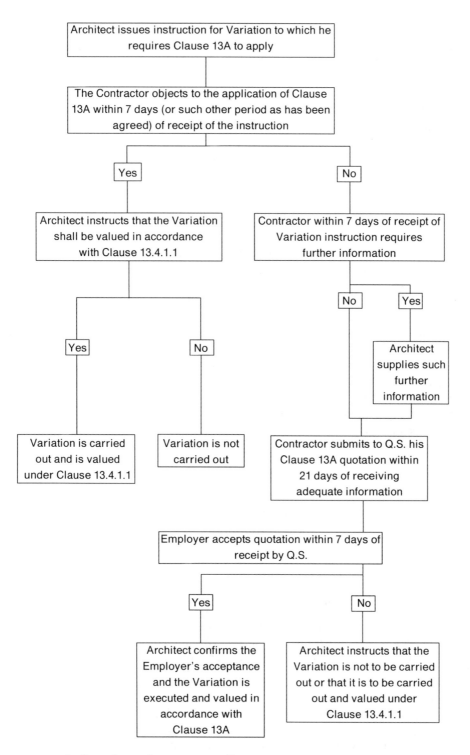

```
┌─────────────────────────────────────────┐
│ Architect issues instruction for Variation to which he │
│          requires Clause 13A to apply          │
└─────────────────────────────────────────┘

┌─────────────────────────────────────────┐
│ The Contractor objects to the application of Clause │
│ 13A within 7 days (or such other period as has been │
│       agreed) of receipt of the instruction        │
└─────────────────────────────────────────┘

         [Yes]                                    [No]

┌───────────────────────┐        ┌───────────────────────┐
│ Architect instructs that the Variation │  │ Contractor within 7 days of receipt of │
│   shall be valued in accordance    │  │    Variation instruction requires    │
│       with Clause 13.4.1.1        │  │          further information         │
└───────────────────────┘        └───────────────────────┘

                                          [No]      [Yes]

    [Yes]              [No]                    ┌──────────────┐
                                               │   Architect    │
                                               │  supplies such  │
                                               │    further     │
                                               │  information   │
                                               └──────────────┘

┌──────────────┐  ┌──────────────┐  ┌───────────────────────┐
│ Variation is carried │  │  Variation is not  │  │ Contractor submits to Q.S. his │
│  out and is valued  │  │  carried out    │  │  Clause 13A quotation within   │
│ under Clause 13.4.1.1│  │                │  │      21 days of receiving      │
└──────────────┘  └──────────────┘  │      adequate information      │
                                     └───────────────────────┘

┌─────────────────────────────────────────┐
│ Employer accepts quotation within 7 days of │
│            receipt by Q.S.             │
└─────────────────────────────────────────┘

         [Yes]                                    [No]

┌──────────────┐              ┌───────────────────────┐
│ Architect confirms the │          │  Architect instructs that the │
│ Employer's acceptance │          │ Variation is not to be carried │
│  and the Variation is  │          │  out or that it is to be carried │
│ executed and valued in │          │   out and valued under      │
│   accordance with    │          │      Clause 13.4.1.1        │
│     Clause 13A      │          └───────────────────────┘
└──────────────┘
```

Figure 3.2 Outline of procedure set out in Clause 13A.

his Quotation to the Quantity Surveyor for acceptance by the Employer. This acceptance must be made by the Employer within 7 days of receipt of the Quotation by the Quantity Surveyor. But for the whole process to be finally effective the Architect must then confirm this acceptance. Note that it is the Employer who accepts the Quotation (which he does by notifying the Contractor in writing); it is the Architect who issues the original instruction and any subsequent extra data and it is the Architect who confirms acceptance of the Quotation; and it is the Quantity Surveyor to whom the Quotation has to be submitted.

The Contractor's Quotation

3.4.5 The Contractor's Quotation must include the following separate components (clause 13A.2):

a) the cost effect of the variation on the Contract Sum (its valuation) including the cost effects of the variation on other work (but excluding any direct loss and/or expense which is to be included elsewhere in the Quotation) as a separate item. Such a valuation of the variation is to be related to the rates in the Contract Bills where relevant;

b) any change required to the Completion Date resulting from the execution of the variation; note that such a change can bring about a reduction in the contract period, i.e. the Date for Completion written into the contract Appendix can be brought forward. This is the only express provision for this to happen (see paras 2.7.21 and 2.9.10);

c) any direct loss and/or expense in respect of the variation which would otherwise have been payable under clause 26;

d) a fair and reasonable amount for preparing the quotation.

And if so required in the original instruction indicative information on:

e) the additional resources required;

f) the method of carrying out the variation.

3.4.6 Each part of these components must contain sufficient data to enable it to be evaluated.

The acceptance of the Quotation

3.4.7 It is for the Employer to accept the Quotation but it is for the Architect to confirm that acceptance before the provisions of clause 13A become effective. The Architect's confirmation of the Employer's acceptance must contain the following components (clause 13A.3.2):

a) a statement that the Contractor is to carry out the variation;

b) the full adjustment to the Contract Sum (i.e. including the cost effect of the variation on other work and any direct loss and/or expense);

c) the revised Completion Date and where relevant the revised periods for the completion of the Nominated Sub-Contract work;

d) where appropriate a statement that the Contractor shall accept such a Quotation from a Nominated Sub-Contractor.

3.4.8 If the Employer does not accept a Quotation submitted under clause 13A the Architect must either instruct that the variation is to be carried out and that it is to be valued in accordance with clause 13.4.1.1 (i.e. the general rules which apply to the valuation of variations) or instruct that the variation is not to be carried out at all. In either event ... *a fair and reasonable amount shall be added to the Contract Sum in respect of the cost of preparation* ... of the Quotation. However, this requirement to pay the cost of preparing the Quotation does not apply if the Quotation was not prepared on a fair and reasonable basis. The fact that the Employer has not accepted a Quotation is not of itself proof that the Quotation was not prepared on a fair and reasonable basis (clause 13A.5).

Variations to work executed under a confirmed acceptance of a clause 13A Quotation

3.4.9 Under clause 13A.8, the Architect may instruct a variation to the work for which there has been a confirmed acceptance of a Quotation under clause 13A. In this event, that variation is not to be valued under clause 13.5 as would be the case with any other variation but is to be valued by the Quantity Surveyor (clause 13A.8). He must do so on a fair and reasonable basis having regard to the accepted Quotation and must include any direct loss and/or expense resulting from that variation materially affecting the regular progress of the Works. It seems anomalous that the strict rules which apply in clause 26 to the ascertainment of direct loss and/or expense arising from variations generally, appear to have been abandoned when it comes to variations to work executed pursuant to an accepted Quotation under clause 13A.

3.4.10 Among such rules, which have been abandoned, are:

a) the need for the Contractor to give written notice of his contention that progress of the Works is being, or is likely to be, materially affected by the variation as soon as this is reasonably apparent;
b) the ability to call for records to be kept by the Contractor;
c) the need for the Architect to be convinced that such direct loss and/or expense has been or will be incurred;
d) the need for the Architect to undertake the ascertainment of the direct loss and/or expense unless he specifically delegates this task to the Quantity Surveyor.

3.4.11 If this reading of the provisions of clause 13A is correct then these are serious matters because rules a) to d) above are the minimum safeguards necessary to ensure a proper and equitable evaluation of this component of cost. In effect the otherwise prescribed detailed ascertainment of direct loss and/or expense incurred has been set aside in favour of a fair valuation thereof.

3.4.12 Finally, it would appear that the valuation of the effect that a variation executed under clause 13A.8 has upon other work is to be valued under clause 13.5.5 and not under clause 13A.8. No reference is made to this aspect of cost in clause 13A.8 and the embargo contained in clause 13A.8 (that clause 13.5 shall not be applied) is in regard to the valuation of the variation itself and not to the changed conditions brought about to other work by it. What is clear is that it should not be valued twice.

3.5 CLAUSE 26 CONSIDERED IN DETAIL

Generally

3.5.1 Clause 26 governs the ascertainment of loss and/or expense reimbursable to the Contractor due to the progress of the Works being affected by the series of events listed therein, including the execution of Architect's instructions requiring a variation. It reads as follows:

---

26 Loss and expense caused by matters materially affecting regular progress of the Works

26.1   If the Contractor makes written application to the Architect stating that he has incurred or is likely to incur direct loss and/or expense (of which the Contractor may give his quantification) in the execution of this Contract for which he would not be reimbursed by a payment under any other provision in this Contract due to deferment of giving possession of the site under clause 23.1.2 where clause 23.1.2 is stated in the Appendix to be applicable or because the regular progress of the Works or of any part thereof has been or is likely to be materially affected by anyone or more of the matters referred to in clause 26.2; and if and as soon as the Architect is of the opinion that the direct loss and/or expense has been incurred or is likely to be incurred due to any such deferment of giving possession or that the regular progress of the Works or of any part thereof has been or is likely to be so materially affected as set out in the application of the Contractor then the Architect from time to time thereafter shall ascertain, or shall instruct the Quantity Surveyor to ascertain, the amount of such loss and/or expense which has been or is being incurred by the Contractor; provided always that:

26.1 .1   the Contractor's application shall be made as soon as it has become, or should reasonably have become, apparent to him that the regular progress of the Works or of any part thereof has been or was likely to be affected as aforesaid; and

26.1 .2   the Contractor shall in support of his application submit to the Architect upon request such information as should reasonably enable the Architect to form an opinion as aforesaid; and

26.1 .3   the Contractor shall submit to the Architect or to the Quantity Surveyor upon request such details of such loss and/or expense as are reasonably necessary for such ascertainment as aforesaid.

26.2   The following are the matters referred to in clause 26.1:

26.2 .1 .1   where an information Release Schedule has been provided, failure of the Architect to comply with clause 5.4.1;

26.2 .1 .2   failure of the Architect to comply with clause 5.4.2;

26.2 .2   the opening up for inspection of any work covered up or the testing of any of the *work, materials or goods in accordance with clause 8.3 (including making good in consequence of such opening up or testing), unless the inspection or test showed that the work, materials or goods were not in accordance with this Contract;*

26.2   .3   *any discrepancy in or divergence between the Contract Drawings and/or the Contract Bills and/or the Numbered Documents;*

26.2   .4   .1   *the execution of work not forming part of this Contract by the Employer himself or by persons employed or otherwise engaged by the Employer as referred to in clause 29 or the failure to execute such work;*

          .4   .2   *the supply by the Employer of materials and goods which the Employer has agreed to provide for the Works or the failure so to supply;*

26.2   .5   *Architect's instructions under clause 23.2 issued in regard to the postponement of any work to be executed under the provisions of this Contract;*

26.2   .6   *failure of the Employer to give in due time ingress to or egress from the site of the Works or any part thereof through or over any land, buildings, way or passage adjoining or connected with the site and in the possession and control of the Employer, in accordance with the Contract Bills and/or the Contract Drawings, after receipt by the Architect of such notice, if any, as the Contractor is required to give, or failure of the Employer to give such ingress or egress as otherwise agreed between the Architect and the Contractor;*

26.2   .7   *Architect's instructions issued*

under clause 13.2 or clause 13A.4.1 requiring a Variation (except for a Variation for which the Architect has given a confirmed acceptance of a 13A Quotation or for a Variation thereto) or

under clause 13.3 in regard to the expenditure of provisional sums (other than instructions to which clause 13.4.2 refers or an instruction for the expenditure of a provisional sum for defined work or of a provisional sum for Performance Specified Work);

26.2   .8   *the execution of work for which an Approximate Quantity is included in the Contract Bills which is not a reasonably accurate forecast of the quantity of work required;*

26.2   .9   *compliance or non-compliance by the Employer with clause 6A.1;*

26.2   .10   *suspension by the Contractor of the performance of his obligations under the Contract to the Employer pursuant to clause 30.1.4 provided the suspension was not frivolous or vexatious.*

26.3   *If and to the extent that it is necessary for ascertainment under clause 26.1 of loss and/or expense the Architect shall state in writing to the Contractor what extension of time, if any, has been made under clause 25 in respect of the Relevant Event or Events referred to in clause 25.4.5.1 (so far as that clause refers to clauses 2.3, 13.2, 13.3 and 23.2) and in clauses 25.4.5.2, 25.4.6, 25.4.8 and 25.4.12.*

26.4   .1   *The Contractor upon receipt of a written application properly made by a Nominated Sub-Contractor under clause 4.38.1 of Sub-Contract Conditions NSC/C shall pass to the Architect a copy of that written application. If and as soon as the Architect is of the opinion that the loss and/or expense to which the said clause 4.38.1 refers has been incurred or is likely to be incurred due to any deferment of the giving of possession where clause 23.1.2 is stated in the*

(continued)

117

26 Loss and expense caused by matters materially affecting regular progress of the Works (continued)

Appendix to apply or that the regular progress of the sub-contract works or of any part thereof has been or is likely to be materially affected as referred to in clause 4.38.1 of Sub-Contract Conditions NSC/C and as set out in the application of the Nominated Sub-Contractor then the Architect shall himself ascertain, or shall instruct the Quantity Surveyor to ascertain, the amount of loss and/or expense to which the said clause 4.38.1 refers.

26.4 .2 If and to the extent that it is necessary for the ascertainment of such loss and/or expense the Architect shall state in writing to the Contractor with a copy to the Nominated Sub-Contractor concerned what was the length of the revision of the period or periods for completion of the sub-contract works or of any part thereof to which he gave consent in respect of the Relevant Event or Events set out in clause 2.6.5.1 (so far as that clause refers to clauses 2.3, 13.2, 13.3 and 23.2 of the Main Contract Conditions), 2.6.5.2, 2.6.6, 2.6.8, 2.6.12 and 2.6.15 of Conditions NSC/C.

26.5 Any amount from time to time ascertained under clause 26 shall be added to the Contract Sum.

26.6 The provisions of clause 26 are without prejudice to any other rights and remedies which the Contractor may possess.

3.5.2 The causes which, if they disrupt the regular progress of the work, will give rise to an entitlement for the Contractor to be reimbursed his direct loss and/or expense are the same as or similar to some but not all of the causes which qualify for an extension of time. Thus reference should be made to the relevant part of Chapter 2 for comment on the text of these causes; for ease of reference this is given in the table at para. 3.5.3.

3.5.3 However, it must be emphasised that in Chapter 2 the causes are considered in the context that they might delay the progress of the work and ultimately delay the Completion Date; in clause 26 they have to be seen from the point of view of the effect if any that they have on the regular progress of the work, and this does not necessarily imply delaying the Completion Date. Thus there can be payment under clause 26 without any extension of time under clause 25. Moreover, there are one or two instances where the wording in clause 26.2 differs marginally, but significantly, from its counterpart in clause 25.4; these differences are considered immediately following this table.

Reference to the text of Relevant Events which also qualify as matters giving rise to a claim for reimbursement of loss and/or expense

Cause	Clause 26 ref.	Clause 25 ref.	Reference to text in this book
Late information	26.2.1.1	25.4.6.1	2.8.29 to 2.8.32
	26.2.1.2	25.4.6.2	2.8.33 to 2.8.44
Inspection and testing	26.2.2	25.4.5.2	2.8.28
Discrepancies	26.2.3	25.4.5.1	2.8.15 to 2.8.17
Work by others	26.2.4.1	25.4.8.1	2.8.51 to 2.8.54

Cause	Clause 26 ref.	Clause 25 ref.	Reference to text in this book
Supply of materials by Employer	26.2.4.2	25.4.8.2	2.8.55
Postponement	26.2.5	25.4.5.1	2.8.24
Failure to give ingress to or egress from the Works	26.2.6	25.4.12	2.8.62 to 2.8.63
Variations	26.2.7	25.4.5.1	2.8.18 to 2.8.22
Approximate quantities	26.2.8	25.4.14	2.8.65 to 2.8.67
Compliance with CDM Regulations	26.2.9	25.4.17	2.8.70
Suspension by the Contractor	26.2.10	25.4.18	2.8.71 to 2.8.75

3.5.4 Although not strictly a *matter* under the terms of clause 26, deferment of site possession (where the parties have opted for this facility) is also a ground for claiming reimbursement of loss and/or expense (clause 26.1) (para. 2.11.2). However, because delayed possession of the site pursuant to clause 23.1.2 does not appear in the items listed in clause 26.2, the issue of the Final Certificate is not conclusive evidence that the reimbursement of direct loss and/or expense is in final settlement of a claim arising out of any such delayed possession. Clause 30.9.1.4 is specifically restricted to such direct loss and/or expense as arises from ... *any of the matters referred to in clause 26.2* ...

3.5.5 It will be seen by comparison between the text of clauses 25 and 26 that generally only discrepancies in or divergence between the Contract Drawings and/or Contract Bills and/or the Numbered Documents are grounds for claiming the reimbursement of loss and expense (clause 26.2.3) whereas discrepancies in and divergences between the following will be grounds for claiming an extension of time (clause 25.4.5.1):

a)   Contract Drawings;
b)   Contract Bills;
c)   Architect's instructions (other than variations);
d)   descriptive schedules and the like;
e)   further drawings or details;
f)   drawings for setting out the Works;
g)   the numbered documents annexed to any Nominated Sub-Contract Agreement

as will any discrepancy or divergence between the Contractor's Statement in respect of Performance Specified Work and any instruction of the Architect issued after receipt by him of the Contractor's Statement.

3.5.6 Further it should be noted that instructions in regard to the expenditure of provisional sums for defined work (see also para. 2.8.19) and for Performance Specified Work as well as those requiring a variation for which the Architect has given a confirmed acceptance of a 13A Quotation cannot be grounds for the reimbursement of loss and/or expense (clause 26.2.7) or grounds for an extension of time (clause 25.4.5.1). The Contractor's tender/13A Quotation are deemed to take both into account.

3.5.7 However, before any matters will give rise to an entitlement to the reimbursement of loss and/or expense:

a) the Contractor must comply with the provisions of the Contract requiring him to give notice and further information; and

b) there must be or have been an effect upon the regular progress of the Works giving rise to loss and/or expense which would not be reimbursed elsewhere under the Contract.

3.5.8 One of the matters which gives the Contractor the ability to claim reimbursement of his direct loss and/or expense is opening up work for inspection or testing where no fault is found (clause 26.2.2). However, this does not extend to such opening up or inspection where it is undertaken pursuant to clause 8.4.4. This clause envisages the possibility that upon discovering a defect in the work the Architect may wish to instruct such further opening up for inspection or testing to establish whether or not there are similar defects elsewhere in the work. Provided that the Architect is reasonable in issuing instructions, and has had regard to the Code of Practice on this matter appended to the Contract, then no sum is to be allowed for such opening up or testing or for any disruption that this may cause, although an extension of time may be granted provided the further inspection reveals that the Work, materials or goods were in accordance with the Contract (see para 2.8.28).

Written application and other data to be provided by the Contractor

3.5.9 The written application from the Contractor to the Architect under clause 26 is a condition precedent to payment; that is, unless it is given, and given in the manner laid down, the Architect cannot act upon it. On the other hand the precise form of an application must be seen in the context of the judgment in *Rees and Kirby -v- Swansea City Council (1985) 30 BLR 1* which, in relation to the form of application (under JCT 63), decided that "... *in the ongoing relationship that exists between a contractor and an Architect carrying out their functions ... a sensible and not too technical attitude must be adopted with regard to the form of such an application*". This does not necessarily mean that the Contractor will lose all entitlement if he fails to comply with the provisions of clause 26.1. Clause 26.6 makes it clear that the Contractor's other rights and remedies are not prejudiced by the remainder of clause 26, thus he could pursue the matter in Arbitration or in the Courts (see para. 1.2.8). But any entitlement would not flow from clause 26 and thus there would be no right to demand payment in interim certificates.

3.5.10 Clause 26 requires that the Contractors' written application be made ... *as soon as it has become, or should reasonably have become, apparent to him that the regular progress of the Works or any part thereof has been or was likely to be affected* ... Thus the Contractor is required to judge the likely effect of the matters listed in that clause upon the regular progress of the Works as soon as they should have become apparent to him (this is in contrast with a similar provision in clause 25 where there is no reference to whom the fact or prospect of a delay must be apparent (para. 2.7.7). The Contractor must scrutinise all instructions and the like to see if they will affect the regular progress of the Works when carried out. If it is reasonably clear that they will and that as a result he is likely to incur direct loss and/or expense which is not recoverable elsewhere, he must give notice in accordance with clause 26.

3.5.11 Application must be made as soon as the likelihood of an effect upon regular progress has become apparent or, more importantly, as soon as it should reasonably have become apparent. Thus the inefficient Contractor, who may be in such a muddle that the disruption due to, say, a variation, is indistinguishable from the general delays, cannot leave his written application until he reviews his final costs on the grounds that it was not until then that the disruption became clear; it is when it **should** reasonably have become apparent to him that such disruption will occur that he must make his application.

3.5.12 The application from the Contractor need only state that:

a) he has incurred or is likely to incur direct loss and/or expense;

b) the loss and/or expense is due to the regular progress of the work being affected by one or more of the matters listed in clause 26;

c) the loss and/or expense would not be reimbursed under any other provision of the Contract (clause 30.6.2 lists those provisions of the Contract that require adjustment to be made to the Contract Sum).

3.5.13 The Contractor's application must be supported by further data. First, such information as will enable the Architect to form an opinion on whether the regular progress of the Works has been or is likely to be affected as claimed and, secondly, such details as will enable the amount of the loss and/or expense to be ascertained.

3.5.14 Both pieces of further information are to be provided to the Architect and in the case of the latter perhaps the Quantity Surveyor ... *upon request* ... not, if requested. Thus it is anticipated that such data will always be necessary with the timing of its supply being determined by the Architect or Quantity Surveyor making his request. The Contractor must therefore always keep such records; he cannot argue that he failed to do so because the Architect or Quantity Surveyor did not request them soon enough. In requesting the provision of such records, the Architect or Quantity Surveyor would be well advised to indicate the sort of data that would enable him to make his judgment or his ascertainment as this should keep misunderstanding to the minimum and speed up the process of determining the Contractor's entitlement.

3.5.15 In April 1998, JCT issued Amendment 18 to JCT 80 which, inter alia, somewhat gratuitously introduced in clause 26.1 a provision whereby ... *the Contractor may give his quantification* ... of the loss and expense incurred and it did so without awarding such quantification any contractual status. It is suggested therefore that when the Contractor opts to provide this quantification (as many did before this amendment) it should be regarded as being "for information only" and should not be accepted in place of any particular further data which may be requested under clause 26.1.2.

The Architect's response to an application for the reimbursement of loss and/or expense

3.5.16 Once the Architect has received the Contractor's application for reimbursement of direct loss and/or expense he has two separate functions to perform. First, to form an opinion on whether the regular progress of the work has been or is likely to be affected in the way suggested by the Contractor to the extent that loss and/or expense will be incurred. Second, if he is of that opinion, he must ascertain the amount of the loss and/or expense directly incurred thereby and which is not reimbursed by any other provision of the Contract. In each case he is required to seek from the Contractor the necessary data to enable him to act. If the Contractor in his application refers to one of the matters listed in clause 26.2 as being the cause of the disruption and the Architect is of the opinion that it was another matter (also listed in that clause) that was the cause, the Architect is not entitled, under the wording of clause 26, to substitute that latter cause and then proceed to ascertain the loss, etc.; he has to decide whether or not the progress was ... *materially affected as set out in the application of the Contractor* ...; see example 3.5.

Example 3.5 Responding to an application for recovery of loss and/or expense where an incorrect cause is given

The Contractor makes written application that he is likely to incur direct loss and/or expense for which he will not be otherwise reimbursed due to late issue of information – the failure of the Architect to comply with clause 5.4.2. The Architect concludes that the issue of the information, although late in relation to the master programme, did not materially affect progress as the Contractor had already dropped behind schedule, owing to an earlier variation which was also an acceptable ground for claiming loss and/or expense. However, the Architect cannot accede to the Contractor's claim as the Architect can only consider the matter ... *as set out in the application of the Contractor* ... Moreover, a similar embargo would apply if, in the opinion of the Architect, the cause complained of did not have the effect upon progress claimed by the Contractor in his application. There is, however, nothing to prevent the Contractor from making a further application if the Architect rejects the original one, provided that it meets the time criteria given in clause 26.1.1. The Contractor may, however, simply specify that the progress of the works have been affected by ... *any one or more matters referred to in clause 26.2* ... In this event the Architect would have to request the data to enable him to form an opinion thereon, one element of that data being the specific matter which the Contractor claims is the cause.

3.5.17 The timing of the Architect's role is not entirely clear. He is first required to consider whether there has been or is likely to be an effect upon the regular progress of the Works. According to clause 26.1 ... *as soon as* ... he has formed this opinion he must from time to time thereafter ascertain the loss and/or expense that has been or is being incurred. The Contractor must give prior notice of a prospective involvement in loss and/or expense but the Architect can only ascertain loss and/or expense that has been or is being incurred. The Architect will in practice therefore, having received the Contractor's application in relation to a future loss and/or expense, have to wait until it is incurred before it can be ascertained. If the Architect can fully and properly ascertain part of such direct loss and/or expense he should do so as soon as possible and the resultant sum should be included in the next interim certificate. However, if he can only roughly calculate any amount due as reimbursement of direct loss and/or expense he should do nothing.

3.5.18 The amount ascertained, from time to time, must be added to the Contract Sum under clause 26.5; in accordance with clause 3, therefore, any amount so ascertained, whether wholly or in part, must be added to the next Interim Certificate. Amounts so included are not subject to Retention (clause 30.2.2.2).

3.5.19 In Chapters 2 and 4, the dangers of relating the calculation of extensions of time to the ascertainment of loss and/or expense are highlighted at section 2.5 and para. 4.4.13 respectively. A direct link between the two has been created by clauses 26.3 and 26.4.2. These require the Architect, insofar as it is necessary for the ascertainment of loss and/or expense, to declare what part of the extensions of time granted both for the main Contract and for any Nominated Sub-Contract are in respect of those causes which also give an entitlement to the payment of loss and/or expense to either the Contractor or Nominated Sub-Contractor as the case may be.

Failure to make written application in good time

3.5.20 A question often posed is: *"If the Contractor fails to make his written application within the specified time does he lose his entitlement completely?"* As the Contractor's entitlement under these clauses arises from the act or default of the Employer (or his agents) the Contractor would be unlikely wholly to lose his rights if he simply failed to comply with an administrative provision, i.e. the giving of a timely written notice. Indeed, clause 26.6 specifically reserves the Contractor's option of pursuing his remedy by other means ... *The provisions of clause 26 are without prejudice to any other rights and remedies which the Contractor may possess* ... In the case of *London Borough of Merton -v- Stanley Hugh Leach Ltd (1985) 32 BLR 51* which was a case under JCT 63 but which contains similar provisions, it was decided that *"... the Contractor is not bound to make an application under clause 24(2). He may prefer to wait until completion of the work..."*

3.5.21 But his remedy in the case of such failure to give the necessary notice would be through Arbitration or the courts, not through the procedure laid down in clause 26. That is to say the matter could not be dealt with by the Architect or Quantity Surveyor nor included in Interim Certificates; and this could be time consuming and costly, and result in a considerable delay in establishing the monetary entitlement. Even if the Contractor does successfully pursue his entitlement other than through the processes laid down in clause 26, it is suggested that the absence of the proper written notice might be to his disadvantage. If he failed to make the written application as required by the Contract and that failure denied the Architect the opportunity of taking action to mitigate the potential loss, then the Contractor might well fail to secure payment to the extent that his losses were exacerbated by his own failure to give notice.

The ascertainment

3.5.22 Clause 26 refers to the ascertainment of direct loss and/or expense and this implies a finite calculation. In practice, however, notwithstanding the availability of detailed cost records or invoices and the like, an element of assessment is invariably involved because:

a) the wording of the Contract is not always as clear as it might be; for example, the boundaries of the term "direct" are difficult to define in practice;

b) when ascertaining loss and/or expense one always has to contrast the costs or losses as recorded with what they might have been had the variation or disruption not occurred thus an element of conjecture becomes inevitable (para. 4.2.7 and Example 4.4 as referred to at para. 4.4.8);

c) often claims are dealt with well after the event and the Contractor is not requested to submit details of his loss and/or expense at the time that it is incurred and the facts will later have become obscured and some of the people most concerned with the events may be difficult to trace.

3.5.23 Nevertheless, this should not be taken as an excuse for abandoning an attempt to make the ascertainment as precise as circumstances permit.

3.5.24 Ascertain means "to make certain of" or "to find out for certain". But it has already been pointed out above that an element of judgment is usually unavoidable in ascertaining any direct loss and/or expense (para. 4.2.7).

3.5.25 Clearly the Architect (or Quantity Surveyor) is unlikely to be able to ascertain the loss and/or expense without the assistance of the Contractor in the provision of cost records and the like; the provision of these data can be demanded under clause 26.1.3 which requires the Contractor to ... *submit to the Architect or to the Quantity Surveyor upon request such details... as are reasonably necessary for such ascertainment ...*

3.5.26 The clause does permit the Architect to delegate the task of ascertainment to the Quantity Surveyor. If he does so, it is suggested, the Architect cannot amend the resulting ascertainment. It is possible for the Architect to review the valuations prepared for the purpose of Interim Certificates and to review a previous Interim Certificate, but clause 3 requires ... *such amount (as has been ascertained) shall be taken into account in the computation of the next interim certificate.* This implies an automatic inclusion of the ascertained sum. Therefore, it would seem that the Architect must include the sum but may advise the Employer to take the matter to Arbitration. In any event, it is most unwise for the Architect wholly to delegate this task to the Quantity Surveyor as it is the Architect who either issues the variation or instructs most of the matters which may be considered under clause 26.

3.5.27 It is the Architect who is administering the Contract and it is therefore necessary for him to be closely associated with the events and progress on site and therefore be the better informed on the effect of variations or instructions. So much so that it is only the Architect who can grant extensions of time under clause 25 (an obligation which he has whether or not the Contractor has notified him of any delay (paras 2.6.8, 2.7.34 and 2.7.35)), and it is only the Architect who is empowered to form an opinion on whether loss and/or expense has been incurred. In ascertaining loss and/or expense it is advisable for the Architect to establish the facts and to assess the extent and nature of any disruption; thereafter he may instruct the Quantity Surveyor to ascertain the actual loss and/or expense directly resulting. In this way the relative skills and experience of each of these professionals are properly recognised and used.

3.5.28 The need to distinguish between items to be valued by reference to Contract Bill prices and those to be ascertained by reference to the Contractor's loss and/or expense is emphasised at paras 3.1.3, 3.1.11 and section 3.7. There is, however, a direct link between the two; if the Contractor's costs appear to exceed the Contract Bill prices in any instance this can be accounted for by one or more of the following factors:

a) disruption under clause 26;
b) inflation;
c) non-reimbursable costs (e.g. caused by Nominated Sub-Contractor's delay);
d) Contractor's inefficiency;
e) remedial work;
f) underestimate in the price.

3.5.29 Of those factors, costs arising at a) would clearly be reimbursable; so would those at b) if the delay was the result of an event envisaged by clause 26 and not reimbursed under any other provision in the Contract. Costs at items c) to e), on the other hand, would not be reimbursable. Item f) presents a problem. As is stated at para. 3.3.17, "errors" in Contract Bill prices cannot be rectified for the purpose of valuing variations. But where there is a direct loss and/or expense as a result of (not as part of) a variation causing disruption as recognised under clause 26 is the Contractor entitled to his total costs even though the item appears elsewhere in his tender at a more optimistic rate? The matter is set out in Example 3.6.

> ## Example 3.6   Where actual expense exceeds the contract rate
>
> A piece of plant is priced in the Contract Bills at £200 per week but the Contractor's records show that the cost is £275 per week. Extra use of the plant as part of a variation is priced at £200 per week in accordance with clause 13. But extra use due to disruption under clause 26 is claimed by the Contractor to be related to cost. He argues that the contract clearly states "direct expense", that the word "expense" is not further qualified and that the £275 per week should therefore prevail. The Architect argues that the Employer has signed a Contract which embodies certain rates and prices; if he requires a variation why should the Employer pay more by way of damages than was envisaged by the parties at the signing of the Contract as being the result of the delay (i.e. the £200 per week) (paras 1.2.16 and 3.5.28).

3.5.30 There is no clear cut answer to this question; it is suggested that the expense must be reasonable and that the Architect or Quantity Surveyor would be entitled to ask the Contractor to demonstrate how, having regard to any lower prices quoted in the Contract Bills, the additional expense now claimed is reasonable. In the event that any part of such additional cost has occurred as a result of items c) to e) in para. 3.5.28 it should be disallowed on the grounds that they are not the **direct** result of the cause relied upon.

3.5.31 Only **direct** loss and/or expense can be reimbursed. Direct damage has been defined as *. . . that which flows naturally from the breach without other intervening cause . . .* (*Saint Line Ltd -v- Richardsons (1940) 2 KB 99*). Such losses or expenses, therefore, must be the direct result of the variation or disruption as the case may be. In the case of *Rees and Kirby -v- Swansea City Council (1985) 30 BLR 1*, the Contractor had claimed recovery of interest charges on losses which arose from disruption by variations and delayed instructions There was a substantial period between his incurring this primary loss and its certification. Part of this time was expended in trying to reach an overall financial settlement by two quite alternative approaches neither of which was related to the original cause of the loss nor to the ascertainment of the amount due. First, there were attempts to settle matters by means of an ex gratia payment but this came to nothing. Secondly, consideration was given to converting the contract from its fixed price basis to one in which the Contractor would be reimbursed any fluctuations in prices. This also came to nothing. In giving its judgment, the Court decided that the interest which had accrued during each of these two periods did not qualify for reimbursement because those periods of time did not directly arise from the variations or delayed instructions. They were therefore indirect rather than direct costs.

3.5.32 The following Examples 3.7 and 3.8 further illustrate this distinction.

Example 3.7 Exclusion of costs because they are *indirect* – I

A Contractor undertakes a Contract containing brickwork priced at £25 per m^2. This rate is arrived at by using a labour rate of £175 per week for a bricklayer. Once work starts on site the Contractor discovers that the prevailing rate, due to local market conditions is, in fact, £200 per week for a bricklayer. Later, the final stage of the brickwork is postponed by an Architect's instruction under clause 23.2. And when this work is recommenced all bricklayers have left the site and a special rate of £215 per week has to be negotiated as the extent of this remaining brickwork is so small, albeit the prevailing rate for bricklayers generally on work of a reasonable volume is still £200 per week, i.e. an extra of £15 per week is related to the nature of the work. In ascertaining the direct loss and/or expense under clause 26 the Architect or Quantity Surveyor should take account of the increase of £15 per hour charged due to the piecemeal nature of the work, but should exclude the £25 increase from £175 to £200 as this extra is not an element of expense which flows directly from the postponement.

Example 3.8 Exclusion of costs because they are *indirect* – II

A Contractor undertakes a fixed price Contract with a fixed Date for Completion. He naturally makes no allowance for the wage award already promulgated to take effect 10 weeks after the Date for Completion. He is delayed by both strikes and variations and is given extensions of time of 12 and 8 weeks, respectively. He claims under clause 26 the effect of the wage award as part of his direct loss and/or expense flowing from the variations. As a compromise he offers to accept 40% of those extra costs (i.e. the proportion of the total extensions of time attributable to variations). However, none of the cost of this wage award is a direct expense; there is an intervening cause, the delay due to strikes, which is not grounds for reimbursement under the Contract. Had the strike not occurred the work, including the variations, would have been completed before the wage award made its impact upon costs.

3.5.33 The last example highlights again the danger of using the extension of time provisions in the Contract as a springboard to establish a claim for monetary entitlement. The extensions of time machinery must be ignored in establishing the quantum of loss and/ or expense. Instead, one must ask the questions "Is the cause one which is recognised by clause 26?" and "What are the costs directly attributable to that cause?" The effect of the strike will then be seen to be irrelevant. For further detailed comment on the process of ascertainment see Chapter 4.

3.6 CONCURRENT DELAYS AND THEIR EFFECT UPON THE ASCERTAINMENT OF DIRECT LOSS AND/ OR EXPENSE

3.6.1 The similarity between the wording of the clauses 25 and 26 often gives rise to a misunderstanding that is sometimes expressed as follows: "If the progress of the works is delayed by, say, exceptionally adverse weather conditions but at the same time there is a delay in information which would have delayed progress anyway, under which sub-clause of clause 25 is the extension granted – clause 25.4.2 or 25.4.6?" The answer to that question is that it does not matter. Either (but not both) can give an entitlement to a proper extension of time. What the questioner really wants to know is "Am I entitled to claim reimbursement of direct loss and/or expense in this situation?", in which event reference must be made to clause 26 and the answer, it is suggested, is "No". The progress of the works is not materially affected by the late information; no work could have proceeded had the information been available. Moreover, there is no direct loss and/or expense due to the late information; any loss and/or expense would have occurred anyway. In any event the Contractor is deemed to have included in his price for all the risks that properly fall on him, e.g. extra costs of delay due to exceptionally adverse weather conditions, delay by statutory undertakers and force majeure. Thus during the period of delay referred to above, the sum deemed to be included in the Contract Sum in respect of those risks is being set against the Contractor's extra costs.

3.6.2 One final point to be made is that there must be a real loss and/or expense for clause 26 to become operative; this clause is not concerned with theoretical entitlements. For example, a piece of plant may have to be standing idle for a period of time in the course of carrying out the contract works. An element of disruption may cause it to be used during this standing time and although there may be a proper entitlement to loss and/or expense under clause 26, it will be limited to the cost of operating that plant, not to hiring it as no extra expense is involved in the latter (paras 4.6.3 and 4.6.4).

3.7 VALUATION UNDER CLAUSE 13 OR ASCERTAINMENT UNDER CLAUSE 26

3.7.1 The need to distinguish between a valuation at prices related to those in the Contract Bills and ascertainment of direct loss and/or expense has already been emphasised (paras 3.1.3 and 3.1.11). The rules for valuing variations are set out in clause 13 and these extend to valuing other work, the conditions of which are changed by the variation (clause 13.5.5). However, they clearly do not extend to valuing the effect that such a variation might have upon the regular progress of the work generally; this must be dealt with under clause 26.2.7 (clause 13.5). What is not clear is where in practical terms one draws the line between clause 13.5.5 – changed conditions created by a variation – and clause 26 – the effect of a variation upon the regular progress of the Works. Example 3.9 illustrates this.

Example 3.9   A range of issues to be valued under clause 13

A variation is issued increasing the extent of a boundary wall. This additional work can be carried out by the Contractor without disrupting the regular progress of the work generally albeit it will delay its completion. The effect of such a variation might be to:

a)   add in some additional blockwork; this might be valued at Contract Bill rates (clause 13.5.1.1) plus a fair allowance for any difference in conditions or quantity if appropriate (clause 13.5.1.2);

b)   add to the period for completing the Works. The cost of the extended preliminary items such as supervision would be valued under clause 13.5.3.3;

c)   require the external paving to be executed in small parcels. This might be valued at Contract Bill rates plus a fair allowance for any difference in conditions if applicable (clauses 13.5.1.2 through to clause 13.5.5);

d)   add further to the period for completing the Works owing to the extra time required in executing the paving under different conditions. The cost of the further extended preliminary items such as supervision would be valued under clause 13.5.3.3 through clause 13.5.5.

3.7.2 If this is a correct interpretation of the application of clause 13 it might be said that it is difficult to see when clause 26.2.7 (loss and expense due to the regular progress of the work being affected by a variation) would apply.

3.7.3 However, the division becomes clearer upon contrasting the relevant part of clause 13 with clause 26. An entitlement under the latter clause is circumscribed by a series of safeguards, which do not pervade an entitlement under clause 13, namely:

a)   the Contractor must make a written application together with supporting data where requested as soon as should reasonably have become apparent that loss and/or expense will occur (clause 26.1.1);

b)   the contractor must, upon request, provide details of the loss and/or expense (clause 26.1.3);

c)   the Architect must be convinced that such a loss and/or expense will occur (or has occurred) and will not be reimbursed by a payment elsewhere under the Contract.

3.7.4 Moreover, it is the Quantity Surveyor who carries out valuations under clause 13 whereas it is the Architect who operates the provisions of clause 26 and if he wishes to delegate this to the Quantity Surveyor he can only delegate the ascertainment (the calculation of the amount). The Architect cannot delegate his responsibility for determining whether there is a liability in principle. In the case of *John Laing*

Construction Ltd -v- County and District Properties Ltd (1982) 23 BLR 1, it was decided that it was the Architect who determined liability and the Quantity Surveyor the quantum.

3.7.5 It seems, therefore, that where the judgment of the reimbursable financial effects of a variation can be made solely by addressing the relevant paperwork (e.g. the variation instruction, the relevant drawings, the measurement of the varied works and of that work affected by the variation, the relevant rates and descriptions in the Contract Bills), the variation will be valued by the Quantity Surveyor by reference to that data. Where, however, the variation, having regard to the special impact that it has on the regular progress of the work, creates elements of direct loss and/or expense which would not be recognised by reference to that data on that particular project, then that loss and/or expense must be ascertained under clause 26 and will properly be subject to the safeguards in that clause. It is as if the Contract is saying "Because of the particular impact that the introduction of this variation has upon this job as a whole in the particular circumstances as they existed on the site at that time, costs were or will be incurred directly as a result of executing that variation which a valuation by the Quantity Surveyor sitting in his office, by reference to the measurements of that varied work and of the work it affects with all the paperwork to hand, could not be aware of."

3.7.6 Examples of a range of calculations of loss and/or expense are given in Chapter 4 of this book

REFERENCES

[1] Davis Langdon & Everest (1999) *Spon's Architects' and Builders' Price Book*, 124th Edition, E. and F. N. Spon, London.

[2] National Joint Consultative Committee for Building (January 1996) *Code of Procedure for Single Stage Selective Tendering*.

[3] Working Group 3 of the Construction Industry Board (May 1997) *Code of Practice for the Selection of Main Contractors*.

[4] *Standard Method of Measurement of Building Works* (Seventh Edition, Revised 1998), The Royal Institution of Chartered Surveyors and the Construction Confederation.

[5] *Definition of Prime Cost of Daywork carried out under a Building Contract* (December 1975 Edition), Royal Institution of Chartered Surveyors and Building Employers Confederation.

[6] *Schedule of Basic Plant Charges* (January 1990 Issue). Royal Institution of Chartered Surveyors.

[7] *Definition of Prime Cost of Daywork carried out under a Heating, Ventilating, Air Conditioning, Refrigeration, Pipework and/or Domestic Engineering Contract* (July 1980 Edition), Royal Institution of Chartered Surveyors and Heating and Ventilating Contractors Association.

[8] *Definition of Prime Cost of Daywork carried out under an Electrical Contract* (March 1981 Edition), Royal Institution of Chartered Surveyors and The Electrical Contractors' Association.

4

ASCERTAINING THE DIRECT LOSS AND/OR EXPENSE

4.1 INTRODUCTION

4.1.1 This chapter is concerned with evaluation and ascertainment of direct loss and/or expense under JCT 98 PWQ. It is not concerned with the amount which the Contractor may require to be paid in lieu of any such ascertainment under either clause 13.4.1.2A2 or clause 13A.2.3. (See paras 3.3.2 to 3.3.6 and section 3.4.) Detailed consideration of the major clauses giving rise to such claims has been undertaken elsewhere in this book as follows:

a)   clause 25 (Extension of time) – Chapter 2;
b)   clause 26 (Loss and/or expense) – Chapter 3.

4.1.2 Claims for direct loss and/or expense under clause 26 are payable only where the Contractor would not be reimbursed by … *a payment under any other provision in this contract* … The main area of potential duplication will occur through payments made in accordance with clause 13 (Variations and provisional sums) and this subject has already been covered in Chapter 3.

4.2 THE ASCERTAINMENT

Generally

4.2.1 Too often, claims for reimbursement of direct loss and/or expense are initiated by the Contractor who submits a seemingly detailed build up of his alleged entitlement at or towards the end of a project. Although clause 26.1 now recognises that … *the Contractor may give his quantification* … it does not require that he shall do so (see para. 3.5.15). It is for the Architect (or Quantity Surveyor if so instructed by the Architect) to ascertain the amount due and to request from the Contractor such details as he believes are necessary to fulfil this obligation. In the case of *London Borough of Merton -v- Stanley Hugh Leach (1985) 32 BLR 51*, it was stated that *"The Contractor must clearly co-operate with the Architect or the Quantity Surveyor in giving such particulars … to enable him to ascertain the extent of that loss and/or expense."* Moreover, the fact that it may be difficult to arrive at an accurate ascertainment cannot be a barrier to making as proper an assessment as the available data permit. Indeed, from the case of *Wood -v- Grand Valley Railway Co (1913) 30 OLR 44*, it is clear that even when it is *"… impossible … to estimate with anything approaching to mathematical accuracy the damages … (this) … cannot relieve the wrongdoer of the necessity of paying damages … and its (the Judge or Jury) conclusion will not be set aside even if the amount of the verdict is a matter of guess work".*

4.2.2 In the case of *F G Minter -v- Welsh Health Technical Services Organisation (1980) 13 BLR 1*, it was emphasised that it was cost, as distinct from the rates and prices given in the Contract Bills, that should be the basis of any ascertainment. However, as has already been pointed out in this book (para. 1.2.16), the damages payable by a party committing a breach should be no more than he anticipated would be the consequence of such a breach at the time he made the contract. Thus, where a Contractor's costs are materially different from the rates quoted in the Contract Bills, there must be a good reason for that difference otherwise any significant unjustifiable increase in costs might have to be disallowed from a claim for reimbursement of direct loss and/or expense on the grounds that to include them would result in the Employer paying more than he anticipated at the time he signed the contract.

4.2.3 Where, however, the Contractor does submit a monied out claim, the Architect's or Quantity Surveyor's task is not to challenge or reduce that submission but, rather, his role is to build up an entitlement quite independently under the terms of the Contract by ascertaining the amounts due, given the Contract Conditions and the established facts. In building up the Contractor's entitlement, the Architect or Quantity Surveyor should start with a comprehensive list of possible headings under which such an entitlement might arise. This list is as follows:

a) materials;
b) labour disruption;
c) attraction money and bonus payment (in certain circumstances);
d) preliminaries and supervision;
e) inflation;
f) head office overheads and profit;
g) interest charges.

4.2.4 Other items often appearing in a Contractor's submission which generally would not be admissible are:

h) cost of accelerating the works;
i) cost of overtime;
j) cost of preparing a claim.

4.2.5 In ascertaining the amounts properly payable as direct loss and/or expense it is advisable to ask the following questions, the answers to which will provide useful background against which to make an evaluation:

a) did the event occur and was it the direct result of the cause claimed?;
b) is the cause one which is recognised by the Contract, i.e. is it either one of the matters listed in clause 26.2 or deferment of possession of the site under clause 23.1.2?;
c) have the conditions laid down by the Contract been complied with, e.g. was the Contractor's application in writing and was it made as soon as it became, or should reasonably have become, apparent to him that the regular progress of the Works had been materially affected?;
d) what are the costs directly attributable to the cause?;
e) is any part of the cost reimbursed elsewhere under the Contract?

4.2.6 Application of some of the above questions may lead one to conclude that the problem is not covered by the Contract terms, in which case neither the Architect nor Quantity Surveyor is empowered to deal with it even if the Contractor has a general right to sue the Employer for breach of Contract. The application of some of the other questions may demonstrate that only part of the costs alleged are reimbursable.

4.2.7 In any event, the Contractor's entitlement is that amount which is ascertained as the difference between the costs properly and directly incurred in carrying out the work, albeit disrupted by the event, and **what those costs would have been had the event not occurred.** In practice this may be very difficult to establish and will always involve an element of judgment rather than established fact: "what the costs would have been" will always be a matter for conjecture (see Example 4.4 as referred to at para. 4.4.8) – a fact not always recognised by those who have to accept the consequences of the ascertainment.

4.2.8 Any such ascertainment will have to ignore:

a) any errors in the level of pricing in the tender; or
b) any extra costs in executing the work not directly associated with the claim, i.e. mismanagement.

An example of such a calculation, of the need for conjecture and of the element of inadmissible costs, is given at Example 4.4 at paras 4.4.6 to 4.4.8.

Global claims

4.2.9 There have been a number of cases in recent years centring around whether claims must be calculated by dealing with each item – each instruction even – or whether a claim can be presented by lumping together all the problems and making a global claim. It seems clear that rarely if ever are the several causes of a claim or their effects indistinguishable one from the other albeit that there will be circumstances where the cost effects of those causes are impossible to isolate item by item. This being the case the response to this question of the ability of the Contractor to make and sustain a global claim may be summarised as follows:

a) a Contractor must give the proper notice on each occasion that an entitlement to make a claim arises;
b) each of these notices must conform fully with the rules written into the Contract about them (e.g. the timing, the need for them to be in writing and to be sent to the Architect);
c) there must be a clear demonstration of the relationship between cause and effect in each case;
d) the Contractor must keep such records as are required of him in relation to every such event;
e) he must make available all the relevant cost data to enable the Architect or Quantity Surveyor to carry out a proper ascertainment; and
f) only where the effect of a series of claims events (e.g. the late issue of instructions or the instructing of variations) is such as to create extra costs which by their nature are indistinguishable one from the other, can an ascertainment on a global basis be entertained. Simply because there is say a series of variation instructions which create overlapping or contiguous disruption it does not follow that all other claims on that project can be lumped into a global claim. The facility to present a global claim should be restricted solely to the events which inevitably create indistinguishable costs.

4.2.10 Thus no Contractor should be penalised by being denied reimbursement simply because he was caused to incur extra costs which could not by their nature be particularised. On the other hand, no Contractor should be allowed to abuse that facility – to make a global claim – by using it to mask facts and figures that could and should properly have been particularised and presented. The notion that a Contractor should be put to quite a burden of proof to particularise his claim was seen in *Crosby & Sons -v- Portland UDC (1967) 5 BLR 121*; in that case, Donaldson J. was considering the award made by an Arbitrator and accepted that, in principle, the loss attributable to each cause should be separately identified and particularised but in those circumstances where such separate identification was difficult, or even impossible, then (and only then) a lump sum award may be made in relation to those circumstances. Whilst the Crosby case has mistakenly been taken as authority for launching so called "global claims" (i.e. claims which do not attempt to establish the nexus between the matters complained of and a line by line explanation of the damages flowing therefrom), another more recent case (*Wharf Properties -v- Eric Cumine Associates (1991) 7 Const LJ 251*) was to breathe life back into the proposition that every effort must be made to separately identify those damages flowing from each of the matters complained of.

The need for the Contractor to use constantly his best endeavours

4.2.11 Clause 25.3.4.1 states that ... *the Contractor shall use constantly his best endeavours to prevent delay in the progress of the Works, howsoever caused, and to prevent the completion of the Works being delayed or further delayed beyond the Completion Date* ... This obligation was considered at para. 2.7.45 where it was suggested that the Contractor is required to expend money in achieving this obligation just as he would do in his own interests. This may come as a surprise to many Contractors. Indeed, there have been occasions where Contractors have averred that they have "used more than our best endeavours"; clearly no level of endeavour can be higher than the best and this emphasises the nature of this stringent obligation. The Contractor, in being obliged to use his best endeavours, is expected to employ those resources which a prudent person would to achieve his own ends (see para. 2.7.45). However, it is suggested that the extra cost of using *best endeavours* pursuant to a disruption resulting from a matter under clause 26 (i.e. one for which the Contractor could claim reimbursement of his loss and/or expense) would form part of that loss and/or expense as it is the direct result of the matter concerned. Where, however, best endeavours are required to be used as a result of any other cause of delay extra expenditure might have to be incurred which would not be reimbursable. These two points are illustrated in Examples 4.1 and 4.2.

Example 4.1   The use of *best endeavours* – I

A Contractor is maintaining his planned progress in carrying out the Works until this progress is disrupted as a result of variations instructed by the Architect under clause 13.2. The Contractor replans his operations and decides that by the introduction of an extra crane he can overcome the delay and still maintain the original Completion Date. This is consistent with the Contractor using ... *his best endeavours to prevent delay in the progress of the Works, howsoever caused, and to prevent the completion of the Works being delayed* ...Thus the extra cost of so doing would have arisen as a direct result of a matter under clause 26.2.7 and would thus be reimbursable under clause 26.1.

135

Example 4.2 The use of *best endeavours* – II

Assume the same conditions prevail as in Example No. 4.1 except that the progress becomes disrupted by the failure of one of the Contractor's Sub-Contractors to perform in accordance with his programme. The Contractor's obligations remain the same and again in pursuit of these obligations the Contractor replans his operations and again decides that the introduction of an extra crane will repair the delay or prospective delay. In these circumstances he must introduce this extra crane but he cannot obtain recompense from the Employer. He may of course be able to obtain redress from his errant sub-contractor but that is another matter.

4.2.12 If the case of *Sheffield District Railway Company -v- Great Central Railway Company (1911) 27 TLR 4512* (see para. 2.7.45), does indeed represent the present law in this matter then there is one way in which the Contractor might determine what level of expenditure the Employer may be prepared to invest "in his own interests". The Employer could be said to have already determined the cost to him of any delay by declaring it in the sum for liquidated and ascertained damages in the Appendix to the Contract Conditions. On this basis, but subject to the caveat which follows, it is suggested that the Contractor is obliged to expend resources up to that level in his pursuit of the Completion Date whether or not he can recover this under clause 26. This is after all what the Employer has declared ... *when he made the contract* ... what he considered his loss to be due to late completion. (See para. 2.13.7.) Whilst there is an attractive simplicity to this proposition, it may not hold good in those circumstances where a contractor's ordinary operating costs far outweigh the liquidated and ascertained damages because, in that situation, the contractor is more likely to have regard to his operating costs than to the liquidated and ascertained damages that he may face were he to be in culpable delay. As a result, best endeavours will more usually have to be read in the context of the particular contract under consideration.

4.2.13 But even where the cause of the delay is a matter listed under clause 26.2 and such an expense is reimbursable as direct loss and/or expense, the Contractor should still be called upon to demonstrate that such expense was effective in achieving its objective, namely recovering the delay and/or preventing further delay.

4.3 MATERIALS

4.3.1 The common elements of a claim for the reimbursement of the extra cost of materials are:

a) materials properly ordered for the project which following the issue of a variation are no longer required; and

b) extra protection of materials during the prolongation of the contract.

Additional waste and the extra cost of purchasing materials in small quantities due to the issue of a variation order would normally be dealt with under clause 13.5 of the Contract (para. 3.3.26).

Surplus materials

4.3.2 If materials properly purchased for the Works become surplus due to a variation then one might reasonably expect that the relevant costs would properly form an element of direct loss and/or expense; except that the cost of removal from site of the materials or goods would be a variation pursuant to 13.1.1.3 and as a consequence this element of cost would fall to be valued under clause 13. It is difficult to think of an example of how rates in the Contract Bills, relating as they do to the carrying out of the Works, immediately lend themselves to determining a fair valuation for dealing with materials rendered surplus due to the issue of a variation. In such a set of circumstances, one might look to valuation under clause 13.5.7 which may well involve adopting principles similar to those used when ascertaining loss and/or expense hence this issue being addressed in this section of the book. The evaluation of the expense should present no difficulty: one would simply look at the invoices, ensure that the amount is reasonable, take account of any credit value and allow for any obligatory payments for cancellation charges. Any cost excess in relation to a Bill rate should, if nothing else, encourage the Architect or Quantity Surveyor to seek further data supporting the charge as the Contractor (or through him the supplier) should not be allowed to exploit a variation.

4.3.3 Such an investigation may reveal an extra charge due to the fact that only a small proportion of the whole order had been supplied – thus a high rate for small quantities might be in order (in these cases any charge for cancellation of the remainder of the order should be comparably less).

4.3.4 The Contractor's claim under this heading may be based upon having ordered his materials from the Contract Bills which, it transpired, contained erroneous quantities. Or the Employer's defence to a claim for abortive material costs might be that the Contractor ordered the materials earlier than was reasonably necessary and had he waited until nearer the date upon which they were required the variation would have been issued omitting them thus avoiding such costs. These two points, therefore, need consideration.

4.3.5 As to the first point there is, it is suggested, no implied warranty as to the accuracy of the Contract Bills for the purpose of constructing the Works only for the purpose of establishing the Contract Sum. The Contractor who placed his order on the basis of the Contract Bills and ignored or failed to notice the fact that the work shown on the Architect's drawings or instructions differed would have no claim for direct loss and/or expense.

4.3.6 Now to the second point – ordering materials at a very early stage – where the Sixth Recital has not been deleted, clause 5.4.1 requires that the Architect releases information in accordance with the Information Release Schedule. In any event, clause 5.4.2 requires that the Architect issues all further drawings, details and instructions ... *as and when from time to time may be necessary* ... It also requires that he shall ... *have regard to the progress of the Works* ... and do so ... *to enable the Contractor to carry out and complete the Works in accordance with the Conditions* ... Thus a Contractor who orders his materials earlier than is reasonably necessary for the regular and diligent progress of the work to achieve the Completion Date, e.g. in advance of the date on the Information Release Schedule for the release of relevant information, runs the risk of not recovering the cost of any materials subsequently not required. This point is reinforced by the clause relating to Interim Certificates (clause 30.2.1.2) which provides that such certificates shall only include the value of materials and goods when ... *they are reasonably, properly and not prematurely so delivered* ... An example relating to surplus materials is given at Example 4.3.

Example 4.3   Ascertaining the cost of materials no longer required

A Contractor orders standard timber windows with special sills to the value of say £6,000 for use in accordance with the Contract Drawings. After these are ordered and the first delivery of windows to site is made the Architect issues a variation changing the specification for all the windows to aluminium. The calculation of direct expense might be as follows:

			£
1.	Account from suppliers:		
	a)	Value of materials already delivered as attached schedule as per attached quotation dated 12.8.99	480.68
	b)	Add for 10% discount previously allowed on bulk order, now cancelled ($\frac{1}{9}$th)	53.41
			534.09
	c)	Loss of profit on remaining part of order	500.00
	d)	Abortive costs (net) incurred on manufacturing special sills after 15% credit for possible reuse of materials	125.00
			1,159.09
2.	Cost from main Contractor:		
	a)	Handling materials 3 hours at £8.00 per hour	24.00
			1,183.09
	b)	Allow credit for resale of standard windows on site; special sills not saleable	(320.00)
			£863.09

4.3.7 From the above it will be noted that the credit available from materials made up specially for the job is much less than that available from standard components.

Extra protection of materials during the prolongation of the contract

4.3.8 Prolongation of a Contract, particularly where the prolongation means working through an additional winter period, can result in materials deteriorating, through rust, bad weather and the like. In consequence, Contractors will often claim the cost of the replacement of such materials; but this is not the correct measure of the expense. The Contractor's obligation in these circumstances is to continue to protect and safeguard the work until Practical Completion. This may entail extending the life of existing protective measures, or replacing or renovating them and such costs would, in principle, be direct expenses. Ascertaining the expenses involved should not provide any difficulty. However, during a delay a Contractor is not entitled to allow the Works to fall into disrepair and to reinstate them at the Employer's cost.

4.4 LABOUR DISRUPTION

Generally

4.4.1 As suggested earlier, the ascertainment of claims is often left until the end of a Contract and probably no head of claim is more difficult to assess long after the event than the loss of labour output due to disruption of the regular progress of the Works. There are four approaches to assessing the disruption of labour, the choice will depend upon the data available and the circumstances. They are:

a) an evaluation of the records of labour output specifically kept for this purpose at the request of the Architect or Quantity Surveyor under clause 26.1.3;
b) a review of labour activity on site at the relevant time;
c) a review of the implications of the extensions of time; and
d) application of a general productivity formula.

4.4.2 Where the data referred to in a) do not exist the assessment becomes one largely of judgment and then it may be necessary to apply more than one of the remaining approaches, where practicable, in order to test the assessment. But in any event, when considering the loss and/or expense under clause 26, it is usually the loss and/or expense arising from **each** variation that has to be considered (para. 4.2.9).

An evaluation of the records of labour output specifically kept for this purpose

4.4.3 JCT 98 PWQ empowers the Architect to demand that records of labour be provided by the contractor of any specific event (para. 3.5.25).

4.4.4 The provision of this data by the Contractor as required by clause 26.1.2 is dependent upon the Architect or Quantity Surveyor requesting them. Therefore where disruption to progress is likely to occur, or where direct loss and/or expense due to a variation is likely, the Architect would be well advised to inform the Contractor immediately what data he expects to be recorded. It would be sensible to ensure that such records were agreed between the Contractor and the Clerk of Works.

4.4.5 If such proper records are kept they will be the best evidence for calculation of loss of output and must be preferable to any subsequent assumptions.

4.4.6 However, there are dangers. First, the request to keep such records must not be mistaken for an acceptance that the Employer will pay all or any of the costs involved whatever they may be; the Contractor may generally be inefficient or part of his labour might be expended on carrying out the work incorrectly or in remedying defective work; to that extent such costs will not be reimbursable and, therefore, they should be isolated. Secondly, steps must be taken to ensure that the records extend only to the work which is or may properly be the subject of a claim. Thus the Architect should agree with the Contractor precisely what is to be recorded and how, and then alert the Clerk of the Works to monitor the manner in which this work is carried out and the efficacy of the cost records. Once these safeguards have been properly implemented, the records should be compared with an assessment of what costs would have been incurred had the delay or disruption not taken place; the difference being reimbursable to the Contractor as the following Examples 4.4 and 4.5 show.

Example 4.4   Ascertaining the loss of output from records

A variation to part of a foundation to a building delays the erection of one bay of the pre-cast concrete frame. This prospect is predicted well before it takes place and the Clerk of Works and the Contractor, at the request of the Architect, agree records of output both before and after the period of disruption. These are as follows:

a)	Recorded cost of an erection gang per week	£3,500
b)	Recorded output of the gang prior to the disruption	1.50 bays per week
c)	Recorded output of the gang during disruption	1.15 bays per week
d)	Output of the gang assumed at tender stage deduced from the Contract Bill prices	1.30 bays per week
e)	After discussion with the Clerk of Works it is decided that the Contractor did not manage the disrupted work as well as he should have done. Had he done so the output at (c) above would have been	1.25 bays per week

Thus the reasonable cost per bay during disruption is (a) ÷ (e)

$$\frac{£3,500}{1.25} = £2,800$$

Whereas the cost per bay before disruption (i.e. that which should have prevailed had disruption not taken place) is (a) ÷ (b)

$$\frac{£3,500}{1.5} = £2,333$$

Therefore the direct loss and/or expense is £467 per bay (£2,800 − £2,333) and this is the Contractor's entitlement.

4.4.7 The following points should be noted:

a) the level of pricing deduced from the Contract Bills is not relevant – at an actual output of 1.5 bays per week the Contractor was beating his tender assumption of 1.3 bays per week and thus reaping an additional profit which should not be eroded;

b) the Contractor can only recover the **direct** loss; that is to say to the extent that mismanagement added to the Contractor's costs (i.e. the reduction in output from

1.25 to 1.15 bays per week) those costs are not reimbursable as they are not the **direct** result of the variation; and

c) whilst this approach may, at first blush, appear to be in conflict with the approach adopted in Example 3.6 (wherein the rates in the Contract Bills were adhered to notwithstanding the fact that the Contractor's records showed that his actual costs for the item in question were greater than the rates in the Contract Bills), it is to be borne in mind that Example 3.6 was directed to valuing a Variation under clause 13 whereas Example 4.4 is directed to ascertaining loss and/or expense under clause 26.

4.4.8 In practice, rarely is one in such a privileged position of having accurate records of the costs before disruption in the form envisaged above; and in any event an element of conjecture is always inevitable as to what the costs would have been had the variation not disrupted the progress of the works, i.e. the costs at e) must almost inevitably be the subject of conjecture. Moreover, it will be noted that neither of the two figures that are available – the Contract Bill prices at d) and the recorded prime costs at c) – are directly relevant to the ascertainment. On the other hand, the Contractor might have priced the Contract Bills on the assumption that the gang would erect only 1.25 bays a week. In the event the 1.5 bays they achieved bettered this; the disruption did no more than restrict him to his original target of 1.25 bays a week. He is nevertheless entitled to the same reimbursement of his loss of output as calculated above for he should not be denied the potential for profit implicit in the rate.

4.4.9 It is the disruption to the regular progress of the frame that is dealt with here in accordance with clause 26 (i.e. by reference to the Contractor's costs). It is assumed that the variation to the foundations would have been dealt with separately under clause 13.5, which sets out the Valuation Rules.

Example 4.5   The cost of out of sequence working from records

A Contractor undertakes work which includes erecting blockwork to form ducts for mechanical services described as isolated casings in the Contract Bills. He programmes his work so that this element can be carried out at the same time as his general blockwork. In the event, instructions related to the mechanical services are delayed and, as a result, the enclosing blockwork has also to be delayed and thus executed out of sequence and in a piecemeal fashion. The Quantity Surveyor requests under clause 26.1.3 that the Contractor keeps records at the time of executing the duct enclosures and agrees them with the Clerk of the Works. The records show that $40\,m^2$ of blockwork in ducts are erected by a gang comprising 2 bricklayers and 1 labourer in 2 days whereas $60\,m^2$ of general blockwork is erected by a similar gang in a similar period. It is judged that this work was as well managed as the circumstances permitted, i.e. that no element of these costs was attributable to poor management. The price in the Contract Bills for blockwork in walls and partitions is £10.50 m^2 and isolated casings is £11.50 m^2 of which the labour element is estimated to represent £6.50 m^2 and £7.50 m^2, respectively. An evaluation of the loss of output might take the following form:

a) Assessment of labour costs actually incurred on 40 m^2 of isolated casings:

	£
Bricklayers, 2 for 2 days at £75 per man per day	300
Labourer, 1 for 2 days at £65 per man per day	130
	430

Therefore labour cost per m^2 = £10.75

b) Assessment of labour costs actually incurred on 60 m^2 of general blockwork:

Total cost as before namely £430

Therefore labour cost per m^2 = £7.17

c) Estimated labour element of the Contract Bill price for general blockwork = £6.50/m^2

d) Summary:

(i)	Labour rate/m^2 of general blockwork in Contract Bills prices	£6.50
(ii)	Ditto but as achieved	£7.17
(iii)	Shortfall in achievement by comparing (i) and (ii) above	10.3%
(iv)	Labour rate/m^2 of blockwork in ducts in Contract Bills prices	£7.50
(v)	Labour rate/m^2 of blockwork in isolated casing as achieved	£10.75
(vi)	Assessed labour rate/m^2 of blockwork in isolated casing as might have been achieved had it not been varied £7.50 m^2 at (iv) + 10.3% at (iii)	£8.27
(vii)	Direct loss/m^2: (v) − (vi)	£2.48

(Inflation is dealt with separately.)

4.4.10 Note again that the comparison is between what the proper costs were and what they would have been had the disruption not taken place; not a comparison between the costs incurred and the price in the Contract Bills. Thus an assessment of likely performance on undisrupted work has to be made. In this example this is achieved by reviewing progress on the general blockwork, i.e. it is assumed that the general shortfall in productivity at (iii) would have applied to the isolated casings in any event. Moreover, it is assumed that the particular delay referred to did not materially affect the regular progress on other work. The Contractor may have sub-let this work at prices dissimilar to those in the Contract Bills with either a greater or lesser profit. This, as has been suggested above, is not relevant because an ascertainment of loss and/or expense should not be used to upset the profit that the Contractor would have earned nor to remedy the loss that he would have incurred.

A review of labour activity on site at the time

4.4.11 If, in the case of example 4.5, no such cost records were available the only reasonable alternative would be to look back at the Contractor's labour level being maintained on the site at the relevant time with particular emphasis on the trades affected; in this instance on the bricklayers and attendant labourers. The difficulty is in isolating those bricklayers and labourers involved with this particular work from those engaged on other activities. Clearly a greater element of judgment is required as Example 4.6 shows.

Example 4.6   The cost of out of sequence working by a review of general labour costs

The facts are as given in Example 4.5 except that detailed cost records are not kept. Upon review of the Contractor's costings the following is revealed.

Total costs of labour for the relevant work = £61,500. It is noted that this amount of labour built a total of $7,500\,m^2$ of blockwork comprising $6,225\,m^2$ of general blockwork and $1,275\,m^2$ of blockwork in duct casings.

The assessed labour element for this work in the Contract Bills was:

	£
$6,225\,m^2$ at £6.50 (see item c) in Example 4.5)	40,462.50
$1,275\,m^2$ at £7.50 (see item d) (iv) in Example 4.5)	9,562.50
	50,025.00

Thus cost exceeded the tender allowance by 23% (£61,500 compared with £50,025). It is recognised that the building of the ducts has been subject to out of sequence working for causes which entitle the Contractor to reimbursement of the loss and/or expense which directly arises. It is also recognised that the Contractor's output generally did not reach that assumed in calculating the Contract Bill prices; losses arising from the latter cause are not reimbursable. However, detailed records do not exist to enable one to differentiate between the two. Therefore all that one can do is to make a series of assumptions and judge whether the implications of the result are reasonable. It is assumed that the general lack of output affected the duct casings and the general blockwork equally. It is postulated that, in addition, the disruption due to the out of sequence working in the duct casings added a further 20% to those particular labour costs.

Thus at the Contract Bill prices and allowing only for the disruption due to out of sequence working on duct casings labour costs for this element would have been:

	£
$6,225\,m^2$ at £6.50	40,462.50
$1,275\,m^2$ at £7.50 + 20%	11,475.00
	51,937.50

If this were the case, and given that the total labour costs incurred were £61,500 (see above), this indicates that the increase in labour costs generally, due to the Contractor's output falling short of target, was 18% (£61,500 is 18% more than £51,937.50). The Architect (assisted by the Quantity Surveyor, if required) must assess whether this 18% loss of output on all blockwork and 20% further disruption on the duct casings represents the right balance bearing in mind the history of the job. If so, then the Contractor's entitlement would be:

20% (particular disruption on duct casings) × 1,275 m (area of
blockwork in duct casings) × £7.50 (rate for undisrupted work)   1,912.50

If, on the other hand, this does not reflect the Architect's judgment alternative assumptions must be tested to achieve the balance that best represents the realities experienced on this site.

A review of the implications of the extensions of time

4.4.12 In most cases, the fixing of a new Completion Date would precede any concomitant direct loss and/or expense. The former is required to be an estimate prepared by the Architect within 12 weeks of receiving adequate details from the Contractor and generally no later than the existing Completion Date (which implies looking forward) whereas the latter is a review of costs (which requires looking back). Thus the evidence used by the Architect in forming his view on extensions of time in principle and the duration arrived at can be useful in determining loss of output; useful but not conclusive. Whilst it may be very dangerous to rely solely upon the extensions to the Completion Date as the prime evidence for calculating loss of output or any other element of claim they cannot be ignored. Indeed, the Architect is required under clause 26.3 and 26.4.2 to disclose what extension of time he has granted in respect of causes which also give rise to a claim for reimbursement of loss and/or expense, if this is necessary for the purposes of ascertainment. The amount of the extension, the reasons for the decision and any element of delay purposely eliminated in the evaluation of the extension can be important guides. This is particularly so in the evaluation of the extended use or hire of plant, site huts and the like to which reference is made later in this chapter.

4.4.13 The main dangers in accepting the evidence of extensions of time for the ascertainment of loss and/or expense are:

a) only delays which affect the Completion Date (i.e. on critical elements in the programme) need to be considered in calculating extensions of time; thus delays on parallel or non-critical activities can be ignored and yet they may all have resulted in loss and/or expense being incurred which may be reimbursable. However, in that the Contractor should always notify the Architect of any delays, if the Architect properly catalogues his detailed responses to all of them, this record can be useful in considering the likely disruption and loss of output on a project where substantial disruption has taken place. To be useful, such a catalogue must contain the Architect's view of the validity of all notified delays, even those that do not affect the Completion Date for these may still rank for consideration in the ascertainment of loss and/or expense. (See Worked Example in Part 4.)

b) there will be occasions where there is an entitlement to an extension of time and no entitlement to reimbursement of loss and/or expense (e.g. exceptionally adverse weather conditions – clause 25.4.2) and vice versa; furthermore, the loss and/or expense is by no means necessarily pro rata to the length of the extension of time;

c) extensions of time should be estimates made by the Architect in advance; loss and/or expense is an amount to be calculated precisely after the event.

4.4.14 However, the following Examples 4.7 and 4.8 serve to illustrate the way in which extensions of time may be usefully consulted as a guide or check on the extent of labour disruption.

Example 4.7   Ascertaining the cost of disruption by reference to extensions of time – I

An Architect grants an extension of time of 5 weeks in a 50-week contract period (i.e. 10% of the contract period) for a series of small consecutive delays, due to a series of minor variations issued consistently but unheralded throughout the contract period which do not significantly change the work content of the project. It is not unreasonable to assume that the cost of disruption will be approaching 100% of the labour content of all critical elements of work during those 5 extra weeks (i.e. around 10% of the total labour content). The Contractor cannot continually hire and fire labour at will and where the delays are many and small, he will have little opportunity of redeploying his resources or otherwise mitigating his losses. However, on the non-critical activities, the output will be less affected. Assume that if the Contract Sum is sub-divided as follows:

Main Contractor	£
Material	180,000
Labour on critical elements (e.g. foundations, frame, roof, etc.)	80,000
Labour on non-critical elements (e.g. partitions, external paving)	40,000
Nominated Sub-Contractor's work	125,000
Preliminaries	75,000
	500,000

4.4.15 Labour disruption could amount to £8,000 (up to 10% of £80,000) plus any specific disruption demonstrated to have been directly caused on the non-critical labour or on the work of a Nominated Sub-Contractor. The effect of inflation, Nominated Sub-Contractor's claims, and loss and/or expense on preliminaries are dealt with elsewhere.

4.4.16 In the above example, the Preliminaries work out at an average of £1,500 per week of the contract period (i.e. £75,000 over a 50-week period). Thus, allowing for a build up of expenditure at the start of the contract and a tail off at the end and allowing for the fact that a proportion of the Preliminaries will not vary with time (para. 3.3.10), the extra cost of Preliminaries for a 5-week extension of time during the course of the contract is unlikely to be less than £7,500 (5 weeks at £1,500) nor more than, say, £8,500 in total. Thus for an extension of time equivalent to 10% of the Contract Period the reimbursable loss and/or expense of these two elements is £16,500 (i.e. £8,000 plus £8,500) or 3.3% of the Contract Sum. It cannot be emphasised too strongly that such a broad approach can never be justified when ascertaining direct loss and/or expense; the detailed approach outlined in this chapter must be followed not least because the above assessment assumes maximum disruption during the whole period of delay – it also assumes that the preliminaries included in the Contract

Sum equate to the direct loss and/or expense suffered. Nonetheless, the rule of thumb in this particular example would suggest that the potential claim might be in the region of one third of the percentage by which the contract period is extended as then applied to the Contract Sum, i.e. for an $x\%$ extension of time, the allowance for direct loss and/or expense is $x/3\%$ of the Contract Sum, thus providing a crude yardstick to compare with the figure initially claimed by the Contractor for direct loss and/or expense and with the figure which is ultimately ascertained in detail. It does not follow, of course, that if either of these figures do not bear comparison with above rule of thumb that they are incorrect, but it will prompt one to investigate whether there are proper grounds for a wide divergence on the particular claim in question.

Example 4.8 Ascertaining the cost of disruption by reference to extensions of time – II

An Architect grants an extension of time of 5 weeks in a 50-week contract period (i.e. 10% of the contract period) due to a single suspension of the works of which he gave the Contractor due warning in advance. In this instance, the loss of output would be at the other end of the scale. The Contractor had the opportunity, and therefore the duty, to mini-mise his loss and would be expected to have re-deployed his resources on other sites or other work to the maximum extent possible.

Figures 4.1–4.3 illustrate Examples 4.7 and 4.8 graphically, but the delay has been exaggerated to 33% of the contract period to illustrate the thesis more clearly.

4.4.17 There will be many projects where the extensions of time granted by the Architect may be in respect of causes which do not entitle the Contractor to reimbursement of loss and/or expense, e.g. *force majeure, exceptionally adverse weather conditions* and also in respect of variations, the extra labour content of which may have been dealt with under clause 13. Nevertheless, again a broad assessment of loss of output can be made as Example 4.9 shows.

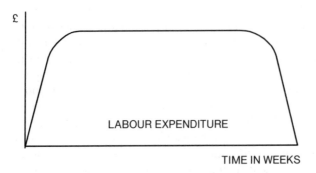

Figure 4.1 Tender programme. Labour resources envisaged in the tender.

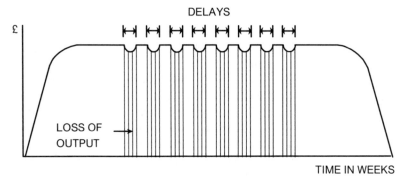

EXTENSION 33% : DISRUPTION 30%

Figure 4.2 Labour resources when many small delays occur.

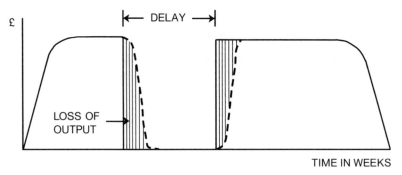

EXTENSION 33% : DISRUPTION 10%

Figure 4.3 Labour resources when one long delay occurs.

Example 4.9 Ascertaining the cost of disruption by reference to extensions of time – III

On a project where the Contract period is 100 weeks the Architect grants an extension of time of 10 weeks. Of this, some 8 weeks is said to be the global effect of variations, late instructions and delay on the part of persons employed directly by the Employer, the remaining 2 weeks are for causes which do not entitle the Contractor to reimbursement of loss and/or expense. The extra work content of those variations considered in this award of extension of time is assessed at $6\frac{1}{2}$ weeks the full extra cost of which has been dealt with in the valuation of variations under clause 13; the remaining $1\frac{1}{2}$ weeks (of the 8 weeks mentioned above) can be said to be the disruptive effect of the causes mentioned.

Application of a general productivity formula

4.4.18 As workmen become more familiar with a task so they become more efficient. The increase in efficiency tails off eventually once workmen reach their maximum

147

potential; this curve of output is known as the learning curve. Very occasionally data on learning curves exist and where one has disruption in the repetitive process these data can be used to assess the change in output. Although such data are hard to come by (and where they exist they rarely wholly suit the purpose in hand) a theoretical learning curve can often be created in order to test the likely loss in output that can result. The following example is based upon data prepared at what was the Building Research Station (now the Building Research Establishment) and published in its Current Papers Construction Series No. 18 entitled *Labour Requirements for House Building. Advantages of Continuity of Work and Experience* [1]. *Figure 4.4* is a copy diagram from this report (reproduced by kind permission of the Controller, HM Stationery Office) and set out below is a table of results deduced from it. Although this paper is now over 30 years old, it illustrates well the potential for using such data in ascertaining direct loss and/or expense.

Labour output for erecting plasterboard in pairs of dwellings

No. of pairs of dwellings	Labout content for last pair (man hours pair)	Average labour constant of whole (man hours pair)
(a) 1	37	37.0
(b) 5	23	27.8
(c) 10	13	21.0
(d) 15	11	18.1

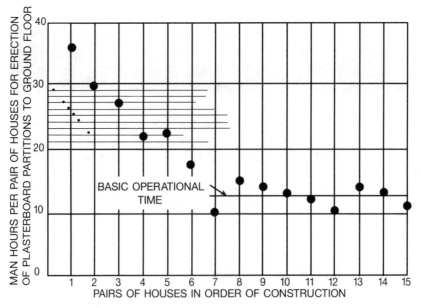

Figure 4.4 Typical "improvement" curve.

4.4.19 Example 4.10, based on these data, illustrates their possible use; the assumptions made are extreme – but nevertheless possible – to highlight the effect of disruption to a repetitive process.

Example 4.10 Ascertaining the cost of loss of output by reference to formulae

A contract includes the erection of plasterboard partitions to 15 pairs of dwellings. Work on the last 5 pairs of dwellings is postponed and re-introduced much later in the Contract after all those tradesmen familiar with the work have left the site and a new gang has to be introduced. The Contractor has based the price in the Contract Bills upon the average labour content for the whole, deduced from table at para. 4.4.18, i.e. 15 pairs × 18.1 man hours (see item (d) of that table) = 272 man hours. However, after the 5 pairs have been postponed, the position would be:

	Man hours
First 10 pairs × 21 man hours (item (c) of the table)	210
Remaining 5 pairs × 27.8 man hours (item (b) of the table)	139
	——
	349

349 man hours expended is 28.3% greater than the 272 man hours estimated; the man hours per dwelling expended on the 5 pairs of postponed dwellings is therefore 54% greater than allowed in the tender (27.8 compared with 18.1).

4.4.20 It must be emphasised here that the above calculation is valid only for the circumstances postulated and is not a general indicator of disruption in any other circumstance but it does illustrate the potential for loss where a highly repetitive process is disrupted. It should be noted, however, that the Contractor would be expected to use those men familiar with the task from their earlier experience if they were reasonably available and to pass on to the Employer the benefit of a far quicker learning curve on the return visit.

Application of an overall percentage for loss of productivity

4.4.21 An overall percentage is often applied to the total labour costs in order to arrive at the reimbursable element of disruption. The dangers of this approach are obvious. Total labour costs will include the following items to which it would clearly be wrong to allocate a percentage as a measure of reimbursable disruption:

a) labour not affected by the disturbance;
b) inefficiency due to bad management;
c) extra labour costs arising from an optimistic view of output at tender stage;
d) the proportion of non-productive labour which is expended in getting to and from the work location on site, breaks in work – for example for taking instructions, consulting drawings and the like; the proportion of a workman's time not expended "at the coal face" as it were;

e) labour expended on remedial work; and

f) extra labour costs due to variations reimbursed under clause 13.

4.4.22 As stated earlier, precise formulae for establishing the extent of disruption of labour output cannot be given; all will depend upon the nature of the project and the nature and extent of the disruption.

4.4.23 Disruption of repetitive processes can result in extensive extra costs being incurred as was seen in Example 4.10. This is also true where pre-fabricated components are involved. Any change resulting in a significant reduction in the number of standard components and a resultant increase in the number of specials or little repeated units can have far reaching consequences. For example, it may become necessary to tailor the manufacturing process more closely to the erection programme, delivery and storage on site may have to be planned more precisely and erection gangs may have to be more particular in the distribution and fixing of the units than might otherwise have been the case. In short, it requires a far more complex and interrelated programme for all aspects of the work and any disruption to any one of the elements of the programme will reverberate through the others to a far greater extent than would have been the case had there been little or no requirement for prefabricated components.

4.4.24 However, proper management will keep the cost of disruption to a minimum. Even on a project that is beset by daily revisions, the tradesman actually working at the "coal face" will spend the large proportion of his day working at his trade quite effectively and productively. The fact that he had expected to be in another more familiar place that day might have affected his learning curve slightly at the outset but it is unlikely to have a dramatic effect upon the majority of his output. A constant stream of variations to the work currently in hand will doubtless add significantly to site and, perhaps, head office overheads and management costs; but it is the task of management to ensure that the actual execution of the work in these circumstances is undertaken with the minimum of disruption (see *Babcock Energy Ltd -v- Lodge Sturtevant Ltd (1994) CILL 981*).

4.4.25 Contractors often complain that a large number of variations results in a high percentage of disruption to the total of their labour resources. Rarely, however, do they involve numerous interruptions to "coal face" activities (e.g. laying bricks or pouring concrete) which comprise the vast majority of the labour costs. In the main, variations are issued in advance of the work being executed – albeit often worryingly close to it. Again, this will doubtless increase the time taken by site management whose task it is to plan the daily work of the operatives but will often have little or no impact upon those operatives.

4.4.26 *A Report on Productivity in the Electrical Contracting Industry* [2] contains in its Appendix a comprehensive list of the possible activities of an electrician undertaking work on a building contract. It contains over sixty separate activities. It is a salutary exercise to look at each of these activities and ask which, if any, would be disrupted by the issue of a particular variation or a late instruction. Alleged disruption of the work people, as opposed to disruption to the work of site managers, is a relatively easy phenomenon to refer to in general, overall terms but extremely difficult to demonstrate as being the direct result of a variation or late instruction and where it is so demonstrated the likely disruption will form a very small proportion of total labour costs.

4.5 ATTRACTION MONEY AND BONUS PAYMENTS

4.5.1 When a Contractor prices Bills of Quantities, he will use rates embodying labour output assumptions which the estimator considers appropriate to the job, its location, its complexity and the like; these he will price at the appropriate rates of pay. Moreover, prior to entering the total resultant sum as the tender amount, adjustments may be made to the figures, possibly amending those assumptions in the light of market conditions. However, once the Contractor is awarded the Contract and work starts on site, a much more detailed appraisal of labour output and costs is usually made – in short a bonus scheme is introduced. This reappraisal is carried out on each parcel of work in order to set targets which the workmen aim to beat in order to reap some personal financial benefit; such targets and achievements are generally operated on a weekly basis.

4.5.2 The targets in such bonus schemes are usually set at a level that will benefit both the Contractor and his workpeople and thus if the regular flow of work is disrupted by, say, delays in instructions, both will lose out. Indeed, the workpeople will justifiably claim in such circumstances that their loss was not of their making and look to their employer (the Contractor) for recompense. On the surface, therefore, the full lost potential of this bonus is, or will become, a loss and/or expense to the Contractor but again some judgment must be made of what would have been achieved under the bonus scheme had the particular delays in the issue of instructions not occurred.

4.5.3 The failure of the Contractor properly to manage the works resulting in a loss of bonus will not of course qualify, nor would the failure to achieve unduly optimistic productivity implicit in any bonus target.

4.5.4 Where the regular flow of work is constantly being interrupted, the Contractor might claim that any attempt at running a bonus scheme had to be abandoned and some form of average weekly payment substituted in lieu became necessary in order to attract men to the site. However, it is difficult to imagine circumstances in which no bonus scheme at all could be operated; the targets are, after all, usually negotiated with the workpeople each week having regard to the particular work then available and to abandon such a process could well be said to be a complete abrogation of management.

4.5.5 It is possible that where the carrying out of work is made more difficult by a disruptive event, the Contractor will have to set output targets lower than might otherwise be the case, but the measure of this difference is unlikely to vary much, if at all, from a straightforward calculation of the effect of the disruption by reference to the labour hours involved. In short, there seems little ground for admitting calculations based upon the operation or restriction of operation of a bonus scheme as a means of calculating disruption; the disruption should be considered on its merits.

4.5.6 Claiming that extra hourly or weekly payments for attraction money had to be made to workpeople, because say the market for obtaining labour has become more difficult than was envisaged at tender stage, is unlikely to be able to be traced as an expense arising directly from a particular disruptive event and is not therefore reimbursable. Normally these payments become necessary simply to match the market rates current in the area. However, one must differentiate between such a general uplift in rates and the need in certain circumstances to negotiate special rates for a particular task to the extent that the peculiarities of that task are a result of it being due to disruption recognised by clause 26 (see Example 3.7 at para. 3.5.32).

4.6 PRELIMINARIES INCLUDING PLANT AND SUPERVISION

Components

4.6.1 The "Preliminaries" section to a Bill of Quantities will list the Contractor's general obligations. The following is a list of those items which may be individually priced or included with the prices for measured work:

a) site administration;
b) remedying defects;
c) insurances;
d) security/safety/protection;
e) facilities/temporary works/services;
f) management and staff;
g) site accommodation;
h) services and facilities including:
> lighting
> power
> water
> telephones
> safety, health and welfare
> removing rubbish
> protection of the Works
> cleaning
> drying the Works
> small plant and tools
i) mechanical plant and transport;
j) temporary works:
> roads, hardstandings, crossings and the like
> scaffolding
> fencing, hoardings, screens, planked footways and the like
> traffic regulations.

4.6.2 Of these items, remedying defects, insurances, lighting, power, water, removing rubbish, drying the Works, small plant and tools are often priced as a percentage of the Contract value. The remainder should be priced in detail and it is important to ensure that a breakdown of the component parts of each item is provided (para. 3.3.10). When ascertaining the Contractor's entitlement to recover any costs incurred in providing any of the facilities in the above list, the calculation must be related to the actual costs incurred at the time the relevant delay occurred. The Worked Example in Part 4 contains such a breakdown and indicates how the reimbursement is calculated.

Plant

4.6.3 To evaluate the proper cost of plant reimbursable as a direct expense, one must first establish that the cause is one that is recognised under clause 26 and that the costs are not reimbursed elsewhere; then it is necessary to consider the costs under a series of headings.

a) **Setting up charges** – If plant is already on site, or is required on site at some time, the cost of delivery, setting in position and commissioning is covered in the Contract Sum and no extra cost can be admitted, but the inflationary effect of a late delivery due to an earlier delay would be. If the use of a piece of plant, not originally required, is made necessary (for example bringing in a small hoist to facilitate increased output to mitigate delay caused by the late release of information) the cost would be recognised as reimbursable loss and/or expense. In contrast, the bringing in of a small hoist to execute isolated work by way of a Variation after a tower crane has left site would be valued under clause 13.5.3.3.

b) **Hire charges** – It may well be that an item of plant put to use in carrying out additional or disrupted work would have remained idle had it not been so used. In these conditions there is no extra cost to the Contractor and therefore no claim against the Employer for hire (although the cost of running the plant would have to be recognised, see c) below). This "idle" or "waiting time" can be substantial in building Contracts but it is difficult to isolate. Nevertheless, in assessing the costs of plant to be charged as a direct expense, an allowance for use, during time that would otherwise have been non-productive, must be made (i.e. no payment for hire charges will be included for this element of the delay). This assessment is achieved by comparing the period that the plant was on the site with the period it would have been on the site had the requirement for the additional work not occurred. If, however, otherwise idle plant is deployed in support of work arising as a result of a variation then the Contractor will collect a "windfall" in that he will be entitled to payment under clause 13.5.3.3.

c) **Running costs** – Extra running costs would be incurred if the plant was used more but this would not apply if work was simply postponed; in that latter event the hire charges might increase but the running costs would remain the same except perhaps for inflation.

d) **Removal costs** – Where the removal of plant is delayed the resultant extra inflation on the cost of removal could form an element of loss and/or expense.

4.6.4 Figure 4.5 and Example 4.11 illustrate the above.

4.6.5 Again it is necessary to emphasise the need to compare whatever Contract Bill prices for plant are available with the cost claimed by the Contractor and seek a reasonable justification for any difference (paras 3.5.22 and 3.5.28 et seq.) and to note that this should only be done to enable comparison between what the proper costs were and what they would have been had no disruption taken place (see para. 4.4.10).

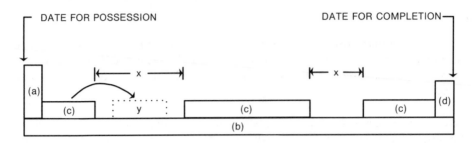

(a) SETTING UP COSTS

(b) HIRE CHARGES   INCREASE IN COST ONLY RECOGNISED IF
                         PLANT IS ON SITE LONGER. NO INCREASE
                         IF DELAY IN USE OCCURS AT "x" WITH
                         NO CONSEQUENT INCREASE IN OVERALL
                         PERIOD OF HIRE

(c) RUNNING COSTS   INCREASE IN COST ONLY RECOGNISED IF
                         USED MORE. NO INCREASE IF USE
                         POSTPONED AS AT "y" EXCEPT FOR INFLATION

(d) REMOVAL COSTS   INCREASE DUE TO INFLATION IF REMOVAL
                         IS DELAYED

Figure 4.5

Example 4.11   The cost of delayed use of plant

Figure 4.6 is a simple bar chart of a programme for a Contract in which the foundations are planned to be completed two weeks before commencement of the superstructure, i.e. there is 2 weeks float. Assume that a 3-week stoppage occurs on part of the foundations whilst schedules of reinforcement are awaited. The Contractor's programme properly reflected the date for the release of these schedules given on the Information Release Schedule (clause 5.4.1). It also envisaged the building being finished 6 weeks before completion of the external works and shows only one Completion Date for the Contract.

The Architect, whilst accepting that a delay has occurred, considers that it will not affect the Completion Date and therefore does not grant an extension of time. The Contractor has plant on site which is costing him £300 per week (although he only allowed £200 per week in his tender for somewhat less sophisticated but nonetheless adequate plant). Because of the delay this plant is on site for an additional week (it would have been standing idle during the 2 week float period anyway). Thus under clause 26.2.1.1 the Contractor is entitled to the cost of 1 week's hire at £200 per week but nothing in respect of running costs as no extra use was required of the plant.

Notes:
1.   Although the Contractor claimed his actual cost of £300 per week for the plant, £100 was disallowed as not being directly attributable to the delay and the difference between the costs claimed and the expectation of those costs from the rates in the Contract Bills was excessive (paras 3.5.22 and 3.5.28 et seq.).

2. Technically 1 week's inflation should have been added to the cost of removing the plant upon completion and 3 weeks' inflation added to the cost of using the plant (e.g. operator, fuels, etc). But it is assumed that the operator's increased costs would be covered by the fluctuations clause and the remaining items are insignificant.

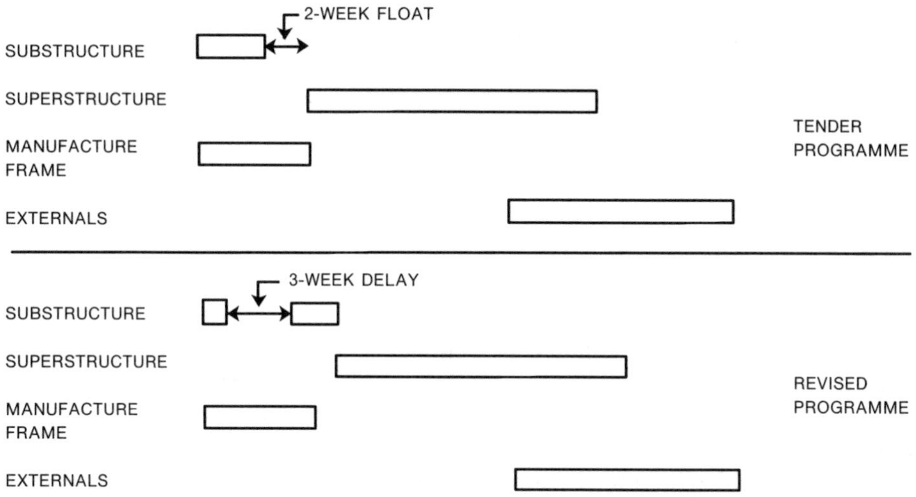

Figure 4.6

Example 4.12 The cost of extra supervision

A building with a precast concrete structure is delayed for 3 months because a discrepancy is found in one structural drawing, which was a Contract Drawing. One bay of precast units is shown on one drawing to be larger than all other bays; moreover, on all other drawings including the setting out drawing that bay size is shown correctly as being of the standard size. The frame is erected and some larger precast units are adapted on site to make them fit, although the large-scale cladding drawings, which show the bay at the correct size, obviously indicate no requirement for such adaptation. The Architect takes the view, in assessing extensions of time, that a diligent Contractor would have discovered the error when the need to adapt the units came to light (the Contractor, in fact, having progressed beyond this stage before discovering the error). The Architect gave an extension of time of 8 weeks, i.e. that necessary to carry out remedial works from the date upon which the Contractor should have discovered the discrepancy (paras 2.8.16 and 2.8.17).

On the basis of the facts and the judgment of the Architect in relation to extensions of time, the Contractor's entitlement to reimbursement of loss and/or expense would be limited to the costs associated with the 8-week period and might be assessed as follows:

	£
Salary and expenses of structural engineer brought to site solely for this purpose, 8 weeks at £650 per week	5,200
Abortive costs in respect of site foreman, 8 weeks at £520 per week	4,160
	———
	9,360

This sum would be reimbursable in full under clause 26.2.3.

Notes:
1. It is assumed that others costs (e.g. extended use of site huts the cost of remedying the work) are ascertained elsewhere.
2. The rates of £650 and £520 are from the Contractor's cost records and appear to be reasonable.

4.7 INFLATION

4.7.1 Strictly speaking, this section should be entitled "fluctuation in prices". But as an increase in price is generally more common than a reduction in price this text refers to inflation only; in any event the mathematical principles are the same.

4.7.2 Rarely do any of the fluctuations clauses fully reimburse the Contractor for the effects of inflation (para. 2.14.3 (Note 3)) but clause 26 does not give the Contractor the right to challenge this inadequacy of the fluctuations clause, this is a risk which he shoulders; it is only the increase in the burden of that risk, due to one of the matters listed in clause 26.2, that is relevant. Care must be taken to ensure that any reimbursement under the fluctuations clause, or any allowance deemed to be included in the Contract Sum to cover fluctuations in prices, is not duplicated when evaluating direct loss and/or expense.

4.7.3 The amount of inflation that qualifies as loss and/or expense is never easy to calculate with precision. That the calculation of an entitlement may present some difficulty is no reason for not making the attempt in whatever manner is most appropriate in the circumstances (para. 4.2.1). Some reference to indices and formulae would appear inevitable in the case of fluctuations calculations.

4.7.4 Before considering examples of the calculation of inflation it is necessary to emphasise the difference between "tender prices" (the cost to the Employer) and "building costs" (the cost to the Contractor).

4.7.5 An index of building costs will be based upon a standard matrix comprising given proportions of different categories of labour and a range of materials at the published rates current at the relevant time.

4.7.6 An index of tender prices will, in addition, reflect current market conditions which, for example, during a period of boom might include the additional cost required to attract labour and possibly to obtain material which may be in short supply together with an increased "mark-up" in respect of overheads and profit. The reverse situation arises during a slump.

4.7.7 The two types of index are given in *Spon's Architects' and Builders' Price Book* [3] and the whole subject is dealt with in detail in *Spon's Construction Cost and Price Indices Cost Handbook* [4]

4.7.8 Over a long period of time, building costs and tender prices keep pace with one another. However, a detailed look at specific periods clearly indicates that there can be a very wide divergence. In the long term, the two sets of indices converge and even cross over, but it is clear that to rely upon the wrong index could give a very misleading result in the short term.

4.7.9 In general, if indices have to be used then the building cost indices are likely to be the more appropriate for ascertaining the cost of inflation as part of loss and/or expense. Any exceptional increase in tender prices is unlikely to qualify as *direct loss and/or expense* (para. 3.5.32).

4.7.10 The range of possibilities for the calculation of inflation can be seen from *Figure 4.7* to which the following indices of building costs apply:

Index at date of tender	100
Index at original Date for Completion	115
Index at (extended) Completion Date	125

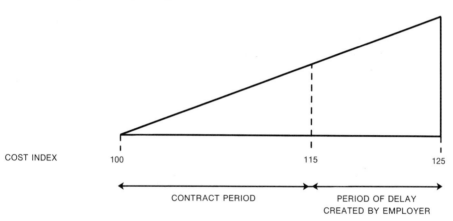

Figure 4.7 Indices of inflation.

4.7.11 The following may be deduced from the above:

a) the index at tender stage is 100 and by the end of the original Contract period the Contractor will be bearing inflation at the rate of 15%. On a fixed price Contract, this is his risk and would accrue progressively throughout the period averaging out at around 7.5%. In fact, on most building projects the greater value of work is carried out in the latter half of the Contract period but taking half the final inflation rate is usually sufficiently reliable at least for illustrative purposes. By the end of the period of extension of time, costs have risen to 25% above those prevailing at tender stage, i.e. the average financial impact on the Contractor will have been 12.5% not the 7.5% referred to above;

b) during the period of delay costs were, on average, at the 120 index level (average of 115 and 125);

157

c) on the basis of these figures, if there are cost records that indicate that over the whole period the average rate charged for plant was say £100 per month, the rate applicable to the extended period only would be:

$$\frac{£100 \times 120}{112.50} = £106.67$$

By the same token, cost records which are only readily available during the original contract period indicated that an average rate of £100 was incurred, the rate applicable to the extended period would then be:

$$\frac{£100 \times 120}{107.50} = £111.63$$

4.7.12 Often it is necessary to recast the record of costs expended over a given period as if they were expended over a different period for comparison.

4.7.13 Examples 4.13, 4.14 and 4.15 illustrate a range of calculations of the effects of inflation. For ease of comparison the following assumptions are made in all cases and are illustrated in *Figure 4.8*:

a) contract period   12 months
b) contract duration   18 months
c) contract value at current rates at tender date (excluding provisional and P C sums)   £500,000
d) monthly rate of inflation in building costs assumed at tender stage   1% simple*
e) actual monthly rate of inflation in building costs   1.5% simple**

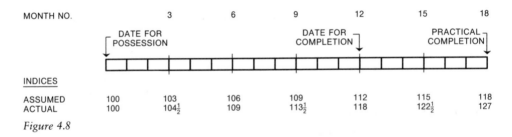

Figure 4.8

Example 4.13   The extra cost of inflation – I

The Contract is let on a fixed-price basis. Of the 6-month delay in Completion, 3 months are due to variations and postponement of parts of the works which affected progress fairly regularly and consistently throughout the Contract; these causes qualify for reimbursement of loss and/or expense. The remaining 3-month delay was in respect of

---

* In practice these rates would vary from month to month and therefore the calculations would be more complex than shown. Moreover, the rates for inflation of 1% and 1.5% per month are considerably higher than are currently being experienced. But this exaggeration does serve to highlight the principle of the calculations.

strikes and bad weather which do not so qualify. The actual cost of disruption to labour that almost certainly would result from such variations and postponement has been assumed to have been dealt with separately so as to clarify the principles involved in calculating the effect of inflation. The range of costs due to inflation are shown below:

a) cost of inflation allowed in tender

$$£500,000 \times 12\% \times \tfrac{1}{2} = £30,000$$

b) cost of inflation had the Contract not been delayed

$$£500,000 \times 18\% \times \tfrac{1}{2} = £45,000$$

c) cost of inflation incurred by practical completion

$$£500,000 \times 27\% \times \tfrac{1}{2} = £67,500$$

d) cost of inflation had the Contract been completed in 15 months (i.e. at end of period of extension of time due to causes which give rise to an entitlement to reimbursement of loss and/or expense)

$$£500,000 \times 22.5\% \times \tfrac{1}{2} = £56,250$$

The Contractor would be entitled to £11,250 in respect of the additional effect of inflation being £56,250 at d) less £45,000 at b).

Notes:

1. Although the Contractor allowed only £30,000 in his tender, it is the cost in excess of £45,000 that becomes his entitlement; this latter figure is what inflation would have cost him had the project finished to time. The fact that actual rate of inflation was greater than he had estimated is his risk; the fact that the actual cost of inflation was exacerbated due to variations and postponement is at the cost of the Employer.

2. The Contractor, in fact, lost a further £11,250 (£67,500 − £56,250) due to the final 3-month delay.

3. It is sometimes said that the proper basis for the calculation of the Contractor's entitlement is $\tfrac{15}{18}$ of the difference between b) and c) on the grounds that the Contractor took 18 months to complete, only 15 months of which were in respect of time for which the Employer had liability. This is incorrect. To allocate the expenditure in this way would result in the extra costs attributable to inflation being increased by matters which are at the Contractor's risk and this element of cost is not a direct expense (para. 3.5.31).

4. The indices of inflation used apply to building work as a whole and do not differentiate between trades. It would be possible to undertake the above calculations in greater detail on the basis, for example, of the individual indices compiled for use with the NEDO formula [5].

4.7.14 These are very simple examples and in practice the facts are likely to be more complex and the calculations more intensive as in the next example.

Example 4.14   The extra cost of inflation – II

The facts are assumed to be as in Example 4.13 except that the 3-month delay due to variations and postponement occurred during the first 6 months on site (i.e. what was planned to be completed in the first 3 months – £110,000 worth of work and which properly should have taken 6 months to complete having due regard to the circumstances, because of strikes and bad weather, took nearly 7 months). The patterns of gross expenditure at Contract prices might reasonably be calculated as shown in the attached tables.

The Contractor would be entitled to a total of £20,401 in respect of the additional effect of inflation, i.e. the difference between the actual inflation for a 12 and 15-month Contract period as shown in the table of monthly costs of inflation. Once again his entitlement is the difference between costs properly incurred in carrying out the work (given the 3-month delay for variations and postponement) and what would have been incurred had that delay/disruption not taken place. The costs estimated at tender stage are not relevant; nor, in this example, are the costs actually incurred as they were further exacerbated by the 3-month delay which was at the Contractor's risk.

If the delays occasioned by the variations and postponement occurred at a different point in time the above figures would have to be recalculated reflecting a different pattern of expenditure.

Gross cumulative expenditure at Contract rates

Month No.	Planned	Actual	If contract completed in 15 months recognising the acceptable proportion of the slow start
	(£000s)	(£000s)	(£000s)
1	26	9	12
2	63	20	26
3	110	32	43
4	163	47	63
5	220	63	85
6	277	81	110
7	334	115	163
8	386	163	220
9	431	208	277
10	467	254	334
11	491	300	386
12	500	345	431
13		386	467
14		423	491
15		454	500
16		478	–
17		494	–
18		500	–

Monthly costs of inflation relevant to Example 4.14

A Month No.	B As estimated at tender stage	C Actual	D Had the contract been completed in: 12 months	E 15 months
	£	£	£	£
1	130	68	195	90
2	555 (1)	248 (2)	833 (3)	315 (4)
3	1,175	450	1,762	638
4	1,855	788	2,783	1,050
5	2,565	1,080	3,847	1,485
6	3,135	1,485	4,702	2,062
7	3,705	3,315	5,558	5,168
8	3,900	5,400	5,850	6,412
9	3,825	5,738	5,738	7,268
10	3,420	6,555	5,130	8,122
11	2,520	7,245	3,780	8,190
12	1,035	7,762	1,552	7,763
13		7,688		6,750
14		7,493		4,860
15		6,743		1,958
16		5,580		
17		3,960		
18		1,575		
	27,820	73,173	41,730	62,131

Difference between £41,730 and £62,131 = £20,401

Notes

(1) Planned expenditure in the second month was £37,000 (£63,000 by month 2 less £26,000 by month 1). The estimated inflation rate for that month was the average of 1% and 2%, i.e. 1.5% which, applied to the planned expenditure of £37,000, is £555.

(2) The actual expenditure in the second month was £11,000 (£20,000 by month 2 less £9,000 by month 1) and the actual inflation rate was the average of 1.5% and 3%, i.e. 2.25% which, applied to the actual expenditure of £11,000 is £247.50 (£248).

(3) Planned expenditure in the second month of £37,000 (see Note 1 above) by the actual inflation rate prevailing in that month of 2.25% equals £832.50 (£833).

(4) Assessed expenditure in the second month if Contract completed in 15 months is £14,000 (£26,000 by month 2 less £12,000 by month 1) which at the average inflation rate prevailing in that month of 2.25% equals £315.

Example 4.15 The extra cost of inflation – III

The facts are again assumed to be as Example 4.13 except that it is a Contract in which the Contractor is to be reimbursed increases in prices that occur after the tender is submitted in accordance with clause 39. In preparing his tender the Contractor estimated that under this clause he would only recover two thirds of the increases which he actually incurred (para. 2.14.3 (Note 3)). The Contract Sum is deemed to include the balance of one-third, and the Contractor, in preparing his tender, allowed accordingly. He estimated that inflation in building costs during the 12-month Contract period would total 6% of the Contract Sum (i.e. 50% of the 12% increase by the end of month 12). Of this he would expect to recover two thirds or 4% under the fluctuations clause and would therefore add the remaining 2% to his tender price. Thus a Contract Sum of £500,000 (excluding p.c. and provisional sums) would include the estimated value of non-reimbursable fluctuations; and the value of work at prices current at the date of tender would be

$$\frac{£500,000 \times 100}{102} = £490,196$$

The range of approximate costs due to inflation are shown below:

a) total estimate at tender stage

$$£490,196 \times 12\% \times \tfrac{1}{2} = £29,412$$

b) non-reimbursable element, one-third of a), included in the tender sum of

$$£500,000 = £9,804$$

c) actual cost of inflation had the Contract not been delayed

$$£490,196 \times 18\% \times \tfrac{1}{2} = £44,118$$

d) actual cost of inflation incurred by practical completion

$$£490,196 \times 27\% \times \tfrac{1}{2} = £66,176$$

e) Actual cost of inflation had Contract been completed in 15 months

$$£490,196 \times 22\tfrac{1}{2}\% \times \tfrac{1}{2} = £55,147$$

The Contractor would be entitled to one-third (the non-reimbursable element) of the difference between c) and e), i.e.

$$\tfrac{1}{3} \times (£55,147 - £44,118) = £3,676.$$

4.7.15 All the above examples deal solely with the effect of inflation due to delay, it being assumed that any costs of disruption and the like would be ascertained separately. However, variations introducing more work would be valued at prices in the Contract Bills which would reflect the fixed price or fluctuating basis of the Contract.

4.8 HEAD OFFICE OVERHEADS AND PROFIT

Generally

4.8.1 An element of Contractors' claims which can prove particularly problematic for those charged with ascertaining the same is that relating to the recovery of head office overheads and loss of profit. It is now accepted practice, on the grounds that it is settled law, that the recovery of head office overheads and loss of profits are allowable heads of claim. It would be a mistake, however, to take the view that any entitlement to loss and/or expense under clause 26 brings with it an automatic right of recovery of such sums. It is common practice amongst Contractors simply to include, in this respect, sums assessed by means of a formula such as Hudson's or Emden's – see paras 4.8.57 et seq. – in the absence of any attempt to establish that, on the merits of the particular circumstances, there is an entitlement to recovery.

4.8.2 In respect of the recovery of head office overheads, a Contractor will generally present his claim in one of two ways, the first relating to disruption and the second to prolongation. Indeed, there will be occasions where a Contractor has suffered the extra cost both of prolongation and of the employment of additional resources during the contract period in which event he may resort to both approaches. These are set out below:

4.8.3 First, the Contractor can contend that the delay or disruption suffered has involved his head office staff in the expenditure of additional time and resources. Having regard to the contractual burden of proof placed upon the Contractor it is necessary (and, indeed, ought to be possible) for the Contractor to provide in support of his claim records of such time and resources expended. However, even having done this, the case of *Babcock Energy Ltd -v- Lodge Sturtevant Ltd (1994) CILL 981* highlights a further test which must be satisfied before the Contractor becomes entitled to the reimbursement of the sums claimed:

> *Managers are of course employed to sort out problems as they arise. If, however, the magnitude of the problem is such that an untoward degree of time is spent on it then their costs are recoverable.*

It would appear from the *Babcock* case that it is incumbent on the Contractor, in order to establish his entitlement to reimbursement, to demonstrate that the problems suffered are not inconsiderable and that the resultant additional head office overheads expended are somewhat more than merely peripheral. Where the Contractor claims the recovery of such costs, i.e. the additional costs brought about by the need for extra resources at a given time, he must support the costs claimed with adequate records related strictly to the project on which they are being claimed, show that adequate management contained these extra resources to a reasonable level in the particular circumstances and show that such costs were the direct consequences of the matter claimed under clause 26.2.

4.8.4 Secondly, and as is more usually the case, the Contractor may present his claim for additional head office overheads on the basis that his site resources have been prevented from earning a contribution to such costs elsewhere by reason of the prolongation of the Contract in question; these are sometimes referred to as unabsorbed overheads. This approach relates more specifically to situations where he has been detained on a site for a longer period than planned and in circumstances where he is entitled to an extension of time. Where the Contractor claims the recovery of such costs, namely his inability to recover contributions to his head office overheads and profit, the use of some formula based approach seems inevitable. Whilst there are a number currently in existence (see below) it is suggested that none of these deals fully with the inevitable complexities. Moreover, it is suggested that in effect circumstances may require two different approaches to the application of formulae. Comments on these published formulae, together with a recommended new formula, are set out in the remainder of this section. Discussing the Canadian case of *Ellis Don -v- The Parking Authority of Toronto (1978) 28 BLR 98*, the respected commentator, the late Vincent Powell-Smith [6], made the following comments:

> *The Canadian Court case also required evidence that the delay prevented the claimant from using his resources elsewhere. That is also the law of England.*

Keating [7] comments in a similar vein but in greater detail:

> *... it is suggested that, in order to succeed, a Contractor has in principle to prove that there was work available which, but for the delay, he would have secured but which in fact because of the delay he did not secure. He might do this by producing invitations to tender which he declined with evidence that the reason for declining was that the delay in question left him insufficient capacity to undertake that other work.*

This is consistent with the decisions in both *Alfred McAlpine Homes (Northern) Ltd -v- Property and Land Contractors Ltd (1995) 76 BLR 59* and *City Axis Ltd -v- Daniel P Jackson (1998) CILL 1382*. As with any aspect of loss and/or expense claimed under clause 26, it is preferable for the Contractor to provide documentary evidence of the actual loss suffered. However, where it is simply not possible to provide the evidence with the degree of particularity envisaged by clause 26, there is now judicial authority to the effect that it is acceptable to establish the measure of the Contractor's entitlement by reference to one or other of the formulae commonly used for such purposes – though it is suggested that the formula approach must be the contractor's last rather than first resort. In the *McAlpine* case, judicial approval was given for the use of Emden's formula (para. 4.8.59) which takes as its percentage for overheads not that allowed for in his tender for the particular contract in question but that actually earned by the Contractor on his turnover according to his accounts. The decision in the McAlpine case was also followed in both *St Modwen Developments -v- Bowmer & Kirkland Ltd (1996) CILL 1203* and *Norwest Holst Ltd -v- CWS (Official Referees' Court Service Website)*; all three cases have a common theme which is that they were all heard on appeal from arbitrators, decisions where each arbitrator had decided on the facts (which are not generally challengeable on appeal) that a loss must have occurred and that Emden's formula may be applied given that the conditions precedent for the application of a formula as decided by case law (which are challengeable on appeal) had been satisfied. What the three cases do not decide is that formulae may be applied indiscriminately – they should only be applied where there is evidence that the Contractor was prevented from recovering a contribution to head office overheads and profit due to the delaying events. When confronted by formulae, it is incumbent upon the Architect or Quantity Surveyor to bear in mind that they are simply a means of measuring, not establishing, the Contractor's entitlement. The latter must be done by reference to the matters discussed in the preceding paragraphs. Accordingly, a claim submitted by a Contractor which does no more than set out a formula based calculation has failed to provide evidence that the Contractor has any entitlement per se capable of being measured. In the *McAlpine* case, the Judge said that ... *a loss must be proved and not left to be a matter of speculation ... A loss in respect of overheads in a period of delay must be proved to be caused by a reduction in turnover which in turn is directly attributable to the delay* ...

4.8.5 In this section head office overheads and profit are considered under the following headings:

a)   the nature of head office overheads and profit generally;
b)   head office overheads;
c)   finance charges;
d)   profit;
e)   the inter relationship between head office overheads and profit;
f)   the development of formulae for ascertaining the recovery of losses on head office overheads and profit;
g)   recommended formulae for ascertaining the recovery of losses on head office overheads and profit;
h)   review of the other formulae currently available.

The nature of head office overheads and profit generally

4.8.6 Head office overheads and profit are usually allowed for in a tender by means of a single percentage on the value (or part of the value) of the Contract. Generally the total of such costs is calculated annually for the coming year and expressed as a percentage on the anticipated turnover for that year. This percentage is then used as the mark-up on all tenders. However, as they each cover quite different elements of the Contractor's resources they should, at least initially, be considered separately.

4.8.7 Head office overheads comprise such items as the cost of renting or purchasing the head office buildings, and running, maintaining and staffing them. Staff based in head office, even if working full time on one project are not often costed individually to that job; they appear as part of the head office facility. An indicative list of the items that fall under the heading of head office overheads is given at para. 4.8.16.

4.8.8 Profit, on the other hand, is less definitive; in monetary terms it is what is left over from the income when all the outgoings have been recognised. Profit is a term which can properly be defined as including the profit or dividend to be paid to equity shareholders. (Somewhat anomalously, the interest on loans or overdrafts – which serve the same purpose as equity payments, namely the provision of the necessary financial resources to run the business – would normally be considered to be an overhead.) If no dividend is paid then after a time the shareholders will remove their investment to alternative enterprises; thus although dividends may be somewhat speculative in amount they are nonetheless a commercial necessity. For this reason also profit should be seen as a recurring payment, i.e. an amount due or required per annum not a single lump sum per project.

4.8.9 One might draw the line between interest charges payable and profits to be distributed on the grounds that the former is a contractual obligation and can be precisely calculated whereas the latter is not. Thus most Contractors will include interest payments as an overhead and not a profit.

4.8.10 It is sometimes suggested that it is important to determine whether the capital, or part of it, is owned by a director or shareholder in a private company out of his own resources or whether it is provided by means of an overdraft: interest charges on the latter being an inevitable expense whereas the former costs nothing to supply. This, of course, is incorrect. In the case of capital owned by a director or shareholder, by having it locked up in the company is to deny him the freedom to invest it elsewhere and thus forego interest or profit on it as the case may be.

4.8.11 A Contract which has been prolonged by the Employer without a commensurate increase in value will result in a reduced cashflow per annum and thus a reduced recovery per annum for profit and overheads where, as is usually the case, these are recovered by the application of a percentage to estimated turnover. A Contractor who finds himself in this position is entitled to argue that but for this delay he could have been using his resources on alternative profitable work bearing in mind that profit should be seen as an annual return.

4.8.12 The case of *Peak Construction (Liverpool) Ltd -v- McKinney Foundations Ltd (1970) 1 BLR 111* contains some interesting comments relating to profit. This was a case which centred around an argument as to the responsibility for a substantial delay (58 weeks) to a building contract. The Court of Appeal in referring the matter back to the Official Referee gave some useful advice on the matter of loss of profit. Lord Justice Edmund Davies said:

> *This outright denial (that no loss of profit had been established), is, in my judgement probably untenable, it being a seemingly inescapable conclusion from such facts as are not challenged that the plaintiffs suffered some loss of profit.*

4.8.13 Lord Justice Salmon said:

> *Moreover, it is possible, I suppose, that an Official Referee might think it useful to have an analysis of the yearly turnover for say 1962 right up to say 1969 (the Contract was signed in February 1964 and the delay occurred between October 1964 and November 1965) so that if the case is put before him on the basis that work lost during 1966 and 1967 by reason of the plaintiffs being engaged upon completion of this block, and therefore not being free to take on other work, he would be helped in forming an assessment of any loss of profit sustained by the plaintiffs.*

4.8.14 It is thus now established that loss of profit and loss in recovery of head office overheads are items which are in principle admissible; the difficulty is the ascertainment. As Lord Justice Edmund Davies said in the *Peak* case (see above) "... the *plaintiffs will need to do some hard thinking on how they should proceed to establish this item of damage".* The matter is therefore considered in some detail below.

4.8.15 In ascertaining the amount of any such loss one must have regard to the following:

a) the percentage amount allowed in the tender for the recovery of head office overheads and profit;

b) the extent to which this planned recovery has been reduced directly as a result of a cause listed in clause 26.2;

c) what recovery the Contractor could have made in the market on alternative projects had his head office resources and capital not been tied up on the current project directly as a result of a cause listed in the contract as grounds for reimbursement of direct loss and/or expense, e.g clause 26.2 in JCT 98 PWQ.

Head office overheads

4.8.16 The description head office overheads will include the following:

a) purchase or rent and rates of offices, plant and yards, etc.;

b) maintenance and running costs of last;

c) furniture and equipment for offices and the like;

d) directors' salaries and expenses;

e) head office technical staff salaries and "on costs", e.g. surveyors, planners, etc.;

f) head office administrative staff salaries and "on costs", e.g. accountants, typists, messengers, cleaners, and maintenance staff;

g) administrative expenses, e.g. postage, printing, stationery, telephones;

h) travelling expenses including the provision of motor cars for directors;

i) legal and professional fees;

j) insurances;

k) finance charges;

l) depreciation.

4.8.17 Most Contractors will include a percentage somewhere in their tenders (either spread over their rates or shown as a separate amount) to allow for the recovery of their head office establishment costs. This percentage is arrived at by reviewing the previous year's head office costs, and projecting them forward to the forthcoming year and relating the resultant figure to their anticipated turnover. In normal market conditions, that is to say where the Industry is not in the middle of a recession or a boom, it usually runs at around 5% to 6%.

4.8.18 Occasionally this is calculated and expressed as a percentage of main Contractor's work only (i.e. excluding the value of Nominated Sub-Contractors) with perhaps a reduced percentage on the value of Nominated Sub-Contractor's work. This creates no difficulty; it is simply necessary to ensure that in calculating any loss the same principles are adopted.

4.8.19 Where a Contract is stretched over a longer period of time with a reduction in turnover, the Contractor can follow one of three courses:

a) increase the percentage to be applied to other contracts remaining to be let in the current year so that the target is achieved;

b) increase the percentage to be applied to a future year or years; or

c) make no adjustment and in effect make a loss on his head office overheads in that year and claim his deficit from the Employer where appropriate.

4.8.20 It is often suggested that a) and b) will mean no ultimate loss therefore there is no entitlement to claim recompense under the Contract. However, this ignores the fact that the allocation of the head office overheads' percentage is only one of the first steps in building up the tender amount; the final overriding decision is the general market level for prices that will be successful in a competitive situation. Example 4.18 at para. 4.8.37 illustrates this.

4.8.21 Having assessed the gross loss in overhead recovery it is necessary to ensure that any recovery by virtue of increases in the value or work through variations and the like is offset so that there is no duplication of payment. This is dealt with in Examples 4.20–4.23 at paras 4.8.48–4.8.54.

Finance charges

4.8.22 The cost of financing an individual project is usually included in head office overheads and thus any adjustment of these overheads will cover these finance charges also. Occasionally, however, finance charges are treated separately; it is therefore considered separately here.

4.8.23 Although most building contracts, including JCT 98 PWQ, make provision for payment for work as it is executed, there is, nevertheless, an inevitable time lag between the time materials are supplied or work executed and the time that payment is received. Moreover, the full value of work and materials is rarely paid immediately; there is an amount retained (usually up to 5%) on building contracts. In general, therefore, money is always outstanding to the Contractor which he has to borrow either from the bank or shareholders or from his suppliers or Sub-Contractors. But insofar as anyone is temporarily held out of funds an interest charge or a return is payable.

4.8.24 The amount outstanding at any one time can vary depending upon the type of expenditure. Under JCT 98 PWQ, monthly certificates are issued up to 7 days after the work and materials have been valued and the Contractor is entitled to payment within 14 days of the issue of the certificate. A Contractor who executes work or who has purchased materials during 1 month will have, in theory, borne the value of those commodities for an average of 2 weeks before they are valued and will not receive payment for a further 3 weeks. There is, therefore, 5 weeks delay on average in receipt of payments. For reasons of cashflow, the Contractor may choose not to pay his suppliers or Sub-Contractors until after he has received the cash – often a long time after. However, it should be noted that following the introduction of clause 19.4.3 in JCT 98 PWQ (relating to Domestic Sub-Contractors) and of clause 4.16.4 in NSC/C (relating to Nominated Sub-Contractors), the Contractor becomes liable for the payment of interest at 5% simple over the Base Rate of the Bank of England on amounts included in certificates but not paid by their final date for payment. On the face of it, the scheme of things contained in JCT 98 PWQ may appear to be somewhat generous when compared with the situation prevailing post *Minter* (see section 4.9) but the situation is instead that, in the absence of the express terms contained in JCT 98 PWQ, the Contractor might potentially become liable for the payment of statutory interest at 8% simple over the Base Rate of the Bank of England on qualifying debts not paid on time and this is due to the introduction on 1 November 1998 of the Late Payment of Commercial Debts (Interest) Act 1998 (hereinafter referred to as the Late Payment Act). The Late Payment Act currently relates only to "small businesses" (as defined in the Act), initially, but the intention is to extend its provisions to all sizes of businesses in the coming years. Whilst the Late Payment Act is aimed at penalising late payers it does not apply where, as is the case with JCT 98 PWQ, the Contract between the parties has its own "substantial remedy" (to quote the Late Payment Act).

In theory, a debtor could deny a creditor his just entitlement on the grounds that, in the fullness of time, the addition to the principal sum of 8% simple over the bank base rate would not be as great as the addition to the principal sum of 1% compound over the bank base rate. A note of caution, however, to anyone disposed to testing this theory; given that the Late Payment Act is aimed at penalising late payers, any attempts that might be made at circumventing it are headed off by another Statute, in this case the Housing Grants, Construction and Regeneration Act 1996 (hereinafter referred to as the Construction Act and, incidentally, implemented on 1 May 1998 being 6 months before the implementation of the Late Payment Act). On its simplest analysis, the Construction Act provides that the aggrieved party (or the "referring party" to quote the Construction Act's own Scheme) will be in a position to have a binding decision made in his favour within 28 days after the date of the referral notice, being the notice sent to the Adjudicator.

Should the other party feel inclined to appeal the Adjudicator's decision, perhaps on the grounds that the sheer speed of the adjudication process led to cogent issues being glossed over, regard should be had to the decisions in *Macob Civil Engineering Ltd -v- Morrison Construction Ltd (1999) CILL 1470* and *Bouygues UK Ltd -v- Dahl-Jensen UK Ltd (CILL 1566)* which provide robust judicial support for the decision of Adjudicators and which, it is suggested, will make appeals against any Adjudicator's decision feasible only in the most exceptional of circumstances, such as jurisdictional challenges as signposted by His Honour Judge Wilcox's decision in *A&D Maintenance -v- Pagehurst (CILL 1518)*. Irrespective of the merits of the Adjudicator's decision in the *Macob* and *Bouygues* cases, Dyson J (the head of the Technology & Construction Court) took the view that to allow the parties to challenge the Adjudicator's decision would undermine Parliament's intention in passing the Construction Act that any such decision should be binding. Further consideration of the Late Payment Act [8] and the Construction Act [9] are beyond the scope of the subject matter in this book but further reading is recommended at the end of this chapter.

Irrespective of the reasons underlying any circumstance in which the Contractor might pay interest to his Sub-Contractors (be they Nominated or Domestic), it is suggested that any such interest is not recoverable by him as direct loss and/or expense under clause 26. If the Contractor fails to pay his Sub-Contractors because he chooses not to, this is a matter between him and them whilst if the Contractor has not paid his Sub-Contractors because he has not himself been paid by the Employer then clause 30.1.1.1 of JCT 98 PWQ provides that he may recover interest from the Employer at the same rate as he is in turn obliged to pay his Sub-Contractors. It is necessary, therefore, to consider each element of expenditure on its merits.

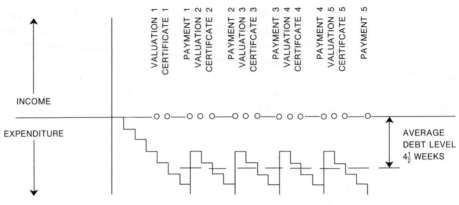

Figure 4.9

4.8.25 *Figure 4.9* illustrates the cash flow in respect of the Contractor's own labour to whom he makes payments at the end of the week which demonstrates that, on average, the Contractor is always financing around 4.50 weeks' costs when monthly valuations are issued.

4.8.26 The time lag between expenditure and income for the various cost elements of a Contract are shown in the following table.

Time lag between expenditure and income assuming monthly variations

A Element of expenditure	B Normal terms of payment	C Approximate delay between expenditure by the Contractor and income from the Employer	D Approximate % of Contract Sum involved	E Weighted delay in receipt of payment (C × D)
		(weeks)	(%)	(weeks)
Contractor's own labour	Paid weekly in arrears	4.50	23	1.035
Materials purchased by Contractor	Paid at the end of the month following that in which they were delivered, i.e. on average 1 week after receiving payment for each certificate	−1.0	33	−0.33
Hire plant	As above	−1.0	2	−0.22

A Element of expenditure	B Normal terms of payment	C Approximate delay between expenditure by the Contractor and income from the Employer	D Approximate % of Contract Sum involved	E Weighted delay in receipt of payment $(C \times D)$
		(weeks)	(%)	(weeks)
Site staff	Paid at end of month, i.e. at valuation date	3.0	6	0.18
Sub-Contractors (whether nominated or not)	Normally paid upon receipt of payment for certificate	–	30	–
Head office overheads	Similar to site staff on average	3.0	6	0.18
			100	1.045

1.045 week's expenditure is approximately 2% of annual turnover.

4.8.27 The outstanding debt is, therefore, 1.045 of a week's expenditure, i.e. in this instance about 2% of the average annual turnover on the site. However, it can vary considerably depending upon:

a) the terms of payment agreed with suppliers and sub-contractors; and
b) the proportion of nominated and other sub-let work.

4.8.28 To this figure must be added the effect of a 3% or 5% retention as the case may be, of the Contract Sum during the Contract period, plus an allowance for reduced retention during the defects liability period. But again the terms of payment agreed with Sub-Contractors are likely to be such that they have to bear the same retention burden as does the main Contractor, and so the retention calculation should be applied to the value of main Contractor's work only; or 70% of the total (i.e. the 100% of expenditure in column D of the table in para. 4.8.26 less the 30% attributable to Sub-Contractors).

4.8.29 Using the data from the table which follows para. 4.8.26, Example 4.16 illustrates the total position.

Example 4.16   The cost of financing

A Contract to the value of £500,000 is to be executed over 2 years. It has provision for retention of 3% and the pattern of expenditure is as given in the table at para. 4.8.26. The defects liability period is six months. The cost of financing the work is as follows:

a) average amount outstanding: £
    (i)   value outstanding during period between payment and receipt

$$\frac{1.045 \text{ weeks (see table)} \times £500,000 \text{ (Contract Sum)}}{104 \text{ weeks (Contract period)}}$$

5,024

    (ii) retention before Practical Completion

$$\frac{3\% \times £500,000 \times 70\% \text{ (Main Contractor's element)}}{2}$$

5,250

_____

10,274

b) finance charges at 10% simple per annum: £
    (i)   £10,274 at 10% per annum × 2 years   2,054.80
    (ii) retention during Defects Liability Period
          1.5% × £500,000 × 70% × $\frac{1}{2}$ year at 10% per annum   262.50

_____

Total cost of financing   2,317.30

4.8.30 Example 4.16 indicates the total cost of financing which the Contractor will build (or will be deemed to have built) into his tender. From the above it will be seen that if the Contract is extended by say one year due to any of the causes listed at clause 26.2 with no comparable increase in value, and if the delays are spread evenly throughout the Contract, the total financing charges will follow the same pattern, but over three years not two years as seen in Example 4.17.

Example 4.17   The cost of financing delay

Details as Example 4.16 but with a delay of 1 year throughout the Contract and no increase in value:

a) average amount outstanding:
    (i)   value outstanding due to delay in payment   £

$$\frac{1.045 \text{ weeks (see table)} \times £500,000 \text{ (Contract sum)}}{156 \text{ weeks (Extended Contract period)}}$$

3,349

    (ii) Retention before Practical Completion (see above)

$$\frac{3\% \times £500,000 \times 70\%}{2}$$

5,250

_____

8,599

b) finance charges at 10% simple per annum: £
    (i)   £8,599 at 10% per annum × 3 years   2,579.70
    (ii) retention during Defects Liability Period as above   262.50

_____

Total cost of financing   2,842.20

This total of £2,842.20 represents an increase of 22.7% on the finance charges that would have prevailed (£2,317.30 (see Example 4.16)) had the delay not occurred.

The original finance charge of £2,317.30 would be built into the Contractor's tender probably as part of the head office overheads element in which event if the increase in cost is to be evaluated and allowed in isolation, as in this example, the original/deemed allowance must be excluded from any overall calculation on loss of overheads.

4.8.31 The above example assumes that the delay is evenly spread throughout the Contract period.

Profit

4.8.32 This is the return the Contractor requires on his enterprise. For the sake of simplicity, it is usually consolidated into tenders as a percentage of annual turnover whereas, in fact, what is required is an annual return on capital resources invested. Taking average figures as a guide, turnover might run at around four times the capital invested. On this basis a Contractor having an annual turnover of £4 million is likely to have £1 million tied up in capital resources. Thus, if he requires a 16% annual return on his capital, this will be achieved by an average mark up of 4% of his annual turnover (whether these proportions are precisely correct is not relevant as it is merely a device to demonstrate the method of calculating loss of profit). As with the recovery of overheads, any reduced turnover will result in a reduced recovery of profit which it is suggested is a real loss to the Contractor (para. 4.8.11).

4.8.33 In calculating such a loss due to delay on a Contract one must have regard to the following questions:

a) could the Contractor have utilised his resources on other Contracts? This can best be ascertained by investigating his workload during the period in question. If the company was working at full capacity, even turning down work or the opportunity to tender, then it can be fairly concluded that he could have used his resources elsewhere had he been free to do so; and

b) what was the level of profitability in the industry generally during the delay; had it changed since the date of his tender; if so by what amount?

4.8.34 There are often two arguments put forward in defence of not reimbursing the Contractor loss of profit. The first is that the Employer does not guarantee that the Contractor will make a profit, far less a profit of a certain amount. The above calculations do not rely upon this; they recognise that the Employer does warrant that he will not, due to his own default, restrict the Contractor's potential to earn profit. The second argument suggests that no additional reimbursement should be made where a contract has been extended in length, albeit that its value has remained unchanged, because the sum originally included for profit has not been diminished. This is to ignore the fact that profit, like interest, is not a lump sum but a time bound return.

4.8.35 What one should not overlook is, that if the variations and/or delay to the Works had been known at tender stage and had formed part of the Contract Documents, the

Contractor would estimate the costs likely to be involved and allow such recovery of profit and overheads on them as he thought fit; and the Employer would not begin to challenge it. Why then should the position be any less advantageous to the unsuspecting Contractor who finds the delay thrust upon him; and why should the Employer think, as he often does, that he is entitled to have extra work carried out, or to cause delay or disruption, on terms which are considerably more advantageous to him than he could have expected at the time the Contract was made? Having entered into a Contract, the Employer does not have a captive Contractor whom he can delay and disrupt nor whose workload he can increase by variations without having to pay a proper price for that facility. The rules for valuing variations and ascertaining loss and/or expense are intended to operate fairly for both parties.

4.8.36 Finally, it is sometimes said, in support of the view that profit should not feature in the ascertainment of loss and/or expense, that if the Employer defaults the Contractor should simply be put back into the position that would have prevailed had the default not occurred but should not be allowed to profit from the default. But "profit" in this latter usage means "take advantage of" whereas profit (in the sense that it is a reimbursable loss and/or expense) means that element of cost that must properly be present in any commercial venture to attract the necessary capital in the same way that wages are paid to attract the necessary workpeople. In that loss and/or expense might be regarded as analogous to general damages for breach of contract, the allowing of profit is on all fours with the dicta from *Robinson -v- Harman (1848) 1 Exch 850* in which Parke B said … *The rule of the common law is, that where a party sustains a loss by reason of a breach of Contract, he is, so far as money can do it, to be placed in the same situation, with respect to damages, as if the Contract had been (properly) performed …*

The interrelationship between head office overheads and profit

4.8.37 As has been demonstrated above, head office overheads and profit are separate components of the financial make up of a Contractor's business. For the purposes of ascertaining any direct loss and/or expense, however, they are closely related and inter-dependent as Example 4.18 shows.

Example 4.18 The extra burden of the cost of head office overheads and profit – I

A Contractor anticipates a net turnover (i.e. before any addition for the recovery of head office overheads and profits) in a given year of £5,000,000. The contract under consideration in this example is planned to start at the beginning of that year and last exactly 52 weeks. It has a net price (i.e. before any addition for head office overhead and profit) of £500,000; therefore 10% of his annual turnover will be committed to this project. His estimated expenditure on head office overheads for the Company in that year is £300,000, i.e. some 6% of his net turnover. Current market conditions indicate to him that he can add 3% to net turnover for profit and still submit a competitive price. He thus submits a tender totalling £545,000 (£500,000 plus 9%, that is 6% for recovery of overheads and 3% for profit). His tender is successful and he secures the contract. During that year, however, the industry moves towards a recession. His turnover drops to £4,000,000; he makes cuts in his head office overheads and reduces its cost to £275,000. This indicates that he should now be adding 6.875% for recovery of his head office costs which, together with 3% for profit, requires a total of 9.875% to be added to tenders. However, he is aware that, with the hardened market, if he adds more than 7.5% for head office overheads and profit (in contrast to the 9% originally added and the 9.875% now contemplated) he would be uncompetitive and not secure further contracts. He thus adds only 7.5% to cover the recovery of both head office overheads and profit on all subsequent projects tendered for in that year. Whether this comprises 6.875% for the recovery of head office overheads and 0.625% for profit is not a relevant distinction. The fact remains that the 7.5% is the going rate for the addition to the net figures to cover the total of **both** components.

4.8.38 As will be seen from subsequent examples in this section, establishing the "going rate" for the recovery of head office overheads and profit – the 7.5% referred to in Example 4.18 – is central to any ascertainment of direct loss and/or expense on head office overheads and profit. The Contractor has an obligation to mitigate any costs he may incur. In ascertaining any amount of direct loss and/or expense due to the Contractor one must, at least in the first instance, undertake any calculation on the basis that he has in fact mitigated such costs whether or not he has done so in practice. With this in mind, it is possible to assume certain levels of redeployment of staff in the event of any significant delay in the works. In practice, however, such theoretical calculations are likely to prove irrelevant if as is suggested above one resorts to the use of the all in rate for recovery of both head office overheads and profit and that this all in rate is the one being dictated by market conditions current at the end of the project; this being the period during which the Contractor could have taken on alternative work had an extension of time not required him to remain on site longer than planned. The loss on head office overheads and profit suffered by a Contractor as a result of being detained on site longer than he had intended is the recovery of head office overheads and profit that would have been available to him in the market at that time.

4.8.39 A contractor may for example consider making some of his staff redundant when there is a significant delay on one project. If he does so he will, at least in the long term, save on his expenditure. However, whether or not he does so will have no effect on the level of head office overheads and profit he can recover in the market at that time. This will be determined by the tendering climate as perceived by all tendering Contractors. And this is the yardstick by which one should judge what recovery of his head office overheads and profit charges he could reasonably expect to recover had he been free to take on new work if the delay on the current contract had not precluded him from doing so.

The development of formulae for ascertaining the recovery of losses on head office overheads and profit

4.8.40 Before embarking upon examples of how to ascertain the direct loss and/or expense of head office overheads and profit as a result of an extended presence on site it is necessary to review the basic rule which governs such calculations. Based on the rule referred to at para. 4.2.7, the Contractor's entitlement is that amount which is ascertained as the difference between:

1. the recovery properly achieved in carrying out the work, albeit disrupted by the event; and
2. the recovery he would have achieved had the event not occurred.

4.8.41 Nowhere is this systematic approach more important than in ascertaining the direct loss and/or expense arising from the extended use of head office overheads and profit. Accordingly, in each of the examples which follow – Examples 4.19 to 4.23 – both calculations are carried out and the difference is the Contractor's entitlement. In each case the background to the contract and to the Contractor's changing financial environment is, unless otherwise stated, as set out in Example 4.18. The word "net" in each example means the various amounts **before** the addition for head office overheads and profit.

Example 4.19   The extra burden of the cost of head office overheads and profit – II

The details are as set out in Example 4.18 save that the project is delayed by late instructions in respect of which the Architect grants an extension of time of 5.2 weeks, i.e. 10% of the original contract period. The Contractor is thus kept on site for 57.2 weeks having originally planned to be there no longer than 52 weeks. It is assumed in the example that it can clearly be demonstrated that had the Contractor not been kept on site for this extra period (i.e. 10% of the contract period) the resources tied up on site could have been released for alternative work valued at 10% of the original contract sum, i.e. he had been denied the opportunity of taking on a net value of work of £50,000. The ascertainment of direct loss and/or expense of head office overheads and profit would be as follows:

1. **Actual recovery of head office overheads and profit over the extended contract period of 57.2 weeks given that the delay occurred:**

	£
Net contract sum £500,000 × 9%	45,000 (1)

2. **Recovery of head office overheads and profit that would have been achieved over the period of 57.2 weeks had there been no delay:**

	£
Year 1 (net value of the original contract works) £500,000 × 9%	45,000 (2)
Year 2 (net value of alternative work that the Contractor was prevented from undertaking during the 5.2 weeks extension of time)	
£50,000 × 7.5% (the rate then prevailing)	3,750
	48,750

The loss in this case is the difference between £48,750 and £45,000, i.e. £3,750.

As the items at (1) and (2) in Example 4.19 are common to both sides of the equation, the formula for ascertaining the loss of recovery for head office overheads and profit could be simplified to read:

$$\frac{\text{Extension of time} \times \text{net Contract Sum} \times \text{MOHP\%}}{\text{contract period}}$$

or put more simply

$$\frac{\text{EOT} \times \text{NCS} \times \text{MOHP}}{\text{CP}}$$

where:

EOT is the extension of time for causes which also carry an entitlement to recover direct loss and/or expense;

NCS is the net Contract sum, i.e. before the inclusion of any amount for head office overheads and profit;

MOHP is the percentage generally recoverable in the market place for head office overheads and profit for alternative work at the relevant time;

CP is the original contract period.

This formula is similar to – but critically different from – the Hudson formula (see para. 4.8.58).

4.8.47 However, as will be seen later, there are dangers in simplifying this calculation in this way.

4.8.48 In the above example it was assumed that there was no change in the scope nor therefore in the valuation of the works. The position is more complicated when the value in the final account differs from the Contract Sum, as it usually does, as Examples 4.20 and 4.21 show.

Example 4.20   The extra burden of the cost of head office overheads and profit – III

The details are as set out in Example 4.19 save that the delay (which amounted to a period of 10% of the original contract period) was not due to late instructions but to a variation which introduced additional work the full value of which (i.e. including the mark up for recovery of head office overheads and profit as set out in the Contract Bills) was £27,250. As this includes the 9% addition for recovery of head office overheads and profit the net value before such additions was £25,000, i.e. 5% of the original net Contract Sum. Thus we have the position where the Contract Sum is increased by 5% whereas the contract period is extended by 10% of the original contract period, both being due to the variation introducing more work. In these new circumstances the ascertainment of direct loss and/or expense due to under recovery of head office overheads and profit would be as follows:

1. **Actual recovery of head office overheads and profit over the extended contract period of 57.2 weeks given that the delay occurred:**

	£
Net contract sum £500,000 × 9%	45,000
Net value of variations £25,000 × 9%	2,250
	47,250

2. **Recovery of head office overheads and profit that would have been achieved by the Contractor generally over the period of 57.2 weeks had there been no delay on this particular project:**

	£
Year 1 (net value of the original contract works) £500,000 × 9%	45,000
Year 2 (net value of alternative work – at 10% of the net value of the original contract works – that the Contractor was prevented from undertaking during the 5.2 weeks extension of time) £50,000 × 7.5%	3,750
	48,750

The loss in this case is the difference between £47,250 and £48,750, i.e. £1,500. This is some £2,250 less than the loss shown at example 4.19 which is due to the 9% recovery the Contractor made on the extra work valued, net, at £25,000 (£25,000 × 9% = £2,250).

4.8.49 The formula postulated at the end of Example 4.19 could now be re-written as follows:

$$\left(\frac{EOT \times NCS \times MOHP}{CP} \right) - (NVV \times COHP)$$

where the definitions are those given at Example 4.19 and where:

NVV is the net value (i.e. before the inclusion of any amount for head office overheads and profit) of variations;

COHP is the percentage allowed in the Contract Sum for the recovery of head office overheads and profit.

4.8.50 But even this approach poses difficulties. There is a possibility that the recovery of head office overheads and profit on the value of variations will exceed any loss of head office overheads and profit incurred by the Contractor as a result of a delay. Clearly if, in the above example, the value of the additional work was say 15% of the original Contract Sum but the extension of time in respect thereof remained at 10% of the original contract period the application of the formula would result in a negative amount, i.e. the Contractor would have made a profit and not a loss. This is set out in Example 4.21.

Example 4.21 The extra burden of the cost of head office overheads and profit – IV

The details are as set out in Example 4.19 save that the full value of the additional work (including the allowance for head office overheads and profit) was £81,250. As this includes the 9% addition for recovery of head office overheads and profit the net value before such additions was £75,000, i.e. 15% of the original net Contract Sum. Thus we have the position where the Contract Sum is increased by 15% whereas the contract period was extended by only 10% of the original contract period but both increases were due to the variation introducing more work. In these new circumstances the ascertainment of direct loss and/or expense due to under recovery of head office overheads and profit would be as follows:

1. **Actual recovery of head office overheads and profit over the extended contract period of 57.2 weeks given that the delay occurred:**

	£
Net contract sum £500,000 × 9%	45,000
Net value of variations £75,000 × 9%	6,750
	51,750

2. **Recovery of head office overheads and profit that would have been achieved over the period of 57.2 weeks had there been no delay:**

	£
Year 1 (net value of the original contract works) £500,000 × 9%	45,000
Year 2 (net value of alternative work that the Contractor was prevented from undertaking during the 5.2 weeks' extension of time) £50,000 × 7.5%	3,750
	48,750

The "loss" in this case is the difference between £51,750 and £48,750, i.e. a profit of £3,000.

4.8.51 It is therefore possible to envisage a situation in which a Contractor might be involved in a loss of say £3,000 on his head office overheads and profit due to disruption caused by the late issue of information; on the same contract he might make an extra profit of £3,000 on the valuation of Variations (see Example 4.21). The Contractor would therefore be in a position where his loss of £3,000 is balanced by an extra profit of £3,000; that is to say on the project overall he would be in balance so far as his head office overheads are concerned. But under JCT 98 PWQ, which contains no provision for the Employer to reclaim extra profit which any of the causes in clause 26.2 might have generated, the Contractor would be entitled to recover the first £3,000 and yet not forfeit the second £3,000, thus being in pocket to the tune of that amount. Example 4.22 illustrates this.

Example 4.22   The extra burden of the cost of head office overheads and profit – V

The details are as set out in Example 4.19 save that the full value of the additional work (including the allowance for head office overheads and profit) was £81,250. As this includes the 9% addition for recovery of head office overheads and profit the net value before such additions was £75,000, i.e. 15% of the original net Contract Sum. The extension of time granted in respect of this additional work was 5.2 weeks, i.e. 10% of the original contract period. However, there has also been the late issue of instructions quite unrelated to the variation referred to above. The Architect has granted an extension of time in respect of these late instructions of a further 5.2 weeks, i.e. a further extension of time amounting to 10% of the original contract period. Thus we have the position where the Contract Sum is increased by 15% whereas the contract period was extended by a total of 20% of the original contract period but as a result of two distinct causes namely the 10% extensions of time due to both the variation and the late instructions. In these new circumstances the ascertainment of direct loss and/or expense due to under recovery of head office overheads and profit would be as follows:

1. **Actual recovery of head office overheads and profit over the extended contract period of 62.4 weeks given that the delays occurred:**

    a) **in respect of the Contract sum and of variations introducing additional work:**

	£
Net contract sum £500,000 × 9%	45,000
Net value of variations £75,000 × 9%	6,750
	51,750 (1)

    b) **in respect of the late instructions:**
    No specific recovery.

2. **Recovery of head office overheads and profit that would have been achieved by the Contractor generally over the period of 57.2 weeks had there not been the first delay on this particular project:**

	£
Year 1 (net value of the original contract works) £500,000 × 9%	45,000
Year 2 (net value of alternative work that the Contractor was prevented from undertaking during the 5.2 weeks' extension of time due to variations) £50,000 × 7.5%	3,750
	48,750

This £48,750 entitlement is less than the £51,750 actual recovery referred to at (1) above by £3,000 and so no loss has resulted; on the contrary, the Contractor has made a profit of £3,000 which he does not have to forfeit.

Year 2 (net value of alternative work that the Contractor was prevented

from undertaking during the 5.2 weeks' extension of time due to late instructions) £50,000 × 7.5%	3,750 (2)
	52,500

The Contractor has made no recovery in respect of the £3,750 forfeited at (2) above; he is entitled therefore to recover this in full as direct loss and/or expense of head office overheads and profit.

4.8.52 Note that had this picture been looked at globally, the Contractor would be entitled to recover a total of £52,500 less the actual recovery of £51,750, i.e. £750 for loss and/or expense in contrast with the £3,750. The difference is the £3,000 profit that he made as a result of executing variations. This has resulted from a combination of earning the full 9% under this contract as opposed to the 7.5% he might have earned in the market for alternative work at that time; and the fact that the extra value was 15% whereas the concomitant delay was only 10%. As suggested above, the Contractor is entitled to this extra profit which should not be eroded by the loss due to extensions of time arising from another cause or event. It must be emphasised, however, that this seeming anomaly, whereby the Contractor does not have to set off the extra profit he has made as a result of one event against losses which arise as a result of another event, only occurs because the contract is assumed to be on the pattern of JCT 98 PWQ which provides for recovery of loss of profit but contains no machinery for passing back to the Employer any increase in profit made by the Contractor.

4.8.53 When operating under such contract provisions therefore it is important to calculate the recovery of loss of profit and overheads on an individual basis. One should not set the benefit of any extra profit that might arise out of the valuation of variations – incorporating as it will the Contract percentage for recovering head office overheads and profit – against the financial recovery to which the Contractor might be entitled as a result of wholly unrelated extensions of time.

4.8.54 It may be of course that the situation will occur where the market conditions for alternative work will have become more favourable to the Contractor during the course of an existing project and that any alternative work could have been carried out at a higher rate, i.e. the recovery provided for in the rates and prices in the Contract Bills may be 5% but shortly thereafter an improvement in market conditions enables the Contractor to increase his rates for recovery on future projects to say 10% yet he is held to his contract percentage of 5% on the current project. This is illustrated in Example 4.23.

Example 4.23   The extra burden of the cost of head office overheads and profit – VI

The details are as set out in Example 4.22 save for the following:

a) the rate added for head office overheads and profit in the Contract is 5% but during its execution prospects in the market improve and during the currency of this project

Contractors find themselves able to increase their addition for the recovery of head office overheads and profit to 10% and still secure the work.

b) the value of the additional work (including the allowance for head office overheads and profit) was £78,750 and as this includes 5% addition for recovery of head office overheads and profit the net value before such additions was £75,000, i.e. 15% of the original net Contract Sum. Thus we have the position where the Contract Sum is increased by 15% whereas the contract period is extended by a total of 20% of the original contract period but as a result of two distinct causes namely 10% due to variations and 10% due to late instructions. The ascertainment of direct loss and/or expense due to under recovery of head office overheads and profit would be as follows:

1. **Actual recovery of head office overheads and profit over the extended contract period of 62.4 weeks given that the delays occurred:**

   a) **in respect of the Contract Sum and of variations introducing additional work:**

	£
Net contract sum £500,000 × 5%	25,000
Net value of variations £75,000 × 5%	3,750
	———
	28,250 (1)

   b) **in respect of the late instructions:**
   No specific recovery.

2. **Recovery of head office overheads and profit that would have been achieved by the Contractor generally over the period of 57.2 weeks had there not been the first delay on this particular project:**

	£
Year 1 (net value of the original contract works) £500,000 × 5%	25,000
Year 2 (net value of alternative work that the Contractor was prevented from undertaking during the 5.2 weeks extension of time due to variations) £50,000 × 10%	5,000
	———
	30,000

   This £30,000 entitlement is £1,250 more than the £28,750 actual recovery referred to at (1) above and so the Contractor has made a loss which he is entitled to recover.

	£
Year 2 (net value of alternative work that the Contractor was prevented from undertaking during the 5.2 weeks extension of time due to late instructions) £50,000 × 10%	5,000 (2)
	———
	35,000

The Contractor has made no recovery in respect of the £5,000 forfeited at (2) above; he is entitled therefore to recover this in full also as direct loss and/or expense of head office overheads and profit.

4.8.55 The anomaly of the position – whereby the contract provides for the recovery of losses by the Contractor but no recovery by the Employer of any extra profit made by the Contractor – can be seen by comparing Examples 4.22 and 4.23. In the former, the Contractor has made a profit which he is entitled to keep and is not required to set this against the loss which arose out of a separate event. In the latter case, he has made a loss on both counts, both of which he is entitled to recover. Examples 4.22 and 4.23 represent a purist approach and serve to demonstrate the application of JCT 98 PWQ to such matters. In practice, however, the reasons behind delays are often difficult to discern with such clinical precision and it is not unusual, therefore, for the contractor's entitlement to the recovery of head office overheads and profit to be viewed globally.

Recommended formulae for ascertaining the recovery of losses on head office overheads and profit

4.8.56 From the above set of examples and formulae it is clear that the formulae to be used in the ascertainment of direct loss and/or expense due to under recovery of head office overheads and profit must be as follows:

For each extension of time for causes which also carry an entitlement to recover direct loss and/or expense, taken individually, in respect of a matter which also entitles the Contractor to an adjustment to the Contract Sum by reference to the rates and prices in the Contract Bills:

$$\frac{(EOT \times NCS \times MOHP)}{CP} - (NVV \times COHP)$$

provided that if the total is a negative amount no deduction from the Contract Sum can be made. For all other extensions of time for causes which also carry an entitlement to recover direct loss and/or expense, taken in aggregate:

$$\frac{EOT \times NCS \times MOHP}{CP}$$

COHP   is the percentage allowed in the Contract Sum for the recovery of head office overheads and profit;

CP   is the original contract period;

EOT   is the extension of time for causes which also carry an entitlement to recover direct loss and/or expense;

MOHP   is the percentage generally recoverable in the market place for head office overheads and profit for alternative work at the relevant time;

NCS   is the net Contract sum, i.e. before the inclusion of any amount for head office overheads and profit;

NVV   is the net value (i.e. before the inclusion of any amount for head office overheads and profit) of variations.

Review of other formulae currently available

4.8.57 There are three formulae which are referred to from time to time in relation to ascertaining the direct loss and/or expense arising from under recovery of head office overheads and profit in building contracts:

    a) Hudson;
    b) Emden;
    c) Eichleay.

4.8.58 The **Hudson** formula appears in *Hudson's Building and Engineering Contracts* [10] and is as follows:

$$\frac{\text{HO/Profit}\% \times \text{Contract Sum} \times \text{Period of delay (weeks)}}{\text{contract period (weeks)}}$$

4.8.59 The **Emden** formula which appears in *Emden's Building Contracts & Practice* is as follows:

$$\frac{\text{HO}\% \times \text{Contract Sum} \times \text{Period of delay (weeks)}}{\text{contract period (weeks)}}$$

4.8.60 The **Eichleay** formula comes from America and is also reviewed in Hudson [11] and is as follows:

1. Allocable overheads =

$$\frac{\text{Contract billings} \times \text{Total of HO overhead for the contract period}}{\text{Total contractor billings for the contract period}}$$

2. Daily contract HO overhead $= \dfrac{\text{Allocable overheads (from 1 above)}}{\text{Days of performance}}$

3. Amount of recovery = Daily contract HO overhead (from 2 above)

$$\times \text{Days of compensable delay.}$$

4.8.61 Emden refers to overheads only whereas Hudson refers to both overheads and profit, moreover, Emden appears to use costs whereas Hudson uses the tender percentages. Beyond these distinctions they are identical in their effect. They apply the relevant percentage (either for overheads or overheads and profit as the case may be) to the Contract Sum and multiply this by the proportion which a qualifying delay (namely a delay which results from a matter which also gives the Contractor the right to claim reimbursement of direct loss and/or expense) bears to the original contract period. Thus an element of double counting is inevitable in that the Contract Sum already includes the amount for head office overheads and profit. This double counting can be avoided when using Eichleay if one excludes from the definition of "billings" any cost of overheads. All three formulae overlook the prospect of double counting when the final account includes the value of variations, which have generated an entitlement to an extension of time, which valuation will already include some reimbursement for the recovery of head office overheads and profit (see Examples 4.20 to 4.23). None of the formulae contain a provision for reflecting the change in the market and/or the requirement for the Contractor to mitigate his costs (see para. 4.8.37 et seq.)

4.9 INTEREST CHARGES

4.9.1 Historically at common law, it was settled that debts do not carry interest. This much was confirmed in *London Chatham and Dover Railway -v- South Eastern Railway*

[1893] AC 429 but their Lordships thought this was unsatisfactory in *President of India -v- La Pintada Compañía Navigación SA [1984] 2 All ER 773*. However, where a creditor is able to establish that he has suffered special damage by, for example, him having to pay interest on an overdraft as a result of late payment then he is entitled to claim that interest as a special damage subject to that damage falling within the "second limb" of *Hadley -v- Baxendale (1854) 9 Ex. 341*. The case of *F. G. Minter Limited -v- Welsh Health Technical Services Organisation (1980) 13 BLR 1*, established that interest is payable as part of direct loss and/or expense on this special damage basis. The Court of Appeal accepted that money spent on meeting loss and/or expense had to be financed from the time that it was incurred until such time as that amount was reimbursed to the Contractor and that the interest payable on such amounts for that period properly formed part of that loss and/or expense. However, because of the very restrictive wording of clauses 11(6) and 24(1) of JCT 63 (the form in use on that project), the right to reimbursement of such interest payments was limited. The Court noted that the written application in that form of contract required by each of these clauses as a condition precedent to the reimbursement of the loss and/or expense, referred to the losses, etc., incurred, in the past tense. Thus, whilst a written application was a prerequisite for any payment, it effectively truncated any continuing element of the loss; the Court suggested that this machinery of making written application for loss and/or expense that had been incurred was hardly apt to deal with continuing losses but accepted the view that an entitlement to reimbursement of a continuing loss could properly be kept alive by lodging a series of written applications, provided of course they were given within the period laid down in the Contract. As to the timing of the initial or subsequent written applications, clause 26.1.1 of JCT 98 PWQ does require that such applications should be made ... *as soon as it has become, or should reasonably have become, apparent to him (the contractor) that the regular progress of the Works or of any part thereof has been or was likely to be affected* ... If nothing else, the prompt issuing of such applications by the Contractor will permit the Architect to consider means of alleviating the problems giving rise to loss and/or expense. The corollary of this, it is suggested, is that the failure by the contractor to observe the requirements of clause 26.1.1 should result in the Contractor's entitlement to loss and/or expense being curbed in some fashion since, but for his failure to observe the contractual requirements upon him, the Architect could have had (and should have taken) the opportunity to control or remove the matters giving rise to the loss and/or expense complained of.

4.9.2 No guidance was offered in the *Minter* case as to whether the interest payments were to be calculated on a simple or compound basis. But, in that the Court simply concluded that in principle interest payments on a loss could be added to that loss, it was perhaps unnecessary. In terms of applying the interest calculation to the Contractor's principal entitlement, Forbes J gave the following guidance in *Tate & Lyle -v- Greater London Council [1982] 1 WLR 149*:

> *I feel satisfied that in commercial cases the interest is intended to reflect the rate at which the plaintiff would have had to borrow money to supply the place of that which was withheld. I am also satisfied that one should not look at any special position in which the plaintiff may have been ... the correct*

thing to do is to take the rate at which plaintiffs in general could borrow money. This does not, however, to my mind mean that you exclude entirely all attributes of the plaintiff other than that he is a plaintiff ... I think that it would always be right to look at the rate at which plaintiffs with the general attributes of the actual plaintiff ... could borrow money as a guide to the appropriate interest rate.

4.9.3 In Keating [12], by reference to case law, it is suggested that *"the rate of interest awarded to established companies is commonly assessed at bank rate or minimum lending rate plus 1 per cent."* In this context, the interest referred to is to be compounded that being the nature of the interest typically payable by companies to their banks.

4.9.4 Moreover, when ascertaining the elements of the loss, it is incumbent upon the Architect and Quantity Surveyor to check the actual costs incurred; thus if the Contractor paid interest on a compound basis (as he almost certainly would) then this should be the basis of ascertainment. This view has been confirmed in *Rees and Kirby -v- Swansea City Council (1985) 30 BLR 1.*

4.9.5 In Example 4.24 it is assumed that there is only one item of direct loss and/or expense to be considered and the interest is assumed only to change at the end of the month. In practice this is unlikely and thus a more complex table and calculation may result.

Example 4.24   The cost of interest charges – I

A Contractor is involved in direct loss and/or expense throughout month 5 of the Contract (May) and this amounts to £23,000. He makes the proper written application under the terms of the Contract to this effect. The reimbursement of interest on a compound basis would be as shown in following table.

Interest accruing on loss and/or expense

Month	Basic loss	Interest Interest rate per annum	Monthly amount	Cumulative amount
	£	%	£	£
May	23,000.00	12	115.00 (1)	115.00
June	23,000.00	12	230.00	345.00
July	23,345.00 (2)	12	233.45	233.45
August	23,345.00	10	194.54	427.99
September	23,345.00	10	194.54	622.53
October	23,345.00	8	155.63	778.16
November	23,345.00	10	194.54	972.70
December	23,345.00	10	194.54	1,167.24

Total payable = £23,345.00 + £1,167.24, i.e. £24,512.24.

186

Interest accruing on loss and/or expense (*continued*)

Notes:

(1)   Average of $\frac{1}{2}$ a month's interest is allowed because the loss was incurred throughout the first month.

(2)   Interest is added to the basic loss at the bank accounting period, here assumed to be every 6 months, i.e. July and January. If interest is rolled up at a more frequent period, say monthly as is usually the case, then the calculation must be adjusted accordingly.

4.9.6 The precise calculation of interest accruing as part of loss and/or expense is best carried out by using a table such as the one referred to in Example 4.25.

Example 4.25   The cost of interest charges – II

A Contractor undertakes a project with a 20-week contract period, during the course of which he is disrupted by variations during week no. 12. The direct loss and/or expense is ascertained by the Quantity Surveyor on the instructions of the Architect at £15,500 with the interest thereon still to be calculated. The proper notices have been given but no payment is included in respect of that direct loss and/or expense. The annual interest rates are as set out in Table 3 to the Worked Example in Part 4. Payment for the direct loss and/or expense is planned for week 26 and the interest charges are calculated by reference to that table by applying the formula:

$$\frac{\text{Accumulator at the time of payment} - \text{Accumulator at the time the expense was incurred}}{52 \text{ weeks}}$$

or from the table of notional interest charges:

$$\frac{239.00 - 105.00}{52} = 2.57\%$$

or £15,500 × 2.57% = £398.35.

4.9.7 The rates of interest given in the table are deemed to be those which, from a review of the Contractor's accounts and of the accounts of other Contractors "with the general attributes" of the Contractor, were payable during the relevant period.

4.9.8 In practice interest charges would, these days, be calculated by computer but the tabular form adopted for this book demonstrates the principles involved.

4.9.9 It could be the case that a "cash rich" Contractor was in an investment rather than an overdraft situation at the time when properly reimbursable loss and/or expense was due to him. If this were the case then it is suggested that the measure of the Contractor's loss would be the amount of interest forgone due to his reduced investment ability rather than the interest charged on his increased borrowing requirements due to the incurred loss and/or expense. In either situation, the calculations would be similar, with just the rates of interest and perhaps the accounting periods varying.

4.9.10 The interest charges discussed in this section are an integral part of the loss and/or expense incurred and are not to be confused with the interest charges mentioned in para. 4.8.24 which are express contractual/statutory rights to charge interest on unpaid debts.

4.10 SOME INADMISSIBLE ITEMS

4.10.1 It is not possible to list all the items that a Contractor might be disposed to claim but which are not admissible. Among the more common, however, are the following:
a) cost of accelerating the works;
b) cost of overtime;
c) cost of preparing a claim.

These are considered in turn.

Cost of accelerating the works

4.10.2 The Contractor may claim that the cost of acceleration should be met by the Employer. But the Contract contains no machinery for instructing the Contractor to accelerate the Works nor any provision for reimbursing him the cost of so doing (the position is different under JCT 98 MC, see Chapter 10). It may be argued that the Architect delayed fixing a new Completion Date and that during this period of indecision the Contractor had no alternative but to speed up the Works to meet the completion date that prevailed at that time. This is incorrect. In the event of a delay by the Architect in fixing a new Completion Date, the Contractor must proceed on the basis that a proper and fair decision on this will be forthcoming – the amount of the extension may be referred to Adjudication, Arbitration or Legal Proceedings if the Contractor's justifiable expectations are not met (para. 2.3.8). This should be contrasted with the financial implications of the obligation upon the Contractor to ... *use constantly his best endeavours* ... (para. 2.7.45).

4.10.3 The Employer may, nevertheless, instruct the Contractor to accelerate the Works and the Contractor may accept such an instruction in which case this will represent a supplementary or additional agreement which should clearly state the additional rights and obligations of the parties. Unless so provided in that agreement, the costs involved cannot be dealt with by either the Architect or Quantity Surveyor as the original Contract does not bestow such powers.

Cost of overtime

4.10.4 Again, this is not a matter on which the Architect is empowered to instruct and therefore no extra costs can be reimbursed. The exception to this rule might be where such costs are incurred pursuant to the Contractor's obligation to "... *use constantly his best endeavours* ..." (see above) and where the delay is the result of a matter listed in clause 26.2. But the Contractor must be able to demonstrate that the working of such overtime was effective in achieving a reduction in that delay. Contract documents often provide that if the Contractor wishes to work overtime he must first obtain the approval of the Architect; such approval when given is not an instruction by the Architect.

Cost of preparing a claim

4.10.5 All too often a Contractor will prepare and submit a voluminous document purporting to set out the basis of his claim and to evaluate the amount of direct loss and/or expense for which he seeks reimbursement. This is not required by the Conditions of Contract although clause 26.1 does now recognise that ... *the Contractor may give his quantification* ... (see para. 3.5.15). The Contractor may therefore feel that the preparation of such a document will be to his ultimate advantage. Nonetheless the Contract Conditions simply require the Contractor to:

a) make a written application that he has been involved in loss and/or expense for which he would not be reimbursed elsewhere (clause 26.1); and

b) keep such information and/or details as are specifically requested under clauses 26.1.2 and 26.1.3;

c) send to the Architect the documentation necessary for the purposes of adjusting the Contract Sum (clause 30.6.1.1).

4.10.6 Therefore the costs of preparing the claim document (quantification) referred to above are not admissible nor are the charges which may be incurred by the Contractor in respect of any consultant he may have engaged for this purpose. But he would, it is suggested, be able to recover the costs properly incurred in making available the necessary data and the like referred to in paragraphs a), b) and c) above.

4.10.7 But in ascertaining that part of the loss the Architect, or Quantity Surveyor, must make sure that:

a) the costs are reasonable in relation to the work involved and no part of the costs expended in preparing any claim submission of the sort referred to earlier is included; and

b) no part of these costs are covered elsewhere in the adjusted Contract Sum, e.g. in any payment for head office overheads included in the value of additions or ascertainment of loss and/or expense.

4.10.8 Finally, whilst on the subject of inadmissible items, two other matters must not be overlooked.

4.10.9 First, an entitlement to claim loss and/or expense is not a mandate to turn the price to be paid for any item, far less the whole of the Contract Sum, into a reimbursement of the Contractor's costs, whatever they may be. The cost of executing any part of the Works is not in any way relevant. What is relevant is that part of the Contractor's costs which comprises the loss and/or expense directly attributable to the variation or disruption. All too often claims are based upon a comparison of the cost of an item with what is purported to have been recovered for it in the final account. The correct entitlement, i.e. the loss and/or expense directly incurred, can only be established by first comparing what an item should properly have cost with what it would have cost had it not been affected by the variation or disruption (para. 4.2.7).

4.10.10 Secondly, claims are often made which may be legitimate at Common Law, but are not in respect of events that are specifically recognised in the Contract Conditions as being matters, the loss and/or expense of which fall to be ascertained by the Architect or Quantity Surveyor. Such matters cannot be dealt with by the Architect or Quantity Surveyor since they must not act outside the terms of reference given to them in the Contract Conditions.

4.11 ANTIQUITIES

4.11.1 Whilst the ascertainment of loss and/or expense arising out of instructions issued in regard to fossils, antiquities and other objects of interest or value found on the site (clause 34.3) has not been specifically mentioned in this chapter, the principles involved in such an ascertainment are the same as those discussed. However, it should be noted that unlike ascertainment under clause 26, the Contractor's application is not a condition precedent to reimbursement under clause 34.3.

REFERENCES

[1] D. Bishop (1965) *Labour Requirements for House Building. Advantages of Continuity of Work and Experience*, Building Research Station Current Papers, Construction Series 18.

[2] Joint Industry Board for the Electrical Industry (1983) *A Report on Productivity in the Electrical Contracting Industry.*

[3] Davis, Langdon & Everest (1999) *Spon's Architects' and Builders' Price Book,* 124th Edition, E. and F. N. Spon, London, pp. 641–644.

[4] Michael C. Fleming and Brian A. Tysoe (1991) *Spon's Construction Cost and Price Indices Cost Handbook*, E and F. N. Spon, London.

[5] Department of the Environment (1990) *Price Adjustment Formulae for Construction Contracts. User's Guide. 1990 Series*, HMSO, London.

[6] Vincent Powell-Smith (1994) *Contract Journal*, 10 November 1994.

[7] *Keating on Building Contracts,* 6th Edition (1995), Sweet and Maxwell, London, p. 230.

[8] Department for Trade & Industry's Better Payment Practice Group (1998) *The Late Payment of Commercial Debts (Interest) Act 1998: A User's Guide*

[9] Tony Francis and Steve Nightingale of Fenwick Elliott (1998) *Adjudication under the Construction Act*, Monitor Press Special Report.

[10] I. N. Duncan Wallace (1995) *Hudson's Building and Engineering Contracts,* 11th Edition, Sweet and Maxwell Ltd, London, p. 1,076 et seq.

[11] I. N. Duncan Wallace (1995) *Hudson's Building and Engineering Contracts,* 11th Edition, Sweet & Maxwell Ltd, London, p. 1,077 et seq.

[12] *Keating on Building Contracts,* 6th Edition (1995), Sweet and Maxwell, London, p. 508.

5

THE CONTROL OF CLAIMS

5.1 INTRODUCTION

5.1.1 When researching to give a lecture on *"How to avoid Contractors' claims"* over two decades ago, Geoffrey Trickey concluded that no other topic in the construction industry was the subject of more misconceptions. The only certain way of avoiding payment of what most practitioners term "Contractors' claims" is to avoid the causes of disruption to regular progress that beget them. One can never prevent a Contractor submitting a claim, however wrongly founded. But the submission of a claim by the Contractor does not of itself create any entitlement to reimbursement; the Contractor must first demonstrate that a real loss and/or expense has been directly caused by one of the matters recognised by the terms of the Contract. Nor should one avoid making proper recompense to the Contractor if the Employer or his agent issues instructions varying the Works or otherwise disrupts the Contractor's progress.

5.1.2 Now, some 25 years after that lecture, it seems that these misconceptions still abound. Looking at some of the evidence gathered for submissions to Sir Michael Latham for his review of the performance of the Industry, Constructing the Team [1], there are frequent references to "the confrontational nature" of certain forms of contract. At the end of the day there is only one way of completely avoiding the prospect of confrontation in a building contract; that is to ask the builder to build what he likes and agree to pay him what he wants. As soon as you insist on a specific commitment – "this is the building I want" or particularly if you go further and say "and I want it completed by a specified date and for a specified price" the seeds of confrontation are inevitably sown. It is not forms of contract or methods of procurement that are of themselves confrontational; it is the parties to a contract who, having been required to undertake commitments, create the confrontation by either failing to achieve them or actively seeking to relieve themselves of them. Certainly ill-prepared Contract Documents, or wanton changes in them thereafter, will produce much fertile ground in which claims will propagate; but to seek to blame particular forms of contract or methods of procurement for the widespread recourse to claims is to tilt at windmills with the result that the real enemy is ignored.

5.1.3 Thus the purpose of this chapter is to lay down some guidelines to assist in minimising the potential for claims and to ensure that where a genuine ground for a claim exists the amount paid in settlement is neither no more nor no less than is contractually proper in the circumstances. Also reviewed are some malpractices that are often adopted in claim submissions in the expectation that their inclusion will enhance the chances of an exaggerated settlement.

5.1.4 Under the terms of a contract such as JCT 98 PWQ, loss and/or expense, being based on fact rather than forecast, will not be ascertained until it has been incurred, i.e. until, during or after the event. This is to be contrasted with some forms of engineering contract, such as NEC/2 (see Chapter 12) which allows for forecasts. As the settlement of such monetary disputes often takes place at the end of the job the Contractor has few tactical weapons with which to bargain. However, issues and facts which are confused or obscured by history may work to the advantage of the Contractor if he can show that he has indeed incurred extra expense notwithstanding the cause.

5.1.5 The provisions of clause 26.1.1 of JCT 98 PWQ require a written application to be submitted by the Contractor ... *as soon as it has become, or should reasonably have become apparent ... that the regular progress of the Works ... has been or was likely* *to be affected* . . . . Often these provisions are not properly implemented and frequently Contractors resort to making applications under clause 26.1 in respect of every conceivable event as a sort of ubiquitous safeguard. In any event the ascertainment of the amount due can, and often does, drag on. Much will depend upon the sufficiency and relevance of the data submitted by the Contractor and on the Architect's (or Quantity Surveyor's) timely response thereto.

5.1.6 Even with the provisions of clause 26.1.1 of JCT 98 PWQ as to the timing of claims applications it still seems to suit many Contractors and the professionals acting under the terms of the Contract to sweep claims under the carpet. Only time will tell if the introduction of clauses 13A.2.3 and 13.4.1.2A1.1 of JCT 98 PWQ under which the Contractor may put forward the amount he requires to be paid for loss and/or expense arising from a variation, in lieu of an ascertainment of the amount under clause 26.1, will do much to change this somewhat entrenched attitude. Such an attitude does not assist the proper administration of the Contract nor does it result in the proper entitlement being established. And it is not only Contractors who seek to exploit such delays. Some Employers, and their advisors, take the view that the longer they hold on to the money the better the deal they can ultimately strike with the Contractor who is hungry for cash.

5.1.7 The only way to ascertain the proper amount due is to bring matters out into the open when they occur, when allegations can be fairly tested against the prevailing facts, and when the real impact of any delay, disruption or the like can be seen and believed. Uncertainty is quite the wrong environment in which to expect even efficient organisations to perform well; and it is fertile ground for the inefficient.

5.1.8 It is unrealistic to attempt to ban variations or design changes. A building project is usually a complicated, unique venture erected largely in the open, on ground the condition of which is never fully predictable, and in weather conditions which are even less so. There is a considerable time span between making the initial decision to build and the completion of works on site. During this period technology, fashion and the client's requirements inevitably change. JCT 98 PWQ recognises these possibilities and attempts to ensure that they are valued fairly in accordance with rules which are known at tender stage, rather than leaving the establishment of those rules until the problems occur at which time one party would most probably be in a more advantageous position than the other.

5.1.9 If the Contractor is to quote a proper price for the project and be kept to his contractual terms he must always know his objectives. Initially, the basis for these will be set out in the Contract documents. The introduction in JCT 98 PWQ of the option for the Employer to provide an Information Release Schedule (clause 5.4.1) should, when used, greatly further assist the Contractor in the formulation of those objectives. They will by its use at least have firm dates on which to expect the release of information and via clauses 25.4.6.1 and 26.2.1.1 they are provided with the appropriate remedies should the Architect fail to abide by the Schedule. Where delays in information, or the extent or timing of variations, eliminate these target dates they should be re-established quickly and fairly using the contractual mechanisms provided. The Contractor's duty will then always be clear; he must achieve those objectives and he should ensure that adequate management is provided to do so. On a Contract that is beset with delays and disruption and where the targets of time and monetary entitlement are temporarily undetermined, it is understandable, if not contractually supportable, for the Contractor to turn his back on his obligations and divert his energies to building a claim rather than the Works.

5.1.10 This is not to say that variations or further instructions should not be given, but where they are necessary the effect both on time and monetary entitlement should be quickly evaluated so that the Contract objectives are repaired. This is where many building projects fail. Albeit that JCT 98 PWQ has had features introduced into it aimed at agreeing the valuation of variations prior to their implementation, all too often Contracts do not provide the necessary procedure for speedy resolution of these matters; and where they do the parties or their representatives do not implement them properly.

5.1.11 There is, however, an intrinsic difficulty in framing the provisions for valuing Variations on the principle adopted by JCT 98 PWQ. This principle is that variations introduced by the Employer shall be paid for at rates analogous to those in the tender; that is to say at the rates that the Contractor might reasonably be expected to have used had those works been incorporated or described in the Tender Documents. Such a provision is an invaluable asset to the Employer; without it he may find himself paying penal amounts for Variations, the value of which would be set by a Contractor well established on site and with all the advantages. However, for such a provision to operate effectively it must operate fairly. Thus JCT 98 PWQ provides for changed conditions or changed character to be taken into account when valuing the Variation; in just the same way that one might presume that the Contractor's estimator would take such conditions and character into account at tender stage had that work been included.

5.1.12 To draft clauses which give these facilities to value Variations at Contract prices and yet be fair to the Contractor in all circumstances is not easy. Provisos are introduced and these can be used by the Contractor to overturn the principle. In JCT 98 PWQ the proviso appears at clause 13.5.7 which entitles the Contractor to a fair valuation (as opposed to using of the rates and prices in the Contract Bills) where, for example, ... *the valuation of any work or liabilities directly associated with a Variation cannot reasonably be effected ... by the application ...* of the detailed provisions of clause 13.5. However equitable and necessary these provisions may be said to be, this lack of commitment to the rules already established in such detail gives either party the opportunity to introduce argument and uncertainty into the price ultimately to be paid; and they play into the hands of any inefficient Contractor who discovers that his Contract may not be turning out as profitably as he had hoped.

5.2 ELIMINATING OR CONTAINING CLAIMS

Generally

5.2.1 As suggested above, and given the traditional provisions for valuing variations at contractually established rates, one cannot wholly eliminate from a form of contract the provision for claims, unless one leaves them to be settled by an Adjudicator, Arbitrator or the Courts.

5.2.2 Without doubt, a number of claims are often successfully pursued which need never have arisen, and many others are settled at levels which are unrealistic given the real facts surrounding them. Such occasions leave one or sometimes both parties dissatisfied and in time they seek extreme contractual procedures for future projects. This cannot be in the long-term interests of the construction industry. It is often claimed that alternative procedures to the competitive lump sum contract historically prove that claims are minimised. Almost invariably this is because such procedures enable the Contractor to build a margin of safety into his price. This of course does not minimise claims, it masks them; and the Contractor keeps the margin whether he is disrupted or not. This penalises the efficient Employer and his design team. Moreover, it does not follow that a profligate price begets a higher profit and a healthier industry; on the contrary, such a "relaxed" price will simply mean the removal of any incentive to strive for efficiency and the resultant picture will be no more edifying than other occasions when a corset is removed. If we are to recreate the view that the construction process is one which represents good value for money, then competitive lump sum contracts must be the bedrock. And for this to be acceptable to clients, for whom reliability of time and cost is usually of paramount importance, the incidence and settlement of claims must not be allowed to reach unacceptable levels.

5.2.3 Listed below therefore are a number of suggestions for avoiding the incidence of claims or containing their potential.

Adequacy of contract documents

5.2.4 It is unrealistic to assume that on every project all the design and specification work will be completed before going to tender; thus it would be fanciful to produce a standard form of contract that is based on this assumption. However, where this can be achieved on any project the contractual clarity, and the absence of any excuses for bad management that will ensue, can be very rewarding. For such documentation to be complete not only must the design team have fully considered and completed all aspects of the building design but they must have also:

a) based the design on a comprehensive survey of ground conditions;
b) eliminated discrepancies;
c) resolved the details of all nominations; and
d) co-ordinated the work of any other Contractors who are directly appointed by the Employer.

5.2.5 Not only will this have removed the majority of the scope for claims but it will also have created a clear backcloth against which the effect of any variations, if they occur, can be precisely measured. All too often the line between a variation and the continuing development of the initial design is impossible to draw.

5.2.6 The Contract is not complete without a Contract Sum and a priced Bill of Quantities; and these have to be supplied by the Contractor. Here again care must be taken to ensure that the documentation is, in all respects, complete and consistent throughout and that full details of the Contractor's costings of his Preliminaries are made available before the Contract is signed. Remember that the rates in the Contract Bills are binding upon both parties as the basis for valuing variations.

Data supplied by the Contractor before commencement

5.2.7 In addition to supplying prices, the Contractor is required under JCT 98 PWQ to provide ... *a master programme for the execution of the Works* (clause 5.3.1.2). This programme will not be a Contract Document, but it would almost certainly be judged to be the best data available to enable the Architect to carry out his function under the Contract particularly in relation to extensions of time and the ascertainment of loss and/or expense.

5.2.8 The Contractor should be required to amplify his programme (paras 2.10.3 et seq.) and produce a written statement on how he intends, in general terms, to undertake the Works. For example, such a statement might include the following "the windows will be erected from the perimeter balconies at each floor level" or "the reinforced concrete stair and lift cores will be erected first and subsequently the steel frame to the remainder of the structure will be erected using these cores as reference points". Whilst the Contractor need not be bound by this it would establish a strategy that both parties could use in determining valid variations to the Information Release Schedule or, where there is not one, in determining the supply of outstanding data and the likely effects of any delays or variations. In the absence of such information it leaves the unscrupulous Contractor with the opportunity to play all the options to his advantage if these programme priorities are allowed simply to unfold in due course.

The supply of further data after commencement and the introduction of variations

5.2.9 There will be projects – probably the majority – where an overlap between the completion of the design and the commencement of the construction is inevitable. In these instances, where design and construction overlap, it would be wise to lay down at tender stage the programme for the supply of outstanding information that will be maintained by the design team by way of the optional Information Release Schedule (clause 5.4.1 of JCT 98 PWQ).

5.2.10 Insofar as the Contract provides for the introduction of Variations such instructions should be issued at the earliest opportunity.

5.2.11 In any event, contrary to the views of many, it is highly desirable that the Contractor be kept in the picture on the prospect and the programme for the supply of any future design development because he will then have greater scope for meeting his obligation to plan to keep any resultant delay to a minimum – see the provisos at clause 25.3.4 of JCT 98 PWQ (para. 2.9.8); and in ascertaining the amount of loss and/or expense due, the Architect or Quantity Surveyor would be entitled to disallow any costs that were avoidable had the Contractor acted on this early warning.

Actions necessary when a cause of delay or loss occurs

5.2.12 The absence of contemporary evidence is probably the single most important reason for the settlement of claims degenerating into an expensive game of poker.

5.2.13 Although JCT 98 PWQ requires notification in advance (clauses 25.2.1.1 and 26.1.1) and the keeping of records (clauses 25.2.1.1, 26.1.2 and 26.1.3), both parties all too often convince themselves that it is to their advantage, for a number of reasons, to delay the resolution of claims for as long as possible. Occasionally this is due to a mistaken belief that matters might become clearer as the Contract progresses or due to the preoccupation with other seemingly more pressing priorities. For example, the Architect who is notified of a delay occurring in the construction of the foundations may understandably be tempted to claim that the inadequacy of the Contractor's particulars and estimates provided under clause 25.3.1 prevents the Architect from determining a new Completion Date. He thus leaves his judgment on extensions of time for, say, 6 months because he feels its effect upon the remainder of the work will have become evident by that time. In practice, nothing is likely to be further from the truth; the occurrence of yet more Variations or instructions, delay due to bad weather or mismanagement will all serve to have obscured the issues rather than have clarified them. The Architect is, after all, only required to **estimate** the delay and whilst this may be no easy task, the Contractor has, in the much more abstract circumstances of the tender period, had to estimate either the whole contract period, or the resources necessary to meet a given contract period.

5.2.14 Another reason offered in support of delaying the granting of extensions of time or ascertaining loss and/or expense is that the Contractor may catch up. So he may but this will not deprive him of his entitlement; the test is to compare what actually happened with what would have happened had the delay or disruption not occurred (para. 4.2.7).

5.2.15 Finally it is sometimes felt that it is advantageous to reject a claim out of hand without regard to its merit or content, and kick the ball conveniently into touch giving the participants a welcome breather whilst the spectators – who always claim to be the experts on both sides – pass it between them.

5.2.16 JCT 98 PWQ requires the Architect to fix a new Completion Date within 12 weeks of receiving notice and reasonably sufficient particulars and estimates of the expected delay from the Contractor. On occasions such particulars and estimates will be far from adequate; on other occasions some Architects may claim that they are inadequate, irrespective of the merits of that claim. Whatever the contractual imperatives, there is no substitute to an extension of time being granted as soon as possible bearing in mind that it is an estimate of the delay in completion with which the Architect is concerned. Equally, there is no substitute for ascertaining the loss and/or expense as it is incurred. Indeed, the first consideration, when the possibility of a disruptive Variation is recognised, is to consider whether the provision, contained in clause 13A, for establishing the value of the variation and any associated loss and/or expense in advance of its sanction, could be implemented.

5.2.17 In any event, the valuation of variations under the detailed rules should be carried out as soon as is possible together with the ascertainment of the direct loss and/or expense so that resolution is achieved and any such ascertainment can, if necessary, be challenged whilst the facts are fresh in the mind. This may now have been assisted, but to what extent only time will tell, by the introduction in JCT 98 PWQ of the primary, optional alternative method for valuing variations (clause 13.4.1.2A) via the Contractor's Price Statement and the separately attached amount required by him, in respect of any resultant loss and/or expense in lieu of any ascertainment under clause 26.1.

5.2.18 Notwithstanding the above exhortation to resolve the matter of establishing the monetary entitlement sooner rather than later, it must follow the event (except in the case of a confirmed clause 13A Quotation or the Contractor's Price Statement), even if only marginally so, as it is the loss and/or expense incurred with which one is concerned. However, the Architect, when considering an extension of time during the currency of work, is required under JCT 98 PWQ to fix such later Completion Date ... *as he then estimates to be fair and reasonable* (clause 25.3.1); this implies looking forward.

5.2.19 In order for such a strategy of timely resolution to work, the Contractor must first give the requisite notice, and again some Contractors, with misplaced gentility, still feel that to give notice of delay or of the incidence of loss is to rock the boat and that as long as such delays or disruption do not occur too often they are best ignored. If such problems do recur it is then difficult to resurrect old skeletons previously buried in a pauper's grave. It is a fundamental misunderstanding of the Contract conditions if either party feels that the submission of a notice of this kind is somehow letting the side down. On the contrary it is, if nothing else, a timely notice which enables the design team so to organise itself as to mitigate the continuing effects of such delays and the like and to require proper records of the facts to be kept.

Keeping proper records

5.2.20 The records necessary for a proper view to be formed of the Contractor's entitlement to extra time or money comprise both those ad hoc records kept specifically as a measure of the effect of particular problems as they occur and those kept regularly throughout all projects.

5.2.21 JCT 98 PWQ contains adequate provisions for demanding ad hoc records to be kept both of delays and of loss and/or expense. The clauses dealing with these matters, predominantly clauses 25 and 26, require the Contractor to give notice as soon as it is reasonably clear that such difficulties will occur (clauses 25.2.1.1 and 26.1.1). Thereafter, the Architect is entitled to particulars of the effect of a delay (clause 25.2.2.1) and the time limit of 12 weeks for granting extensions of time during construction only commences when such details are sufficient for that purpose (clause 25.3.1). Moreover, the Architect or Quantity Surveyor may demand under clauses 26.1.2 and 26.1.3 information and details of the loss and/or expense which the Contractor notifies that he is incurring or will incur. Thus a well administered Contract under JCT 98 PWQ should result in all necessary contemporary records being available. In any event, the Architect should require the Clerk of Works to keep or should otherwise himself keep such detailed records as will be beneficial when it comes to settling these issues, for it is the Architect who has eventually to fix a Completion Date having agreed to all Relevant Events, whether or not notified to him by the Contractor.

5.2.22 It is appropriate, in making reference to Clerks of Works, to move on to the subject of regular data kept in relation to a project. Ascertaining a Contractor's entitlement to the reimbursement of loss and or/expense after the event would be a veritable nightmare if it were not for Clerk of Works' reports. Hence the need for a Clerk of Works to be appointed on all but the simplest of projects (see *Gray -v- TP Bennett & Sons (1987) 43 BLR 63* as regards the duties of a Clerk of Works).

5.2.23 The Clerk of Works' weekly reports contain vital historic information that can be invaluable when it comes to settling claims, most particularly if each of these weekly reports is amplified to record the various activities on which each trade is occupied each week. They contain pure facts – the number of men on site, the amount of stoppages, lists of information received and required, progress achieved, delays experienced and the like. Whilst they are not countersigned by the Contractor these facts are generally indisputable, the more so if the major points made are aired at site meetings.

5.2.24 Reports of site meetings are probably the second most useful source of data when trying to settle the facts after the event although, conscious of this potential, it is not unknown for the whole of a site meeting to be spent disputing the minutes of the previous one.

5.2.25 Correspondence between the Contractor and the design team is also relevant when looking back over a project in order to establish the facts. There is always the temptation, particularly for the Architect, not to respond to correspondence; he is, after all, responsible for the management of the project from the Employer's point of view and this can be very time consuming and pressing. The Contractor, on the other hand, may feel he has a direct monetary interest in pursuing relentlessly any potential for a claim. However tedious, therefore, all correspondence must be dealt with at the time and in the proper degree of detail. Often such correspondence, from either side, is written with the potential Adjudicator, Arbitrator or Judge in mind rather than to achieve current objectives, and such a tactic may rebound heavily if it is responded to quickly and resolutely with facts. Nothing concentrates the mind better on the realities than a recognition that the other party is master of the project and is prepared to devote the necessary resources to establish a solution in accordance with the terms of the Contract.

5.2.26 It is always useful if there can be an agreement at the outset of a contract as to the form in which a notice of delay or disruption will be submitted; the more so if such an agreement is written into the contract conditions and if all such notices are numbered sequentially and the responses thereto are cross referenced. There can then be no argument as to what represents a written notice and what represents a formal reply to it. Too often either side will claim, often well after the event, that a somewhat casually worded letter or note was in reality a written notice issued under a particular clause of the contract conditions – indeed, some lawyers draft letters for their contractor clients in such a casual style in order that the contractor may rely on them later, so it is important that contractors' letters are read carefully. In these days of Word Processors, agreeing upon an approach to the notification of claims need not be cumbersome. The Architect can imprint on the reverse of every notice by the Contractor of a delay or disruption recording his response in a structured fashion; this can save a lot of cross referencing and additional correspondence. Indeed there have been projects where, without resorting to the use of Word Processors, a large rubber stamp has been produced for endorsing the back of any such notice or similar data.

5.2.27 Examples 5.1 and 5.2 outline the sort of data that might be covered in proformae to be agreed between the Architect and Contractor for communicating notice of delay and application for reimbursement of direct loss and/or expense and the responses thereto.

199

Example 5.1 Agreed pro forma for notice of a delay in progress under clause 25 and the response thereto

Project Title ... *Primary School at Meadow Lane Newton* ...

Date of notice ... **27 October 1999** ...

Serial No. of Notice ... **N7** ....

To ...... **R Green and Partners** .... (Architect)

cc **Nominated Subcontractor (where applicable)**
... *Not applicable* ....
(Contractor)

From .... **A Brown & Co.** ....

Contractor's notice

Architect's response dated 3 November 1999

Item Ref.	Cause of Delay	Relevant Event where applicable	Material circumstances and expected effects	Estimated extent of delay beyond Completion Date	Response
N7.1	*Late instruction for partitioning*	*Clause 25.4.6.2*	*We advised the Architect on 7th June 1999 that these drawings were needed by us by no later than 4th October 1999. They were not supplied until 25th October 1999 – a delay of 3 weeks. However we have fallen behind by one week in this area by that date. We previously notified you of this prospective delay in our Notice No 3.*	*2 weeks*	*We agree with your assessment and will be fixing a new Completion Date shortly.*
N7.2	*Additional work to the kitchens and associated area. Variation Order No 23 refers*	*Clause 25.4.5.1*	*This variation as yet does not show the full extent of the extra work required in this area. It is sufficiently detailed however for us to know that a delay beyond the Completion Date will occur.*	*It is impracticable for us to estimate the extent of this delay. We shall notify you of our estimate as soon as it is possible for us to prepare it.*	*We await your further notification*
N7.3	*It has not been possible for us to secure the necessary drainage goods for the South-East corner of the site in time to meet our programme.*	*Clause 25.4.10.2*	*Our Master Programme shows clearly that this work is critical to our progress and to achieving the Completion Date. Accordingly we will inevitably be delayed by 2 weeks*	*2 weeks*	*Our information is that the position with regard to the supply of drainage goods has certainly not deteriorated since the Base Date and accordingly this is not a Relevant Event under clause 25.4.10.2.*

NB The Architect's response does not constitute the fixing of a new Completion Date which, where applicable, will be executed by the issue of a Notice in writing clearly identified as such.

Example 5.2 Agreed pro forma for application for reimbursement of Loss and/or Expense under clause 26 and the response thereto

Project Title ... *Primary School at Meadow Lane Newton* ...

Date of Application ... **27 October 1999** ...

Serial No. of Application ... **A5** ....

To .... *R Green and Partners* .... (Architect)

From ... *A Brown & Co.* .... (Contractor)

Contractor's application

Architect's response dated 3 November 1999

Item Ref.	Matter creating material delay	Clause reference	Details of the matter in question and the reason(s) it is claimed to result in direct loss and/or expense	Response	Amount ascertained £
A5.1	*Late instructions for partitioning in Hall*	*Clause 26.2.1.2*	*We advised the Architect on 7th June 1999 that these drawings were needed by us by no later than 4th October 1999. They were not supplied until 25th October 1999 - a delay of 3 weeks. However we have fallen behind by one week in this area by that date. We previously notified you of this prospective delay in our Notice No N3. This will create a delay in the progress of other trades resulting in loss and/or expense for which we require reimbursement.*	*In our opinion, having referred to your Master Programme, the work of other trades will not be materially affected.*	*Nil.*
A5.2	*Additional work to the kitchens and associated area. Variation Order No 23 refers*	*Clause 26.2.7*	*This variation as yet does not show the full extent of the extra work required in this area. It is clear however that loss and/or expense will be incurred.*	*Your comments are noted. It is too soon to form a view on whether or not loss and/or expense will be incurred. Please liaise with the C of W so that proper records are kept of labour output associated with this work*	*Further details to be provided in due course*

5.3 TACTICS

5.3.1 No book on the subject of claims would be complete without some reference to the tactics sometimes resorted to by one or both parties in either the presentation, response or negotiation of a claim. Not all the following devices are employed all of the time; nor does everyone resort to them some of the time. In the same way as a medical textbook, crammed with the real or potential physical maladies of the body, is not typical of the majority of the human race, the following paragraphs are not included because they are typical of the building industry. But it would be reckless not to counsel readers on their potential and accordingly listed below are some of the tactics which have been encountered over the years.

The fatality of the first figure (or the "relentless wedge")

5.3.2 This rarely fails. Either party gives the other an exaggerated view of his expectations; the Employer or Architect pitches the figure in payment of claims very low, or the damages for delay very high, or both; or the Contractor pitches his alleged loss and/or expense very high. Of course the Contractor would not normally copy the Employer with the correspondence he has with the Architect, but a copy of a claim submission sent to the Employer "for information" may be a very worthwhile investment for the Contractor. The party in receipt of such figures seeks consolation in the thought that at least he is aware of the worst position and is curiously grateful for the removal of uncertainty; he steels himself for a disappointing outcome and the more he becomes resigned to this the less objectivity and resolve he has for supporting what might be a long struggle through to a just settlement; this is the thin end of a relentless wedge.

5.3.3 Architects and Quantity Surveyors faced with the need to deal with a claim submission from a Contractor should in no way be influenced by the global figures put forward by him. The Contractor's duty in the ascertainment of loss and/or expense is simply to establish the facts and to make available such cost records as are called for; it is, under the terms of the Contract, the job of the Architect or, if delegated to him, the Quantity Surveyor to ascertain the amount due. But almost without exception one finds that Contractors are only too keen to money out their claims as they judge that it suits their cause. Such action by the Contractor is of course now recognised in JCT 98 PWQ where, in clause 26.1, it states in respect of direct loss and/or expense ... *of which the Contractor may give his quantification* ... but see our comments in para. 3.5.15 on this somewhat gratuitous provision. Equally, the recipient will often be unable to resist the temptation of peeping at the final total and being seduced into accepting that it has some relevance (even if only as an Aunt Sally) rather than undertaking a wholly objective ascertainment. Irrespective of the revised wording of clause 26.1 referred to above it is still the Architect's or Quantity Surveyor's task to build up the Contractor's entitlement, not to knock down his claim. Thus one should start by setting out the seven heads of claim referred to in Chapter 4, namely:

a) materials;
b) labour disruption;
c) attraction money and bonus payments;
d) preliminaries including plant and supervision;

e) inflation;
f) head office overheads and profit;
g) interest charges

and then proceed to evaluate them referring only to the Contractor's submission where any of the facts supplied are relevant. However, if it is judged that the Contractor's monetary evaluation is exaggerated, it is unrealistic to pretend that this will not have had some impact upon the Employer (who will almost certainly wish to be kept informed of developments) and he may find himself in a dilemma if in contrast he is to be faced with what might appear to be a very low assessment in response from his advisers. To prevent such a credibility gap appearing, then, in the first instance an immediate and approximate ascertainment is necessary from the Architect or Quantity Surveyor in order to counter any prejudices forming in the Employer's mind as to the validity of the Contractor's figure; in doing so, the Architect or Quantity Surveyor might also bear in mind the rule of thumb given in para. 4.4.16.

5.3.4 Example 5.3 contains the summary of a Contractor's claim and the associated table illustrates the Architect's preliminary initial response to it.

Example 5.3   The outline of a Contractor's claim

A contractor submits a claim the summary page of which is as follows:

Item	£
Loss of output on labour	33,878
Attraction money	14,000
Extended preliminaries	28,324
Excessive waste on materials	17,832
Inflation	14,282
Overheads and profit	17,057
Cost of preparing the claim	5,000
TOTAL EVALUATION OF DIRECT LOSS and/or EXPENSE	130,373

The Architect and Quantity Surveyor quickly consult each other over this submission as a result of which they prepare an initial response in the form set out in table 5.1.

5.3.5 In table 5.1, the Contractor's claim and the Architect's preliminary ascertainment are shown side by side, and although the difference between the two is not relevant contractually, notes designed to explain these differences are included for the Employer's peace of mind. In this table, it is assumed that the Contractor did not complete on site until 13 weeks after the (extended) Completion Date.

5.3.6 Having perused the analysis in table 5.1, the Employer will have a more balanced view of the matters in dispute and be less inclined to accept that where a Contractor's submission is exaggerated, it is any proper indication of entitlement.

Table 5.1. Reconciliation between Contractor's claims and Architect's or Quantity Surveyor's preliminary ascertainment.

Item	Amount claimed by Contractor (£)	Approximate entitlement assessed by Architect (£)	Difference accounted for, say (£)	Comments upon the difference
Loss of output on labour	33,878	7,500	26,500	The Contractor has claimed the difference between his alleged labour costs and what he reckons to have recovered to date. This is incorrect, in particular it ignores: Say £ a) Remedying defective work — 6,000 b) Inefficient work (he was 13 weeks late finishing) — 13,000 c) Costs which are his risk (bad weather, strikes, Nominated Sub-Contractor's delays) — 3,000 d) Unaccountable balance (perhaps due to underestimating at tender stage) — 4,500 —— 26,500
Attraction money	14,000	–	14,000	This was paid in order to attract labour on to the job and is not reimbursable as direct loss and/or expense.
Extended Preliminaries	28,324	8,500	18,000	The Contractor has claimed either on the basis of his Bill rates and/or his costs as it has suited him and irrespective of the reasons for the expense; only his costs properly and directly incurred are admissible; this results in a difference of around £8,000. He has also claimed extended Preliminaries when the works were delayed due to a strike on the grounds that information would not have been available anyway; and also for his 13 weeks default in completion, a difference of some £10,000.

Excessive waste on materials	17,832	1,500	The Contractor has enclosed some theoretical calculation purporting to demonstrate that he has only recovered £150,000 for materials in the final account whereas his costs records show an expenditure of £167,832. He blames the rest on extra waste due to disruption. There may be a small entitlement due to extra wastage when materials had to be moved due to variations but this is unlikely to amount to much more than £1,500. There was little supervision and inadequate watching and losses of up to 5% on his materials could well have resulted – £8,000: the material cost records appear to include consumable stores and formwork – approximate cost £7,500.	15,500
Inflation	14,282	3,500	The Contractor has claimed the difference between his estimate of inflation at tender stage with that actually experienced; this is his risk. His entitlement is limited to the increased impact of inflation due to delays caused by the Employer or his agent.	10,500
Overheads and profit	17,057	2,000	This claim is based upon a formula and its application makes no allowance for: a) Mitigation of costs that should have been made – £3,000 b) Contractor' default in completion – £10,000 c) Additional value for overheads covered in the value of variations – £2,000.	15,000
Cost of preparing the claim	5,000	–	This detailed claim submission was not at our request and is not required by the Contract conditions. If and when we call for data to be provided we will consider meeting the cost of its preparation provided that it is not covered elsewhere.	5,000
	130,373	23,000		104,500

The Solomon Syndrome

5.3.7 At the root of too many negotiations is the bland acceptance that, if two contrary views are expressed on a claim, there must be some merit in both and the real truth lies somewhere in between; moreover, the nearer the middle that the answer is judged to be, the more sensitive and sophisticated has been the adjudicator. This, of course, is nonsense. But the more widely it is practised the more it is exploited, with each side stretching ever wider the boundaries of their imagination in order to establish a base in as extreme a position as can be tolerated without totally losing credibility. Before concluding that "there is no smoke without fire" one must be able to differentiate between smoke and hot air, and in any event to recognise that the amount of smoke is no measure of the real intensity of the fire. The amount of the entitlement is by no means pro rata to the amount of paper generated; often the converse is the case.

5.3.8 This predisposition to split the difference can be particularly distressing when attempting to reconcile the Contractor's and the Architect's or Quantity Surveyor's view on entitlement to reimbursement of loss and/or expense. Their respective roles in the run up to any such confrontation are quite different and this difference is often not recognised by the Employer nor, regrettably, by his legal advisors. In arriving at his figure, the Architect or Quantity Surveyor will have been under a duty to operate the terms of the Contract as best he can as between Employer and Contractor. The Contractor, on the other hand, has no such constraint and is perfectly entitled to express every item in dispute in a way which is to his advantage. Given such a situation it is quite clearly wrong to assume that the view of each side has equal merit, because the Architect or Quantity Surveyor has been representing the Contract, whereas the Contractor has been representing himself. Any temptation to split the difference should be resisted. The Contractor and the Architect or Quantity Surveyor are not at opposite ends of a tug-of-war rope; the latter are standing where they judge the centre marker to be.

Miscellaneous manoeuvres

5.3.9 Finally there is a series of manoeuvres open to any party when submitting his claim and some of these are given below so that the recipient is primed to recognise them. The examples given appear to suggest that it is only the Contractor who exploits these potential areas. If this is so, it is almost entirely because Employer's claims are usually limited to liquidated and ascertained damages which, once the extensions of time are established, permit no room for manoeuvre in calculating the maximum entitlement.

The clandestine cost plus

5.3.10 In this type of claim the text will present undeniable facts relating to the running of the Contract, trace the remedy to the correct clause in the Contract, and then finish with a calculation of loss and/or expense which turns out to be no more than a comparison between what the Contractor alleges his costs to be and what he claims to have recovered through the Contract to date; he is in effect seeking reimbursement of all his costs under this heading which is not the test to be applied however just his cause or claim (para. 4.2.7). Indeed, this approach is often camouflaged by a distracting detour through fascinating formulae. Converting the alleged labour content of the final account into a percentage of the total value and describing this as the "tender output", relating the alleged labour costs incurred to the same value and describing this as the "achieved output", comparing the two and describing the difference as the "disruptive loss" is a typical route as illustrated in Example 5.4

Example 5.4 Claims based upon the labour cost incurred

			£
1.	Value of final account		4,586,327
2.	Less Preliminaries		(432,313)
3.	Subtotal		4,154,015
4.	Tender labour content (see Appendix 7)	38.43%	
5.	Actual labour cost incurred		2,103,296
6.	Therefore achieved labour output (Item 5/Item 3)	50.63%	
7.	Therefore loss of output (Item 6 less Item 4)	12.20%	
8.	Resultant disruption of labour 12.20% × £4,154,015		506,790

5.3.11 This approach is quite unacceptable because:

a) it makes no attempt to relate cause to effect;
b) it does not seek to ascertain the direct loss and/or expense attributable to each matter listed in clause 26.2 of JCT 98 PWQ;
c) it applies a global formula to total labour costs; and
d) the global formula is deduced entirely from the total costs said to have been incurred on the scheme and is not an ascertainment of the effect upon specific labour elements of the individual items of disruption.

5.3.12 A nice touch is often added by making a modest deduction for fluctuations charged elsewhere by the introduction of sophisticated but irrefutable calculations. But at the end of the day it is no more nor less than a claim for reimbursement of all labour costs, however they were caused, and this is irrelevant in the ascertainment of direct loss and/or expense.

The veneer of precision

5.3.13 No one is impressed by a claim of such generality as £25,000; one's suspicions are aroused as to its authenticity. On the other hand a total of £24,787.32 does concentrate the mind wonderfully. Peeling back the veneer, however, the details shown in Example 5.5 may be revealed.

Example 5.5   The veneer of precision

Item 7 Claims by domestic Sub-Contractors	£
a) A Smith & Co.'s claim documented at Appendix 5	4,787.32
b) Provisional allowance for claims likely to be forthcoming from other Sub-Contractors, say	20,000.00
To summary	24,787.32

"I'm not clairvoyant!"

5.3.14 This approach is usually expressed in less challenging terms in which the Contractor suggests that it is sufficient for him to be able to demonstrate beyond reasonable doubt that some event has occurred which is detrimental to his interests and which he could not possibly have known about at the outset and therefore had made no allowance for in his tender. However, this shows a basic misconception of the contractual obligations in the matter of time and price. In principle, any Contractor's entitlement under a lump sum contract with a contracted time for completion (whether in the building industry or not) is as shown at para. 1.2.14, and any occurrence not being due to the act or default of the Employer nor being covered by one of the other grounds listed in the Contract is at the Contractor's risk for which due allowance is deemed to have been made in his price. And even then damages due to the consequences of the act or default of the Employer will have to be pursued through proceedings at common law if their resolution is not specifically provided for in the Contract Conditions.

The heart-rending narrative

5.3.15 Building is at best a difficult process and an accurate record of the unfolding of even a well-run contract with few variations would strike fear into the hearts of many from other industries. The scope that exists therefore for exaggerating the picture of events surrounding a claim is almost limitless. On the other hand severe difficulties may face the Employer in meeting the sudden cost of inflation, in finding a market for his building upon completion during a depression, or trading from it should a recession descend upon the country just as he takes possession of the finished building.

5.3.16 Neither of these attitudes has any place in establishing contractual entitlements which should not be put on a par with handing out charity.

Duplicated costs

5.3.17 There is always a danger, when allocating parts of global costs, of duplicating figures. For example, calculations of disruption arrived at by applying percentages to total labour costs may well result in the fluctuations element or extra payments made under the valuation of variations being included twice. The only safe solution is to avoid such overall calculations and build up the ascertainment step by step.

5.4 THE APPROACH TO A SETTLEMENT

Generally

5.4.1 For reasons set down earlier (para. 4.2.7) some element of conjecture is inevitable in the settlement of a claim and thus some form of negotiation may have to take place if a formal dispute (be that by way of Litigation, Arbitration or Adjudication) is to be avoided. Notwithstanding the fact that the vast majority of claims made by Contractors are satisfactorily dealt with by the Architect or Quantity Surveyor there will also be those occasions when the Contractor will press for a payment substantially in excess of what can be ascertained.

Settlement meetings

5.4.2 In these circumstances the Employer will often press for meetings in an attempt to reach a settlement and avoid the legal process. Such meetings will often include representatives from the Employer's organisation, no doubt supported by the Architect or Quantity Surveyor and legal advisor. In this, the penultimate section of this chapter, advice is given on how such meetings should be approached.

5.4.3 For the Employer such a meeting is likely to be a new, or infrequent, experience for unless he is a Property Developer construction is unlikely to be his main preoccupation in life. For the Contractor this may be his meat and drink; certainly he is likely to be much more in command of the legal and technical intricacies. It is for this reason that the comments contained in the following paragraphs seem to be aimed more at the Employer and his advisors than at the Contractor and his.

Preparatory work

5.4.4 There may well be grey areas (of fact and of law) in which the Contractor is to be offered the benefit of the doubt in an attempt to reach a settlement. Before embarking upon such an offer it is important to:

a) ensure that the Employer has given his support to any such concession; and
b) make clear to the Contractor that any such concession is made on a "Without Prejudice" basis solely to achieve a settlement and if this is not achieved it will be withdrawn.

5.4.5 The priority at this juncture is to persuade the Contractor to accept the amount on offer and in doing so remind him that the concessions will be withdrawn if agreement is not reached.

5.4.6 Thoroughly plan any meeting well in advance so that from the outset the following are clear in the minds of the team:

a) precisely what you are trying to achieve;
b) the limit of your authority; and
c) how the meeting will be handled and by whom.

Each of these is referred to in greater detail below.

5.4.7 Probably the biggest and easiest mistake to make is to start exchanging views on figures, whether in detail or in total, without first being crystal clear on precisely what those figures are to cover. It is impossible to set out here a fully comprehensive list of such matters as much will depend upon the particular project and its history. However, the following list will indicate the range of items that might be considered, so that their status in any settlement is made clear:

a) amount paid;
b) amount due but not paid;
c) Nominated Sub-Contractors' claims (including claims under Nominated Sub-Contractors' Warranties);
d) legal and professional costs incurred to date in seeking a settlement;
e) outstanding work;
f) remedial work;
g) retention;
h) liquidated and ascertained damages.

5.4.8 Preferably any settlement should include the resolution of all matters. To reach a settlement in which say Nominated Sub-Contractors are excluded to be dealt with separately is far from ideal because:

a) the Employer will still have no clear idea what overall settlement is in prospect; and
b) the line between what is covered and what is left to be resolved is often difficult to draw and can lead to further disputes. For example, the status of such matters as the main Contractor's profit, attendance and discount and any claims and counterclaims between the Contractor and the Nominated Sub-Contractors.

To leave any matter unresolved is to provide fertile ground for misunderstanding and even puts the partial settlement in jeopardy.

5.4.9 Be clear also whether the monies under discussion:

a) are in addition to the final account and at what level that account has been finalised (e.g. what sum, if any, is already ascertained and included for direct loss and/or expense);
b) are in addition to the amounts already certified;
c) are in addition to the amounts already paid;
d) take into account the payment of liquidated and ascertained damages; and
e) include or exclude Valued Added Tax.

It is all the more important to clarify the above if the Employer himself is playing a leading role in any negotiations as he may not otherwise be aware of their respective values.

5.4.10 It is important to know the limit of your authority and not to stray beyond it. Never use it as a barrier to reaching a settlement, lest the other side interpret this as evidence that you have been convinced of the justice of their case. In establishing the level of your authority bear in mind the advice given in para. 5.4.9. The Employer's Board for example may have in mind conceding a payment of £25,000 more than has been paid; this may be quite different from a payment of £25,000 in addition to what has been established as due.

5.4.11 Rehearse how a meeting is to be run and predetermine what each member is to contribute in each likely eventuality. Establish an agenda in advance of the meeting; this will help to avoid the otherwise inevitable roaming over other related (and sometimes unrelated!) topics. Also, establish who is to head your team, the extent of their mandate and on what subject and to whom they are to defer. Ensure that everyone is aware of the current facts and that your team has a "ready reckoner" so that as the meeting develops and various amounts in settlement are canvassed these can be put into context. Example 5.6 illustrates these points.

Example 5.6   The hazards of settling on a figure

A project, the Contract Sum for which was £715,057.00, has been completed on site and the Form of Contract is JCT 98 PWQ. The final account has been agreed (save for the ascertainment of direct loss and/or expense) at £748,057.94. The loss and/or expense has been ascertained at £32,534.90 but this is disputed by the Contractor who claims that this should be £75,350.00; to date only part – £25,500 – of the ascertained amount has been included in interim certificates.

The project finished 8 weeks later than the Date for Completion entered into the Appendix to the Contract Conditions. The Architect has granted an extension of time of 4 weeks and has certified under clause 24.1 that the Contractor has failed to complete by the Completion Date. The Employer has deducted liquidated and ascertained damages for the remaining 4 weeks at £1,500 per week (i.e. a total of £6,000.00) from the amount due under interim certificate No. 38 (some 2 months ago) having first complied with the provisions of clause 24.2.1.2. The Employer is unhappy that any extension of time has been granted at all and believes that his entitlement to liquidated and ascertained damages should have been for the full 8-week delay – a total for liquidated and ascertained damages of £12,000.00. The Architect on the other hand is now minded to grant a further extension of time, but only of a further two weeks.

The Employer has been advised by the Quantity Surveyor that the final cost of the building contract is likely to be in the region of £782,000.00. This figure is the current limit of expenditure approved by the Board and is made up as follows:

	£
Measured account including adjustment of all PC and Provisional Sums	748,057.94
Present estimate of maximum amount due as direct loss and/or expense	37,000.00
Subtotal say	785,000.00
Less liquidated and ascertained damages (based on 2 weeks because out of an 8-week delay, the Architect is of the view that the Contractor is entitled to an extension of time of 6 weeks)	(3,000.00)
Current estimate to final cost of building contract	782,000.00

Valuation No. 40 has been issued by the Architect based on the following figures:

	£
Work executed as agreed final account	748,057.94
Less work not in accordance with the Contract	(5,000.00)
	743,057.94
Less retention at 1.50% on last	(11,145.87)
	731,912.07
Partial payment for direct loss and/or expense	25,500.00
	757,412.07
Less total of previous certificate No. 39	730,472.60
Amount now due	26,939.47

However, at an earlier stage the Employer has made a deduction from the amount due from him to the Contractor of £12,531.00 in respect of Nominated Sub-Contractors which the Contractor had failed to discharge, although it was included in certificates.

Before interim certificate No. 40 is paid a meeting takes place with a view to settling outstanding matters. At this meeting the Contractor makes a Without Prejudice offer saying *"I will reduce my present claim considerably with a view to settling this dispute so that all you have to do is to agree to a further £25,000.00 to cover my loss and/or expense. Of course I would expect you to drop your counterclaim for liquidated damages."* The Employer does a quick calculation on the back of an envelope and concludes that the total cost to him on this basis will be only £736,941.60 and that this is comfortably within the £782,000 of which he had been advised by the Quantity Surveyor; the Contractor on the other hand sees his offer totalling a settlement of £805,592.84. A staggering difference of £68,651.24! The differing totals are arrived at as follows:

The Employer's calculation:		£
Amount of Certificate No. 39		730,472.60
Less:	£	
Amounts paid direct to Nominated Sub-Contractors	12,531.00	
Liquidated and Ascertained Damages (still based on a cupable delay of 4 weeks even though the Architect is now minded to grant a further extension of time of 2 weeks which will reduce the culpable delay to 2 weeks).	6,000.00	(18,531.00)
		711,914.60
Add Contractor's offer to settle		25,000.00
Total out turn cost to compare with latest advice of £782,000.00		736,941.60

The Contractor's calculation:		£
Agreed final account		748,057.94
Ascertained amount of direct loss and/or expense		32,534.90
Offer to settle		25,000.00
Total settlement		805,592.84
There is therefore a disparity of £68,651.24, made up as follows:		£
The Employer, in reaching his conclusion, has omitted to include:	£	
The amount due in Certificate No. 40	26,939.47	
Repayment of amount previously deducted for unsatisfactory work when it is put right	5,000.00	
Retention money to be released in due course	11,145.87	
Repayment of liquidated and ascertained damages as requested by the Contractor as part of the deal	6,000.00	49,085.34
The Contractor, in reaching his conclusion had also included:	£	
The difference between direct loss and/or expense as ascertained (£32,534.90) and as Certified (£25,500.00)	7,034.90	
The amount paid direct to the Nominated Sub-Contractors by the Employer	12,531.00	19,565.90
Disparity between Employer's and Contractor's Calculations		68,651.24

5.4.12 Had the proper preparations been made before the meeting, both sides should have been armed with the following analysis of the current amounts committed:

Paid to the Contractor:	£	£
Total of Certificate No. 39	730,472.60	
Less paid direct to Nominated Sub-Contractors	(12,531.00)	
Less Liquidated and Ascertained Damages	(6,000.00)	711,941.60
Paid direct to Nominated Sub-Contractors		12,531.00
Certificate No. 40 in the pipeline		26,939.47
Retention to be released		11,145.87
Loss and/or Expense Ascertained but not yet Certified	7,034.90	
Repayment of amount deducted for work not in accordance with the Contract once it has been rectified		5,000.00
Total amount currently committed (with the option to the Employer of deducting Liquidated and Ascertained Damages in respect of a culpable delay of 2 weeks)		774,592.84

5.4.13 The Employer should also be reminded of the estimated out turn cost of £782,000 of which he had previously been advised.

5.4.14 The employer might even be armed with a ready reckoner, based on the above figures, so that he can readily judge the total financial effect of any proposition; it might be as set out in the following table.

Total out-turn cost resulting from a range of options

Basic financial data:

- Contract sum   £715,057.00
- Amount currently committed, i.e. the agreed final account plus amount currently ascertained for loss and/or expense (claims) less amount presently deducted for delay in completion – £774,592.84 say   £775,000,00

  (see figure boxed below)

- Amount paid to date   £724,472.60

Effect of a range of options:

Additional amounts for direct loss and/or expense beyond the £32,534.90 already ascertained (£000's)

Adjustment to existing extensions of time (E o T)	0	2	4	6	8	10	12	14	16	18	20
				Total out turn cost for a range of options (£000's)							
Reduction of 2 weeks E o T – i.e. 2-week extension	772	774	776	778	780	782	**784**	**786**	**788**	**790**	**792**
Reduction of 1 week E o T – i.e. 3-week extension	774	776	778	780	782	**784**	**786**	**788**	**790**	**792**	**794**
4 weeks E o T as at present	775	777	779	781	783	**785**	**787**	**789**	**791**	**793**	**795**
Increase of 1 week E o T – i.e. 5-week extension	777	779	781	**783**	**785**	**787**	**789**	**791**	**793**	**795**	**797**
Increase of 2 weeks E o T – i.e. 6-week extension	778	780	782	**784**	**786**	**788**	**790**	**792**	**794**	**796**	**798**

NB Figures in **bold** exceed the presently notified out-turn cost as approved by the Board of £782,000.

The negotiation

5.4.15 It is impossible to set out in advance all the detailed rules for running a meeting aimed at reaching a negotiated settlement. There are, however, a few rules that one should always follow.

5.4.16 Be aware that one is trying to achieve a figure that best represents the proper amount due under the Contract. It is not a car boot sale in which bluff and counterbluff are the order of the day. There is always the fear that if a negotiation fails then formal proceedings will follow together with all the costs and fees which that might entail. But the same threat hangs over both sides and such a fear should therefore only have a neutral role to play in any negotiation. In any event if one constantly gives way to the threat of legal costs one will set the pattern for settlements which will always favour the other party and the more this predisposition is recognised the more it will be exploited.

5.4.17 Never be induced into believing that because the other side has "made a concession" it is now your turn to do so. However, if an obvious error is revealed in your case concede it quickly and graciously. Beware of those who, having so grossly overstated their case, are shamed into modifying it and in doing so claim it as a genuine attempt to settle and then ask you to make the next move.

5.4.18 There is a widely held theory that the first one to break the deadlock of silence in a negotiation will lose the battle; and it is particularly unrewarding to be involved in a final negotiation when both protagonists subscribe to this view. However, this ploy of keeping silent is likely to find favour only with those who are short of real facts; those who have a well documented case can calmly but relentlessly pursue it.

5.4.19 It is often the case that negotiation is required on a number of items; some of which will be easier to resolve than others. Failure to agree a particular item should not prevent progression to the next, albeit that some item(s) will be left temporarily unagreed. Once all items have been considered, one can total up the unagreed items by ascribing to them the amounts that you are prepared to accept in respect of each. It may well be that the resulting total is acceptable to the other side; or at least very much nearer their aspirations than may otherwise have been thought. If nothing else it may encourage both parties to continue the quest where without such a review they might have given up.

5.4.21 Never forget that at the end of the day an Adjudicator, Arbitrator or the Courts may have to review the facts and if you genuinely believe in the strength of your case, as you should, there is no reason why others should feel differently. Bear in mind that the views being so vociferously put forward by the other side might, in any event, be flying in the face of received advice. Remember the text of the Vicar's sermon which was annotated *"Argument weak here, so shout!"*

5.4.22 If the negotiation fails then before resorting to the Adjudicator, Arbitrator or the Courts consider using one of the methods of Alternative Dispute Resolution procedures that have come into being in recent years, e.g. mediation.

5.5 CONCLUSION

5.5.1 The word "claim" is an emotive one in the construction industry for a variety of reasons and much criticism is levelled at Forms of Contract (not least the JCT Forms) that recognise claims – even if such forms do not mention them by name. At one end of the scale are those Employers or design teams who feel that, once appointed, the Contractor is their unpaid servant to do their bidding at no extra cost. At the other end are those Contractors who act as if buildings are built out of mountains of paper and apply their utmost energy in imagining or inventing some claim potential out of every sigh of the design team.

5.5.2 Such views have done a great disservice to the industry. Increasingly they tempt Employers to seek alternative routes to lump sum Contracts, particularly those routes which offer a tempting alternative to competition. In a competitive lump sum contract, the Contractor is more readily brought face to face with the choice of either seeking an efficient solution to the construction process or of muddying the clarity of his contractual obligations by purporting to have a justified claim at every turn of events. The desire to launch such claims is born out of a misconception, namely that claims have only to be registered to become an entitlement. They do not. Certainly it is open to a Contractor to pursue claims off the back of an inefficient design team. But to succeed, claims have to be rooted in fact and the amount payable will be a proper reflection of the loss and/or expense directly resulting therefrom. Inevitably, the amounts payable under many of the appealing alternative contractual arrangements have a margin built into them to mask the incidence of what otherwise would appear as a claim. These alternative arrangements are based on a short-sighted policy which not only contains less incentive to be efficient (on both sides) but also results in the Employer paying a premium for claims whether or not they occur. This encourages an inefficient industry and makes the decision to build progressively less viable; it is a short term palliative only.

5.5.3 It is hoped that this book will go some way towards taking the mystique out of claims, the settlement of which is, after all, simply the process of ensuring that the Employer pays only a fair price for interfering with the Contractor in the execution of the work; too often, claims are seen as the excuse for undermining the financial certainty of the project. If this book succeeds to any degree, it should help to restore the confidence of both sides of the industry in the lump sum Contract mainly sought in competition as representing the best value for money and the best environment for growing efficiency on all sides.

REFERENCE

[1] Department of the Environment and the Construction Industry (1994) Constructing the Team (The Latham Report), HMSO.

Part 3

FORMS OF CONTRACT OTHER THAN JCT 98 PWQ

6

JCT NOMINATED SUB-CONTRACT CONDITIONS 1998 EDITION

6.1 INTRODUCTION

General

6.1.1 Clause 35.1 of JCT 98 PWQ provides that where ... *the Architect has, whether by the use of a prime cost sum or by naming a sub-contractor, reserved to himself the final selection and approval of the sub-contractor to the Contractor ... the sub-contractor so named or to be selected and approved shall be nominated in accordance with the provisions of clause 35 and ... shall be a Nominated Sub-Contractor for all the purposes of this Contract ...*

6.1.2 Thus, this chapter deals with the form of sub-contract required to be used for Nominated Sub-Contracts let under JCT 98 PWQ. The current version of this form is the JCT Standard Form of Nominated Sub-Contract Conditions (NSC/C) 1998 Edition incorporating Amendments 1 and 2 hereinafter referred to as NSC/C. Notwithstanding the JCT's decision to update NSC/C to the 1998 version, it has proved to be a Sub-Contract which has been somewhat underutilised since the late 1980s when a trend began, still very much in evidence, towards amending Main Contracts in a manner which would entitle the Employer to have Sub-Contractors of his choice yet require Contractors to engage those Sub-Contractors on a domestic basis. The advantage of proceeding down the domestic Sub-Contract path, as perceived by clients and those advising them, is that the Contractor is directly responsible for the performance (both as to financial stability and progress of the Sub-Contract Works) of his domestic Sub-Contractor without any sharing of that risk as is provided for under NSC/C.

6.1.3 NSC/C is referred to in the Articles of Nominated Sub-Contract Agreement (Agreement NSC/A) Article 1.3 and is incorporated by that reference in the Nominated Sub-Contract. NSC/C, which must be used in conjunction with NSC/A, is compatible with JCT 98 PWQ dealt with in this book.

6.1.4 Since the Nominated Sub-Contractor is required to enter into a Sub-Contract with the Contractor (not the Employer), it is of some importance that the Sub-Contract is wholly compatible with the Main Contract. To this end, the rights and obligations contained in JCT 98 PWQ as between the Employer and Contractor are largely mirrored in NSC/C as between the Contractor and Sub-Contractor and this intention is further underpinned by clause 1.10.1 which provides that:

> ... *The Sub-Contractor shall ... observe, perform and comply with all the provisions of the Main Contract on the part of the Contractor to be observed, performed and complied with so far as they relate and apply to the Sub-Contract Works (or any portion of the same). Without prejudice to the generality of the foregoing, the Sub-Contractor shall observe, perform and comply with the following provisions of the Main Contract Conditions: clauses 6, 7, 9, 16, 32, 33 and 34 ...*

6.1.5 The review which follows will concentrate in large part upon an analysis of the differences which exist between the provisions of JCT 98 PWQ and those of NSC/C, and of the way in which such differences are essentially a reflection of the circumstances within which the two forms operate. In this respect, the essential difference between these two forms is that the latter document regulates contractual relationships as between the Contractor and the Sub-Contractor while the former does so as between the Contractor and the Employer. The manner in which this impacts upon the provisions contained within NSC/C, insofar as they relate to extensions of time and direct loss and/or expense, will form the subject matter of the discussion in Sections 6.2, 6.3 and 6.4. By way of an introduction, however, to the nature of the relationship between the Contractor and the Sub-Contractor the remainder of this section, 6.1, sets out a number of features of NSC/C which serve to regulate this relationship under the following heads:

- limitations on Contractor's liability;
- "set-off";
- Contractor and Sub-Contractor defaults.

Limitations on Contractor's liability

6.1.6 The Contractor's obligations to the Employer in respect of the Sub-Contractor's undertakings are limited in two key respects, thus:

(i) The design of, the selection of the kinds of materials and goods for, the satisfaction of any performance specification for and the provision of information drawings and details of the Sub-Contract Works – clause 35.21 of JCT 98 PWQ provides that:

The Contractor shall not be responsible to the Employer for:
.1 the design of any nominated Sub-Contract Works insofar as such Nominated Sub-Contract Works have been designed by a Nominated Sub-Contractor;
.2 the selection of the kinds of materials and goods for any nominated Sub-Contract Works insofar as such kinds of materials and goods have been selected by a Nominated Sub-Contractor;
.3 the satisfaction of any performance specification or requirement insofar as such performance specification or requirement is included or referred to in the description of any nominated Sub-Contract Works included in or annexed to the numbered tender documents enclosed with any NSC/T Part 1;
.4 the provision of any information required to be provided pursuant to Agreement NSC/W in reasonable time so that the Architect can comply with the provisions of clauses 5.4.1 and 5.4.2 in respect thereof.
Nothing in this clause 35.21 shall affect the obligations of the Contractor under this Contract in regard to the supply of workmanship, materials and goods by a Nominated Sub-Contractor.

The responsibilities of the Nominated Sub-Contractor to the Employer are contained in the 1998 version of the JCT Standard Form of Employer/ Nominated Sub-Contractor Agreement (referred to in this book as Agreement NSC/W). By virtue of entering into Agreement NSC/W the Sub-Contractor warrants under clause 2.1 thereof that he has exercised and will exercise all reasonable skill and care in:

.1 the design of the Sub-Contract Works insofar as the Sub-Contract Works have been or will be designed by the Sub-Contractor; and

221

.2 *the selection of the kinds of materials and goods for the Sub-Contract Works insofar as such kinds of materials and goods have been or will be selected by the Sub-Contractor; and*

.3 *the satisfaction of any performance specification or requirement insofar as such performance specification or requirement is included or referred to in the description of the Sub-Contract Works included in or annexed to the numbered tender documents enclosed with NSC/T Part 1 ...*

Nothing in clause 2.1 shall be construed so as to affect the obligations of the Sub-Contractor under the Sub-Contract entered into by the execution by the Main Contractor and the Sub-Contractor of Agreement NSC/A in regard to the supply under that Sub-Contract of workmanship, materials and goods ...

(ii) Extensions of time – clause 25.4.7 of JCT 98 PWQ provides the Contractor with grounds for an extension of time in the event that there is ... *delay on the part of Nominated Sub-Contractors ... which the Contractor has taken all practicable steps to avoid or reduce ...* The occurrence of such a Relevant Event would therefore permit the Contractor to finish the Works later than the Date for Completion and yet not be liable to the Employer in respect of liquidated and ascertained damages. In these circumstances the Employer's right of redress is provided for by way of clause 3.3.2 of Agreement NSC/W which reads as follows:

... The Sub-Contractor shall so perform the Sub-Contract that the Main Contractor will not become entitled to an extension of time for completion of the Main Contract Works by reason of the Relevant Event in clause 25.4.7 of the Main Contract Conditions ...

Even though the Contractor can gain relief from the imposition of liquidated and ascertained damages when there is a ... *delay on the part of Nominated Sub-Contractors ...* he is not entitled to recover from the Employer any direct loss and/or expense so arising since there is not a corresponding matter in clause 26 of JCT 98 PWQ. However, the Contractor does have means of recovering costs directly from the Sub-Contractor and this is addressed in the remainder of this section.

"Set-off"

6.1.7 In addition to the Sub-Contractor being required to comply with the Sub-Contract itself, obligations are introduced in relation to the Main Contract by virtue of clause 1.10.2 of NSC/C which requires the Sub-Contractor to:

... indemnify and save harmless the Contractor against and from:

.1 *any breach, non-observance or non-performance by the Sub-Contractor or his servants or agents or Sub-Sub-Contractors of any of the provisions of the Main Contract; and*

.2 *any act or omission of the Sub-Contractor or his servants or agents or Sub-Sub-Contractors which involves the Contractor in any liability to the Employer under the provisions of the Main Contract ...*

6.1.8 If by the act, default or omission of the Sub-Contractor there is incurred by the Contractor loss and/or expense then such loss and/or expense may be "set-off" from any monies due to the Sub-Contractor. Whilst the right of "set-off" may be implied as a common law right, NSC/C recognises the right at clause 4.16.1.2 wherein it provides:

> ... Not later than 5 days before the final date for payment ... the Contractor may give a written notice to the Sub-Contractor which shall specify any amount proposed to be withheld and/or deducted ... the ground or grounds for such withholding and/or deduction and the amount of the withholding and/or deduction attributable to each ground ...

6.1.9 This is a much simpler but nonetheless effective provision than that which pertained prior to the issue of Amendment 7 to NSC/C in April 1998.

6.1.10 If the Sub-Contractor disagrees the amount of "set-off" notified by the Contractor, he may contest it at Adjudication or Arbitration or in Legal Proceedings under Section 9 of NSC/C.

6.1.11 There will be occasions where the Contractor and Sub-Contractor cause delay and disruption to one another for reasons unconnected with the acts or omissions of the Employer or his agents – whilst the Employer will not be immediately concerned with such occurrences (in the sense that he will not have a financial liability to either party), the Sub-Contract provides for the rights of redress that the Contractor and Sub-Contractor have against one another.

6.1.12 The Sub-Contractor's rights against the Contractor are set out in clause 4.39 whilst those of the Contractor against the Sub-Contractor are set out in clause 4.40. As provided in clause 4.41, such rights as the Sub-Contractor and Contractor might have against one another are without prejudice to any other rights or remedies which they may possess. This means that the existence of remedies set down within the contract (such as those contained in clauses 4.39 and 4.40) does not preclude either party from pursuing a remedy outside the terms of the contract – e.g. an action for common law damages for breach of contract.

6.1.13 The text of clauses 4.39 to 4.41 reads as follows:

> 4.39   *If the regular progress of the Sub-Contract Works (including any part thereof which is Sub-Sub-Contracted) is materially affected by any act, omission or default of the Contractor (including where the Contractor is the Principal Contractor, any omission or default in the discharge of his obligations as the Principal Contractor), or any person for whom the Contractor is responsible (see clause 6.3.1), the Sub-Contractor shall within a reasonable time of such material effect becoming apparent give written notice thereof to the Contractor and the agreed amount of any direct loss and/or expense thereby caused to the Sub-Contractor shall be recoverable by the Sub-Contractor from the Contractor as a debt. Provided always that:*
>
>   *.1   the Sub-Contractor's application shall be made as soon as it has become, or should reasonably have become, apparent to him that the regular progress of the Sub-Contract Works or of any part thereof has been or was likely to be affected as aforesaid; and*

.2 *the Sub-Contractor, in order to enable the direct loss and/or expense to be ascertained, shall submit to the Contractor such information in support of his application including details of the loss and/or expense as the Contractor may reasonably require from the Sub-Contractor.*

4.40 *If the regular progress of the Works (including any part thereof which is sub-contracted) is materially affected by an act, omission or default of the Sub-Contractor or any person for whom the Sub-Contractor is responsible (see clause 6.3.1), the Contractor shall within a reasonable time of such material effect becoming apparent give written notice thereof to the Sub-Contractor and the agreed amount of any direct loss and/or expense thereby caused to the Contractor (whether suffered or incurred by the Contractor or by Sub-Contractors employed by the Contractor on the Works from whom claims under similar provisions in the relevant Sub-Contracts have been agreed by the Contractor, Sub-Contractor and the Sub-Contractor) may be deducted from any monies due or to become due to the Sub-Contractor or may be recoverable from the Sub-Contractor as a debt. Provided always that:*

.1 *the Contractor's application shall be made as soon as it has become, or should reasonably have become, apparent to him that the regular progress of the Works. (including any part thereof which is sub-contracted) has been or was likely to be affected as aforesaid; and*

.2 *the Contractor, in order to enable the direct loss and/or expense to be ascertained, shall submit to the Sub-Contractor such information in support of his application including details of the loss and/or expense as the Sub-Contractor may reasonably request from the Contractor.*

4.41 *The provisions of clauses 4.38 to 4.40 are without prejudice to any other rights or remedies which the Contractor or the Sub-Contractor may possess ...*

6.1.14 The content of the foregoing clauses may be summarised in the following way:

a) The party claiming to have suffered direct loss and/or expense is required both to notify the defaulting party to that effect and to provide the defaulting party with supporting details as to the sums claimed.

b) Unlike clause 26 of JCT 98 PWQ any amounts recovered are to be the subject of agreement between the two parties rather than ascertainment by a third party. Should it not prove possible to reach agreement then the matter must be referred to Adjudication, Arbitration, or Legal Proceedings.

c) Clauses 4.39 and 4.40 give essentially the same rights to either party in the event of default by the other. The exception is that, while the Sub-Contractor must recover as a debt from the Contractor any sums due to him by operation of clause 4.39, the Contractor may deduct any sums due to him by operation of clause 4.40 from monies otherwise due to the Sub-Contractor.

The operation of clauses 4.39 and 4.40 is discussed further at point 6.2.3 below.

6.2   EXTENSIONS OF TIME

General

6.2.1 The detailed provisions of JCT 98 PWQ which relate to the granting of extensions of time are dealt with in Chapter 2. As noted above, the section which follows deals primarily with those aspects in which the provisions of NSC/C differ from those of JCT 98 PWQ. It does so under the following subject headings:

- procedures;
- Relevant Events;
- fluctuations.

Procedures

6.2.2 Table 6.1 provides references to the text within this book dealing with JCT 98 PWQ which is also relevant to the following review of NSC/C.

Table 6.1. Extension of time procedures

Provision in Contract	Clause in NSC/C	Equivalent clause in JCT 98 PWQ	Explanatory text in this book
Requirement for the Contractor to give notice of delay and to give further details to assist the Architect in establishing any extension of time.	2.2.1	25.2.1	2.7.7
Requirement for the Contractor to give, in respect of each Relevant Event identified in the notice of delay, particulars of the expected effects thereof together with an estimate of expected delay and to update such information as and when necessary.	2.2.2	25.2.2 and 25.2.3	2.7.8 and 2.7.9
Requirement for the Architect, in certain circumstances, to grant extensions of time during the construction period or to notify the Contractor that an extension of time is not, in his view, justified.	2.3	25.3.1	2.7.11 to 2.7.31
Requirement for the Architect to review the Completion Date following Practical Completion.	2.5	25.3.3	2.7.32 to 2.7.35
The effect of omissions upon extensions of time.	2.4 and 2.5.2	25.3.2 and 25.3.3.2	2.7.36 to 2.7.41
General obligations imposed upon the Architect and Contractor in relation to their conduct.	2.5	25.3.4, 25.3.5 and 25.3.6	2.7.42 to 2.7.45

6.2.3 The procedures by which extensions of time are requested and granted are basically similar as between JCT 98 PWQ (clauses 25.2 and 25.3 apply) and NSC/C (clauses 2.2 to 2.5 apply), albeit there are a number of differences which are a product of the rather different circumstances under which the NSC/C Conditions operate.

6.2.4 As regards the giving of written notice of any anticipated delay, clause 2.2.1 of NSC/C states that such notice must be given by the Sub-Contractor to the Contractor, who must then inform the Architect of such anticipated delays and must also pass on to the Architect any supporting details supplied by the Sub-Contractor. It is incumbent upon the Sub-Contractor to ensure that such supporting details are kept up to date (clause 2.2.2.3 refers), and any such further information as the Sub-Contractor provides in this respect must be forwarded timeously by the Contractor to the Architect (clause 2.2.3 refers).

6.2.5 The principal reason for maintaining the Architect's involvement in the process is that, whilst any extension to the Sub-Contract period will be granted by the Contractor, the Contractor can only grant such extension of time as has first been approved in writing by the Architect. In this respect both the Contractor and the Sub-Contractor must jointly request the Architect's consent to an extension of the Sub-Contract period. It is then for the Architect to form an opinion as to a fair and reasonable extension, if any, to the Sub-Contract period, which opinion will be binding upon both the Contractor and the Sub-Contractor. These matters are dealt with in clause 2.3 of NSC/C.

6.2.6 As regards the written notice required of the Sub-Contractor, NSC/C differs slightly from JCT 98 PWQ in that it requires such notice to be provided in the event that delay is anticipated not only to the progress of the Works but also to either its commencement or completion (clause 2.2.1 of NSC/C refers). Whilst this may on first perusal appear to be a matter of semantics it in practice reflects the fact that, as a result of the interaction of the operations of the Sub-Contractor with those of the Contractor or any other sub-contractors, there may be events which impact upon the eventual progress of the Sub-Contractor's work but which predate the Sub-Contractor's involvement on site.

6.2.7 A further difference between the two forms relates to the requirement in both to identify the cause or causes of any delay. Clause 25.2.1.1 of JCT 98 PWQ requires the Contractor to identify in its notice to the Architect such of the alleged causes of delay as in the Contractor's opinion are Relevant Events, these being the only causes of delay for which the Architect is able to grant an extension of time. Clauses 2.2.1 and 2.3.1 of NSC/C reproduce this requirement but add to it the further requirement that such of the alleged causes of delay as may not be Relevant Events but are instead the fault of the Contractor are to be identified as such (i.e. separately from any Relevant Events) by the Sub-Contractor in his notice. The reason for this is that, in addition to Relevant Events, the Sub-Contractor is also entitled to an extension of the sub-contract period in respect of any delays caused by events in respect of which the Contractor is at fault.

6.2.8 In the normal course of events, should the Architect consent to an extension of the sub-contract period this would, in the event that the main contract works were as a result also delayed, give rise to a like extension to the main contract period by virtue of the analogous Relevant Event under JCT 98 PWQ. However, the ability of the Architect to consent to extensions of the sub-contract period to cover delays which are caused by some default of the Contractor gives rise to a situation whereby the Sub-

Contractor may be granted an extension to the sub-contract period but the Contractor is not granted a corresponding extension to the main contract period. Should the Sub-Contractor require an extension to the sub-contract period in these circumstances he will therefore initially be faced with the problem of requesting the Contractor to join with him in seeking the Architect's consent to a measure which may ultimately involve the Contractor in the payment of liquidated and ascertained damages to the Employer. Interestingly this does not apply to claims for reimbursement of direct loss and/or expense whereby, if the Sub-Contractor submits an application to the Contractor, the Contractor is bound to request that the Architect investigate the matter without first being requested to join with the Sub-Contractor in so doing (clause 4.38.1 of NSC/C applies).

6.2.9 With the exception of the above, the procedural rules in respect of the awarding of extensions of time under NSC/C are more or less the same as those under JCT 98 PWQ, with the proviso that those matters which relate to the timing of the actions of the various parties are set out by reference to the sub-contract rather than the main contract period.

Relevant Events

6.2.10 Table 6.2 provides references to the text within this book dealing with JCT 98 PWQ which is also relevant to the following review of NSC/C.

Table 6.2. Relevant Events

Relevant Event	Clause in NSC/C	Equivalent clause in JCT 98 PWQ	Explanatory text in this book
Force majeure	2.6.1	25.4.1	2.8.2 to 2.8.3
Exceptionally adverse weather conditions	2.6.2	25.4.2	2.8.4
Loss or damage due to Specified Perils	2.6.3	25.4.3	2.8.5
Civil commotion	2.6.4	25.4.4	2.8.6 and 2.8.7
Compliance with Architect's instructions	2.6.5.1	25.4.5.1	2.8.8 to 2.8.27
Opening up or testing	2.6.5.2	25.4.5.2	2.8.28
Late information	2.6.6	25.4.6	2.8.29 to 2.8.44
Nominated Sub-Contractors/ Suppliers	2.6.7	25.4.7	2.8.45 to 2.8.50
Works by others	2.6.8.1	25.4.8.1	2.8.51 to 2.8.54
Supply of materials by Employer	2.6.8.2	25.4.8.2	2.8.55
Statutory power restrictions	2.6.9	25.4.9	2.8.56 to 2.8.57
Inability to secure labour	2.6.10.1	25.4.10.1	2.8.58 to 2.8.60
Inability to secure materials	2.6.10.2	25.4.10.2	2.8.58 to 2.8.60
Work by local authority/statutory undertaker	2.6.11	25.4.11	2.8.61

(continued)

Table 6.2 (cont.)

Relevant Event	Clause in NSC/C	Equivalent clause in JCT 98 PWQ	Explanatory text in this book
Site access/egress restrictions	2.6.12	25.4.12	2.8.62 and 2.8.63
Suspension by the Contractor	2.6.13.1	25.4.18	2.8.71 to 2.8.75
Suspension by the Sub-Contractor	2.6.13.2	–	–
Deferment of date of possession	2.6.14	25.4.13	2.8.64
Approximate quantities	2.6.15	25.4.14	2.8.65 to 2.8.67
Terrorism	2.6.16	25.4.16	2.8.69
CDM Regulations	2.6.17	25.4.17	2.8.70

6.2.11 As with procedural matters, the Relevant Events by virtue of which extensions of time may be granted are much the same across both JCT 98 PWQ (clause 25.4 refers) and NSC/C (clause 2.6 refers). The differences which are worthy of note can be summarised as follows:

1. Compliance with Architect's instructions
    This qualifies as a Relevant Event (clause 2.6.5 of NSC/C refers) whether compliance with such instructions is required of the Contractor only or of the Sub-Contractor only or of both. It is of course feasible that compliance by the Contractor with Architect's Instructions which are not necessarily directly connected with the Sub-Contract Works can nevertheless affect the programming of them. Hence this is recognised as a Relevant Event by NSC/C.
2. Performance Specified Work
    Clause 2.6.5.1 of NSC/C lists (as does clause 25.4.5.1 of JCT 98 PWQ) the particular matters in connection with which Architect's Instructions may be issued and which give rise to an entitlement to an extension of time. Unlike JCT 98 PWQ, there is no reference in NSC/C in this respect to instructions issued in connection with Performance Specified Work. Performance Specified Work is itself dealt with in clause 42 of JCT 98 PWQ, clause 42.18 stating the following:

    ... Performance Specified Work pursuant to clause 42 shall not be provided by a Nominated Sub-Contractor under a Nominated Sub-Contract ...

    As Performance Specified Work does not exist within the confines of NSC/C it is accordingly not mentioned as a matter in relation to which grounds for an extension of time can arise. This is also the case in relation to clause 25.4.15 of JCT 98 PWQ, which deals with the effect on Performance Specified Work of changes in statutory requirements.
3. Opening up for testing
    Clause 2.6.5.2 of NSC/C refers (as does clause 25.4.5.2 of JCT 98 PWQ) to instructions issued in respect of the opening up for inspection of work previously covered up or the testing of work, materials or goods. JCT 98 PWQ states in this respect that if such opening up and inspection reveals work which has been constructed in accordance with the Contract then the Contractor is entitled to an

extension of time in respect of any delay thereby caused to the progress of the Works. The principle behind this clause is that the Contractor should not be penalised should it prove necessary to delay the progress of the Works as a result of the Architect's requirement to inspect work which is subsequently shown to have been installed in accordance with the Contract. Similarly, clause 2.6.5.2 of NSC/C permits an extension to the Sub-Contract period in similar circumstances relating to the sub-contract works. Whilst this clause of NSC/C recognises as a Relevant Event any delay to the Sub-Contractor's work as a result of inspections carried out in respect of the Contractor's work which inspections reveal that the work was in accordance with the contract it does not permit an extension to the sub-contract period in circumstances in which the Sub-Contractor's work is delayed as a result of inspections carried out in respect of the Contractor's work, which inspections reveal work which has not been constructed by the Contractor in accordance with the Contract. In this eventuality it would be necessary for the Sub-Contractor to apply for an extension to the Sub-Contract period on the grounds, not of a Relevant Event, but as an act, omission or default by the Contractor. Should this be consented to by the Architect then the Sub-Contractor would further have grounds for recovery of direct loss and/or expense from the Contractor by reference to clause 4.39 of NSC/C (see para. 6.4.2).

4. Late information

    If the Architect fails to provide instructions or information in due time which are necessary for maintaining the progress of the Sub-Contract Works then the Sub-Contractor will become entitled to an extension of time. In the event that the Architect forwards such information to the Contractor but the Contractor does not forward the same in due course to the Sub-Contractor then the Sub-Contractor will remain entitled to an extension of the Sub-Contract period regardless of the fact that the Contractor will not be entitled to an extension of the Main Contract period under the terms of JCT 98 PWQ. Such an extension to the Sub-Contract period would be granted not by reference to the Relevant Events set out in clause 2.6.6 of NSC/C but as a result of a default by the Contractor as described in clause 2.3.1 thereof. This distinction becomes important in the context of a related claim in respect of direct loss and/or expense (see para. 6.4.2) in that such a claim would be dealt with under clause 4.39 of NSC/C, whereby the amount of loss and/or expense is a matter for agreement between the Contractor and the Sub-Contractor rather than under clause 26.4 of JCT 98 PWQ (via clause 4.38.1 of NSC/C) whereby the amount due would be ascertained by the Architect or the Quantity Surveyor under the Main Contract.

5. Inability to secure labour and materials

    The fact that problems affecting the progress of the Main Contract Works can have consequential effects upon the progress of the Sub-Contract Works is again recognised in clause 2.6.10 of NSC/C. This clause permits an extension to the Sub-Contract period by reason not only of the Sub-Contractor being unable to obtain the necessary labour and/or materials to progress the Works but also in respect of the Contractor being similarly unable in relation to the Main Contract Works. As with compliance with Architect's Instructions (discussed in point 1 above), it is feasible that the inability on the part of the Contractor to obtain resources which are not necessarily directly connected with the Sub-Contract Works could nevertheless affect the programming of them.

6. Suspension by the Contractor
   If the Contractor suspends the performance of his obligations pursuant to clause 30.1.4 of JCT 98 PWQ (see paras 2.8.71 to 2.8.75) it would inevitably delay the Sub-Contract Works. Although this eventuality is recognised as a Relevant Event under clause 2.6.13.1 of NSC/C it is not a matter giving rise to recovery of any direct loss and/or expense under clause 4.38 of NSC/C. Thus any such loss and/or expense suffered by the Sub-Contractor would have to be pursued under clause 4.39 of NSC/C (see also point 4 above).

7. Suspension by the Sub-Contractor
   Clause 2.6.13.2 of NSC/C has no JCT 98 PWQ equivalent since it is dealing with the situation in which the Sub-Contractor is entitled to suspend the Sub-Contract Works as a result of the improper withholding of payment from the Sub-Contractor by the Contractor. In such situations, JCT 98 PWQ does not, unsurprisingly, entitle the Contractor to an extension of the Main Contract period. The above referred to clause of NSC/C does, however, allow the Architect to consent to an extension of the Sub-Contract period in such circumstances. Any direct loss and/or expense suffered by the Sub-Contractor as a result of such suspension is recoverable from the Contractor under clause 4.21.3 of NSC/C. As suspension by the Sub-Contractor is not a matter to which clause 4.38 of NSC/C applies it is suggested that the amount of any direct loss and/or expense has to be agreed between the Contractor and the Sub-Contractor as opposed to being ascertained by the Architect or the Quantity Surveyor under the Main Contract.

Fluctuations

6.2.12 Clauses 4A, 4B and 4C of NSC/C (all of which are optional) set out the mechanisms by which the amount due to the Sub-Contractor is adjusted in the event of fluctuations in cost occurring during the sub-contract period. By virtue of clauses 4A.4.7, 4B.5.7 and 4C.7.1 of NSC/C the adjustment to the Sub-Contractor's entitlement does not encompass increased costs arising during a period of culpable delay on the part of the Sub-Contractor. It is worthy of note in this respect that this only continues to be the case should the printed text of clauses 2.2 to 2.7 of NSC/C (which incorporates the provisions relating to extensions of time) remain unamended and provided that the Architect has, in respect of every request by the Contractor and Sub-Contractor under clause 2.2 of NSC/C consented or not consented to a revision of the period for completion of the Sub-Contract Works. Any amendment whatsoever to this printed text or any failure of the Architect properly to respond to the request of the Contractor and Sub-Contractor, where either clause 4A, 4B or 4C of NSC/C apply, will result in the Sub-Contractor becoming entitled to increased costs which arise even during a period of culpable delay on the part of the Sub-Contractor.

6.2.13 Clauses 4A.4.7, 4B.5.7 and 4C.7.1 of NSC/C reflect clauses 38.4.7, 39.5.7 and 40.7.1 of JCT 98 PWQ respectively and the reader is referred to sections 2.12 and 4.7 in this connection.

6.3 VARIATIONS AND DISRUPTION

General

6.3.1 The detailed provisions of JCT 98 PWQ which relate to the valuation of variations, and the means by which the Contractor may recover additional costs arising from disruption, are dealt with in Chapter 3. The section which follows deals only with those aspects in which the provisions of NSC/C differ from those of JCT 98 PWQ, and does so under the following subject headings:

- Variations;
- Procedures for the ascertainment of direct loss and/or expense;
- List of matters.

Variations

6.3.2 Table 6.3 provides references to the text within this book dealing with JCT 98 PWQ which are also relevant to the following review of NSC/C.

Table 6.3. Valuation of variations

Valuation rule	*Clause in NSC/C*	*Equivalent clause in JCT 98 PWQ*	*Explanatory text in this book*
Valuation of variations – Alternative A Within 21 days (17 days for Sub-Contractor) of receipt of instruction/commencement of work for which an Approximate Quantity is given the Contractor **may** submit a Price Statement for such work.	4.4.2A1	13.4.1.2A1	3.3.4
Contractor/Sub-Contractor **may** separately attach to the Price Statement: • any amount required in lieu of direct loss and/or expense; • any adjustment required to the Completion Date.	4.4.2A1	13.4.1.2A1	3.3.5
Acceptance of Price Statement (in whole or in part) by Quantity Surveyor within 21 days of receipt from Contractor (within 24 days of receipt in the case of Sub-Contract).	4.4.2A2	13.4.1.2A2	3.3.3
Non-acceptance of Price Statement (in whole or in part) – the Quantity Surveyor to supply amended Price Statement for Contractor/Sub-Contractor to accept or reject within 14 days.	4.4.2A4	13.4.1.2A4	3.3.3
When the Price Statement or amended Price Statement are neither accepted nor referred to Adjudication valuation of the work shall be by Alternative B.	4.4.2A6	13.4.1.2A6	3.3.2

(continued)

231

Table 6.3 (cont.)

Valuation rule	Clause in NSC/C	Equivalent clause in JCT 98 PWQ	Explanatory text in this book
Where attachments to the Price Statement are not accepted direct loss and/or expense and revisions to the Completion Date are to be dealt with as provided for elsewhere in the Contract.	4.4.2A7	13.4.1.2A7	3.3.6
Valuation of variations – Alternative B Valuation determined by the rates and prices in the Contract Bills/priced document where the additional or substituted work is of similar character, is executed under similar conditions and does not significantly change the quantity of work.	4.6.1.1	13.5.1.1	3.3.26
Fair allowance on the above in the event that work of a similar character is not executed under similar conditions and/or there is a significant change in the quantity of the work.	4.6.1.2	13.5.1.2	3.3.29 3.3.31 3.3.34 3.3.43
Valuation at fair rates or prices in the event that the additional or substituted work is not of similar character to that in the Contract Bills/priced documents.	4.6.1.3	13.5.1.3	3.3.29 3.3.31 3.3.43
Valuation determined by the rate or price entered against an approximate quantity where such quantity is a reasonably accurate forecast of the work required.	4.6.1.4	13.5.1.4	3.3.34
Fair allowance on the above in the event that the approximate quantity is not a reasonably accurate forecast of the work required.	4.6.1.5	13.5.1.5	3.3.29
Valuation of omitted works determined by the rates or prices in the Contract Bills/priced document.	4.6.2	13.5.2	–
In valuing in accordance with the above: – measurement is to be in accordance with the same convention as is applicable to the preparation of the Contract Bills; – allowance is to be made for percentage/lump sum adjustments in the Contract Bills; – allowance is to be made for any addition to or reduction of preliminary items.	4.6.3	13.5.3	3.3.23
Valuation by reference to dayworks where the additional or substituted work cannot properly be valued by measurement.	4.6.4	13.5.4	3.3.35 to 3.3.42
Valuation of other works as if they were variations in the event that a variation substantially changes the conditions under which such other works are executed.	4.7	13.5.5	3.3.43

Valuation rule	Clause in NSC/C	Equivalent clause in JCT 98 PWQ	Explanatory text in this book
Fair valuation to be made of work which cannot be valued in accordance with the above rules.	4.6.5	13.5.7	3.3.29 3.3.34
Right of Contractor/Sub-Contractor to be present where it is necessary to measure work for the purpose of Valuation.	4.8	13.6	3.3.23
Submission of Quotations for Variations before Work Instructed Instruction to provide sufficient information for Contractor/Sub-Contractor to provide quotation.	3.3A.1.1	13A.1.1	3.4.2
Quotation to be submitted by Contractor within 21 days (by the Sub-Contractor in 17 days) and thereafter remain open for acceptance for 7 days (14 days in the case of the Sub-Contract).	3.3A.1.2	13A.1.2	3.4.3 and 3.4.4
Variation only to be carried out after acceptance of quotation.	3.3A.1.3	13A.1.3	3.4.7
Quotation to separately comprise:	3.3A.2	13A.2	3.4.5 and 3.4.6
– cost of carrying out variation; – any adjustment to the Completion Date; – any amount in lieu of direct loss and/or expense; – cost of preparing quotation; – method statement (if requested).			
Acceptance of quotation by Architect.	3.3A.3	13A.3	3.4.7
Non-acceptance of quotation – Variation either to be valued in accordance with normal rules or not to be carried out.	3.3A.4	13A.4	3.4.8

6.3.3 Instructions giving rise to variations are dealt with in clause 3.3 of NSC/C. In this respect, the requirements of NSC/C are much the same as those set out in JCT 98 PWQ except that the Sub-Contractor is required to comply not only with instructions issued by the Architect but also with any reasonable directions of the Contractor. Clause 3.3A of NSC/C, furthermore, deals with variations which are the subject of a quotation in compliance with the instruction from the Sub-Contractor in much the same way that clause 13A of JCT 98 PWQ deals with those which are the subject of a quotation in compliance with the instruction from the Contractor. In the former situation, however, the quotation must be submitted by the Sub-Contractor not to the Quantity Surveyor under the Main Contract but, instead, to the Contractor. It is for the Contractor to accept the quotation, albeit the Quantity Surveyor under the Main Contract must be kept informed of the same and may have views as to its acceptability which would inevitably colour those of the Contractor.

6.3.4 Definitions of that which constitutes a Variation – clause 1.4 of NSC/C and clause 13.1 of JCT 98 PWQ, the manner in which variations are to be valued – clause 4.4 in

NSC/C and clause 13.4.1 in JCT 98 PWQ and the Valuation Rules – clause 4.6 in NSC/C and clause 13.5 in JCT 98 PWQ are essentially the same in the two forms of contract. Hence the detailed consideration given to these JCT 98 PWQ clauses in sections 3.2 and 3.3 applies equally to their NSC/C counterparts and consequently there is no need for further comment here except to note that whilst it is correct under NSC/C (as under JCT 98 PWQ) to include additional preliminary items in the valuation of a variation, costs which arise as a result of disruption to the progress of the Sub-Contract Works must be dealt with by reference to clauses 4.38 to 4.41 of NSC/C (direct loss and/or expense). The exceptions to this under NSC/C (as under JCT 98 PWQ) are the acceptance of an attachment to a Price Statement under clause 4.4.2A7 of NSC/C or of a 3.3A Quotation which include amounts to be paid in lieu of direct loss and/or expense.

Procedures for the ascertainment of direct loss and/or expense

6.3.5 Table 6.4 provides references to the text within this book dealing with JCT 98 PWQ which is also relevant to the following review of NSC/C.

Table 6.4. Direct loss and/or expense procedures

Direct loss and/or expense procedures	*Clause in NSC/C*	*Equivalent clause in JCT 98 PWQ*	*Explanatory text in this book*
Contractor/Sub-Contractor to make a written application regarding direct loss and/or expense.			3.5.9
Architect to form an opinion as to the Contractor's (Sub-Contractor) entitlement.	4.38.1 (plus, by reference, clause 26.4 of JCT 98 PWQ)	26.1	3.5.16 and 3.5.17
Architect (or Quantity Surveyor) to ascertain the amount of the Contractor's/Sub-Contractor's entitlement.			3.5.22 to 3.5.33
Contractor's/Sub-Contractor's written application must be made timeously.	4.38.1.1	26.1.1	3.5.10 3.5.11 3.5.20
Contractor/Sub-Contractor is to submit such information as will enable the Architect to form an opinion as to entitlement and the Architect or Quantity Surveyor to ascertain any amount due.	4.38.1.2 4.38.1.3	26.1.2 26.1.3	3.5.13 and 3.5.14
Amounts ascertained as being due are to be added to the Contract Sum.	4.38.3	26.5	3.5.18
The above provisions are without prejudice to any other rights or remedies which the Contractor/Sub-Contractor may possess.	4.4.1	26.6	3.5.9 3.5.21

6.3.6 Except insofar as it is included in an accepted 13A Quotation or an accepted attachment to a Price Statement such direct loss and/or expense as the Contractor considers himself entitled to recover in respect of the Main Contract Works must be ascertained either by the Architect or by the Quantity Surveyor in accordance with clause 26.1 of JCT 98 PWQ. Should the Sub-Contractor consider himself similarly entitled to recover direct loss and/or expense in respect of disruption to the Sub-Contract Works, and make written application to the Contractor to this effect, clause 4.38.1 of NSC/C requires the Architect to put into effect clause 26.4 of JCT 98 PWQ. This latter clause requires the Architect to form an opinion as to whether the Sub-Contractor has incurred direct loss and/or expense in relation to the Sub-Contract Works and, if so satisfied, to ascertain (or to require the Quantity Surveyor to ascertain) the amount of such loss and/or expense. Clause 26.4.1 of JCT 98 PWQ is analogous in form and content to clause 26.1 of that form, the effect of this being to subject claims for direct loss and/or expense made on behalf of the Sub-Contractor to the same process of ascertainment by the Architect or Quantity Surveyor as claims made by the Contractor in respect of the Main Contract Works. Similarly, in the same way that the Contractor under clause 26.1.3 of JCT 98 PWQ shall submit to the Architect or Quantity Surveyor such details as they may reasonably require in order to carry out such ascertainment so the Sub-Contractor under clause 4.38.4 of NSC/C is required to comply with any reasonable directions of the Contractor in this respect.

6.3.7 There are, however, three instances in which the above does not hold good, these being covered by clauses 4.21.3, 4.39 and 4.40 of NSC/C. These three clauses cover the occasions when either the Contractor or the Sub-Contractor incur direct loss and/or expense as a result of some default by the other (rather than by reason of the matters listed either in clause 26.2 of JCT 98 PWQ or in clause 4.38.2 of NSC/C). In such circumstances the settlement of the amount due is a matter for agreement between the Contractor and the Sub-Contractor and not for ascertainment by the Architect or Quantity Surveyor. It is worthy of note in respect of the Sub-Contractor's entitlement under clauses 4.21.3 and 4.39 of NSC/C that, the Sub-Contractor will be unable to recover any sums, even on an interim basis, in the absence of a final agreement with the Contractor. Should such an agreement not be reached then the Sub-Contractor would have to refer the matter to Adjudication, Arbitration or Legal Proceedings.

List of matters

6.3.8 Table 6.5 provides references to text within this book dealing with JCT 98 PWQ which is also relevant to the following review of NSC/C.

6.3.9 The list of matters which give rise to an entitlement on the part of the Sub-Contractor to recover direct loss and/or expense and which are set out in clause 4.38.2 of NSC/C are, without exception, analogous to those which similarly entitle the Contractor in respect of the Main Contract and which are set out in clause 26.2 of JCT 98 PWQ. Hence the detailed consideration given to JCT 98 PWQ clause 26.2 matters in Sections 2.8 and 3.5 applies equally to their NSC/C counterparts and consequently there is no need to comment further here.

Table 6.5. List of matters

Cause	NCS/C clauses		JCT 98 PWQ clauses		References to text in this book
	Relevant event	Matter	Relevant event	Matter	Paras
Late information	2.6.6.1 2.6.6.2	4.38.2.1.1 4.38.2.1.2	25.4.6.1 25.4.6.2	26.2.1.1 26.2.1.2	2.8.29–32 2.8.33–44
Inspection or testing	2.6.5.2	4.38.2.2	25.4.5.2	26.2.2	2.8.28 3.5.8
Discrepancies	2.6.5.1	4.38.2.3	25.4.5.1	26.2.3	2.8.15–17
Work by others	2.6.8.1	4.38.2.4	25.4.8.1	26.2.4.1	2.8.51–54
Supply of materials by Employer	2.6.8.2	4.38.2.4	25.4.8.2	26.2.4.2	2.8.55
Postponement	2.6.5.1	4.38.2.5	25.4.5.1	26.2.5	2.8.24
Failure to give ingress to or egress from the Works	2.6.12	4.38.2.6	25.4.12	26.2.6	2.8.62–63
Variations	2.6.5.1	4.38.2.7	25.4.5.1	26.2.7	2.8.18–22
Approximate quantities	2.6.15	4.38.2.8	25.4.14	26.2.8	2.8.65–67
Compliance with CDM Regulations	2.6.17	4.38.2.9	25.4.17	26.2.9	2.8.70

6.4 ASCERTAINING THE DIRECT LOSS AND/OR EXPENSE

6.4.1 The matters discussed in Chapter 4, as regards the heads under which claims for direct loss and/or expense may be submitted and the principles which must be brought to bear in assessing the Contractor's entitlement, are equally applicable to loss and/or expense arising under clause 4.38 of NSC/C as under clause 26 of JCT 98 PWQ. They will therefore not be the subject of further discussion here.

6.4.2 Any entitlement accruing either to the Sub-Contractor under clauses 4.21.3 and 4.39 of NSC/C or to the Contractor under clause 4.40 thereof is not, however, subject to the requirement of ascertainment. As stated above, such sums are to be the subject of agreement between the two parties concerned. Notwithstanding this it is suggested that, in the context of claims relating to clauses 4.21.3, 4.39 and 4.40 of NSC/C, the most equitable course as between the two parties is nevertheless to apply those principles which apply to ascertainment of direct loss and/or expense under clause 4.38 and which are discussed in detail in Chapter 4.

7

JCT AGREEMENT FOR MINOR BUILDING WORKS 1998 EDITION

7.1 INTRODUCTION

7.1.1 This Chapter deals with the JCT Agreement for Minor Building Works, 1998 Edition incorporating Amendments MW1 and MW2 hereinafter referred to as JCT 98 MW. As in section 2, the term "Architect" has been used throughout the text of this chapter and such references are deemed to include references to "Architect/Contract Administrator".

7.1.2 Practice Note M2, issued by the JCT in 1981 and revised and incorporated in JCT 98 MW, provides guidance on the use of JCT 98 MW. It is recommended that the form be used where the following conditions apply:

- the value of the contract does not exceed £70,000 (at 1992 prices);
- the works are to be the subject of an agreed lump sum;
- an Architect (or Contract Administrator) has been appointed to represent the Employer;
- a price has been obtained based on drawings and/or a specification and/or schedules but without either detailed measurements or bills of quantities having been prepared;
- the period required for the work is likely to be such as not to require full labour and materials fluctuations;
- Employer selection of sub-contractors is not required.

7.1.3 As with all recommendations issued by the JCT as to the use of its various standard forms, the proposed maximum contract value of £70,000 ought not to be (and, in practice, is not) taken too literally. By comparison with JCT 98 PWQ, many clauses do not appear and others have been reduced to no more than the basic requirements. The detailed legal provisions and, particularly, administrative procedures present in JCT 98 PWQ do not occur in JCT 98 MW. Much of the "gap filling" which is necessary as a result must be done on the basis of agreement between the Employer (or his Architect) and the Contractor; this will require a more flexible and consensual approach than is envisaged by the detailed provisions of JCT 98 PWQ. Inasmuch that the stresses imposed on the parties by the need to complete complex schemes in accordance with critical programme requirements militates against such a consensual approach, the key requirement in any scheme which is to be the subject of JCT 98 MW is simplicity.

7.1.4 The Contracts in Use Survey, prepared for the RICS in November 1996 by Davis Langdon & Everest, indicated that the 1980 Edition of JCT 98 MW was used on some 23% of all contracts by number (albeit, because of the nature of the form, only just over 2% of all contracts by value). The fact that the JCT Agreement for Minor Building Works is so regularly employed, and yet has to date given rise to a comparatively small body of case law, would appear to demonstrate that the flexible and consensual approach recommended above is in evidence in practice. Another factor, of course, is that relatively small sums of money are involved which do not generally merit the use of litigation as a means of resolving disputes.

7.2 EXTENSIONS OF TIME

Procedures

7.2.1 The procedures by which extensions of time are granted are set out in clause 2.2 of JCT 98 MW and, in terms of administrative detail, are in accordance with the minimalist philosophy apparent throughout the form.

7.2.2 Clause 2.2 of JCT 98 MW comes into effect upon it becoming apparent, first, that the Works will not be completed by the most recently fixed date for completion and, secondly, that this is for reasons beyond the control of the Contractor. In such circumstances, and only in such circumstances, the Contractor is required to submit a written notice to the Architect. This differs from clause 25.2.1.1 of JCT 98 PWQ in the following two respects:

(i) JCT 98 PWQ requires written notice to be given upon it becoming reasonably apparent ... *that the progress of the Works is being or is likely to be delayed* ... regardless of the likely effect, if any, on the completion date. In the case of complex schemes of the type for which JCT 98 PWQ is intended, it is not uncommon for the Contractor to experience delays to non-critical activities which will not, in the final analysis, be translated into corresponding delays to the overall completion date. In such circumstances, the Contractor is required nevertheless to provide written notification to the Architect. Clause 2.2 of JCT 98 MW, however, requires such notice to be given only when it becomes apparent that the Works will not be completed by the completion date. The logic behind such a requirement is that, on simple schemes of the type for which JCT 98 MW is intended, it ought to be reasonably apparent as to whether a delay to progress will impact upon the completion date.

(ii) JCT 98 PWQ also requires notice to be given by the Contractor regardless of whether the cause of the delay is one which gives rise, prima facie, to an entitlement to an extension of time. The Contractor is required to identify in his written notice the cause or causes of the alleged delay including, but not restricted to, those which qualify as Relevant Events. Clause 2.2 of JCT 98 MW, by contrast, requires no more than a simple notice that the completion date will not to be achieved for reasons (which the contract does not require be specified) beyond the Contractor's control. Should the anticipated failure to meet the completion date be the result of matters within the Contractor's control, no notification is required, such matters include ... *any default of the Contractor or of others employed or engaged by or under him for or in connection with the Works or of any supplier of goods or materials for the Works* ...

7.2.3 Unlike clause 25.2.2 of JCT 98 PWQ, clause 2.2 of JCT 98 MW does not require the Contractor to support his written notice either with particulars of the expected effects of the Relevant Events identified or with an estimate of the extent to which the completion date is likely to be delayed. Once the Contractor has submitted the (largely unspecific) notice required under clause 2.2 of JCT 98 MW, it is for the Architect to then arrive at what he considers to be a reasonable extension of time.

7.2.4 The requirement of JCT 98 PWQ that the Contractor provide information in support of his written notice is aimed at ensuring that the Architect is sufficiently apprised of the relevant details when forming a view as to a fair and reasonable extension of time. It follows that the Architect ought to be similarly apprised when forming the same view under JCT 98 MW. Accordingly, even though the agreement does not expressly require supporting information from the Contractor, it would be advisable for the parties to agree that reasonable requests made by the Architect in this respect will be complied with by the Contractor (it is, after all, in the latter's interest that the Architect be able to form a reasoned opinion). In a similar vein, clause 2.2 of JCT 98 MW expressly requires the Architect to provide the Contractor with written notification only when he does in fact grant an extension of time. Should the Architect decide that no extension of time is warranted, it would appear from the wording of this clause that he is under no duty to communicate this fact to the Contractor. However, the wording of clause A4.4.4.2 of JCT 98 MW which states ... *the Architect/the Contract Administrator has, in respect of every written notification by the Contractor under clause 2 of this Agreement, fixed such completion date as he considered to be in accordance with that clause* ... suggests otherwise. It is therefore recommended that the Architect should in practice always respond to the Contractor's notices, whatever the outcome of his deliberations, and in doing so either fix a new completion date or confirm the existing one. This will leave the Contractor in no doubt as to the current completion date.

7.2.5 Unlike JCT 98 PWQ, JCT 98 MW sets down no express time limit within which the Architect is required to reach a decision concerning the granting of an extension of time. This being the case, it is suggested that it is an implied term of the agreement that the Architect act within a reasonable time frame. Such a period ought generally to be capable of being agreed by the Architect and the Contractor.

7.2.6 Reference is made in paras 9.2.5 to 9.2.8 to clause 2.3 of JCT 98 IFC, which gives the Architect an express power to grant extensions of time in respect of what are known as "acts of prevention" (i.e. those events which may delay the Contractor and which are within the control of the Employer or his agents) which occur during a pre-existing period of delay for which the Contractor is culpable. There is no equivalent provision either in JCT 98 PWQ or in JCT 98 MW. This being the case it had been considered that, should the Employer hinder the Contractor's progress during a period of culpable delay, the Architect was not empowered to grant an extension and time therefore became at large (i.e. the Contractor would thereafter have a reasonable period during which the Works could be completed and the Employer would be debarred from levying liquidated and ascertained damages). However, as noted at para. 9.2.7, the case of *Balfour Beatty Building Ltd -v- Chestermount Properties Ltd [1993]* confirmed that the Architect is in fact empowered to grant an extension of time during a period of culpable delay, thereby preserving the Employer's right to recover liquidated and ascertained damages. Although the *Balfour Beatty* case was concerned with a contract let under JCT 80, it is suggested that the decision is also applicable to JCT 98 MW.

7.2.7 JCT 98 MW has no equivalent of clause 25.3.3 of JCT 98 PWQ, which requires the Architect to carry out a post-completion review of the Completion Date. As with JCT 98 IFC, therefore, this suggests that all decisions of the Architect as to extensions of time are effectively irreversible. This makes it that much more important that such decisions are made on an informed basis and, therefore, that the parties agree a procedural framework involving the exchange of information above and beyond the bare administrative details expressly set out in the contract.

7.2.8 Unlike JCT 98 PWQ the requirement in clause 2.2 of JCT 98 MW for the Contractor to give written notice that the Works will not be completed by the completion date is a condition precedent to the fixing of a revised completion date; that is unless the notice is given in the manner prescribed the Architect cannot act upon it. Further as there is no express reservation of the other rights and remedies which the Contractor may possess attached to clause 2.2 of JCT 98 MW, as there is for example at clause 7.3.4 thereof, it is suggested that in default of providing a sufficient notice the Contractor would not have a remedy in Arbitration or in the Courts.

Relevant events

7.2.9 JCT 98 MW does not contain a prescribed list of events which entitle the Contractor to an extension of time. There is no equivalent of the list of Relevant Events set out at clause 25.4 of JCT 98 PWQ against which the Contractor's entitlement can be judged. Clause 2.2 of JCT 98 MW simply entitles the Contractor to an extension of time if the Works will not be completed by the completion date … *for reasons beyond the control of the Contractor, including compliance with any instruction of the Architect/ the Contract Administrator under this Agreement whose issue is not due to a default of the Contractor* … Were such a phrase to have been appended to a short list of the type of events which would give rise to an entitlement, then it is likely – following the decision in *Wells -v- Army & Navy Co-operative Society (1902)* – that the ejusdem generis rule would come into effect and only events of the type set out in the list would give rise to an extension of time. However, there is no such list and the terminology of JCT 98 MW is therefore such that the ejusdem generis rule would not apply and the phrase … *reasons beyond the control of the Contractor* … will instead be given a very wide interpretation. Accordingly, anything which can reasonably be said to be outwith the control of the Contractor may give rise to an entitlement to an extension of time should it delay the completion of the Works. In this connection, the final sentence of clause 2.2 of JCT 98 MW serves to make clear that defaults on the part of sub-contractors or suppliers to the Contractor are not to be regarded as matters beyond the Contractor's control.

7.2.10 Those practitioners who deal largely with contracts in the JCT family other than JCT 98 MW will be unfamiliar with an approach to this particular issue which is not prescriptive. In order to highlight some of the differences which may be encountered when dealing with a blanket provision such as that on which clause 2.2 of JCT 98 MW is based, paras 7.2.11 to 7.2.12 identify two instances whereby events which may not give rise to the fixing of a revised completion date under JCT 98 PWQ may do so under JCT 98 MW.

Adverse weather

7.2.11 Clause 25.4.2 of JCT 98 PWQ entitles the Contractor to an extension of time where delays have been caused by … *exceptionally adverse weather conditions* … A certain degree of adverse weather is to be anticipated on any Contract (particularly during the winter) and the Contractor is expected to make some allowance for this in his programme. He will therefore only be entitled to an extension of time in circumstances where the weather conditions are so exceptionally adverse that he could not reasonably be expected to have anticipated them and made due allowance. Under JCT 98 MW, however, it is arguable that any adverse weather may entitle the Contractor to an extension of time on the grounds that it is outwith his control.

Deferment of possession of the site

7.2.12 Clause 23.1.2 of JCT 98 PWQ, which entitles the Employer to defer giving possession of the site to the Contractor for up to 6 weeks, is optional. If it does not apply then the Architect is not empowered to give an extension of time in this respect. Should the Employer nevertheless defer giving possession of the site to the Contractor, this would constitute an 'act of prevention' for which the Contractor would have no entitlement to an extension of time. Under the JCT 98 PWQ regime, therefore, time would thereafter become at large and the Contractor would have a reasonable period in which to carry out and complete the Works, with the Employer losing any entitlement to recover liquidated and ascertained damages. JCT 98 MW, however, would entitle the Contractor to an extension of time in such circumstances as the causative event would lie outwith the Contractor's control. Consequently, the Employer would retain his right to recover liquidated and ascertained damages in respect of any further delays which were caused by the Contractor.

7.2.13 Whilst it would appear that, by comparison with JCT 98 PWQ, the Contractor has a far wider entitlement under JCT 98 MW, it is suggested that the terminology of the latter form does allow the Architect to exert a measure of control notwithstanding the blanket nature of clause 2.2. This clause also empowers the Architect to grant ... *such extension of time for completion as may be reasonable* ... It is suggested that the Architect may consider it reasonable that the Contractor make due allowance for non-exceptional adverse weather in his planning and programming. In such circumstances, the Architect may consider it unreasonable to grant an extension of time in the absence of weather conditions which can be said to be *exceptionally* adverse.

7.2.14 Whilst the contract machinery of JCT 98 MW is somewhat more flexible than that set out in JCT 98 PWQ, it is considered that the two forms endeavour to achieve the same result – that the Contractor will be awarded an extension of time in circumstances where this is reasonable. JCT 98 PWQ seeks to achieve this by prescribing those events which entitle the Contractor to an extension of time and requiring the Architect to award such an extension only where those events can be demonstrated to have caused the delay. JCT 98 MW, on the other hand, is not prescriptive as to the events which may give rise to an entitlement, but instead gives the Architect a degree of discretion to award only such extensions of time as are reasonable in the circumstances. As with many aspects of JCT 98 MW, a measure of goodwill and flexibility on both sides is required in order to ensure that the contract runs in accordance with its intended principles.

Fluctuations

7.2.15 Part A of the Supplementary Memorandum to JCT 98 MW, which operates by virtue of optional clause 4.6, sets out the mechanism by which the Contractor's entitlement to payment of increased costs arising during the contract period is to be computed. Unlike most of the other forms in the JCT family, however, JCT 98 MW permits such adjustment only in respect of contribution, levy and tax changes. There is no provision in JCT 98 MW for any form of full labour and materials fluctuations. By virtue of clause A4.4.4.1, the Contractor's entitlement to payments in respect of increased costs does not encompass those which arise during a period of culpable delay. As with the equivalent provisions in JCT 98 PWQ, clause A4.4.4.2 of JCT 98 MW dictates that this only continues to be the case should the printed text of clause 2 of JCT 98 MW (which contains the provisions relating to extensions of time) remain unamended and provided that the Architect has, in respect of each notification of the Contractor under clause 2, fixed a completion date. Any amendment whatsoever to the printed text or any failure on the part of the Architect to so fix a completion date will result in the Contractor becoming entitled to increased costs which arise during a period of culpable delay – unless, of course, clause A4.4.4.2 of JCT 98 MW is deleted (for further comments on this subject see paras 2.12.1 to 2.12.5).

7.2.16 Clause A4.4.4.1 of the Supplementary Memorandum to JCT 98 MW reflects clause 38.4.7 of JCT 98 PWQ and the reader is referred to 2.12.1 to 2.12.5 in this connection. As noted above, there is no equivalent in JCT 98 MW of JCT 98 PWQ clauses 39 (labour and materials cost and tax fluctuations) and 40 (use of price adjustment formulae).

7.3   VARIATIONS AND DISRUPTION

Variations

7.3.1 Clause 3.5 of JCT 98 MW empowers the Architect to issue instructions to the Contractor. This power is somewhat wider and less strictly defined than the equivalent power under JCT 98 PWQ. Clause 4.1.1 of JCT 98 PWQ permits the Architect to issue instructions only where expressly empowered to do so elsewhere in the contract, whilst there is no such restriction on the Architect under JCT 98 MW. Accordingly, a provision, such as that at clause 4.2 of JCT 98 PWQ, permitting the Contractor effectively to challenge the Architect's right to issue an instruction would be superfluous under JCT 98 MW and so there is no equivalent in that form.

7.3.2 As with JCT 98 PWQ, clauses 1.2 and 3.5 of JCT 98 MW require that all instructions from the Architect be in writing. Should instructions be issued orally, clause 3.5 of JCT 98 MW requires that they be confirmed in writing by the Architect within 2 days. This regime is somewhat stricter than that set out in JCT 98 PWQ in that, in addition to the somewhat harsh time limit imposed on the Architect, there is no provision for the Contractor himself to confirm oral instructions in writing, as he is permitted to do under clause 4.3.2 of JCT 98 PWQ. Should the Architect neglect to issue written confirmation within the time allowed, the Contractor would appear to have no effective remedy in the event of a dispute as to the authenticity of an instruction. Measures aimed at reducing the scope for such an eventuality could include the setting up of administrative procedures which would ensure compliance with the contractual machinery (e.g. putting work in hand only after receipt of a written instruction) or, alternatively, the implementation of a more flexible approach aimed at protecting the interests of both sides in the event of a problem (e.g. by encouraging the routine written confirmation by the Contractor of instructions received orally).

7.3.3 Written instructions must be complied with "forthwith" by the Contractor. Whilst this term is not defined, should the Contractor not have complied within 7 days of the date of a notice requiring the Contractor to comply with an instruction the Employer may pay someone else to carry out the necessary work and recover such cost as he thereby incurs from the Contractor.

7.3.4 Clause 3.6 of JCT 98 MW deals with variations, which are defined rather more loosely than in clause 13.1 of JCT 98 PWQ. JCT 98 MW defines a variation as ... *an addition to or omission from or other change in the Works or the order or period in which they are to be carried out* ... Whilst the thrust of this definition is not dissimilar to that in clause 13.1 of JCT 98 PWQ, its looser framework and more general nature could result in instances whereby changes which would not be regarded as Variations under JCT 98 PWQ are regarded as such under JCT 98 MW. The most obvious example of this relates to the ability of the Architect under JCT 98 MW to alter the contract period by means of an instruction. By virtue of clause 3.6, any such instruction would be classed as a variation. Conversely, the application of the ejusdem generis rule may well serve to exclude from the JCT 98 MW definition those matters set out in clause 13.1.2 of JCT 98 PWQ (the imposition or removal of restrictions on access to or over the site, etc.) as these do not give rise to a ... *change in the Works* ...

7.3.5 The definition of that which constitutes a variation is particularly important in the case of JCT 98 MW, as variations are the only means within the agreement by which the Contractor becomes entitled to reimbursement of direct loss and/or expense (an issue which is discussed in more detail in paras 7.3.10 to 7.3.25 below). It is suggested that, as elsewhere in the form, a flexible and reasonable approach is required of the Architect in interpreting the definition contained in clause 3.6. Should a Court ever be required to arrive at its own interpretation, it is possible that this will result from an inflexible and unreasonable approach to this question on the part of one or both parties.

7.3.6 Although JCT 98 MW contains provision in the Fourth Recital for the naming of a Quantity Surveyor, there are no duties expressly required of him by the Contract. Accordingly, the valuation of variations under clause 3.6 is stated as being a task for the Architect. In practice, however, where a Quantity Surveyor has been appointed it is likely to be he who carries out such valuations. In this respect, JCT 98 MW contains none of the detailed rules for the valuation of Variations set out in clause 13.5 of JCT 98 PWQ. Nonetheless, the fundamental principle adopted is the same – that variations shall be valued on a fair and reasonable basis by reference where possible to a priced contract document. As with the requirement (under clause 3.5) for the Architect to award a reasonable extension of time in the absence of detailed contractual provisions which dictate the Contractor's entitlement, so the Architect or Quantity Surveyor is required to arrive at a reasonable valuation of a variation in the absence of detailed contractual guidance as to the mechanics of such a valuation.

7.3.7 The main feature which distinguishes JCT 98 MW from JCT 98 PWQ, insofar as the valuation of variations is concerned, is that under the former such valuation is to include the amount of any direct loss and/or expense which results from compliance with the instruction. Under JCT 98 PWQ, of course, all such direct loss and/or expense is expressly excluded from consideration under clause 13.5 (valuation rules) and is instead dealt with in accordance with clauses 13.4.1.2A1.1, 13A.2.3 and 26.1. The effect which this may have on the Contractor's entitlement under JCT 98 MW and the means by which such entitlement is assessed are discussed in more detail in paras 7.3.10 to 7.3.13.

7.3.8 Clause 3.6 of JCT 98 MW expressly provides, as an alternative to valuation, for the Architect and the Contractor to reach prior agreement on a price for the variation. Such a price would include any direct loss and/or expense arising from the variation. Whilst such a procedure is not expressly provided for in JCT 98 PWQ it is nonetheless possible to adopt by virtue of the words ... *unless otherwise agreed by the Employer and the Contractor* ... in clause 13.4.1.1 thereof (see para. 3.1.2b).

7.3.9 There is no equivalent in JCT 98 MW of either clause 13.4.1.2A (valuation of Variations – Contractor's Price Statement – see section 3.3) or clause 13A (Variation instruction – Contractor's quotation in compliance with the instruction – see section 3.4) of JCT 98 PWQ. This is not surprising having regard to the procedurally detailed and cumbersome nature of both clauses.

Procedures for the ascertainment of direct loss and/or expense

7.3.10 Unlike the other forms of contract in the JCT family, JCT 98 MW has no detailed provisions dealing specifically with the Contractor's entitlement to reimbursement of direct loss and/or expense. As noted above, the Contractor will be reimbursed the direct loss and/or expense arising from a variation as part of the valuation of that variation by the Architect. In this respect, no written application is required from the Contractor of the type specified in clause 26.1 of JCT 98 PWQ. Having said this, it is most likely in practice that the Contractor will be quick to advise the Architect of any direct loss and/or expense he has suffered and considers reimbursable.

7.3.11 JCT 98 MW is not entirely clear as to how the reimbursable amount of the Contractor's direct loss and/or expense is to be valued. Clause 26 of JCT 98 PWQ is quite clear that the Contractor's entitlement in this respect is to be ascertained.

7.3.12 Because the sums to which the Contractor becomes entitled under clause 26 of JCT 98 PWQ must be established as a matter of fact, as opposed to judgment, there is a requirement in clause 26.1.3 that the Contractor provide such information as is requested and necessary to enable the Architect or Quantity Surveyor to establish such entitlement. JCT 98 MW, by contrast, provides for the reimbursement of direct loss and/or expense as part of the valuation of a variation – albeit, somewhat confusingly, clauses 4.2 and 4.3 of JCT 98 MW refer to the payment to the Contractor of sums ... *either ascertained or agreed under clauses 3.6 and 3.7* ... The term "valuation" has a particular meaning in this context and, unlike "ascertainment", may involve the use of professional judgment to arrive at a fair and reasonable result. In accordance with this philosophy, JCT 98 MW does not expressly require the Contractor to provide information to the Architect to assist in establishing his entitlement.

7.3.13 It is suggested that it is preferable that the principles of ascertainment be applied to the direct loss and/or expense element of the valuation of a variation under clause 3.6 of JCT 98 MW and that the Contractor endeavour to provide the Architect with such information as the latter may reasonably require in this respect. Should it prove neither practicable nor economic to proceed on this basis, the provisions of JCT 98

MW may still be satisfied by a more flexible approach based on use of the professional and commercial judgment of the parties.

List of matters

7.3.14 As noted at para. 7.3.10 JCT 98 MW contains no detailed provisions dealing with the Contractor's entitlement to direct loss and/or expense. Consequently, JCT 98 MW has no equivalent of the list of matters set out at clause 26.2 of JCT 98 PWQ. Instead, as noted above, the Contractor is entitled within the agreement to reimbursement of only that direct loss and/or expense which can be said to have arisen as the result of an instruction requiring a variation. An examination of the list of matters which give rise to an entitlement under JCT 98 PWQ (see para. 3.5.3) demonstrates the more restrictive nature of the Contractor's right to reimbursement under JCT 98 MW.

Late information or instructions (JCT 98 PWQ clauses 26.2.1.1 and 26.2.1.2)

7.3.15 It is suggested that the Contractor would not be entitled to reimbursement of direct loss and/or expense purely by virtue of information or an instruction having been received late. However, if the late information or instruction was also a variation as defined by clause 3.6 of JCT 98 MW then he would be entitled to such reimbursement.

The opening up for inspection of work which proves to be in accordance with the Contract (JCT 98 PWQ clause 26.2.2)

7.3.16 There is no equivalent in JCT 98 MW of clause 8.3 (Inspection – tests) of JCT 98 PWQ. Hence the Architect can only instruct the Contractor to open up the Works for inspection under clause 3.5 of JCT 98 MW. If, following such an instruction, the inspection reveals that the Works are in accordance with the Agreement then it is suggested that such an instruction falls to be valued under clause 3.6 of JCT 98 MW in that it is in effect … *a change in the Works* … ordered by the Architect. The Contractor would therefore be entitled to reimbursement of direct loss and/or expense.

Correcting discrepancies within/between documents (JCT 98 PWQ clause 26.2.3)

7.3.17 By virtue of clause 4.1 of JCT 98 MW, any such correction would be regarded as a variation and would entitle the Contractor to reimbursement of direct loss and/or expense under clause 3.6.

Work carried out and/or materials or goods supplied by the Employer (JCT 98 PWQ clauses 26.2.4.1 and 26.2.4.2)

7.3.18 Unless the work to be carried out and/or the materials or goods to be supplied by the Employer had previously been omitted from the Works to be carried out by the Contractor, it is difficult to envisage this falling within the definition of a variation under clause 3.6 of JCT 98 MW. Accordingly, no entitlement to reimbursement of direct loss and/or expense would arise.

Instructions to postpone the work (JCT 98 PWQ clause 26.2.5)

7.3.19 As these could be said to be instructions changing ... *the order or period in which* [the Works] *are to be carried out* ... they would fall within the definition of a variation under clause 3.6 of JCT 98 MW and would thereby entitle the Contractor to reimbursement of direct loss and/or expense.

Failure of the Employer to give timely ingress/egress to or over the site (JCT 98 PWQ clause 26.2.6)

7.3.20 Again, it is difficult to envisage this falling within the definition of a variation under clause 3.6 of JCT 98 MW. Accordingly, no entitlement to reimbursement of direct loss and/or expense would arise.

Instructions requiring Variations (JCT 98 PWQ clause 26.2.7)

7.3.21 This would, of course, give rise to an entitlement to reimbursement of direct loss and/or expense under clause 3.6 of JCT 98 MW.

Inaccurately forecast Approximate Quantities (JCT 98 PWQ clause 26.2.8)

7.3.22 JCT 98 MW does not cater for Approximate Quantities. This provision of JCT 98 PWQ, therefore, has no equivalent in JCT 98 MW and therefore the question of entitlement to reimbursement of direct loss and/or expense does not arise.

Compliance or non-compliance with CDM regulations (JCT 98 PWQ clause 26.2.9)

7.3.23 This is expressly recognised by clause 3.6. Accordingly, the Contractor would be entitled to reimbursement of direct loss and/or expense.

Suspension by the Contractor (JCT 98 PWQ clause 26.2.10)

7.3.24 Whilst clause 4.8 of JCT 98 MW gives the Contractor the same right, under certain circumstances, to suspend the performance of his obligations under the Agreement as does clause 30.1.4 of JCT 98 PWQ it does not provide for the recovery by the Contractor of any direct loss and/or expense incurred as a result of such suspension, nor does clause 3.6 of JCT 98 MW.

7.3.25 It is apparent from paras 7.3.15 to 7.3.24 that a number of situations which would give rise to an entitlement to direct loss and/or expense under JCT 98 PWQ do not give rise to a similar entitlement under JCT 98 MW. One view of this situation is that the Contractor nevertheless remains entitled to pursue an extra-contractual remedy in damages at common law. However, as noted at para. 7.2.8, it is most likely that the remedies set out within the agreement are in fact exhaustive and that no recourse can be had to such extra-contractual remedies. This view is supported by the fact that clause 26.6 of JCT 98 PWQ expressly reserves the Contractor's common law rights, whilst such an express provision is absent from JCT 98 MW except in clause 7.3.4. It is suggested therefore that the Contractor should be aware of such risks and be deemed to have included for them in his price (see also para. 3.6.1).

7.4 ASCERTAINING THE DIRECT LOSS AND/OR EXPENSE

7.4.1 Insofar as it is practicable to ascertain the Contractor's entitlement to reimbursement, the discussion in Chapter 4 – as regards the heads under which claims for direct loss and/or expense may be considered and the principles which must be brought to bear in assessing the Contractor's entitlement – is equally relevant in the context of JCT 98 MW. Such matters are not, therefore, the subject of further discussion here. Insofar as it is impracticable to carry out an exercise of ascertainment (see para. 7.3.13) then arriving at a fair and reasonable assessment of the Contractor's entitlement will be a matter requiring the exercise of the commercial and professional judgment of the parties. Having regard to the fact that significant sums ought not to be involved on a project of the type for which JCT 98 MW was drafted, such an exercise ought to be capable of being carried out in accordance with the flexible and consensual approach to the administration of this form of agreement described above.

8

JCT STANDARD FORM OF BUILDING CONTRACT WITH CONTRACTOR'S DESIGN 1998 EDITION

8.1   INTRODUCTION

8.1.1 This chapter deals with the JCT Standard Form of Building Contract with Contractor's Design 1998 Edition incorporating Amendments 1 and 2 hereinafter referred to as JCT 98 WCD.

8.1.2 JCT 98 WCD differs fundamentally from JCT 98 PWQ in that the Contractor's responsibilities encompass not only the construction of the Works but also their design. The Contractor's obligations in this respect are set out in clause 2.1 of JCT 98 WCD in the following terms:

> ... *The Contractor shall upon and subject to the Conditions carry out and complete the Works referred to in the Employer's Requirements, the Contractor's Proposals (to which the Contract Sum Analysis is annexed), the Articles of Agreement, these Conditions and the Appendices in accordance with the aforementioned documents and for that purpose shall complete the design for the Works including the selection of any specifications for any kinds and standards of the materials and goods and workmanship to be used in the construction of the Works so far as not described or stated in the Employer's Requirements or Contractor's Proposals* ...

8.1.3 A consequence of the Contractor having responsibility under JCT 98 WCD for the design of the Works, and one which impacts considerably upon the Contract itself, is the fundamentally altered nature of the relationships between the Employer, the Contractor and the various members of the design team as compared with the situation pertaining under JCT 98 PWQ. Under JCT 98 WCD the design team will be appointed and retained by the Contractor (although they may have originally been appointed by the Employer and subsequently had their appointments novated to the Contractor). A consequence of this is that the managerial and quasi-arbitral roles performed by the Architect and Quantity Surveyor under JCT 98 PWQ (particularly as regards the functions of valuation and certification) do not exist under JCT 98 WCD. Indeed, neither the Architect nor the Quantity Surveyor receive any mention in JCT 98 WCD itself.

8.1.4 The functions referred to above are to be performed, if at all, by the Employer under JCT 98 WCD although, insofar as such functions call for the application of professional or technical expertise, they are more likely to be performed by the Employer's Agent. This individual is named in Article 3 of JCT 98 WCD and his function is essentially to maintain a watching brief upon the Contractor and his design team and to perform on the Employer's behalf those duties which fall to the Employer under the various provisions of the contract. This function is set out in the following terms in Article 3:

> ... *save to the extent which the Employer may otherwise specify by written notice to the Contractor, for the receiving or issuing of such applications, consents, instructions, notices, requests or statements or for otherwise acting for the Employer under any other of the Conditions* ...

8.1.5 It is important to understand that the Employer's Agent under JCT 98 WCD does not stand in the position of the Architect under JCT 98 PWQ. In the latter situation the Architect acts when required as an impartial certifier and quasi-arbitrator between the two parties to the Contract whereas the Employer's Agent is acting in the Employer's stead and is therefore himself in the position of a party to the Contract as opposed to acting as between the two parties. This distinction is of some significance and is discussed in further detail at section 8.2.

8.1.6 A further fundamental distinction between JCT 98 PWQ and JCT 98 WCD lies in the documentation upon which the Contract is based and which is referred to in clause 2.1 quoted above. Unlike the situation pertaining under JCT 98 PWQ, whereby the design and quantum of the Works is indicated to the Contractor in the form of Contract Drawings prepared by the Architect and Contract Bills prepared by the Quantity Surveyor, the only document issued by the Employer to the Contractor under JCT 98 WCD is that known as the Employer's Requirements. This document will set the parameters within which the scheme will be designed by the Contractor and will vary in nature from project to project. It is not unknown for the Employer's Requirements to adopt an essentially prescriptive approach and to take the form of fulsome designs supported by detailed specifications and quantities. However, it is suggested that JCT 98 WCD works best when the Employer's Requirements are kept to the bare minimum of detail commensurate with achieving the Employer's aims, thereby affording the maximum degree of flexibility to the Contractor in completing the design.

8.1.7 The means by which the Contractor aims to satisfy the Employer's Requirements will be set out in a document known as the Contractor's Proposals, which itself will be supported by a Contract Sum Analysis. The Contractor's Proposals will generally take the form of a statement of the design fundamentals (supported by drawings) together with a more or less detailed specification of the Works. The Contract Sum Analysis will generally be in less fulsome detail than typical bills of quantities in use under a JCT 98 PWQ contract and is intended more as a general indication of the way in which the Contractor's price breaks down rather than as a detailed financial control document although it is, contractually, the reference base for the valuation of additional, substituted and omitted work.

8.1.8 Should discrepancies occur within the contract documentation, the Contractor's Proposals will generally prevail. However, any discrepancies within the Contractor's Proposals themselves must be rectified by the Contractor without any adjustment to the Contract Sum. In this way it is quite possible, unlike the situation which generally pertains under JCT 98 PWQ, for the design of the Works to be changed without any consequent adjustment to the Contract Sum or any consideration required as to the question either of an extension of time or of direct loss and/or expense. In general, only where such a change is the result of a Change in the Employer's Requirements (as defined in clause 12.2 of JCT 98 WCD) will these latter considerations apply.

8.1.9 The principles set out in the foregoing paragraph are taken from clauses 2.4.1, 2.4.2 and 12.2.1 of JCT 98 WCD which provide as follows:

> 2.4 .1 *Where there is a discrepancy within the Employer's Requirements (including any Change issued in accordance with clause 12.2) the Contractor's Proposals shall prevail (subject always to compliance with the Statutory Requirements) without any adjustment of the Contract Sum. Where the Contractor's Proposals do not deal with*

any discrepancy within the Employer's Requirements (including any Change issued in accordance with clause 12.2) the Contractor shall inform the Employer in writing of his proposed amendment to deal with the discrepancy and the Employer shall either agree the proposed amendment or himself decide how the discrepancy shall be dealt with; such agreement or decision shall be notified in writing to the Contractor and such notification shall be treated as a Change in the Employer's Requirements.

2.4 .2 *Where there is a discrepancy within the Contractor's Proposals the Contractor shall inform the Employer in writing of his proposed amendment to remove the discrepancy; and (subject always to compliance with Statutory Requirements) the Employer shall decide between the discrepant items or otherwise may accept the Contractor's proposed amendment and the Contractor shall be obliged to comply with the decision or acceptance by the Employer without cost to the Employer.*

12.2 .1 *The Employer may subject to the proviso hereto and to clause 12.2.2 and to the Contractor's right of reasonable objection set out in clause 4.1.1 issue instructions effecting a Change in the Employer's Requirements. No Change effected by the Employer shall vitiate this Contract. Provided that the Employer may not effect a Change which is, or which makes necessary, an alteration or modification in the design of the Works without the consent of the Contractor which consent shall not be unreasonably delayed or withheld.*

JCT 98 WCD is silent on the situation where there is a discrepancy as between the Employer's Requirements and the Contractor's Proposals but what is ostensibly an unusual omission may be prompted by the Third Recital which provides that ... *the Employer has examined the Contractor's Proposals and ... is satisfied that they appear to meet the Employer's Requirements.*

8.1.10 Against the backdrop of the foregoing comments, the differences between those provisions of JCT 98 PWQ and JCT 98 WCD which deal with extensions of time, the valuation of additional and/or varied work and the reimbursement of direct loss and/or expense are essentially to be explained by reference to the differing nature both of the relationships between the various parties and the documentation which passes between them. The discussion which follows will hopefully prove to be the more illuminating with this simple fact in mind.

8.2   EXTENSIONS OF TIME

Procedures

8.2.1 Table 8.1 provides references to the text within this book which is also relevant to the following review of JCT 98 WCD.

Table 8.1. Extension of time procedures

Provision in Contract	Clause in JCT 98 WCD	Equivalent clause in JCT 98 PWQ	Explanatory text in this book
Requirements for the Contractor to give notice of delay and to give further details to assist in establishing any extension of time.	25.2.1*	25.2.1	2.7.7
Requirement for the Contractor to give, in respect of each Relevant Event identified in the notice of delay, particulars of the expected effects thereof together with an estimate of expected delay and to update such information as and when necessary.	25.2.2* and 25.2.3*	25.2.2 and 25.2.3	2.7.8 to 2.7.10
Requirement for the Architect (Employer) in certain circumstances, to grant extensions of time during the construction period or to notify the Contractor that an extension of time is not, in his view justified.	25.3.1*	25.3.1	2.7.11 to 2.7.31
Requirement for the Architect (Employer) to review the Completion Date following Practical Completion.	25.3.3*	25.3.3	2.7.32 to 2.7.41
The effect of omissions upon extensions of time.	25.3.2* and 25.3.3.2*	25.3.2 and 25.3.3.2	2.7.36 to 2.7.41
General obligations imposed upon the Architect (Employer) and Contractor in relation to their conduct.	25.3.4* and 25.3.5*	25.3.4, 25.3.5 and 25.3.6	2.7.42 to 2.7.45

* The provisions of JCT 98 WCD may be modified by the application of the Supplementary Provisions. Provision S6 requires the Contractor to submit estimates of extensions of time consequent upon Changes to the Employer's Requirements. Such estimates may be accepted or rejected in whole or in part by the Employer or used as the basis for negotiations.

8.2.2 The procedures by which extensions of time are requested and granted under clause 25 of JCT 98 WCD are basically the same as those set out in clause 25 of JCT 98 PWQ albeit with certain exceptions.

8.2.3 The most notable of these exceptions is that applications for extensions of time made by the Contractor under JCT 98 WCD are dealt with by the Employer rather than the Architect. The reason for this is that, under JCT 98 WCD, there is no Architect named in the contract to undertake, inter alia, the managerial and quasi-arbitral role provided for in JCT 98 PWQ. The granting of extensions of time is a duty under JCT 98 PWQ which particularly requires the Architect to act as quasi-arbitrator in that it requires that he exercise skill and independent judgment in order to resolve a matter as between two parties to a contract (i.e. the Contractor and the Employer) and it is for this reason that clause 25.3.1 of JCT PWQ makes reference to the Architect forming an ... *opinion* ... as to the Contractor's entitlement to an extension of time.

8.2.4 JCT 98 WCD, on the other hand, makes no such reference to the need for the Employer to form an "opinion" as to the Contractor's entitlement to an extension of time. Clause 25.3.1 of JCT 98 WCD simply requires the Employer to award such extension of time as is fair and reasonable in the circumstances. As the Employer is himself a party to the contract he cannot act as quasi-arbitrator in this respect (see the case of *Stevenson -v- Watson (1879) 4 CPD 148*) and must instead make a decision by reference to a supposed objective standard rather than by the exercise on his own part of any professional skill or judgment. In this sense, the Employer may be seen as a judge in his own cause but this is nonetheless subject to the strictures of the Contract if challenged by adjudication, arbitration or the courts. No guidance is given in the Contract as to where such an objective standard may be found, but in practice it is expected that the Employer will be guided in such matters by the Employer's Agent who will invariably be drawn from the ranks of the construction professionals.

8.2.5 As a result of the case of *J F Finnegan -v- Ford Sellar Morris Developments Ltd [1991] 53 BLR 42* in which it was held that an Employer's notice of non-completion issued under clause 24 of the JCT Standard Form of Building Contract With Contractor's Design 1981 Edition, **but disputed by the Contractor**, would be of no effect until settled at arbitration, whereas an Architect's notice under clause 24 of JCT 80 (the predecessor to JCT 98 PWQ) would be binding unless and until set aside by an arbitrator, the wording of clause 2.1 of JCT 98 WCD now reads:

> ... *The Contractor shall comply with any instruction and be bound by any decision of the Employer issued or made under or pursuant to the Conditions and any such instruction or decision shall have effect except to the extent that any such instruction or decision is varied pursuant to the provisions of clause 39A or 39B or 39C ...*

8.2.6 The effect of this wording is to place the Employer under JCT 98 WCD in essentially the same position as the Architect under JCT 98 PWQ in that his decisions are binding upon the Contractor unless and until they are overturned at Adjudication or Arbitration or in Legal Proceedings.

8.2.7 To facilitate the process of agreeing upon appropriate extensions of time, provision S6 of the Supplementary Provisions requires the Contractor (unless the Employer instructs otherwise or the Contractor raises reasonable objection to doing so) to submit estimates of extensions of time to which he considers himself entitled as a result of changes in the Employer's Requirements. The Employer is not bound by any such submission and is at liberty to proceed on the basis of his own assessment or refer the matter for adjudication.

8.2.8 A further distinction between clause 25 of both JCT 98 PWQ and JCT 98 WCD is that the latter document makes no mention of Nominated Sub-Contractors. Whilst the Supplementary Provisions of JCT 98 WCD make reference to Named Sub-Contractors, Nominated Sub-Contractors do not exist under JCT 98 WCD and cannot therefore fall to be considered under the extension of time (or indeed any) provisions of that form of contract.

Relevant Events

8.2.9 Table 8.2 provides references to the text within this book which is also relevant to the following review of JCT 98 WCD.

Table 8.2. Relevant Events

Relevant Event	Clause in JCT 98 WCD	Equivalent clause in JCT 98 PWQ	Explanatory text in this book
Force majeure	25.4.1	25.4.1	2.8.2 and 2.8.3
Exceptionally adverse weather conditions	25.4.2	25.4.2	2.8.4
Loss or damage due to Specified Perils	25.4.3	25.4.3	2.8.5
Civil commotion	25.4.4	25.4.4	2.8.6 and 2.8.7
Compliance with Architect's (Employer's) instructions	25.4.5.1	25.4.5.1	2.8.8 to 2.8.27
Opening up or testing	25.4.5.2	25.4.5.2	2.8.28
Late information from Architect (Employer)	25.4.6	–	8.2.10.3
Late permission or approval of any statutory body	25.4.7	–	–
Works by others	25.4.8.1	25.4.8.1	2.8.51 to 2.8.54
Supply of materials by Employer	25.4.8.2	25.4.8.2	2.8.55
Statutory power restrictions	25.4.9	25.4.9	2.8.56 and 2.8.57
Inability to secure labour	25.4.10.1	25.4.10.1	2.8.58 to 2.8.60
Inability to secure materials	25.4.10.2	25.4.10.2	2.8.58 to 2.8.60
Work by local authority/statutory undertaker	25.4.11	25.4.11	2.8.61
Site access/egress restrictions	25.4.12	25.4.12	2.8.62 and 2.8.63
Change in Statutory Requirements	25.4.13	25.4.15	2.8.68
Deferment of date of possession	25.4.14	25.4.13	2.8.64
Terrorism	25.4.15	25.4.16	2.8.69
CDM Regulations	25.4.16	25.4.17	2.8.70
Suspension by the Contractor	25.4.17	25.4.18	2.8.71 to 2.8.75

8.2.10 The Relevant Events by virtue of which the Contractor becomes entitled to an extension of time are in large part the same in both JCT 98 PWQ and JCT 98 WCD. Those differences which are worthy of note can be summarised as follows:

1. *The effect of strikes, lock-outs, civil commotion, etc.*
    Clause 25.4.4 of JCT 98 WCD extends this to cover those engaged not only in constructing the Works but also those designing same, on the grounds that under JCT 98 WCD design is the responsibility of the Contractor.

2. *Instructions*

Under JCT 98 WCD, these are Employer's (as opposed to Architect's) Instructions and differ in a number of other respects from those which can give rise to an extension of time under JCT 98 PWQ:

- Under clause 2.3 of JCT 98 PWQ an instruction issued by the Architect with a view to correcting discrepancies within and between the Contract Documents is a Relevant Event. The analogous instruction under clause 2.3.1 of JCT 98 WCD (which is a Relevant Event by virtue of clause 25.4.5.1 of JCT 98 WCD) makes reference only to any discrepancy concerning the definition of the site boundary. As there are no Contract Documents under JCT 98 WCD in the sense anticipated by JCT 98 PWQ (the only document provided by the Employer being the Employer's Requirements) any discrepancies between such clearly cannot arise.

- Clause 2.4.1 of JCT 98 PWQ requires the Architect to issue Instructions in respect of discrepancies relating to Performance Specified Work, which Instruction will be a Relevant Event under clause 25.4.5.1 thereof. There is no analogous Instruction under JCT 98 WCD principally because, by its nature, all of the work under a JCT 98 WCD contract is likely to be specified by reference to performance criteria.

- Clause 13.2 of JCT 98 PWQ enables the Architect to issue instructions with regard to Variations, any of which instructions may be a Relevant Event under clause 25.4.5.1 thereof. The analogous provision in JCT 98 WCD is clause 12.2, which refers to Employer's instructions with regard to Changes in the Employer's Requirements, any of which instructions may be a Relevant Event under clause 25.4.5.1 of JCT 98 WCD.

- Clause 13.3 of JCT 98 PWQ permits the Architect to issue Instructions with regard to the expenditure of provisional sums in the Contract Bills. Under the provisions of SMM7 such provisional sums may be in respect of defined or undefined work. Only Instructions in respect of provisional sums for *undefined* work, however, can constitute a Relevant Event under clause 25.4.5.1 of JCT 98 PWQ. Whilst there are no Contract Bills under JCT 98 WCD there may well be provisional sums included in the Employer's Requirements. However, as this latter document will not necessarily have been prepared in accordance with the rules contained in SMM7 there is no distinction between provisional sums for defined work and those for undefined work. Accordingly there is no distinction between the two in clause 25.4.5.1 of JCT 98 WCD, by which any Employer's Instruction in respect of the expenditure of a provisional sum can constitute a Relevant Event.

- Clause 23.2 of JCT 98 PWQ permits the Architect to issue Instructions regarding the postponement of any part of the work to be executed, any of which Instructions may be a Relevant Event under clause 25.4.5.1 thereof. These provisions are replicated at clause 23.2 in JCT 98 WCD and extend also to any Instruction to postpone the *design* work (for which the Contractor is, of course, responsible).

- Architect's Instructions issued under clauses 35 (Nominated Sub-Contractors) and 36 (Nominated Suppliers) of JCT 98 PWQ may constitute a Relevant

Event under clause 25.4.5.1 thereof. As there are no Nominated Sub-Contractors or Suppliers under JCT 98 WCD there is no analogous provision under clause 25 thereof.

3. *Late instructions, information, etc.*
   Clause 25.4.6 of JCT 98 WCD differs from its counterpart clause in JCT 98 PWQ in the following fundamental respects:
   - If the Architect fails to provide information in accordance with the Information Release Schedule, clause 25.4.6.1 of JCT 98 PWQ provides that this can constitute a Relevant Event. As there is no role for the Architect under JCT 98 WCD it follows that there is no need for an Information Release Schedule hence there is no analogous Relevant Event under clause 25 thereof.
   - If the Architect fails to provide all drawings, details and instructions in such time as to enable the Contractor to carry out and complete the Works in accordance with the Conditions, whether or not such information has been requested by the Contractor, clause 25.4.6.2 of JCT 98 PWQ provides that this can constitute a Relevant Event. By contrast clause 25.4.6 of JCT 98 WCD provides that the failure of the Employer (the role of Architect is not recognised under this form of contract) to provide necessary instructions, decisions, information and consents ... *in due time* ... can constitute a Relevant Event **but** only in those circumstances when the Contractor has made specific written application for such information at a date ... *which having regard to the Completion Date was neither unreasonably distant from nor unreasonably close to the date when it was necessary for him to receive the same* ... In this context ... *in due time* ... means "in a reasonable time" and not "in time to avoid delay"; see *Percy Bilton -v- Greater London Council [1982] 1 WLR 794* in which the materially similar provisions of clause 23(f) of JCT 63 were considered. Further, it is worth noting that the Contractor must make his written application for outstanding information in sufficient time for the Employer to be able to respond adequately so as not to delay the Works but not so far in advance of the need for the information that the request loses its potency. It is suggested that to comply with these provisions there is no substitute for the Contractor constantly to review his requirements to enable him to give proper and adequate notice to the Employer of his up to date requirements and for the Employer to respond promptly.

4. *Nominated Sub-Contractors and/or Suppliers*
   Delay on the part of Nominated Sub-Contractors and/or Suppliers may be a Relevant Event under clause 25.4.7 of JCT 98 PWQ. As neither exist under the terms of JCT 98 WCD there is no analogous Relevant Event under clause 25 thereof.

5. *Late permission or approval of any statutory body*
   By reason of clause 25.4.7 of JCT 98 WCD, delays in obtaining these can constitute a Relevant Event, although this is not the case under JCT 98 PWQ. This particular provision of JCT 98 WCD would apply to situations whereby the Contractor, for reasons beyond his control, was unable to secure (for instance) planning permission in due time in respect of the whole or part of the design of the

scheme. Such a situation would not pertain, of course, under JCT 98 PWQ whereby such matters are the responsibility of the Employer. Accordingly no provision is required under JCT 98 PWQ in order to grant the Contractor an extension of time in this respect.

6. *Changes in statutory requirements*

    Clause 25.4.13 of JCT 98 WCD permits the Contractor to receive an extension of time by reason of changes in statutory requirements or by reason of decisions of a statutory body (for example a planning authority) which in turn necessitate a change in the Contractor's Proposals. JCT 98 PWQ clause 25.4.15 contains a similar provision but it is restricted to Performance Specified Work.

7. *Approximate quantities*

    Clause 25.4.14 of JCT 98 PWQ entitles the Contractor to an extension of time in circumstances in which an Approximate Quantity contained in the Contract Bills proves not to be a reasonably accurate forecast of the work involved and the Contractor is thereby delayed. As there are no Contract Bills (and thereby no Approximate Quantities) under JCT 98 WCD, there is no corresponding Relevant Event under clause 25 thereof.

Fluctuations

8.2.11 Clauses 36, 37 and 38 of JCT 98 WCD (all of which are optional) set out the mechanisms by which the Contractor's entitlement to payment of increased costs arising during the contract period can be computed. By virtue of clauses 36.4.7, 37.5.7 and 38.6.1, the Contractor's entitlement to such payments does not encompass increased costs arising during a period of culpable delay on the part of the Contractor. It is worthy of note in this respect that this only continues to be the case should the printed text of clause 25 of JCT 98 WCD (which incorporates the provisions relating to extensions of time) remain unamended and provided that the Employer has, in respect of every written notification by the Contractor under clause 25 of JCT 98 WCD, fixed a new or confirmed the existing Completion Date. Any amendment whatsoever to this printed text or any failure of the Employer to fix a new or confirm the existing Completion Date, where either clause 36, 37 or 38 apply, will result in the Contractor becoming entitled to increased costs which arise even during a period of culpable delay on the part of the Contractor.

8.2.12 Clauses 36.4.7, 37.5.7 and 38.6.1 of JCT 98 WCD reflect clauses 38.4.7, 39.5.7 and 40.7.1 of JCT 98 PWQ respectively and the reader is referred to in sections 2.12 and 4.7 in this connection.

8.3 VARIATIONS AND DISRUPTION

Changes

8.3.1 Table 8.3 provides references to the text within this book which is also relevant to the following review of JCT 98 WCD.

Table 8.3. Valuation of changes (variations)

Variation Rule	Clause in JCT 98 WCD	Related clauses in JCT 98 PWQ	Explanatory text in this book
Valuation of Changes and provisional sum work (variations) – Alternative A Within 21 days of receipt of instruction the Contractor **may** submit a Price Statement for such work.	12.4.2A1	13.4.1.2A1	3.3.4
Contractor **may** separately attach to the Price Statement:			3.3.5
• any amount required in lieu of direct loss and/or expense; • any adjustment required to the Completion Date.	12.4.2A1.1 12.4.2A1.2	13.4.1.2A1.1 13.4.1.2A1.2	
Acceptance of Price Statement (in whole or in part) by Employer (Quantity Surveyor) within 21 days of receipt from Contractor.	12.4.2A2	13.4.1.2A2	3.3.3
Non-acceptance of Price Statement (in whole or in part) – the Employer (Quantity Surveyor) to supply amended Price Statement for Contractor to accept or reject.	12.4.2A4	13.4.1.2A4	3.3.3
When the Price Statement or amended Price Statement are neither accepted nor referred to Adjudication valuation of the work shall be by Alternative B.	12.4.2A6	13.4.1.2A6	3.3.2
Where attachments to the Price Statement are not accepted direct loss and/or expense and revisions to the Completion Date are to be dealt with as provided for elsewhere in the Contract.	12.4.2A7	13.4.1.2A7	3.3.6
Valuation of Changes and provisional sum work (variations) – Alternative B Valuation of work shall be consistent with values in the Contract Sum Analysis, with due allowance made for differences in conditions and/or quantities. Where no similar work in Contract Sum Analysis a fair valuation to be made.	12.5.1	13.5.1.1 to 13.5.1.3	3.3.26 3.3.29 3.3.31 3.3.34 3.3.43

(continued)

Table 8.3. (cont.)

Variation Rule	Clause in JCT 98 WCD	Related clauses in JCT 98 PWQ	Explanatory text in this book
Valuation of omitted work by reference to values in Contract Sum Analysis.	12.5.2	13.5.2	–
Valuation to include necessary allowance for items of a preliminary nature.	12.5.3	13.5.3	3.3.23
Valuation by reference to dayworks where this would be an appropriate basis of a fair valuation of the additional or substituted work.	12.5.4	13.5.4	3.3.35 to 3.3.42
Valuation of other works as if they were variations in the event that a variation substantially changes the conditions under which such other works are executed.	12.5.5	13.5.5	3.3.43
Fair valuation to be made of work which cannot be valued in accordance with above rules.	12.5.6	13.5.7	3.3.29 3.3.34
Submission of Estimates (Quotations) in respect of Changes and provisional sum work (Variations) – Supplemental Provisions Estimate (Quotations) to be provided within 14 days (21 days in case of Quotation).	S6.2*	13A.1	3.4.3 and 3.4.4
Change (Variation) only to be carried out after agreement of Estimate (acceptance of Quotation).	S6.2*	13A.1.3	3.4.7
Extimate shall comprise (Quotation shall separately comprise):	S6.3*	13A.2	3.4.5 and 3.4.6
• cost of carrying out the Change (Variation);	S6.3.1*	13A.2.1	
• additional resources required (if requested in case of Quotation);	S6.3.2*	13A.2.5	
• method statement (if requested in case of Quotation);	S6.3.3*	13A.2.6	
• any adjustment to the Completion Date required;	S6.3.4*	13A.2.2	
• any amount of lieu of direct loss and/or expense.	S6.3.5*	13A.2.3	
Note: JCT 98 PWQ also requires that the cost of preparing the Quotation to be stated – JCT 98 WCD does not so provide.	–	13.A.2.4	

(continued)

Variation Rule	Clause in JCT 98 WCD	Related clauses in JCT 98 PWQ	Explanatory text in this book
Agreement of Estimate (Acceptance of Quotation by Architect)	S6.4*	13A.3	3.4.7
Failure to agree Estimate (Non-acceptance of quotation) – Change (Variation) either to be valued in accordance with normal rules or not to be carried out.	S6.5*	13A.4	3.4.8

* Note that these provisions are optional and dependent upon the application of the supplementary provisions.

8.3.2 Instructions to the Contractor are dealt with in clause 4 of JCT 98 WCD. These provisions are generally the same as those set out in clause 4 of JCT 98 PWQ except that, under JCT 98 WCD, the Contractor receives instructions from the Employer rather than from the Architect.

8.3.3 Whereas clause 13 of JCT 98 PWQ deals with Variations and provisional sums, the equivalent clause 12 in JCT 98 WCD deals with Changes in the Employer's Requirements and provisional sums. The definition of a change in clause 12.1 of JCT 98 WCD, however, is analogous to that of a Variation in clause 13.1 of JCT 98 PWQ (which subject is discussed in detail in Section 3.2). The only additional qualification in JCT 98 WCD is that, in order to qualify as a "Change", an Employer's instruction must result in some material alteration to the Employer's Requirements. Furthermore, the Contractor has a right of reasonable objection to a "Change" only inasmuch that it necessitates altering the design of the Works.

8.3.4 The manner in which Changes in the Employer's Requirements and provisional sum work are to be valued is set out in clause 12.4.1 of JCT 98 WCD and this, like its counterpart, clause 13.4.1, in JCT 98 PWQ provides for the valuation of such work to be made either by way of a Contractor's Price Statement – Alternative A (if the Contractor so chooses) or by way of specific Valuation rules – Alternative B.

8.3.5 The provisions in JCT 98 WCD relating to valuation Alternative A are the same as those in JCT 98 PWQ except that it is the Employer rather than the Quantity Surveyor and Architect who receives and deals with the Contractor's Price Statement. Hence the detailed consideration given to these JCT 98 PWQ clauses in section 3.3 applies equally to their JCT 98 WCD counterparts and consequently there is no need to comment further here.

8.3.6 However, the specific Valuation rules set out in clause 12.5 of JCT 98 WCD are more truncated than those in clause 13.5 of JCT 98 PWQ. The reason for this is that, generally speaking, the rules in JCT 98 PWQ make reference to the Contract Bills and the detailed pricing information contained therein whereas the rules in JCT 98 WCD relate to the less detailed Contract Sum Analysis which is generally the only pricing document which will be in the possession of the Employer under JCT 98 WCD. Notwithstanding the fact that the valuation rules in JCT 98 WCD are truncated in comparison with those in JCT 98 PWQ, they are nonetheless based on essentially the same set of principles – that is, wherever possible evaluation is to take place by reference to a priced Contract document or otherwise to be by way of a fair and reasonable approach.

8.3.7 As with clause 13.5 of JCT 98 PWQ so clause 12.5 of JCT 98 WCD requires that, whilst due allowance be made in the valuation of a Change for additional items of a preliminary nature (site administration, facilities etc.), any direct loss and/or expense arising out of disruption to the progress of the work must be reimbursed under the terms of clause 26 of JCT 98 WCD. The exceptions to this are the acceptance of an attachment to a Contractor's Price Statement under clause 12.4.2A7 of JCT 98 WCD, this is the same as under clause 13.4.1.2A7 of JCT 98 PWQ and where it is stated in the Appendix to the Contract that the Supplementary provisions are to apply since, as previously observed, clause S6 permits the Contractor to submit to the Employer, in much the same way as clause 13A of JCT 98 PWQ, an estimate which includes not only the cost of giving effect to the Change but also any direct loss and/or expense which results from compliance with the instruction. This estimate may then be accepted by the Employer, adjusted following negotiation between the Employer and Contractor or rejected. In the latter case, any direct loss and/or expense would then be reimbursed pursuant to the provisions of clause 26 of JCT 98 WCD.

Procedures for the ascertainment of direct loss and/or expense

8.3.8 Table 8.4 provides references to the text within this book which is also relevant to the following review of JCT 98 WCD.

Table 8.4. Direct loss and/or expense procedures

Direct loss and/or expense procedures	*Clause in JCT 98 WCD*	*Related clause in JCT 98 PWQ*	*Explanatory text in this book*
Contractor to make a written application regarding direct loss and/or expense.	26.1*	26.1	3.5.9
Contractor's written application must be made timeously.	26.1.1*	26.1.1	3.5.10 3.5.11 3.5.20
Contractor is to submit such information as the Employer reasonably requires in order to make a payment.	26.1.2*	26.1.2 26.1.3	3.5.13 and 3.5.14
Amounts ascertained as being due are to be added to the Contract Sum.	26.3	26.5	3.5.18
The above provisions are without prejudice to any other rights or remedies which the Contractor may possess.	26.4	26.6	3.5.9 3.5.21

* The provisions of JCT 98 WCD may be modified by the application of the Supplementary Provisions. Clauses S6 and S7 permit the Contractor to submit estimates of direct loss and/or expense. Such estimates

may be accepted or rejected by the Employer or used as the basis for negotiations.

8.3.9 Reimbursement of direct loss and/or expense incurred by the Contractor is, subject to an accepted attachment to a Contractor's Price Statement and to the operation of clause S6, dealt with in accordance with the provisions of clause 26 of JCT 98 WCD as modified by clause S7 where it is stated in the Appendix to the Contract that the Supplementary Provisions are to apply. This differs in a number of respects from clause 26 of JCT 98 PWQ. Principally there is no requirement for such loss and/or expense as has been incurred to be ascertained by an independent party (such as the Architect or Quantity Surveyor under JCT 98 PWQ). Instead, as set out in clause 26.1 of JCT 98 WCD, it is incumbent upon the Employer to reimburse those amounts that have actually been incurred by the Contractor. In making reimbursement the Employer is entitled to such information from the Contractor as is reasonably necessary in order to arrive at the correct figure. In effect, the Employer will be making his own ascertainment of the Contractor's entitlement and may well refer the matter to the Employer's Agent (who, as noted above, will generally be drawn from the ranks of the construction professionals).

8.3.10 The comments in section 8.2 concerning the Employer's position as an interested party to the contract rather than an impartial certifier apply equally in the context of reimbursement of loss and/or expense. Furthermore, if the Supplementary Provisions are incorporated into the contract then, under clause S6, the Contractor may (as with extensions of time) submit to the Employer an estimate of the direct loss and/or expense which he claims to have incurred as a result of Changes to the Employer's Requirements. Likewise, as with extensions of time, the Employer may choose not to be guided by such an estimate. If the alleged loss and/or expense arises other than by reason of a Change in the Employer's Requirements then clause S7 applies. By virtue of clauses S7.2 and S7.3 the Contractor is to provide the Employer with ongoing estimates of such direct loss and/or expense as the Contractor claims to have incurred. The Employer, under clause S7.4, may then either accept or reject the estimate or use it as a basis for negotiations with the Contractor. In the case of both clauses S6 and S7, should the Contractor fail to comply with the requirements of the contract regarding the submission of estimates in due time then any assessment made by the Employer in the absence of an estimate from the Contractor will not be included in any Interim Payments to the Contractor but instead will only be reimbursed as part of the final adjustment of the Contract Sum and then it will be without any addition in respect of loss of interest or financing charges suffered or incurred by the Contractor on such direct loss and/or expense prior to the issue of the Final Statement and Final Account. Interestingly in this context it would appear that, whilst the Employer is entitled to withhold payment of direct loss and/or expense from interim payments in circumstances in which the Contractor fails to submit an estimate (or fails to do so in due time) he cannot do so where the Contractor does submit such an estimate but fails to support the same with necessary details and particulars as required by clause S7.4.

8.3.11 Where the Supplementary Provisions do not apply any decision of the Employer in respect of direct loss and/or expense will (by virtue of clause 2.1 of JCT 98 WCD) be binding upon the Contractor unless and until it is overturned by an Adjudicator or Arbitrator or in Legal Proceedings.

8.3.12 A further difference between JCT 98 PWQ and JCT 98 WCD as to the procedural aspects of the reimbursement of direct loss and/or expense lies in the fact that clause 26.4 of JCT 98 PWQ, which requires the Architect or Quantity Surveyor to ascertain the amount of any loss and/or expense incurred by Nominated Sub-Contractors, has (for reasons already noted above) no equivalent in JCT 98 WCD.

List of matters

8.3.13 Table 8.5 provides references to the text within this book which is also relevant to the following review of JCT 98 WCD.

Table 8.5. List of matters in JCT 98 WCD

Cause	JCT 98 WCD clauses		JCT 98 PWQ clauses		Reference to text in this book
	Relevant Event	Matter	Relevant Event	Matter	Paras
Inspection or testing	25.4.5.2	26.2.1	25.4.5.2	26.2.2	2.8.28 3.5.8
Late information (from statutory body)	25.4.7	26.2.2	–	–	8.2.10.5
Work by others	25.4.8.1	26.2.3.1	25.4.8.1	26.2.4.1	2.8.51–54
Supply of materials by Employer	25.4.8.2	26.2.3.2	25.4.8.2	26.2.4.2	2.8.55
Postponement	25.4.5.1	26.2.4	25.4.5.1	26.2.5	2.8.24
Failure to give ingress to or egress from the Works	25.4.12	26.2.5	25.4.12	26.2.6	2.8.62–63
Changes (variations)	25.4.5.1	26.2.6	25.4.5.1	26.2.7	2.8.18–22
Late information (from Employer)	25.4.6	26.2.7	–	–	8.2.10.3
Compliance with CDM Regulations	25.4.16	26.2.8	25.4.17	26.2.9	2.8.70
Suspension by the Contractor	25.4.17	26.2.9	25.4.18	26.2.10	2.8.71–75

8.3.14 The situation concerning relevant clause numbers is somewhat complicated by the fact that the matters listed in clause 26.2 of JCT 98 WCD are inexplicably ordered differently from those in clause 26.2 of JCT 98 PWQ. The remaining differences between the two forms in respect of these matters essentially replicate those concerning Relevant Events for which extensions of time may be granted (as noted at section 8.2). These differences can be summarised as follows:

1. *Late information (from Statutory Body)*

Clause 26.2.2 of JCT 98 WCD entitles the Contractor to reimbursement of loss and/or expense incurred as a result of delays in receiving ... *any permission or approval for the purposes of Development Control Requirements* ... These will in large part consist of the obtaining of planning approvals which, under JCT 98 WCD, are the responsibility of the Contractor (who is also the designer). There is no equivalent provision in clause 26 of JCT 98 PWQ although should the Architect's inability to secure such approvals lead to a delay on his part in issuing the Contractor with the necessary drawings and/or instructions then the Contractor would have recourse to clause 26.2.1 of JCT 98 PWQ (failure of the Architect to provide necessary instructions, etc. in due time) in order to found a claim for reimbursement of direct loss and/or expense.

2. *Changes (Variations)*
   Whereas under clause 26.2.7 of JCT 98 PWQ the Contractor is entitled to reimbursement of direct loss and/or expense caused by Architect's Instructions relating to Variations and/or the expenditure of provisional sums for undefined work, under clause 26.2.6 of JCT 98 WCD such reimbursement relates to Employer's instructions regarding Changes to the Employer's Requirements and/or the expenditure of provisional sums per se. Apart from this difference, which is essentially cosmetic and is a result of the different relationships which pertain between the parties under JCT 98 WCD, this Matter is the same in both of the contract forms.

3. *Late information (from Employer)*
   As with clause 25.4.6 (Relevant Events see para. 8.2.10.3) such late information will originate from the Employer (as opposed to the Architect) and will not cover drawings, details, etc. which under JCT 98 WCD are the responsibility of the Contractor.

4. *Approximate quantities*
   There are no Contract Bills under JCT 98 WCD. Accordingly, as there can be no approximate quantities under JCT 98 WCD then inaccuracies in approximate quantities cannot be a ground for the Contractor to claim reimbursement of direct loss and/or expense.

5. *Discrepancies between Contract Documents*
   As noted at para. 8.1.6, there are no Contract Documents as defined in JCT 98 PWQ (i.e. Contract bills, Contract Drawings, etc.) as such under JCT 98 WCD. Discrepancies between them, which can give rise to an entitlement to reimbursement of direct loss and/or expense under clause 26.2.3 of JCT 98 PWQ, cannot therefore arise under JCT 98 WCD. The only contractual document with which the Contractor will be provided by the Employer under JCT 98 WCD will be the Employer's Requirements. Discrepancies within the Employer's Requirements are dealt with in clause 2.4 of JCT 98 WCD. By virtue of this clause if such discrepancies have been dealt with within the Contractor's Proposals then the latter document will prevail and no entitlement to reimbursement of any direct loss and/or expense will arise. Where discrepancies have not been dealt with in the Contractor's Proposals then such measures as are taken to correct the discrepancies will constitute a Change in the Employer's Requirements and, if the Contractor suffers disruption as a result, entitlement to reimbursement of direct

loss and/or expense will arise from the application of clause 26.2.6 of JCT 98 WCD.

9

JCT INTERMEDIATE FORM OF BUILDING CONTRACT FOR WORKS OF SIMPLE CONTENT 1998 EDITION

9.1 INTRODUCTION

General

9.1.1 This chapter deals with the latest edition of the JCT Intermediate Form of Building Contract for works of simple content 1998 Edition incorporating Amendments 1 and 2 hereinafter referred to as JCT 98 IFC.

9.1.2 In 1984, JCT introduced its Intermediate Form of Building Contract, known as IFC 84, and later declared it as suitable for use on those contracts in the range between those for which JCT 80 and MW 80 had been issued, namely above £70,000 in value but not exceeding £280,000 (all at 1992 prices) and where the contract period was not more than 12 months. The endorsement on the back cover of IFC 84 and JCT 98 IFC describes the form as being suitable where the proposed building works are:

1. of a simple content involving the normally recognised basic trades and skills of the industry; and
2. without any building services installations of a complex nature, or other specialist work of a similar nature; and
3. adequately specified, or specified and billed, as appropriate prior to the invitation of tenders.

9.1.3 JCT Practice Note 20 (August 1993 Revision) adds that ... *the Intermediate Form may however be suitable for somewhat larger or longer contracts provided the three criteria referred to in the endorsement are met* ... The JCT is not in a position to lay down absolute rules covering those situations in which any of its family of contracts must be used and where their use is prohibited; this is entirely a matter for the parties. Nevertheless, it must be borne in mind that JCT 98 IFC is less detailed than JCT 98 PWQ and, where it is applied to larger and/or longer contracts, it is possible that it may not be able to cater equitably for all of the eventualities which may arise in such circumstances.

9.1.4 The parties under JCT 98 IFC, and the relationships between them, are generally as under JCT 98 PWQ in that the Architect (or Contract Administrator if not a registered architect) and the Quantity Surveyor are engaged by the Employer and are allocated similar duties to those set out in JCT 98 PWQ. Consistent with its application to simpler schemes, however, JCT 98 IFC does contain rather less in the way of procedural detail and also envisages contract documentation of a less comprehensive nature than that required under JCT 98 PWQ (drawings and specification, for instance, as an optional alternative to fully measured bills of quantities).

Named persons as sub-contractors

9.1.5 One notable respect in which JCT 98 IFC differs from JCT 98 PWQ is in the absence from the former document of any reference to Nominated Sub-Contractors. Instead, in accordance with the provisions of clause 3.3 of JCT 98 IFC, work can be required to be executed by a named person who is to be employed as a sub-contractor by the Contractor. Whilst this situation may prima facie appear to be not dissimilar to nomination (in that the Employer or Architect can exercise control over the Contractor's choice of sub-contractor) the allocation of risk is such that in many respects the named person is in effect a domestic sub-contractor to the Contractor. As this has important implications for the contract provisions dealing with, for instance, extensions of time and direct loss and/or expense it is necessary at this point to provide a brief explanation as to the nature of the "named person" and the contractual framework into which he fits.

9.1.6 JCT Practice Note IN/1 explains the procedure for naming a person to be employed by the Contractor as a sub-contractor in the following terms:

> ... *The provisions in clause 3.3 require the Contractor to employ as a sub-contractor for execution of part of the Works a person who may be named in one of the following ways:*
> – *Procedure One: in the Specification/Schedules of Work/Contract Bills for pricing by the Contractor in respect of work described therein; or*
> – *Procedure Two: in an instruction as to the expenditure of a provisional sum included in the Specification/Schedules of Work/Contract Bills*
> *In either case, the Contractor must be given a full description of the work, in the Specifications/Schedules of Work/Contract Bills under Procedure One or in the instruction under Procedure Two ...*

9.1.7 Also, in either case, the named person must have submitted a bona fide tender for the Works using the prescribed JCT Form of Tender and Agreement NAM/T which document will provide the basis for the sub-contract between the Contractor and the named person.

9.1.8 In the case of Procedure One described above, the Contractor is required to execute a sub-contract with the named person within 21 days of execution of the main contract between the Employer and the Contractor. In the case of Procedure Two described above, the Contractor has 14 days following the date of issue of the Architect's instruction within which to register an objection following which he is required to enter into a sub-contract with the named person. Clause 3.3 of JCT 98 IFC deals with the allocation of risk between the Employer and the Contractor, in terms of both time and money, should the Contractor find himself unable to execute a sub-contract with the named person or should it be necessary, having executed such a sub-contract, subsequently to determine that sub-contract. As these provisions differ from those in respect of Nominated Sub-Contractors set out in JCT 98 PWQ, and as they impact upon the subject matter of this book, they are explained below.

Inability to execute a sub-contract

9.1.9 If Procedure One is followed and the Contractor is unable to execute a sub-contract with the named person because certain of the particulars in the Contract Documents are preventing agreement between the Contractor and the named person then clause 3.3.1 of JCT 98 IFC sets out the following options for the Architect:

(a) *change the particulars so as to remove the impediment to such execution; or*

(b) *omit the work; or*

(c) *omit the work from the Contract Documents and substitute a provisional sum ...*

9.1.10 In the case of options (a) or (b) above, any such instruction by the Architect shall be classed as a Variation (as defined in clause 3.6) and may entitle the Contractor both to an extension of time (via clause 2.4.5) and to the reimbursement of direct loss and/or expense (via clause 4.12.7). In the case of option (c) above, an instruction will later be required from the Architect to expend the provisional sum which instruction may entitle the Contractor both to an extension of time and to reimbursement of loss and/or expense in the same way as with a Variation. In any event, therefore, the Employer assumes the risk (in terms of both time and money) should the particulars in the Contract Documents prevent the execution of a sub-contract between the Contractor and the proposed named person.

9.1.11 Should it not prove possible to enter into a sub-contract with the proposed named person for any reason other than the particulars in the Contract Documents proving an impediment then it would appear that the attendant risk remains with the Contractor by reason of clause 3.3.9 as follows:

> ... *Save as otherwise expressed in the Conditions the Contractor shall remain wholly responsible for carrying out and completing the Works in all respects in accordance with clause 1.1 notwithstanding the naming of a sub-contractor for the execution of work described in the Specifications/Schedules of Work/Contract Bills* ...

9.1.12 Should Procedure Two be adopted, the Contract contains no equivalent provisions in the event that the Contractor and the named person are unable to execute a sub-contract. In such circumstances clause 3.3.2 gives the Contractor a right of reasonable objection within 14 days of receiving the instruction to expend the provisional sum. The contract is, however, silent as to what should happen in the event that the Contractor exercises the right.

9.1.13 It would appear likely that the Architect, faced with an objection from the Contractor say on the grounds that the named person could not comply with the Contractor's programme, would be obliged to issue a further instruction naming an alternative person. Alternatively, he could allow the Contractor to enter into a sub-contract with the person originally named in the knowledge that the progress of the Works may be delayed and that the Contractor may incur direct loss and/or expense. In either event, it is suggested that the Contractor may be entitled to an extension of time (via clause 2.4.5) and to reimbursement of any loss and/or expense incurred (via clause 4.12.7).

9.1.14 Should it not prove possible to execute a sub-contract between the Contractor and the proposed named person, therefore, the allocation of risk as between the Employer and the Contractor would appear to be the same whichever procedure is followed. What the above discussion relative to Procedure Two does illustrate is the less comprehensive nature of JCT 98 IFC as compared to JCT 98 PWQ and how it is inappropriate to apply the former contract to an unduly complex situation.

Determination of the Sub-Contract

9.1.15 Clauses 27.2, 27.3 and 27.4 of NAM/SC (the conditions of sub-contract between the Contractor and the named person) empower the Contractor to determine the employment of the named person in the event of the named person's default, insolvency or corrupt act. In such a situation the allocation or risk, in terms of time and money, between the Contractor and the Employer is dealt with at clauses 3.3.3 to 3.3.5 of JCT 98 IFC.

9.1.16 Should the Contractor find it necessary to determine the employment of the named person under clauses 27.2, 27.3 or 27.4 of NAM/SC then clause 3.3.3 of JCT 98 IFC sets out the following options for the Architect:

(a) *name another person to execute the work or the outstanding balance of the work …*

(b) *instruct the Contractor to make his own arrangements for the execution of the work or the outstanding balance of the work …*

(c) *omit the work or the outstanding balance of the work …*

9.1.17 The precise allocation of risk is dependent both upon the option pursued and upon the means by which the named person was appointed.

9.1.18 Where the named person was appointed under Procedure One, clause 3.3.4 deals with the allocation of risk. In this event, should it cost more or less to complete the work than was originally envisaged then the Contract Sum is adjusted (in the case of all three options set out above) to account for the difference. In the case of options (b) and (c), any instruction from the Architect is to be treated as a Variation also entitling the Contractor (in principle) to an extension of time and to reimbursement of direct loss and/or expense. Under option (a), however, any such entitlement on the part of the Contractor is restricted to an extension of time, with no entitlement arising to reimbursement of direct loss and/or expense.

9.1.19 The situation where the named person has been appointed in accordance with Procedure Two is set out in clause 3.3.5 of JCT 98 IFC. In the case of all of options (a) to (c) above, any instruction issued by the Architect is to be treated as an instruction issued with regard to the expenditure of the original provisional sum. Such instruction is to be valued in accordance with clause 3.7 and may entitle the Contractor both to an extension of time (via clause 2.4.5) and to the reimbursement of any direct loss and/or expense incurred (via clause 4.12.7).

9.1.20 It would therefore appear that, with the exception of direct loss and/or expense incurred as a result of the Architect being required to name another person in circumstances where the original named person was appointed under Procedure One, the risk of time and money attaching to the determination by the Contractor of the named person's employment because of the default, insolvency or corrupt act of the named person lies with the Employer. If the employment of the named person is determined otherwise than in accordance with the provisions of clauses 27.2, 27.3 or 27.4 of NAM/SC (e.g. at the instigation of the named person because of some default by the Contractor under clause 28 of NAM/SC) then the risk lies with the Contractor in that he is entitled to no increase in the Contract Sum, no extension of time and no reimbursement of direct loss and/or expense suffered as a result.

9.1.21 With the exception of these two instances, i.e.:

(i) the inability to execute a sub-contract between the Contractor and the named person for reasons connected with particulars in the Contract Documents; and

(ii) the determination of the named person's employment by reason of his default or insolvency;

the risks of time and money attaching to the employment of named persons under JCT 98 IFC lies with the Contractor. In this way the treatment of named persons under JCT 98 IFC differs from that of Nominated Sub-Contractors under JCT 98 PWQ. There is, for instance, no provision under JCT 98 IFC which entitles the Contractor to an extension of time in respect of delay on the part of a named person such as there is in respect of delay on the part of a Nominated Sub-Contractor by way of clause 25.4.7 of JCT 98 PWQ.

9.2 EXTENSIONS OF TIME

Procedures

9.2.1 Table 9.1 provides references to the text within this book which is also relevant to the following review of JCT 98 PWQ.

Table 9.1. Extension of time procedures

Provision in Contract	Clause in JCT 98 IFC	Equivalent clause in JCT 98 PWQ	Explanatory text in this book
Requirement for the Contractor to give notice of delay and to give further details to assist the Architect in establishing any extension of time.	2.3	25.2.1	2.7.7
Requirement for the Architect, in certain circumstances, to grant extensions of time during the construc-ion period or to notify the Contractor that an extension of time is not, in his view, justified.	2.3	25.3.1	2.7.11 to 2.7.31
Requirement for the Architect to grant extensions of time in respect of acts of prevention occurring during a period of culpable delay.	2.3	–	7.2.6 9.2.5 to 9.2.8
Requirement for the Architect to review the Completion Date following Practical Completion.	2.3	25.3.3	2.7.32 to 2.7.35
General obligations imposed upon the Architect and Contractor in relation to their conduct.	2.3	25.3.4, 25.3.5 and 25.3.6	2.7.42 to 2.7.45

9.2.2 The procedures by which extensions of time are granted under clause 2.3 of JCT 98 IFC are similar to those set out in clause 25 of JCT 98 PWQ, albeit with less in the way of detail.

9.2.3 It is a requirement under both forms that the Contractor must, as soon as a problem becomes reasonably apparent, notify the Architect of the cause of any delay. Unlike clause 25.2.1.1 of JCT 98 PWQ, however, there is no requirement on the Contractor under clause 2.3 of JCT 98 IFC to identify separately in his written notice those causes which are Relevant Events. Furthermore, the requirement at clause 25.2.2 of JCT 98 PWQ for the Contractor to provide, in respect of each and every Relevant Event identified in his notice, full particulars of the expected effects thereof and an estimate of the expected delay to the Completion Date effectively becomes in clause 2.3 of JCT 98 IFC no more than a requirement to provide such information as is reasonably necessary to enable the Architect to assess the Contractor's entitlement to an extension of time. Clause 25.3 of JCT 98 PWQ is also quite specific in setting down the timescale within which the Architect is to respond to the Contractor's request for an extension of time – i.e. at most within 12 weeks of receipt of adequate particulars and estimate from the Contractor. Clause 2.3 of JCT 98 IFC, on the other hand, simply requires the Architect to act ... *so soon as he is able* ...

9.2.4 Section 9.1 explains the position under JCT 98 IFC of named persons employed by the Contractor as sub-contractors, and distinguishes between this and the position of Nominated Sub-Contractors under JCT 98 PWQ. With the exception of the two instances discussed at para. 9.1.21, all issues relating to delay on the part of the named person are a matter as between the Contractor and the named person and do not directly involve or affect either the Employer or the Architect. There is no requirement under clause 12 of NAM/SC, as there is under clauses 2.2 to 2.5 of NSC/C (discussed at section 6.2), for the Architect to be notified of any anticipated delay to the sub-contract works nor for the Architect's approval to be sought in respect of any extension proposed to the sub-contract period. Similarly the requirement, under clause 25 of JCT 98 PWQ, for Nominated Sub-Contractors to receive from the Architect and the Contractor copies of relevant notices, details and awards of extensions of time is absent in respect of named persons under clause 2.3 of JCT 98 IFC.

9.2.5 One matter which is dealt with expressly under clause 2.3 of JCT 98 IFC, although not so under clause 25 of JCT 98 PWQ, concerns delays occasioned by the Employer during a pre-existing period of delay for which the Contractor is responsible (also known as "culpable delay"). It was previously argued by contractors that JCT contracts did not empower the Architect to grant an extension of time in respect of events occurring after the Completion Date set down in the contract had passed and that, should the Contractor's progress be delayed by the Employer during this period, the Completion Date would cease to apply and time would become at large. The practical effect of such an argument is that, from the point in time at which the Contractor is delayed by the Employer, the Contractor has a reasonable period of time in which to complete the Works and the Employer is debarred from levying liquidated and ascertained damages against the Contractor in respect of any delay already caused by the Contractor. It would therefore be possible to envisage a situation in which the Contractor, despite having himself caused delays amounting to several months, would be released from any liability to pay liquidated and ascertained damages in respect of such delays merely because, on the final day of the contract, he is requested by the Architect to apply an additional coat of paint to a door.

9.2.6 Clause 2.3 of JCT 98 IFC seeks to avoid such an eventuality by giving the Architect an express power to grant extensions of time in respect of what are known as "acts of prevention" which occur during a period of culpable delay. Acts of prevention can be defined as those of the events set out in clause 2.4, such as late and/or varied information, which are within the control of the Employer or his agents (as opposed to those, such as exceptionally adverse weather conditions, which are not). They will be found to correspond with those matters set out at clause 4.12 of JCT 98 IFC which give rise to an entitlement to reimbursement of direct loss and/or expense.

9.2.7 Although clause 25 of JCT 98 PWQ contains no equivalent provision, the decision of the Commercial Court in the case of *Balfour Beatty Building Limited -v- Chestermount Properties Limited [1993]* confirms that an Architect is in fact empowered under JCT 98 PWQ to grant extensions of time in respect of Relevant Events which occur during a period of culpable delay. This case also confirms that the extension granted by the Architect should equate to the extent of the delay caused by the Relevant Event in question and that this extension ought to run from the previously fixed Completion Date. By way of explanation of this point, one can imagine a contract with a Completion Date of 31 December 1998 but which, due to the fault of the Contractor, is still incomplete on 30 June 1999. At this point the Contractor is instructed to carry out a Variation which causes a further 2 weeks of delay in respect of which he is entitled to an extension of time. Following the decision in *Balfour Beatty*, the Completion Date ought now to be refixed as 14 January 1999, notwithstanding the fact that the Contractor was not instructed to carry out the Variation until 31 June 1999.

9.2.8 The position at law in respect of such matters would therefore appear to be identical as between both JCT 98 PWQ and JCT 98 IFC, despite the difference in wording.

9.2.9 Where the two forms do differ is in respect of the process by which the Architect is permitted to review previous decisions regarding extensions of time. Once the Completion Date has passed clause 25.3.3 of JCT 98 PWQ allows the Architect to review his earlier decisions and further, provides that he must do so within 12 weeks following the date of Practical Completion. It is also envisaged that this review process may lead to the Architect fixing a Completion Date earlier than that previously fixed provided that, following an earlier award of an extension of time, an instruction has been issued omitting works. Under clause 2.3 of JCT 98 IFC, however, the Architect is allowed, but not required to review any previous decision at any time up to 12 weeks after the date of Practical Completion but he is not allowed to fix a Completion Date earlier than that previously fixed.

Relevant Events

9.2.10 Table 9.2 provides references to the text within this book which is also relevant to the following review of JCT 98 IFC.

Table 9.2. Relevant Events

Relevant Event	Clause in JCT 98 IFC	Equivalent clause in JCT 98 PWQ	Explanatory text in this book
Force majeure	2.4.1	25.4.1	2.8.2 to 2.8.3
Exceptionally adverse weather conditions	2.4.2	25.4.2	2.8.4
Loss or damage due to Specified Perils	2.4.3	25.4.3	2.8.5
Civil commotion	2.4.4	25.4.4	2.8.6 and 2.8.7
Compliance with Architect's instructions	2.4.5	25.4.5.1	2.8.8 to 2.8.27

Relevant Event	Clause in JCT 98 IFC	Equivalent clause in JCT 98 PWQ	Explanatory text in this book
Opening up or testing	2.4.6	25.4.5.2	2.8.28
Late information	2.4.7	25.4.6	2.8.29 to 2.8.44
Works by others	2.4.8	25.4.8.1	2.8.51 to 2.8.54
Supply of materials by Employer	2.4.9	25.4.8.2	2.8.55
Inability to secure labour	2.4.10	25.4.10.1	2.8.58 to 2.8.60
Inability to secure materials	2.4.11	25.4.10.2	2.8.58 to 2.8.60
Site access/egress restrictions	2.4.12	25.4.12	2.8.62 and 2.8.63
Work by local authority/statutory undertaker	2.4.13	25.4.11	2.8.61
Deferment of date of possession	2.4.14	25.4.13	2.8.64
Approximate quantities	2.4.15	25.4.14	2.8.65 to 2.8.67
Terrorism	2.4.16	25.4.16	2.8.69
CDM Regulations	2.4.17	25.4.17	2.8.70
Suspension by the Contractor	2.4.18	25.4.18	2.8.71 to 2.8.75

.2.11 The events which entitle the Contractor to an extension of time are set out in clause 2.4 of JCT 98 IFC and are in large part identical to those in clause 25.4 of JCT 98 PWQ (save that, in JCT 98 IFC, they are not referred to as "Relevant Events" – they are simply referred to as "events"). There are some differences of substance between the two forms, however, and these can be summarised as follows:

1. *Instructions*

   Clause 2.4.5 of JCT 98 IFC is identical in substance to clause 25.4.5.1 of JCT 98 PWQ save that instructions issued under clause 3.3 of JCT 98 IFC, relate to named sub-contractors whilst those issued under clauses 35 and 36 of JCT 98 PWQ relate respectively to Nominated Sub-Contractors and Nominated Suppliers. In addition, there is no equivalent in JCT 98 IFC to clause 13A of JCT 98 PWQ which empowers the Architect to instruct the Contractor to provide a quotation in respect of a variation instruction before it is implemented. Nor is there an equivalent in JCT 98 IFC to clause 34 of JCT98 PWQ which empowers the Architect to issue instructions in the event of a find of antiquities or fossils. This suggests that JCT 98 IFC may be unsuitable for contracts undertaken in areas of historical interest, as the Architect is not empowered to grant an extension of time should excavations reveal the remains of, for instance, a Roman settlement. He is, however, able to issue instructions requiring a valuation in respect of such remains under clauses 3.6.1 and 3.6.2, both of which could entitle the Contractor to an extension of time.

2. *Nominated Sub-Contractors and/or Suppliers*

   Whilst provision is made for named sub-contractors, there is no equivalent in JCT 98 IFC to clause 25.4.7 of JCT 98 PWQ, which entitles the Contractor to an extension of time in the event of delay on the part of Nominated Sub-Contractors and Suppliers. This is quite simply because, as previously noted, Nominated Sub-Contractors and Suppliers do not exist under JCT 98 IFC. Delay on the part of named sub-contractors does not of itself entitle the Contractor to an extension of time under JCT 98 IFC and this does not, therefore, appear as an event under clause 2.4 thereof.

3. *Exercise of Statutory Powers*

There is no equivalent in JCT 98 IFC to clause 25.4.9 of JCT 98 PWQ, which entitles the Contractor to an extension of time should any governmental action affect the progress of the Works by restricting the availability of labour or materials. As this eventuality is somewhat remote in the foreseeable future, and as JCT 98 IFC is not generally recommended for contracts of greater than 12 months' duration such a provision as this appears to have been regarded as superfluous.

4. *Inability to secure labour and materials*

Clauses 2.4.10 and 2.4.11 of JCT 98 IFC and clause 25.4.10 of JCT 98 PWQ deal with the same issue, that is the Contractor's entitlement to an extension of time if, for reasons beyond his control, he is unable to secure the necessary labour and/or material to progress the Works. Under JCT 98 PWQ these provisions are compulsory whereas, by contrast, under JCT 98 IFC they are optional – in the latter case it is necessary to state in the Appendix to the contract whether or not clauses 2.4.10 and 2.4.11 are to apply. To require the parties to positively choose one or other option would be both unnecessary and confusing given that the relevant clause simply states ... *where this clause is stated in the Appendix to apply* ...

5. *Performance Specified Work*

There is no equivalent in JCT 98 IFC to clause 25.4.15 of JCT 98 PWQ, which entitles the Contractor to an extension of time by reason of any change in Statutory Requirements which has an effect upon any Performance Specified Work. The reason for this is that there is in JCT 98 IFC no equivalent to clause 42 of JCT 98 PWQ, which deals with Performance Specified Work.

Fluctuations

9.2.12 Supplemental Conditions C and D to JCT 98 IFC (which are optional) set out the mechanisms by which the Contractor's entitlement to payment is adjusted in the event of there being increases or decreases in prices arising during the contract period. By virtue of conditions C4.7 and D12, the Contractor's entitlement to extra payment does not encompass increased costs arising during a period of culpable delay on the part of the Contractor. It is worthy of note in this respect that this only continues to be the case should the printed text of clauses 2.3, 2.4 and 2.5 of JCT 98 IFC (which contain the provisions relating to extensions of time) remain unamended and provided that the Architect has, in respect of every written notification by the Contractor under clause 2.3 of JCT 98 IFC, stated in writing what extension of time, if any, he has granted. Any amendment whatsoever to the printed text or any failure of the Architect to state in writing what, if any, extension of time he has granted where either Supplemental Conditions C or D apply, will result in the Contractor becoming entitled to increased costs which arise even during a period of culpable delay on his part.

9.2.13 Supplemental Conditions C4.7 and D12 to JCT 98 IFC reflect clauses 38.4.7 and 40.7.1 of JCT 98 PWQ respectively and the reader is referred to sections 2.12 and 4.7

in this connection. There is no equivalent in JCT 98 IFC to clause 39 of JCT 98 PWQ (Labour and Materials Cost and Tax Fluctuations).

9.3 VARIATIONS AND DISRUPTION

Variations

9.3.1 Table 9.3 provides references to the text within this book which is also relevant to the following review of JCT 98 IFC.

Table 9.3. Valuation of Variations

Valuation Rule	Clause in JCT 98 IFC	Equivalent clause in JCT 98 PWQ	Explanatory text in this book
Valuation of Variations and provisional sum work – Alternative A Within 21 days of receipt of instruction or from the commencement of work for which an Approximate Quantity is included in the Contractdocuments the Contractor **may** submit a Price Statement for such work.	3.7.1.2A1	13.4.1.2A1	3.3.4
Contractor **may** separately attach to the Price Statement:			3.3.5
• any amount required in lieu of direct loss and/or expense.	3.7.1.2A1.1	13.4.1.2A1.1	
• any adjustment required to the Completion Date.	3.7.1.2A1.2	13.4.1.2A1.2	
Acceptance of Price Statement (in whole or in part) by Quantity Surveyor within 21 days of receipt from Contractor.	3.7.1.2A2	13.4.1.2A2	3.3.3
Non-acceptance of Price Statement (in whole or in part) – the Quantity Surveyor to supply amended Price Statement for Contractor to accept or reject within 14 days.	3.7.1.2A4	13.4.1.2A4	3.3.3
When the Price Statement or amended Price Statement are neither accepted nor referred to Adjudication valuation of the work shall be by Alternative B.	3.7.1.2A6	13.4.1.2A6	3.3.2
Where attachments to the Price Statement are not accepted direct loss and/or expense and revisions to the Completion Date are to be dealt with as provided for elsewhere in the Contract.	3.7.1.2A7	13.4.1.2A7	3.3.6

(continued)

Table 9.3. (cont.)

Valuation Rule	Clause in JCT 98 IFC	Equivalent clause in JCT 98 PWQ	Explanatory text in this book
Valuation of Variations and provisional sum work – Alternative B Valuation by reference to the Contract Bills/priced document where the additional or substituted work is of similar character, is executed under similar conditions and does not significantly change the quantity of work.	3.7.4(a)	13.5.1.1	3.3.26
Fair allowance on the above in the event that work of a similar character is not executed under similar conditions and/or there is a significant change in the quantity of work.	3.7.4(a)	13.5.1.2	3.3.29 3.3.31 3.3.34 3.3.43
Valuation at fair rates or prices in the event that the additional or substituted work is not of similar character to that in the Contract Bills/priced document.	3.7.5	13.5.1.3	3.3.29, 3.3.31 3.3.34 3.3.43
Valuation by reference to the value entered against an approximate quantity where such quantity is a reasonably accurate forecast of the work required.	3.7.4(b)	13.5.1.4	–
Fair allowance on the above in the event that the approximate quantity is not a reasonably accurate forecast of the work required.	3.7.4(b)	13.5.1.5	3.3.29
Valuation of omitted works by reference to the rates or prices in the Contract Bills/priced document	3.7.3	13.5.2	–
In valuing in accordance with the above allowance is to be made for any addition to or reduction of preliminary items.	3.7.7	13.5.3	3.3.23
Valuation by reference to dayworks where this would be the appropriate basis of a fair valuation of the additional or substituted work.	3.7.6	13.5.4	3.3.35 to 3.3.42
Valuation of other works as if they were variations in the event that a variation substantially changes the conditions under which such other works are executed.	3.7.9	13.5.5	3.3.43
Fair valuation to be made of work which cannot be valued in accordance with above rules.	3.7.5 3.7.10	13.5.7	3.3.29 3.3.34

9.3.2 Instructions to the Contractor are dealt with in clause 3.5 of JCT 98 IFC. The provisions in JCT 98 IFC in this respect differ from those in JCT 98 PWQ only in that there is no equivalent in JCT 98 IFC to clause 4.3.2 of JCT 98 PWQ, which enables oral instructions from the Architect to be confirmed subsequently by the Contractor in writing. All instructions issued under JCT 98 IFC must be in writing by the Architect although, by virtue of clause 3.6 of JCT 98 IFC (which replicates clause 13.2.4 of JCT 98 PWQ), the Architect may retrospectively sanction in writing an uninstructed variation already carried out by the Contractor since the clause provides that the Architect may ... *sanction in writing any Variation made by the Contractor otherwise than pursuant to an* [earlier written] *instruction* ...

9.3.3 Both clause 3.6 of JCT 98 IFC and clause 13.1 of JCT 98 PWQ define Variations in essentially the same way. The manner in which Variations, provisional sum work and work covered by an Approximate Quantity are to be valued is set out in clause 3.7.1 of JCT 98 IFC and this, like its counterpart (clause 13.4.1 in JCT 98 PWQ) provides for the valuation of such work to be made either by way of a Contractor's Price Statement – Alternative A (if the Contractor so chooses) or by way of specific Valuation rules – Alternative B.

9.3.4 The provisions in JCT 98 IFC relating to valuation Alternative A are the same as those in JCT 98 PWQ. Hence the detailed consideration given to these JCT 98 PWQ clauses in section 3.3 applies equally to their JCT 98 IFC counterparts and consequently there is no need to comment further here.

9.3.5 However, the specific valuation rules set out in clause 3.7 of JCT 98 IFC are somewhat shorter than those in clause 13.5 of JCT 98 PWQ albeit the general principles involved remain the same. Indeed, the rules for the valuation of Variations in clause 3.7 of JCT 98 IFC are not dissimilar to those in clause 12.5 of JCT 98 WCD (discussed in section 8.3). This is a reflection of the somewhat less detailed pricing document likely to be in use in respect of those two contract forms as compared with that to be expected in connection with a contract let on JCT 98 PWQ. Again, in common with the other JCT forms JCT 98 IFC requires the valuation of any Variation to include an amount in respect of additional preliminary items which attach to such Variation (clause 3.7.7) but stipulates (at clause 3.7.8) that costs which arise as a result of disruption to the progress of the Works must be recovered by reference to clause 4.11 (ascertainment of direct loss and/or expense).

9.3.6 Both JCT 98 IFC and JCT 98 PWQ allow for the machinery for the valuation of Variations to be set aside in favour of a sum agreed between the Employer and the Contractor. In this respect, clause 3.7.1.1 of JCT 98 IFC makes specific reference to the fact that such agreement must be made prior to the Contractor complying with the instruction. JCT 98 PWQ is not specific in this regard and it would appear that, using this form of contract, the Employer and the Contractor can agree between themselves the value of Variations even after such Variations have been carried out and completed by the Contractor. It is suggested that the approach taken by JCT 98 IFC is to be preferred, in that certainty as to the Employer's financial liability to the Contractor may be determined from the outset.

9.3.7 We have explained the operation of clause 13A of JCT 98 PWQ in section 3.4. This clause permits the Architect to obtain in advance from the Contractor a quotation in respect of a proposed Variation. This quotation will set out not only the cost of carrying out the work but also details of any adjustment to the time required for

279

completion of the Works and the amount required to be paid in lieu of any ascertainment of direct loss and/or expense as a result of carrying out the Variation. There is no equivalent to clause 13A in JCT 98 IFC and therefore, there are no corresponding implications in respect of any of the clauses dealing with variations, extensions of time and direct loss and/or expense.

Procedures for the ascertainment of direct loss and/or expense

9.3.8 Table 9.4 provides references to the text within this book which is also relevant to this review of JCT 98 IFC.

Table 9.4. Direct loss and/or expense procedures

Direct loss and/or expense procedures	Clause in JCT 98 IFC	Equivalent clause in JCT 98 PWQ	Explanatory text in this book
Contractor to make a written application regarding direct loss and/or expense.	4.11	26.1	3.5.9
Architect to form an opinion as to the Contractor's entitlement.			3.5.16 and 3.5.17
Architect (or Quantity Surveyor) to ascertain the amount of the Contractor's entitlement.			3.5.22 to 3.5.33
Contractor's written application must be made timeously.		26.1.1	3.5.10, 3.5.11 and 3.5.20
Contractor is to submit such information as will enable the Architect to form an opinion as to entitlement and the Architect or Quantity Surveyor to ascertain any amount due.		26.1.2 and 26.1.3	3.5.13 and 3.5.14
Amounts ascertained as being due are to be added to the Contract Sum.		26.5	3.5.18
The above provisions are without prejudice to any other rights or remedies which the Contractor may possess.		26.6	3.5.9 and 3.5.21

9.3.9 The procedures by which the Contractor is able to recover direct loss and/or expense from the Employer are set out in clause 4.11 of JCT 98 IFC. In common with the philosophy behind JCT 98 IFC the style is somewhat simpler than the equivalent clause 26.1 of JCT 98 PWQ but, in substance, the two clauses are essentially the same. The Contractor is required to make a timeous application for reimbursement following which the Architect is required to form an opinion as to the Contractor's entitlement in principle. Should the Architect be of the opinion that the Contractor is so entitled then the amount of the loss and/or expense is to be ascertained either by the Architect or by the Quantity Surveyor. The Contractor for his part is obliged to provide the Architect and/or the Quantity Surveyor with such information as is necessary to establish both the fact of the Contractor's entitlement and its quantum.

9.3.10 Procedurally, JCT 98 IFC does differ from JCT 98 PWQ in two respects. The first of these relates to clause 26.3 of JCT 98 PWQ which has no equivalent in JCT 98 IFC. This clause requires the Architect, insofar as it is necessary for the ascertainment of direct loss and/or expense, to notify the Contractor in writing as to those extensions of time which have been granted in respect of Relevant Events which also qualify as Matters under clause 26.2. Section 2.5 contains a discussion of the relationship between extensions of time and direct loss and/or expense and concludes that, whilst there is of course a relationship between time and money, there are dangers in assuming too close a correlation between delays to the Completion Date and entitlement to reimbursement of direct loss and/or expense. Extensions of time can be granted without any accompanying entitlement to recovery of direct loss and/or expense, and vice versa. Clause 26.3 of JCT 98 PWQ has been criticised by some commentators as encouraging the view that extensions of time and direct loss and and/or expense are directly interdependent. As noted above, this clause has no equivalent in JCT 98 IFC.

9.3.11 The second difference between the two forms in respect of procedural matters concerns clause 26.4 of JCT 98 PWQ (which is discussed in detail at paras 3.5.19 and 6.3.6). This deals with the situation whereby a Nominated Sub-Contractor considers himself to be entitled to recovery of direct loss and/or expense and involves the Architect and Quantity Surveyor directly in ascertaining both the fact and the quantum of the Nominated Sub-Contractor's entitlement. Any entitlement of a named sub-contractor under JCT 98 IFC in this respect is essentially a matter for agreement between the Contractor and the sub-contractor and is dealt with purely by reference to clause 14 of NAM/SC. Of course, if the Contractor feels himself entitled to reimbursement by the Employer of sums agreed between himself and the sub-contractor further to the operation of clause 14 of NAM/SC then this will be dealt with as between the Contractor and the Employer in accordance with clause 4.11 of JCT 98 IFC and the sums involved will then become subject to ascertainment by the Architect or the Quantity Surveyor. In this way it is quite possible (unlike the operation of clause 26.4 of JCT 98 PWQ) for the sum awarded to the Contractor under clause 4.11 of JCT 98 IFC to be different to the sum agreed between the Contractor and the named sub-contractor under clause 14 of NAM/SC.

List of matters

9.3.12 Table 9.5 provides references to the text within this book which is also relevant to the following review of JCT 98 IFC.

Table 9.5. List of matters

Cause	JCT 98 IFC clauses		JCT 98 PWQ clauses		Reference to text in this book
	Relevant Event	*Matter*	*Relevant Event*	*Matter*	*Paras*
Late information	2.4.7.1	4.12.1.1	25.4.6.1	26.2.1.1	2.8.29–32
	2.4.7.2	4.12.1.2	25.4.6.2	26.2.1.2	2.8.33–44
Inspection and testing	2.4.6	4.12.2	25.4.5.2	26.2.2	2.8.28, 3.5.8
Work by others	2.4.8	4.12.3	25.4.8.1	26.2.4.1	2.8.51–54
Supply of materials by Employer	2.49	4.12.4	25.4.8.2	26.2.4.2	2.8.55
Postponement	2.4.5	4.12.5	25.4.5.1	26.2.5	2.8.24
Failure to give ingress to or egress from the Works	2.4.12	4.12.6	25.4.12	26.2.6	2.8.62–63
Variations (including those relating to inconsistencies)	2.4.5	4.12.7	25.4.5.1	26.2.3 26.2.7	2.8.18–22
Approximate quantities	2.4.15	4.12.8	25.4.14	26.2.8	2.8.65–67
Compliance with CDM Regulations	2.4.17	4.12.9	25.4.17	26.2.9	2.8.70
Suspension by the Contractor	2.4.18	4.12.10	25.4.18	26.2.10	2.8.71–75

9.3.13 The list of matters in respect of which the Contractor is entitled to reimbursement of direct loss and/or expense is set out in clause 4.12 of JCT 98 IFC. There is only one respect in which this list differs from that set out in clause 26.2 of JCT 98 PWQ and this is again in respect of named Sub-Contractors. By operation of clause 4.12.7 of JCT 98 IFC, the contractor may be entitled to the recovery of direct loss and/or expense arising as a result of instructions issued by the Architect in connection with a named Sub-Contractor. In this respect, clause 4.12.7 refers in turn to clause 3.3 which sets out the following instructions in respect of which the Contractor may be entitled to reimbursement of direct loss and/or expense:

- instructions pursuant to the Contractor's inability to execute a Sub-Contract (clause 3.3.1 refers);
- instructions as to the expenditure of a provisional sum (clause 3.3.2 refers);
- instructions pursuant to the determination of the Sub-Contractor's employment, except where the instruction in question names an alternative person to complete the balance of the work in a situation in which the original named person was appointed in accordance with Procedure One (clauses 3.3.4 and 3.3.5 refer).

These matters are discussed in detail in section 9.1.

9.4 ASCERTAINING THE DIRECT LOSS AND/OR EXPENSE

9.4.1 The matters discussed in Chapter 4, as regards the heads under which claims for direct loss and/or expense may be submitted and the principles which must be brought to bear in assessing the Contractor's entitlement, are equally applicable to loss and/or expense arising under clause 4.11 of JCT 98 IFC as under clause 26 of JCT 98 PWQ. They will not therefore be the subject of further discussion here.

10

JCT STANDARD FORM
OF MANAGEMENT CONTRACT
1998 EDITION

10.1 INTRODUCTION

General

10.1.1 This chapter differs from Chapters 6–9 in that it considers separate but related Contracts, namely a Management Contract and a Works Contract. These forms were first published by JCT in 1988 in order to provide a standard contract form for management contracting which, at the time, had become an increasingly popular method of procurement. The philosophy behind this method of procurement can be summarised as follows:

- to isolate the **management** of the construction work as a discrete function and place it in the hands of a Management Contractor who is not responsible for physically carrying out any of the work on site;
- to bring the expertise of the Management Contractor to bear during the pre-construction phase and to encourage co-operation between the Management Contractor and the Professional Team;
- to divide the Works as a whole into discrete work packages to be tendered for by and let to individual Works Contractors employed by the Management Contractor under separate Works Contracts thereby permitting a greater degree of overlap between the design and construction phases of the project.

10.1.2 This chapter deals with the JCT Standard Form of Management Contract 1998 Edition incorporating Amendments 1 and 2, hereinafter referred to as JCT 98 MC, the Works Contract (Works Contract/1) 1998 Edition incorporating Amendments 1 and 2, hereinafter referred to as WC/1 and the Works Contract Conditions (Works Contract/2) 1998 Edition incorporating Amendments 1 and 2, hereinafter referred to as WC/2.

10.1.3 The Management Contractor's primary obligation to the Employer is recorded in Article 1.2 of JCT 98 MC and is to ... *set out, manage, organise, supervise and secure the carrying out and completion of the Project* ... In order to achieve this the Management Contractor is required to engage Works Contractors to carry out the physical work on the site. Contracts between the Management Contractor and the various Works Contractors are subject to the terms and conditions set out in WC/1 and WC/2. Each Works Contractor is responsible to the Management Contractor for the Works described in his own Works Contract documentation, whilst the Management Contractor is responsible to the Employer for the Project as a whole.

10.1.4 The Contract Conditions set out in WC/2, which regulate the relationship between the Management Contractor and the Works Contractor, are not dissimilar to those in NSC/C which regulate the relationship between Contractor and Nominated Sub-Contractor in the JCT 98 PWQ arrangement. However, the structure and content of JCT 98 MC differs from that of JCT 98 PWQ in order to reflect the approach adopted in management contracting. The discussion which follows may, as a consequence, be somewhat more enlightening if preceded by a brief explanation of the general mechanics of JCT 98 MC.

10.1.5 One of the more obvious differences between JCT 98 MC and JCT 98 PWQ is that the former is entered into by the parties before the design of the Project has been completed and well in advance of a decision to proceed with construction. Consequently the Project timescale is divided into two distinct periods, with the option available to the Employer to determine the Management Contractor's employment after the first period. They are:

(a) the Pre-Construction Period – this is analogous to the pre-contract period for a JCT 98 PWQ Contract, during which the design and planning of the project is developed;

(b) the Construction Period – this is analogous to the contract period for a JCT 98 PWQ contract and begins on the day when the Management Contractor is given possession of the site.

10.1.6 A detailed list of all services required to be performed by the Management Contractor during both of these periods is set out in the Third Schedule to JCT 98 MC. During the Pre-Construction Period these services are primarily concerned with preparing project and construction programmes, advising on buildability, layout of site facilities and Works Contracts packages, agreeing the Contract Cost Plan, preparing lists of Works Contractors, assisting in the procurement of Works Contractors and suppliers, and agreeing terms for Project Insurance. During the Construction Period the services are mainly concerned with planning and programming, monitoring the progress of design work and drawing production, cost control and payment disciplines, labour relations and procedures, site management and quality control, and controlling and reporting on the performance of Works Contractors.

10.1.7 The Professional Team as defined at clause 1.3 of JCT 98 MC comprises the Architect, the Quantity Surveyor and such other consultants as are appointed by the Employer. The Management Contractor himself is not a member of the Professional Team, notwithstanding the fact that the management contracting ethos regards the function which he performs as a professional discipline in its own right. Unlike JCT 98 PWQ, however, there is via Article 1 and clause 1.4 of JCT 98 MC, a specific requirement upon the Management Contractor to co-operate with the Professional Team throughout the course of the Project. It is anticipated that such co-operation will manifest itself, among other things, in the Management Contractor's input into the design on questions of buildability, the preparation of the Project programme for acceptance by the Professional Team and the agreement of the Contract Cost Plan with the Quantity Surveyor.

10.1.8 Unlike the position under a JCT 98 PWQ contract, the Management Contractor under JCT 98 MC gives no overall commitment to a price for the Project. Prior to proceeding with the construction of the Project, a Contract Cost Plan will be agreed between the Management Contractor and the Quantity Surveyor. This amounts to no more than the parties' best estimate of the likely overall cost of the Project and is in no way binding upon the Management Contractor, since his actual remuneration will instead comprise the following elements:

Element 1 – Prime Cost

10.1.9 This represents the actual cost to the Management Contractor of carrying out the work on site and is defined in the Second Schedule to JCT 98 MC. The Prime Cost itself can be sub-divided into the following elements:

(a) amounts payable by the Management Contractor to the various Works Contractors;

(b) amounts incurred directly by the Management Contractor in respect of his own on-site staff, facilities and other sundry costs, on-site labour, materials and goods, plant, consumable stores and services.

In general, the Management Contractor will be reimbursed his actual costs in respect of the above save where such costs are incurred as a result of his negligence, in which case they do not form part of the Prime Cost and must be met by the Management Contractor himself. Alternatively, lump sums can be agreed between the Employer and the Management Contractor in respect of all or any of those items comprising (b) above, together with a basis upon which such lump sums will be adjusted in the event of compliance by the Management Contractor with the Architect's Instructions.

Element 2 – Management Fee

10.1.10 In respect of the management services which the Management Contractor provides, he is paid a Management Fee. In order to reflect the manner in which the Project timescale is divided, the Management Contractor's fee is similarly divided into:

- the Pre-Construction Period Management Fee is recorded in Appendix Part 1 and Part 2 to JCT 98 MC and will either be a non-adjustable lump sum or, alternatively, an amount to be calculated by reference to a Schedule agreed by the parties (generally on the basis of resources employed by the Management Contractor).

- the Construction Period Management Fee is recorded in Appendix Part 2 to JCT 98 MC and is a lump sum which is adjusted by reference to the formula set out in clause 4.10 of JCT 98 MC should the final total of the Prime Cost vary from the Contract Cost Plan Total by a margin of more than 5% (or such other percentage as is given in the Appendix to JCT 98 MC).

Should the Management Contractor's employment be determined before the Construction Period then pursuant to clause 7.22 of JCT 98 MC the Employer's liability to the Management Contractor is restricted to an appropriate proportion of the Pre-Construction Period Management Fee.

Element 3 – Sums arising out of Works Contractors' default

10.1.11 As noted at para. 10.1.24, should sums become payable by the Management Contractor to a Works Contractor as a result of a breach of contract by another Works Contractor then the Management Contractor is entitled to recover from the Employer any part of such sums which he is unable to recover from the defaulting Works Contractor. According to para. 1.2 of Part 2 of the Second Schedule of JCT 98 MC these amounts do not form part of the Prime Cost or the Management Fee and must therefore be considered as a separate element or the Management Contractor's overall potential remuneration.

10.1.12 As noted at para. 10.1.3, upon acceptance of a tender in respect of a particular work package the Management Contractor is required to execute a contract with the successful Works Contractor using forms WC/1 and WC/2. There is, therefore, no direct contractual link between the Employer and the Works Contractors. The Employer's only contractual link in respect of the construction work is with the Management Contractor, whose primary obligation is to secure the carrying out and completion of the Project by means of contracts entered into with the various Works Contractors. The exception to this situation is where the Employer and the Works Contractor execute between them the JCT Standard Form of Employer/ Works Contractor Agreement (hereinafter referred to as WC/3). This constitutes a collateral contract between the Employer and the Works Contractor whereby the latter provides a direct but limited warranty to the Employer in respect of certain matters, predominantly related to specialist design work, selection of materials and satisfaction of performance specification requirements.

10.1.13 Because of the manner in which the various contract forms interrelate, it would be misleading and uninformative to discuss any one of these in isolation. Many of the issues with which we are concerned will involve sets of circumstances which encompass the Employer, the Professional Team, the Management Contractor and the Works Contractors – a fact which is acknowledged by the way in which the contract forms are drafted. The approach taken in the following section, therefore, will be to recognise the inseparable nature of the Management and Works Contracts and to review the way in which those issues which form the subject matter of this book are dealt with in both JCT 98 MC and WC/2.

10.1.14 Before proceeding with this review, however, it will be helpful to set out briefly a number of further aspects of JCT 98 MC which particularly distinguish it from JCT 98 PWQ and which also impact upon the way in which claims under the management form of contract arise and are dealt with. This will be done under the following headings:

(a) Acceleration;
(b) Breach on the part of Works Contractors;
(c) "Set-off".

Acceleration

10.1.15 JCT 98 MC and its associated documents are unique in the JCT family of contracts in that they provide, albeit by optional clauses, for the acceleration or alteration of sequence or timing of the Works. This may be done either as a means of recovering delays already incurred or as a means of completing the Project at a date earlier than the Date for Completion set out in the Appendix to the contract. In this latter respect, JCT 98 MC differs from other JCT contracts which do not empower the Architect to fix a completion date earlier than the Date for Completion set down in the Appendix to the contract – even should the Architect issue an instruction omitting half of the work originally covered by the Contract – except by virtue of accepting an attachment to a Contractor's Price Statement (for example pursuant to clause 13.4.1.2A7.1.2 of JCT 98 PWQ) or accepting a Contractor's Quotation (e.g. pursuant to clause 13A of JCT 98 PWQ).

10.1.16 As noted above, such situations are likely to involve repercussions across all the different parties to the Project. Accordingly, it is necessary for our purposes to look at the manner in which the rights and responsibilities of these parties are dealt with in both JCT 98 MC and WC/2. For the purposes of this discussion we have assumed that the Employer's decision to accelerate is not prompted by the need to counteract the effects of some earlier default on the part of one of the Works Contractors. This latter eventuality is discussed in paras 10.1.24 to 10.1.32.

10.1.17 Acceleration is dealt with under clause 3.6 of JCT 98 MC. This clause is optional and it is necessary to have indicated in the Appendix to the Contract at the time it was signed whether the clause is to apply. The same issue is dealt with by clauses 3.4.2 to 3.4.7 of WC/2, which will only apply where it is stated in Section 1 of WC/1 that clause 3.6 of JCT 98 MC is in operation. Both JCT 98 MC and WC/2 deal with the issue of acceleration in much the same way but from a different perspective.

10.1.18 Should the Employer wish to accelerate the progress of the Project, either by requiring the Completion Date to be brought forward in time or by reducing or cancelling a prospective extension of time, then the Architect must issue a Preliminary Instruction to that effect to the Management Contractor (clauses 3.6.2 and 3.6.3 of JCT 98 MC refer). The Management Contractor must then forward this Preliminary Instruction to any Works Contractor affected by the proposed acceleration together with a request for details of any reasonable objection which the Works Contractor may have to the proposals (clause 3.4.4 of WC/2 refers). Both the Management Contractor, and any Works Contractor affected may object to the proposals, which objections will be passed on to the Architect by the Management Contractor, and will require the Architect either to withdraw or amend his proposals (clause 3.6.4 of JCT 98 MC refers).

10.1.19 Having overcome any objections to the Preliminary Instruction, clause 3.4.6 of WC/2 requires each Works Contractor affected to submit to the Management Contractor what is, in effect, a quotation stating (wherever practicable) the cost and programming implications of the Architect's proposals. Clause 3.6.5 of JCT 98 MC then requires the Management Contractor to submit to the Architect what is effectively a consolidated quotation in this respect, covering the costs to be incurred by all the Works Contractors affected.

10.1.20 The contract does not, however, allow for the Management Contractor's costs in this respect to be similarly communicated to the Architect and a question arises as to how the Management Contractor is to be reimbursed in the event that a decision to accelerate exposes him to additional costs over and above those to be paid to Works Contractors. Inasmuch as those costs are included in element 1(b) of the Prime Cost (see para. 10.1.9) they will be reimbursed **either** as they have been incurred by the Management Contractor **or** by the adjustment of any pre-agreed lump sums. Inasmuch as they form part of a fixed Management Fee the Management Contractor will not be entitled to any additional recompense unless the Prime Cost exceeds the Contract Cost Plan Total by more than 5% (see para. 10.1.10). In this case, the Management Fee will be adjusted in line with the formula set out at clause 4.10 of JCT 98 MC.

10.1.21 Should the Employer decide to accept the Management Contractor's/Works Contractors' proposals in full, the Architect then issues an instruction to this effect which instruction must be complied with by the Management Contractor and the Works Contractors affected and the Date for Completion is refixed in accordance with the details confirmed on the Architect's Instruction (clause 3.6.6 of JCT 98 MC refers).

10.1.22 It is worthy of note that both JCT 98 MC (at clause 3.6.5.1) and WC/2 (at clause 3.4.6.1) require the Management Contractor/Works Contractors to quote a lump sum figure for reimbursement of acceleration costs only inasmuch as it is reasonably practicable to do so. Should this not be practicable then the costs attaching to any acceleration would fall to be ... *ascertained in accordance with all the relevant Works Contract Conditions* ... As to how exactly such ascertainment should be carried out WC/2, unfortunately, does not provide a clear answer. Whilst the final sentence of clause 3.4.7 of WC/2 provides ... *The Works Contractor shall be paid in accordance with the Works Contract Conditions the amount stated or referred to in clause 3.4.6.1 (or as may have been agreed between the Management Contractor and the Works Contractor) for compliance with the Instruction* ... it is suggested that the words in brackets are in conflict with clause 3.4.6.1 which requires ... *that the cost of compliance ... be ascertained under section 4* ... Costs associated with such action would normally be expected to take the form of direct loss and/or expense suffered as a result of the regular progress of the Works having been disrupted (albeit that more usually disruption is associated with a delay to, rather than an acceleration of, the Works). This is dealt with at clauses 4.45 to 4.51 of WC/2 and the list of matters which give rise to an entitlement to reimbursement is set out at clause 4.46. Instructions to accelerate the progress of the Works, issued under clause 3.6.6 of JCT 98 MC, do not qualify as a matter giving rise to reimbursement of direct loss and/or expense under the relevant provisions of WC/2.

10.1.23 The only alternative means of arriving at a figure in respect of acceleration costs is via clause 4.4 of WC/2, whereby the Quantity Surveyor is empowered to value ... *all Instructions requiring a Variation* ... It is unlikely, however, that an Instruction to accelerate could be termed a Variation as defined at clause 1.3 of WC/2. It is suggested, therefore, that the Quantity Surveyor is in somewhat of a quandary should he be called upon to ascertain the sums payable to Works Contractors in respect of acceleration costs under clause 3.6.5.1 of JCT 98 MC.

Breach on the part of Works Contractors

10.1.24 Clause 3.21 of JCT 98 MC sets out the rights and responsibilities of the various parties in the event of ... *any breach of, or non-compliance with a Works Contract by a Works Contractor* ... In this regard, the Management Contractor is prima facie responsible to the Employer in the following terms (clause 1.7 of JCT 98 MC refers):

> ... *subject to clause 3.21 the Management Contractor shall be fully liable to the Employer for any breach of the terms of this Contract including any breach occasioned by the breach by any Works Contractor of his obligations under the relevant Works Contract* ...

10.1.25 It is clear from clause 3.21 of JCT 98 MC, however, that notwithstanding the gratuitous, propagandist undertaking that the ... *Management Contractor shall be fully liable to the Employer* ... any liability on the part of the Management Contractor in the event of a breach by a Works Contractor is severely curtailed and is largely underwritten by the Employer (see *Copthorne Hotels (Newcastle) Ltd -v- Arup Associates CILL 1318* which discussed this issue). This is because, on it becoming apparent that a Works Contractor is in breach of the terms of his Works Contract, clause 3.21 of JCT 98 MC requires the Management Contractor, in consultation with the Architect and the Employer, to take all reasonable steps to achieve the following (clause 3.21.1 of JCT 98 MC refers):

(a) recover all sums due from the Works Contractor to the Management Contractor as a result of the breach;
(b) ensure the satisfactory completion of the Project, if necessary by appointing a new Works Contractor to replace the defaulting Works Contractor;
(c) meet all claims submitted by other Works Contractors as a result of the breach.

10.1.26 Whilst this would appear to be an onerous set of obligations placed upon the Management Contractor, a further reading of clause 3.21 demonstrates this not to be the case. For instance, clause 3.21.2.1 requires the Employer to meet all **expenses** properly incurred by the Management Contractor in meeting obligations (a) and (b) in para. 10.1.25. Furthermore, in the event that the breach in question causes a delay to the Project as a whole, clause 3.21.2.2 of JCT 98 MC provides that the Employer may only keep an amount of any liquidated and ascertained damages as would have been recoverable had the delay been due to the default of the Management Contractor and his recovery is limited to those amounts which the Management Contractor is in turn able to recover from the defaulting Works Contractor as part of his obligation under (a) in para. 10.1.25.

10.1.27 This particular element of clause 3.21 gives rise to a rather interesting legal problem should the Management Contractor be required to resort to litigation in order to pursue recovery from the defaulting Works Contractor. This problem arises from the fact that the Management Contractor is pursuing the Works Contractor for recovery of a loss which he has not in fact incurred – in accordance with clause 3.21.2.2 of JCT 98 MC the Management Contractor has not been obliged to pay or allow to the Employer those liquidated and ascertained damages in respect of which he is seeking recompense from the Works Contractor. In any litigation, it is a defence to an action for damages to claim that the plaintiff has in fact suffered no loss as a result of the defendant's alleged conduct. Such a defence would be open to a defaulting Works Contractor were it not for clause 1.6.2 of WC/2, whereby the Works Contractor gives an undertaking not to contend in any subsequent proceedings that the Management Contractor has suffered no loss and that the Works Contractor's liability ought as a result to be reduced or extinguished. The "no loss defence" is therefore not an option available to the defaulting Works Contractor. Notwithstanding this, it is suggested that a Court may well award the Management Contractor only nominal damages on the grounds that (regardless of any arguments put forward by the parties in this respect) the Management Contractor has indeed suffered no loss. In such an event, the Employer would similarly forfeit any substantial recovery of the liquidated and ascertained damages foregone as a result of the application of clause 3.21.2.2 of JCT 98 MC. The issue has yet to be tested in the Courts.

10.1.28 Any **costs** (as distinct from expenses) that the Management Contractor incurs in respect of (b) and (c) in para. 10.1.25 are, prima facie, to be recovered by him from the defaulting Works Contractor under (a) in para. 10.1.25. To the extent that the Management Contractor is unable so to recover such sums (due, e.g., to the insolvency of the defaulting Works Contractor) then he is entitled to be reimbursed by the

Employer. In this way it is the Employer, rather than the Management Contractor, who bears the financial risk attaching to a breach by a Works Contractor of the terms of WC/2 – despite the Employer not himself being a party to that contract. The burden which remains with the Management Contractor is that of bearing the costs associated with (b) and (c) in para. 10.1.25 until such time as it becomes apparent that recovery from the defaulting Works Contractor will not be possible. Bearing in mind that part of the Management Contractor's obligation under (a) in para. 10.1.25 is to pursue recovery, if necessary, via Adjudication, Arbitration or Litigation, the period during which the Management Contractor is required to finance the sums involved before obtaining recompense from the Employer could be quite extensive. Even so, the Management Contractor would probably be successful in arguing that the cost of financing such sums is in itself recoverable as part of his expenses under clause 3.21.2.1 of JCT 98 MC.

10.1.29 Clause 3.22 of JCT 98 MC provides a further codicil in this respect in the context of allegations of breach against the Management Contractor by a Works Contractor. To the extent that such allegations prove to be unfounded the Employer is required to underwrite the Management Contractor in respect of the latter's costs incurred in defending the allegations.

10.1.30 It is worthy of note that clause 3.21 of JCT 98 MC only applies in respect of ... *any breach of, or non-compliance with, a Works Contract by a Works Contractor ...* It is possible to envisage situations whereby a Works Contractor may cause delay and disruption to the Project as a whole but technically may not be in breach of the Works Contract. For instance, at para. 2 of WC/3 the Works Contractor warrants to the Employer in the following terms:

> ... *The Works Contractor shall so supply the Architect/the Contract Administrator with information (including drawings) in accordance with any agreed programme or at such time as the Architect/the Contract Administrator may reasonably require so that the Architect/the Contract Administrator will not be delayed in issuing the necessary instructions and/or drawings and/or other documents to the Management Contractor in accordance with the Management Contract ...*

10.1.31 Unless specifically required by the particular tender documents referred to at (f) in section 1 of WC/1, there is no equivalent provision in either WC/1 or WC/2 by which the Management Contractor could require the Works Contractor to supply information along the lines described above. This is notwithstanding the potential for disruption and delay to the Project should such information not be supplied timeously by the Works Contractor. Accordingly, such delay or disruption could occur to the Project without the Works Contractor being in breach of the terms of the Works Contract.

10.1.32 In such circumstances it would still be open to the Management Contractor to seek to recover any consequent loss and/or expense by the application of clause 4.50 of WC/2 which renders the Works Contractor liable to the Management Contractor in respect of ... *any act, omission or default of the Works Contractor ...* (i.e. not restricted to breaches of contract). In such a situation, however, any sums which the Management Contractor was unable to recover from the Works Contractor would not be reimbursable by the Employer under clause 3.21 of JCT 98 MC since they will not have arisen as the result of a breach of the terms of the Works Contract.

10.1.33 Should a Works Contractor be in breach of a Works Contract in the manner envisaged by clause 3.21 of JCT 98 MC, or in default in a manner not amounting to a breach of contract, there are a number of provisions contained in WC/2 which serve to regulate the position as between the Management Contractor and the Works Contractor.

10.1.34 By virtue of clause 1.8 of WC/2, for instance, the Works Contractor indemnifies the Management Contractor against the effects of ... *any negligence or act or omission or default of the Works Contractor* ... which involves the Management Contractor in liability towards the Employer. Note that this liability is somewhat wider than that resulting simply from a breach of contract on the part of the Works Contractor. Furthermore, clause 2.12 of WC/2 requires that the Works Contractor reimburse the Management Contractor in respect of any direct loss and/or expense suffered by the latter should the Works Contractor fail to complete the Works on time. Such loss and/ or expense includes (but is not restricted to) any liquidated and ascertained damages which the Management Contractor may be required to pay to the Employer under the provisions of JCT 98 MC should the Project as a whole be delayed. In this respect it is worthy of recollection that clause 3.21.2.2 of JCT 98 MC allows the Employer to recover such liquidated and ascertained damages only to the extent that the Management Contractor is able to recover them from the defaulting Works Contractor. It is also worthy of recollection that this only applies in cases where the Works Contractor is in breach of contract. In the example quoted in paras 10.1.30 to 10.1.32 in which the Works Contractor does not provide information timeously, the Management Contractor may not be entitled to an extension of time because any delay would result from a ... *default, whether by act or omission* ... (clause 2.13.1 of JCT 98 MC refers) of the Works Contractor, albeit not amounting to a breach of the Works Contract. In such a situation, the Employer would be entitled to levy liquidated and ascertained damages against the Management Contractor, who would himself have to bear the cost of those amounts which he could not recover from the defaulting Works Contractor.

"Set-off"

10.1.35 There is a high degree of similarity between the approach adopted by WC/2 as regards the question of default between the Management Contractor and the Works Contractor and that adopted in NSC/C as between the Contractor and the Nominated Sub-Contractor (discussed in detail in section 6.1). For instance, clauses 4.49 and 4.50 of WC/2 effectively replicate clauses 4.39 and 4.40 of NSC/C in providing for recovery by either party of **agreed** sums in respect of direct loss and/or expense caused by the default of the other. Should it not be possible to agree the amounts owed between the parties, then it would be necessary to refer the matter to Adjudication, Arbitration or Legal Proceedings under section 9 of WC/2.

10.1.36 It is an implied common law right that any amounts owing to the Management Contractor in respect of direct loss and/or expense caused by the default of the Works Contractor can be "set-off" by the Management Contractor against sums owing to the Works Contractor. JCT 98 MC recognises this right at clause 4.26.2 wherein it provides ... *Not later than 5 days before the final date for payment ... the Management Contractor may give a written notice to the Works Contractor which shall specify the amount proposed to be withheld and/or deducted ... the ground or grounds for such withholding and/or deduction and the amount of the withholding and/or deduction ...*

10.1.37 As with Amendment 7 to NSC/C this is a much simpler but nonetheless effective provision than that which pertained prior to the issue of JCT 98 MC.

10.1.38 Whilst dealing with the subject of "set-off" it is worth noting that clause 3.21.3 of JCT 98 MC entitles the Management Contractor to deduct (set-off) from amounts certified due to a Works Contractor who is in breach, the amount of direct loss and/or expense to which the other Works Contractors are entitled together with any costs incurred by the Management Contractor as a result of the breach.

10.2   EXTENSIONS OF TIME

General

10.2.1 Delays to the progress of the work on site are, first and foremost, a matter concerning the Works Contractors as it is they who are responsible for physically carrying out the work. Accordingly, the provisions of WC/2 deal with the issue of delays and extensions of time in some considerable detail and in a manner similar to that adopted in JCT 98 PWQ as between the Contractor and the Employer. As the Management Contractor carries an overall responsibility to the Employer for securing the carrying out and completion of the Project, the issue is also dealt with by JCT 98 MC (albeit in truncated fashion). The approach adopted in the following paragraphs will be, first, to examine the way in which the issues are dealt with as between the Management Contractor and the Works Contractor (by reference to WC/2) and, secondly, to examine the means by which this is translated up the line as between the Management Contractor and the Employer (by reference to JCT 98 MC). The issues which arise are dealt with under the following headings:

- Procedures;
- Relevant Events;
- Fluctuations.

Procedures

10.2.2 Table 10.1 provides references to the text within this book dealing with JCT 98 PWQ which is also relevant to the following review of WC/2 and JCT 98 MC.

Table 10.1. Extension of time procedures

Provision in Contract	Clause in WC/2	Equivalent clause in JCT 98 PWQ	Explanatory text in this book
Requirement for the Works Contractor (Contractor) to give notice of delay and to give further details to assist the Management Contractor (Architect) in establishing any ex-tension of time.	2.2.1	25.2.1	2.7.7

Provision in Contract	Clause in WC/2	Equivalent clause in JCT 98 PWQ	Explanatory text in this book
Requirement for the Works Contractor (Contractor) to give, in respect of each Relevant Event identified in the notice of delay, particulars of the expected effects thereof together with an estimate of expected delay and to update such information as and when necessary.	2.2.2	25.2.2 and 25.2.3	2.7.8 to 2.7.10
Requirement for the Management Contractor (Architect), in certain circumstances, to grant extensions of time during the construction period or to notify the Works Contractor (Contractor) that an extension of time is not, in his view, justified.	2.3	25.3.1	2.7.11 to 2.7.31
Requirement for the Management Contractor to notify the Architect of any proposed Extension of Time to a Works Contract and to notify the Works Contractor of any dissent expressed by the Architect.	2.3	–	–
The effect of omissions upon extensions of time.	2.6	25.3.2 and 25.3.3.2	2.7.36 to 2.7.41
Requirement for the Management Contractor (Architect) to review the Date for Completion following practical completion of the Works Contract.	2.7	25.3.3	2.7.32 to 2.7.41
General obligations imposed upon the Management Contractor (Architect) and the Works Contractor (Contractor) in relation to their conduct.	2.8 2.9	25.3.4, 25.3.5 and 25.3.6	2.7.42 to 2.7.45

Provision of Contract	Clause in JCT 98 MC	Equivalent clause in JCT 98 PWQ	Explanatory text in this book
Requirement for the Management Contractor (Contractor) to advise of likely delay to Project.	2.12.1	25.2.1	2.7.7
Requirement for the Architect either to grant a fair and reasonable extension of time or to notify the Management Contractor (Contractor) that an extension of time is not justified.	2.12.1	25.3.1	2.7.11 to 2.7.31
The effect of omissions upon extensions of time.	2.12.2	25.3.2 and 25.3.3.2	2.7.36 to 2.7.41

(continued)

295

Table 10.1. (cont.)

Provision of Contract	Clause in JCT 98 MC	Equivalent clause in JCT 98 PWQ	Explanatory text in this book
Requirement for the Management Contractor to notify the Architect of any proposed extension of time to a Works Contract and for the Architect to notify the Management Contractor of any dissent to such proposal.	2.14	–	–

10.2.3 The means by which the Works Contractor must apply to the Management Contractor for an extension of time are dealt with by clause 2.2 of WC/2. This clause effectively replicates clause 25.2 of JCT 98 PWQ except that the Works Contractor and the Management Contractor under WC/2 take the place respectively of the Contractor and the Architect under JCT 98 PWQ.

10.2.4 The consideration of the Works Contractor's application and the granting by the Management Contractor of an extension of time are dealt with under clause 2.3 of WC/2. This clause differs in a number of relatively minor respects from clause 25.3 of JCT 98 PWQ. These differences are a reflection of the tripartite nature of the management contracting environment and in this respect are similar to those found in clause 2.3 of NSC/C which also deals with the tripartite relationship between the Employer, Contractor and Nominated Sub-Contractor (discussed at section 6.2) – albeit under NSC/C it is the Architect, not the Contractor, who actually grants the extension of time whereas, under WC/2, this function is performed by the Management Contractor.

10.2.5 For instance, in addition to the occurrence of a Relevant Event, the Works Contractor can also become entitled to an extension of time by reason of . . . *an act, omission or default of the Management Contractor . . . or any person for whom the Management Contractor is responsible . . .* (clause 2.3.1.1 of WC/2 refers). Such default by the Management Contractor would include any delay occasioned by a proper suspension of the Works by the Works Contractor caused by the improper withholding of payment by the Management Contractor (clauses 4.28 and 2.3.2 of which WC/2 refer).

10.2.6 Furthermore, while the issue of extensions of time is contractually one as between the Works Contractor and the Management Contractor the latter is required by clause 2.3.1 of WC/2 to notify the Architect as to any proposed extension to be granted to the Works Contractor. This is also dealt with under clause 2.14 of JCT 98 MC which permits the Architect to register his dissent to any such proposed extension of time. Should the Management Contractor be notified of any such dissent by the Architect then clause 2.3.1 of WC/2 states the following:

> . . . *If under clause 2.14 of the Management Contract Conditions the Architect has expressed dissent from the decision of the Management Contractor on giving or not giving an extension of time or the length thereof the Management Contractor shall notify in writing the terms of such dissent to the Works Contractor . . .*

10.2.7 Both JCT 98 MC and WC/2 are unclear as to the precise effect of the Architect's dissenting from the Management Contractor's proposed extension of time. The Management Contractor's obligation under WC/2 is to give such extension of time ... *as the Management Contractor then estimates to be reasonable* ... It would appear that, whilst the Management Contractor would no doubt regard the Architect's dissent as highly persuasive (not least because it may well affect any extension of time granted to the Management Contractor arising out of the same events) the Management Contractor is not bound by such dissent and must, in the final analysis, arrive at his own opinion as to what is fair and reasonable. This view is supported by JCT Practice Note MC/2 [1] which states that:

> ... *The Architect is **not** required to consent to the Management Contractor's decision ... Where there is dissent by the Architect this puts the Management Contractor on notice that where the grounds for this extension are also a Project Extension Item the Architect is in no way bound by the Management Contractor's decision under the Works Contract when deciding under clause 2.12 [of JCT 98 MC] on any extension of the Project Completion Date ...*

10.2.8 In other respects, the procedural aspects of the granting of extensions of time under the WC/2 are similar to those under clause 25 of JCT 98 PWQ. Generally, the same time limits apply within which the Management Contractor is required to act, the same powers of review apply to the Management Contractor's decisions and the same restrictions apply to the fixing of a Completion Date earlier than the Date for Completion set out in the Contract Documents (except inasmuch as clause 3.6 of JCT 98 MC – acceleration – applies).

10.2.9 The situation as between the Employer and the Management Contractor, as set out in clauses 2.12 to 2.14 of JCT 98 MC is somewhat more loosely defined. The Management Contractor is simply required to ... *advise* ... the Architect should it become clear that the Completion Date is unlikely to be or has not been achieved. Upon receiving such advice the Architect then grants what he considers to be a fair and reasonable extension of time. There are no time limits within which the Architect is required to act and no express provisions dealing with the degree of detail which the Management Contractor is required to submit in support of his advice. It is suggested, however, that, as the overall effect of events upon the progress of the Project will be an amalgam of events as they affect the individual Works Contracts, the timescale and level of detail of the actions required under clause 2.12 of JCT 98 MC will be very much driven by the requirements of WC/2 as they relate to individual Works Contracts which are in delay. In this respect, it is important for the Architect to distinguish between those extensions to Works Contracts which arise from the default of either the Management Contractor or a Works Contractor and those which do not. Only the latter will entitle consideration to be given to an extension of the Project Completion Date (albeit that, in the case of breach by a Works Contractor, the Employer's right to recover liquidated and ascertained damages is limited by the application of clause 3.21 of JCT 98 MC – see paras 10.1.24 et seq.).

Relevant Events

10.2.10 Table 10.2 provides references to the text within this book dealing with JCT 98 PWQ which is also relevant to the following review of WC/2 and JCT 98 MC.

Table 10.2. Relevant Events/Project Extension Items

Relevant Event	Clause in WC/2	Equivalent clause in JCT 98 PWQ	Explanatory text in this book
Force majeure	2.10.1	25.4.1	2.8.2 to 2.8.3
Exceptionally adverse weather conditions	2.10.2	25.4.2	2.8.4
Loss or damage due to Specified Perils	2.10.3	25.4.3	2.8.5
Civil commotion	2.10.4	25.4.4	2.8.6 and 2.8.7
Compliance with instructions	2.10.5.1 and 2.10.5.2	25.4.5.1	2.8.8 to 2.8.27
Opening up or testing	2.10.5.3	25.4.5.2	2.8.28
Late information	2.10.6	–	10.2.11.3
Other works contractors/Nominated Suppliers	2.10.7	25.4.7	2.8.45 to 2.8.50
Works by others	2.8.10.1	25.4.8.1	2.8.51 to 2.8.54
Supply of materials by Employer	2.10.8.2	25.4.8.2	2.8.55
Statutory power restrictions	2.10.9	25.4.9	2.8.56 and 2.8.57
Inability to secure labour	2.10.10.1	25.4.10.1	2.8.58 to 2.8.60
Inability to secure materials	2.10.10.2	25.4.10.2	2.8.58 to 2.8.60
Work by local authority/statutory undertaker	2.10.11	25.4.11	2.8.61
Site access/egress restrictions	2.10.12	25.4.12	2.8.62 and 2.8.63
Deferment of date of possession	2.10.13	25.4.13	2.8.64
Approximate quantities	2.10.14	25.4.14	2.8.65 to 2.8.67
Terrorism	2.10.15	25.4.16	2.8.69
Effect on Performance Specified Work of change in Statutory Requirements	2.10.16	25.4.15	2.8.68
CDM Regulations	2.10.17	25.4.17	2.8.70
Suspension by Works Contractor	2.10.18	25.4.18	2.8.71 to 2.8.75
Project Extension Items	*Clause in JCT 98 MC*	*Equivalent clause in JCT 98 PWQ*	*Explanatory text in this book*
Any cause which impedes the Management Contractor including (but not restricted to):			
– default by Employer		–	–
– late information	2.13	–	–
– deferment of date of possession		25.4.13	2.8.64
– any Relevant Event as per clause 2.10 of WC/2 (except clause 2.10.7.1 of WC/2)		As above	As above
Always provided it is not caused by the default of either the Management Contractor or a Works Contractor.			

10.2.11 The Relevant Events in respect of which the Works Contractor may be entitled to an extension of time are set out in clause 2.10 of WC/2. They are broadly similar to those in clause 25.4 of JCT 98 PWQ but, as with clause 2.6 of NSC/C, there are a number of differences which reflect the changed circumstances in which the conditions operate. Those differences which are worthy of note can be summarised as follows:

1. *Compliance with Instructions*

   This qualifies as a Relevant Event (clause 2.10.5 of WC/2 refers) whether compliance is required of the Management Contractor or of the Works Contractor or of both, and recognises the fact that the progress of the Works in respect of one particular Works Contract may well be affected by the issue of Instructions not directly concerned with that Works Contract. One point worthy of note, however, is that this Relevant Event only covers Instructions issued by the Architect to the Management Contractor. Clause 3.3 of WC/2 empowers the Management Contractor not only to pass on to the Works Contractor Instructions from the Architect but also to issue Directions of his own to the Works Contractor. Any Direction of the Management Contractor to the Works Contractor does not qualify as a Relevant Event under clause 2.10 of WC/2 and will not of itself entitle the Works Contractor to an extension of time. In this respect, the definition of a Direction at clause 1.3 of WC/2 is worthy of note:

   > *... reasonable requirements in writing of the Management Contractor regulating for the time being the due carrying out of the Project ...*

   It would appear from this that Directions issued by the Management Contractor are intended to be of an essentially regulatory nature, and not of themselves to give rise to changes such as would adversely affect a Works Contractor's progress towards the Date for Completion of the Works Contract. Hence they do not give rise to an entitlement to an extension of time.

2. *Opening up or testing*

   Clause 2.10.5.3 of WC/2 provides the same rights and remedies as between the Works Contractor and the Management Contractor as does clause 2.6.5.2 of NSC/C between the Nominated Sub-Contractor and the Contractor. The reader is referred to para. 6.2.11.3 wherein this matter is discussed in detail. When making this reference it should be noted that it is clause 4.49 in WC/2 that entitles the Works Contractor to recover direct loss and/or expense incurred as a result of any act, omission or default of the Management Contractor.

3. *Late information*

   Clause 2.10.6 of JCT 98 MC differs from its counterpart clause in JCT 98 PWQ in the following fundamental respects:

   - If the Architect fails to provide information in accordance with the Information Release Schedule, clause 25.4.6.1 of JCT 98 PWQ provides that this can constitute a Relevant Event. As JCT 98 MC does not provide for an Information Release Schedule there is no analogous Relevant Event under clause 2.10 thereof.
   - If the Architect fails to provide all drawings, details and instructions in such time as to enable the Contractor to carry out and complete the Works in accordance with the Conditions, whether or not such information has been

requested by the Contractor, clause 25.4.6.2 of JCT 98 PWQ provides that this can constitute a Relevant Event. By contrast clause 2.10.6 of JCT 98 MC provides that the failure of the Professional Team to provide necessary Instructions, drawings, details and levels ... *in due time* ... can constitute a Relevant Event **but** only in those circumstances when the Management Contractor or the Works Contractor through the Management Contractor has made specific written application for such information at a date ... *which having regard to the Completion Date or the period or periods for completion of the Works was neither unreasonably distant from nor unreasonably close to the date when it was necessary for the Management Contractor or the Works Contractor to receive the same* ... In this context ... *in due time* ... means "in a reasonable time" and not "in time to avoid delay" (see also para. 8.2.10.3). Further, it is worth noting that the Management Contractor or the Works Contractor through the Management Contractor must make written application for outstanding information in sufficient time for the Professional Team to be able to respond adequately so as not to delay the Works but not so far in advance of the need for the information that the request loses its potency. It is suggested that to comply with these provisions there is no substitute for the Management Contractor and the Works Contractor to constantly review the information needed to enable them to give proper and adequate notice to the Professional Team of their up to date requirements and for the Professional Team to respond promptly. In the event that the Professional Team forwards information to the Management Contractor but the Management Contractor does not forward the same in due course to the Works Contractor then the Works Contractor will remain entitled to an extension of time regardless of the fact that the Management Contractor will not be entitled to an extension of the Project Completion Date under the terms of JCT 98 MC. Such an extension to the Works Contract period would be granted not by reference to the Relevant Events set out in clause 2.10 of WC/2 but as a result of a default by the Management Contractor as described in clause 2.3.1 thereof. This distinction becomes important in the context of a related claim in respect of direct loss and/or expense in that such a claim would be dealt with under clause 4.49 of WC/2 whereby the amount of direct loss and/or expense is a matter for agreement between the Management Contractor and the Works Contractor rather than under clause 4.45 of WC/2 whereby the amount due would be ascertained by the Architect or the Quantity Surveyor albeit in consultation with the Management Contractor.

4. *Delay on the part of other Works Contractors*
   Should other Works Contractors delay the progress of the Works Contractor, for example as a result of their work not having been constructed in accordance with their contract, then the non-defaulting Works Contractor is entitled to an extension of time by virtue of clause 2.10.7.1 of WC/2. However, clause 2.13 of JCT 98 MC does not entitle the Management Contractor to an equivalent or any extension of time in such circumstances. Notwithstanding this, as discussed at para. 10.1.26, the Employer will only be able to recover liquidated and

ascertained damages from the Management Contractor in respect of such delays to the extent that the Management Contractor can recover such sums from the defaulting Works Contractor (clause 3.21 of JCT 98 MC refers).

10.2.12 As with procedural matters, the position set out in JCT 98 MC as between the Employer and the Management Contractor is more loosely defined than under WC/2. According to clause 2.12 of JCT 98 MC, the Management Contractor is entitled to an extension of time in respect of delays caused by one or more of the Project Extension Items. These are defined at clause 2.13 of JCT 98 MC as being ... *any cause which impedes the proper discharge by the Management Contractor of his obligations under this Contract* ... Such matters include (but are not restricted to) the following:

- compliance or non-compliance of the Employer with the CDM Regulations;
- any default of the Employer or his agents;
- the Management Contractor not having received information in due time from the Professional Team. It is suggested that the Management Contractor's obligations relating to the timeous written application for information needed from the professional team required by this clause is in addition to and not part of his obligations relating to planning and programming under items 16 and 17 of the Third Schedule to JCT 98 MC;
- the deferment of the date of the Employer giving possession of the site (where clause 2.3.2 of JCT 98 MC applies);
- any of the Relevant Events as defined in WC/2 except delay caused by one Works Contractor to another.

10.2.13 The Management Contractor does not, of course, receive an extension of time from the Architect by reason of delays caused by his own default and this is by virtue of the proviso at the end of clause 2.13 of JCT 98 MC which states that no extension of time will be given to the Management Contractor to the extent that the delay was ... *caused or contributed to by any default* ... *of the Management Contractor* ... Nonetheless, the Management Contractor may be required to grant extensions of time in this respect to Works Contractors under the provisions of WC/2.

10.2.14 When one considers the varying obligations of the parties, a situation whereby the Management Contractor is required to grant extensions of time to Works Contractors which are incompatible with, and perhaps of different duration to, those to which he himself becomes entitled under JCT 98 MC is not difficult to envisage.

Fluctuations

10.2.15 Clauses 4A, 4B and 4C of WC/2 (all of which are optional) set out the mechanisms by which the amount due to the Works Contractor is adjusted in the event of fluctuations in cost occurring during the period for carrying out the Works. By virtue of clauses 4A.4.7, 4B.5.7 and 4C.7.1 of WC/2 the adjustment to the Works Contractor's entitlement does not encompass increased costs arising during a period of culpable delay on the part of the Works Contractor. It is worthy of note in this respect that this only continues to be the case should the printed text of clauses 2.2 to 2.10 of WC/2 (which incorporates the provisions relating to extensions of time) remain unamended and provided that the Management Contractor has, in respect of every written notice by the Works Contractor under clause 2.2 of WC/2 given his decision in writing to revise or not to revise the period or periods for completion of the Works. Any amendment whatsoever to this printed text, or any failure of the Management Contractor to properly respond to the notice of the Works Contractor, where either clause 4A, 4B or 4C of WC/2 apply, will result in the Works Contractor becoming entitled to increased costs which arise even during a period of culpable delay on the part of the Works Contractor.

10.2.16 Clauses 4A.4.7, 4B.5.7 and 4C.7.1 of WC/2 reflect clauses 38.4.7, 39.5.7 and 40.7.1 of JCT 98 PWQ respectively and the reader is referred to sections 2.12 and 4.7 in this connection.

10.2.17 The situation as it pertains to the Management Contractor under JCT 98 MC is slightly different. If the Management Fee (Element 2 of the Management Contractor's remuneration – see para. 10.1.10) is a lump sum then it is not adjustable in any event by virtue of increased costs to which the Management Contractor may be exposed. This is not true, however, of that part of the Prime Cost which relates to the Management Contractor's directly incurred costs (Element 1(b) of the Management Contractor's remuneration – see para. 10.1.9). Para. 2 of Part 1 of the Second Schedule to JCT 98 MC provides that, if the Management Contractor is in culpable delay, he shall be reimbursed his directly incurred Prime Cost on the basis **either** of rates and prices applicable immediately before the Completion Date **or** the actual costs incurred, whichever is the **lesser** amount. This safeguard does not exist in respect of delay on the part of Works Contractors. In this way, while the Works Contractor under the terms of WC/2 can take advantage of any price falls during a period of his culpable delay (i.e. he will continue to be reimbursed at the earlier, higher rate), the Management Contractor is unable to do so in respect of his own costs under the terms of JCT 98 MC. A further distinction between JCT 98 MC and WC/2 is that the restrictions on the Management Contractor's ability to recover increased costs incurred during a period of culpable delay continue to apply whether or not the printed text of the extension of time clauses in JCT 98 MC is amended.

10.3 VARIATIONS AND DISRUPTION

Variations

10.3.1 Table 10.3 provides references to the text within this book dealing with JCT 98 PWQ which is also relevant to the following review of WC/2.

Table 10.3. Valuation of Variations

Valuation rule	Clause in WC/2 where WC/1 Section 3 Article 2.1 applies – Lump Sum Contract	Clause in WC/2 where WC/1 Section 3 Article 2.2 applies – Remeasurement Contract	Equivalent clause in JCT 98 PWQ	Explanatory text in this book
Valuation generally Valuation by reference to	4.7.1.1	4.15.1.1	13.5.1.1	3.3.26

Valuation rule	Clause in WC/2 where WC/1 Section 3 Article 2.1 applies – Lump Sum Contract	Clause in WC/2 where WC/1 Section 3 Article 2.2 applies – Remeasurement Contract	Equivalent clause in JCT 98 PWQ	Explanatory text in this book
the Contract Bills/priced document where the additional or substituted work is of similar character, is executed under similar conditions and does not significantly change the quantity of work.				
Fair allowance on the above in the event that work of a similar character is not executed under similar conditions and/or there is a change in the quantity of the work.	4.7.1.2	4.15.1.2	13.5.1.2	3.3.29 3.3.31 3.3.34 3.3.43
Valuation at fair rates or prices in the event that the additional or substituted work is not of similar character to that in the Contract Bills/priced document.	4.7.1.3	4.15.1.3	13.5.1.3	3.3.29 3.3.31 3.3.34 3.3.43
Valuation by reference to the rate or price entered against an approximate quantity where such quantity is a reasonably accurate forecast of the work required.	4.7.1.4	–	13.5.1.4	–
Fair allowance on the above in the event that the approximate quantity is not a reasonably accurate forecast of the work required.	4.7.1.5	–	13.5.1.5	3.3.29
Valuation of omitted works by reference to the rates or prices in the Contract Bills/priced document.	4.7.2	–	13.5.2	–
In valuing in accordance with the above:				
• Measurement is to be in accordance with the same convention as is	4.7.3.1	4.15.2.1	13.5.3.1	3.3.23

(continued)

303

Table 10.3. (cont.)

Valuation rule	Clause in WC/2 where WC/1 Section 3 Article 2.1 applies – Lump Sum Contract	Clause in WC/2 where WC/1 Section 3 Article 2.2 applies – Remeasurement Contract	Equivalent clause in JCT 98 PWQ	Explanatory text in this book
applicable to the preparation of the Contract Bills/priced document.				
• Allowance is to be made for percentage/ lump sum adjustments in the Contract Bills/ priced document.	4.7.3.2	4.15.2.2	13.5.3.2	3.3.23
• Allowance for any addition to or reduction of preliminary items.	4.7.3.3	4.15.2.3	13.5.3.3	3.3.23
Valuation generally Valuation by reference to dayworks where the additional or substituted work cannot properly be valued by measurement.	4.7.4	4.15.3	13.5.4	3.3.35 to 3.3.42
Valuation of Performance Specified Work.	4.7.5	–	13.5.6	3.3.1
Fair valuation to be made of work which cannot be valued in accordance with the above rules.	4.7.6	4.15.4	13.5.7	3.3.29 3.3.34
Valuation of other works as if they were variations in the event that a variation substantially changes the conditions under which such other works are executed.	4.8	4.16	13.5.5	3.3.43
Right of Works Contractor (Contractor) to be present when it is necessary to measure the work for the purpose of valuation.	4.9	4.13	13.6	3.3.23
Submission of Quotations for Variations before Work Instructed Instruction to provide sufficient information for	3.14.1.1	3.14.1.1	13A.1.1	3.4.2

Valuation rule	Clause in WC/2 where WC/1 Section 3 Article 2.1 applies – Lump Sum Contract	Clause in WC/2 where WC/1 Section 3 Article 2.2 applies – Remeasurement Contract	Equivalent clause in JCT 98 PWQ	Explanatory text in this book
Works Contractor (Contractor) to provide quotation.				
Quotation to be submitted by Works Contractor within 17 days (by Contractor within 21 days) and thereafter remain open for acceptance for 14 days (7 days in the case of the Contractor).	3.14.1.2	3.14.1.2	13A.1.2	3.4.3 3.4.4
Variation only to be carried out after acceptance of quotation.	3.14.1.3	3.14.1.3	13A.1.3	3.4.7
Quotation to separately comprise:	3.14.2	3.14.2	13A.2	3.4.5 and 3.4.6
• Cost of carrying out the Variation. • Any adjustment to the time required for completion of the Works. • Any amount in lieu of direct loss and/or expense. • Cost of preparing quotation. • Method statement (if requested).				
Acceptance of quotation by Management Contractor (Architect)	3.14.3	3.14.3	13A.3	3.4.7
Non-acceptance of quotation – Variation either to be valued in accordance with normal rules or not to be carried out.	3.14.4	3.14.4	13A.4	3.4.8

0.3.2 Instructions from the Architect to the Management Contractor are dealt with at clauses 3.3 to 3.6 of JCT 98 MC. The means by which such Instructions may be handed down by the Management Contractor to the Works Contractor are dealt with at clauses 3.3 to 3.7 of WC/2. In addition, the Management Contractor may also, on his own behalf, issue directions to the Works Contractor – this is also dealt with at clauses 3.3 to 3.7 of WC/2 (for an explanation of the difference between Instructions and Directions see 10.2.11.1).

10.3.3 The provisions which empower the Architect to issue instructions under JCT 98 PWQ relate specifically to particular matters (e.g. clause 2.4.1 requires the Architect to issue instructions in regard to any discrepancy or divergence between certain specified documents and clause 34.2 requires that the Architect issue instructions in regard to what is to be done concerning a find of antiquities) but the Architect is not empowered to issue Instructions otherwise than in accordance with these particular provisions. In contrast, clause 3.3 of JCT 98 MC is significantly less restrictive in this respect as it empowers the Architect to issue ... *such Instructions as are reasonably necessary to enable the Management Contractor properly to discharge his obligations* ... There are also a number of specific instances in which the Architect may issue Instructions to the Management Contractor. These relate to:

- Project Change (as defined at clause 1.3 of JCT 98 MC) – clause 3.4.1;
- Works Contract Variation (as defined at clause 1.3 of JCT 98 MC and at clause 1.3 of WC/2) – clause 3.4.1;
- the expenditure of provisional sums – clause 3.4.1;
- postponement of work – clause 3.5;
- acceleration of sequence or timing – clause 3.6;
- opening up of work for inspection or testing – clause 3.10;
- work not in accordance with the Contract – clause 3.11;
- making good defects – clause 3.12; and
- antiquities found – clause 3.27.

10.3.4 The Works Contractor is entitled, by virtue of clauses 3.3.3 of JCT 98 MC and 3.7 of WC/2 to request that the Architect specify under which provision of the Contract he is empowered to issue any particular instruction. This requirement is analogous to that set out in clause 4.2 of JCT 98 PWQ. Whilst this may well on occasion tax the ingenuity of the Architect under JCT 98 PWQ, clause 3.3 of JCT 98 MC is so widely drawn that it may of itself provide a simple answer to any request on the part of the Works Contractor made pursuant to clause 3.7 of WC/2. It ought to be noted in this respect that the Management Contractor is not likewise empowered to request this information from the Architect, presumably on the grounds that (in general) it is the Works Contractor and not the Management Contractor who will be directly affected by the Instruction.

10.3.5 As with JCT 98 PWQ, all Instructions from the Architect are required to be in writing, although clause 3.3.2 of JCT 98 MC permits either the Architect or the Management Contractor to confirm oral instructions within 7 days. Similarly, clause 3.5 of WC/2 permits either the Management Contractor or the Works Contractor to confirm oral Instructions or Directions within 7 days.

10.3.6 As noted above, the Architect is empowered under clause 3.4 of JCT 98 MC to issue Instructions in respect of Project Changes. These are defined at Clause 1.3 of JCT 98 MC in the following terms:

> ... *the alteration or modification of the scope of the Project as shown and described generally in the Project Drawings and the Projects Specification* ...

Project Changes are therefore **general** in nature and apply to the **scope** rather than the detail of the Project. They are not intended to cover detailed amendments within individual works packages. These latter are the subject of Works Contract Variations. As noted at para. 10.3.3, Works Contract Variations are defined, in the same terms, both at clause 1.3 of JCT 98 MC and at clause 1.3 of WC/2 which definitions are analogous to those relating to Variations as set out at clause 13.1 of JCT 98 PWQ.

10.3.7 The rules for the valuation of Works Contract Variations are set out at clauses 4.4 to 4.16 of WC/2. Clauses 4.4 to 4.10 are applicable when the Works Contract has been let on a lump sum basis and is not subject to remeasurement while clauses 4.11 to 4.16 are applicable to Works Contracts which are subject to full remeasurement. Except for those relating to approximate quantities, omissions and performance specified work (which are not necessary in a remeasurement contract) the rules for the valuation of Works Contract Variations are essentially the same in both situations.

.3.8 Unlike JCT 98 PWQ there is no provision in either JCT 98 MC or WC/2 for the valuation of Variations by way of a Contractor's Price Statement. All Works Contract Variations, except those for which a Works Contractor's Quotation (known as a "3.14 Quotation") has been accepted under clause 3.14 of WC/2 are to be valued by the Quantity Surveyor in accordance with the rules set out in clause 4.7 of WC/2 on lump sum contracts and clause 4.15 of WC/2 on remeasurement contracts. These clauses are virtually identical to the rules for the valuation of Variations set out at clause 13.5 of JCT 98 PWQ which are considered in detail at paragraphs 3.3.21 to 3.3.46 and, as such require no further comment here. However, there are two aspects of the rules for the valuation of Works Contract Variations which are worthy of note. These are:

(a) Where any Instruction issued by the Architect to the Management Contractor under JCT 98 MC substantially changes the conditions under which part of the Works Contractor's operations are carried out, even though it is not a Variation relating to the particular Works Contract, then this too shall be treated as a Works Contract Variation and valued accordingly. This is a recognition of the fact that Instructions may be issued to the Management Contractor in respect of Works Contracts other than that of a particular Works Contractor and yet still have an effect on the work of that particular Works Contractor.

(b) Clauses 4.5 and 4.12 of WC/2 concern Instructions which are issued ... *arising out of any negligence or default, whether by act or omission, of the Management Contractor ... or of any other works contractor ...* Such Instructions (whether or not they give rise to a Works Contract Variation) are to be valued in accordance with the rules set out at clauses 4.7 or 4.15 of WC/2, albeit that they are not required to be valued by the Quantity Surveyor and the sums involved are not to be included in the Management Contractor's remuneration (although the Management Contractor is required to pay such amounts to the Works Contractor). Where such Instructions have been made necessary as a result of default on the part of the Management Contractor it is he who is ultimately responsible for bearing the resultant costs. However, where such Instructions have been made necessary as a result of default on the part of a Works Contractor the possibility exists (by virtue of clause 3.21 of JCT 98 MC – see paras 10.1.24 to 10.1.34) that the Employer may ultimately have to bear the costs arising. In such an event it is unfortunate that the Quantity Surveyor is not expressly empowered, either under clauses 4.5 and 4.12 of WC/2 or clause 3.21 of JCT 98 MC, to approve the basis upon which payment is made by the Management Contractor to the Works Contractor of sums which may ultimately fall to be paid by the Employer. This subject is discussed further at section 10.4.

Procedures for the ascertainment of direct loss and/or expense

10.3.9 Table 10.4 provides references to the text within this book dealing with JCT 98 PWQ which is also relevant to the following review of WC/2 and JCT 98 MC.

Table 10.4.

Direct loss and/or expense procedures	Clause in WC/2	Equivalent clause in JCT 98 PWQ	Explanatory text in this book
Works Contractor (Contractor) to make a written application to Management Contractor (Architect) regarding direct loss and/or expense.	4.45	26.1	3.5.9
Architect (or Quantity Surveyor) to ascertain the amount of the Works Contractor's (Contractor's) entitlement.	4.45	26.1	3.5.22 to 3.5.33
Works Contractor's (Contractor's) written application must be made timeously.	4.45.1	26.1.1	3.5.10, 3.5.11 and 3.5.20
Works Contractor (Contractor) is to submit such information as is requested by the Architect or Quantity Surveyor.	4.45.2	26.1.2 and 26.1.3	3.5.13 and 3.5.14
Amounts ascertained as being due are to be added to the Works Contract Sum (Contract Sum).	4.47	26.5	3.5.18
Works Contractor must comply with all reasonable directions of the Management Contractor to enable ascertainment.	4.48	–	–
The above provisions are without prejudice to any other rights or remedies which the Works Contractor (Contractor) may have.	4.51	26.6	3.5.9 and 3.5.21

Direct loss and/or expense procedures	Clause in JCT 98 MC	Equivalent clause in JCT 98 PWQ	Explanatory text in this book
Management Contractor to pass Works Contractor's written application to Architect with comments and to collaborate with Quantity Surveyor in ascertaining loss and/or expense.	8.5	–	–

Direct loss and/or expense procedures	Clause in JCT 98 MC	Equivalent clause in JCT 98 PWQ	Explanatory text in this book
Architect to form an opinion as to Works Contractor's (Contractor's) entitlement.	8.5	26.1	3.5.16 3.5.17

.3.10 The way in which the Works Contractor must apply to the Management Contractor for reimbursement of direct loss and/or expense is set out in clauses 4.45 to 4.51 of WC/2. In addition, clause 8.5 of JCT 98 MC details how such matters are to be dealt with as between the Architect, the Quantity Surveyor and the Management Contractor. The provisions of WC/2 in this respect largely replicate those in clause 26 of JCT 98 PWQ except that the Works Contractor's application must be made to the Management Contractor. The latter is then required to pass the Works Contractor's application (together with his comments on the same) to the Architect for his consideration (clause 8.5 of JCT 98 MC refers). It is for the Architect to form an opinion as to the Works Contractor's entitlement in principle and for the Architect/Quantity Surveyor to ascertain the quantum of the Works Contractor's entitlement. Clause 8.5 of JCT 98 MC requires the Management Contractor to co-operate in this process. As part of this process, the Management Contractor may issue Directions to the Works Contractor (presumably as regards the provision of supporting information by the Works Contractor) with which the Works Contractor must comply in order to preserve his right to reimbursement.

.3.11 The above provisions apply to situations in which the Works Contractor has incurred loss and/or expense by reason of the regular progress of the Works or a part thereof having been materially affected as a result either of the Employer having deferred giving possession of the site to the Management Contractor or due to one or more of the Matters set out in clause 4.46 of WC/2. The situation is different where either the Management Contractor or the Works Contractor claims to have incurred direct loss and/or expense as a result of the default of the other (clauses 4.49 and 4.50 of WC/2 refer). Such direct loss and/or expense is a matter for agreement between the Management Contractor and the Works Contractor and not a matter for ascertainment by the Architect/Quantity Surveyor (see paras 10.1.24 to 10.1.34).

.3.12 This is also the case where a Works Contractor claims to have suffered direct loss and/or expense as a result of the default of another Works Contractor. Reimbursement of such direct loss and/or expense is dealt with under clause 4.49 of WC/2 rather than clause 4.45 and the Architect/Quantity Surveyor is therefore not required to ascertain the Works Contractor's entitlement – notwithstanding the fact that, by virtue of clause 3.21 of JCT 98 MC, the Employer may ultimately have to pay the sum involved. This subject is discussed in further detail at section 10.4.

List of matters

.3.13 Table 10.5 provides references to the text within this book dealing with JCT 98 PWQ which is also relevant to the following review of WC/2.

Table 10.5. List of matters (WC/2)

Cause	WC/2 clauses		JCT 98 PWQ clauses		Reference to text in this book
	Relevant Event	Matter	Relevant Event	Matter	Paras
Late information	2.10.6	4.46.1	–	–	10.2.11.3
Inspection or testing	2.10.5.3	4.46.2	25.4.5.2	26.2.2	2.8.28 3.5.8
Discrepancies	2.10.5.2	4.46.3	25.4.5.1	26.2.3	2.8.15–17
Postponement	2.10.5.1	4.46.4	25.4.5.1	26.2.5	2.8.24
Work by others	2.10.8.1	4.46.5	25.4.8.1	26.2.4.1	2.8.51–54
Supply of materials by Employer	2.10.8.2	4.46.5	25.4.8.2	26.2.4.2	2.8.55
Failure to give ingress to or egress from the Works	2.10.12	4.46.6	25.4.12	26.2.6	2.8.62–63
Variations	2.10.5.1	4.46.7	25.4.5.1	26.2.7	2.8.18–22
Approximate quantities	2.10.14	4.46.8	25.4.14	26.2.8	2.8.65–67
Compliance with CDM Regulations	2.10.17	4.46.9	25.4.17	26.2.9	2.8.70
Suspension by Works Contractor (Contractor)	2.10.18	4.46.10	25.4.18	26.2.10	2.8.71–75

10.3.14 The list of matters which give rise to an entitlement on the part of the Works Contractor to recover direct loss and/or expense are set out at clause 4.46 of WC/2. This list of matters is broadly similar to that contained in clause 26.2 of JCT 98 PWQ except in the following respects:

1. *Late information*

    As with clause 2.10.6 of WC/2 (Relevant Events – see para. 10.2.11.3) such late information may have been caused by either the default of the Professional Team in not having responded timeously to the proper request for information by the Management Contractor, or by the Works Contractor through the Management Contractor, or by the default of the Management Contractor who, having received the information fails to forward it to the Works Contractor in time. In the former case the Works Contractor will be entitled to reimbursement of such direct loss and/or expense as is ascertained by the Architect or the Quantity Surveyor pursuant to clause 4.45 of WC/2. Whereas in the latter case the Works Contractor is entitled to reimbursement of such direct loss and/or expense as he is able to agree with the Management Contractor pursuant to clause 4.49 of WC/2.

2. *Postponement*

    This item extends to cover postponement of work in respect of other Works Contracts which may have the effect of disrupting progress on the Works Contract under consideration.

310

3. *Instructions*

It is worthy of note that this matter does not include Instructions made necessary as a result either of the Management Contractor's default or that of a Works Contractor. Direct loss and/or expense arising from such Instructions must be agreed between the Management Contractor and the Works Contractor in accordance with clause 4.49 of WC/2 rather than being ascertained by the Architect or the Quantity Surveyor in accordance with clause 4.45 of WC/2.

.3.15 As there are no matters equivalent to those listed at clause 4.46 of WC/2 in JCT 98 MC the Management Contractor is not contractually entitled to claim direct loss and/or expense from the Employer for such occurrences. However, all amounts to which the Works Contractors are entitled pursuant to clause 4.45 in WC/2 will be passed on to the Employer as part of the Prime Cost payable to the Management Contractor.

10.4 ASCERTAINING THE LOSS AND/OR EXPENSE

General

0.4.1 The matters discussed in Chapter 4, as regards the heads under which claims for direct loss and/or expense may be submitted and the principles which must be brought to bear in ascertaining the Contractor's entitlement under clause 26 of JCT 98 PWQ, are equally applicable in the case of direct loss and/or expense arising under clause 4.45 of WC/2. They will not, therefore, be the subject of further discussion here.

0.4.2 However, a number of issues do arise as regards the manner in which sums in respect of direct loss and/or expense are incurred, ascertained and reimbursed under a management contract and these are worthy of further comment here. As all payments from the Employer under a management contract are made to the Management Contractor, it is worth considering the means by which any loss and/or expense is reimbursed. We will do this by examining each of the three elements of the Management Contractor's remuneration as described at section 10.1.

Element 1 – Prime Cost

0.4.3 There are two means by which direct loss and/or expense may be reimbursed under this head, namely that incurred by the Works Contractors and that incurred directly by the Management Contractor.

Amounts due to Works Contractors

0.4.4 As noted above, direct loss and/or expense may be incurred by Works Contractors as a result of one of the matters listed at clause 4.46 of WC/2, in which case it falls to be ascertained by the Quantity Surveyor (in accordance with the principles set out in Chapter 4) by application of clause 4.45 of WC/2. All such sums as are ascertained in this way will be added to the Prime Cost and will form part of the Management Contractor's remuneration.

10.4.5 However, Works Contractors may also incur direct loss and/or expense as a result of some default on the part either of the Management Contractor or of other Works Contractors. In this event the quantum of such direct loss and/or expense is not required to be ascertained by the Quantity Surveyor but is instead a matter for agreement between the Management Contractor and the Works Contractor under clause 4.49 of WC/2 and does not form part of the Management Contractor's remuneration from the Employer. Where the loss and/or expense arises as a result of a breach of contract on the part of another Works Contractor then the Employer may find himself having to reimburse such sums by virtue of clause 3.21 of JCT 98 MC. This situation is discussed further below in relation to Element 3 of the Management Contractor's remuneration.

Sums incurred directly by the Management Contractor

10.4.6 Parts 3 and 4 of the Second Schedule to JCT 98 MC set out those of the Management Contractor's own directly incurred costs (site staff, facilities, etc.) which form part of the Prime Cost and are, in the absence of agreement to the contrary, to be reimbursed as incurred. The sums involved are to be ascertained by the Quantity Surveyor on the basis of information provided by the Management Contractor.

10.4.7 It is quite possible that a proportion of such amounts as are ascertained by the Quantity Surveyor under this head may have arisen as a result of some disruption to the regular progress of the Project and would, under a traditional lump sum contract, be more correctly classified as direct loss and/or expense. Although, in the case of a Management Contract, it is still required that such sums are ascertained by the Quantity Surveyor, there does not appear to be the additional safeguard, as with the ascertainment of direct loss and/or expense under JCT 98 PWQ, that a direct casual link be established between the event complained of and the loss alleged to have been incurred.

10.4.8 The Second Schedule of JCT 98 MC does provide as an alternative for the Employer and the Management Contractor to agree a lump sum basis for the reimbursement of those costs set out in Parts 3 and 4 thereof. If this alternative applies then the parties are required to attach to the Second Schedule a memorandum which sets out the basis upon which such sums will be adjusted as a result of compliance with Instructions. There is no guidance given as to the form which such a memorandum may take, but it is notable that it is to be employed only in the event of compliance with Instructions and not simply where additional costs arise as a result of delay. The implication is that, should the Management Contractor be exposed to additional site-based direct costs as a result of delays to the Project which have occurred independently of an Instruction, he will be required to bear those costs himself.

10.4.9 It would appear, therefore, that the Management Contractor's right to reimbursement of loss and/or expense incurred as part of his direct site-based costs may be somewhat more circumscribed where these costs are reimbursed on the basis of pre-agreed lump sums (as opposed to on the basis of costs actually incurred) and is certainly more restrictive than the equivalent right held by Works Contractors under clauses 4.45 to 4.48 of WC/2.

Element 2 – Management Fee

10.4.10 As events during the Construction Period can only affect the Construction Period Management Fee, it is this particular element of the overall Management Fee which will be considered here.

10.4.11 The Construction Period Management Fee will be the lump sum stated in the Appendix to JCT 98 MC. It is only adjustable should the Prime Cost vary from the Contract Cost Plan Total by more than 5%, or such other percentage as is inserted in the Appendix to JCT 98 MC. As the Prime Cost will include, among other things, any direct loss and/or expense incurred by Works Contractors as ascertained under clause 4.45 of WC/2 then, to the extent that this serves to increase the Prime Cost by more than the applicable percentage, it could be argued that the Management Contractor is by extension being reimbursed additional costs, whether incurred or not, via his Management Fee. However, any costs actually incurred will not be subject to ascertainment by the Quantity Surveyor because the Management Fee is only adjusted in accordance with the formula set out at clause 4.10 of JCT 98. It is again possible, therefore, that the Management Contractor may be reimbursed in respect of costs more properly described as direct loss and/or expense without the need to establish a causal link in this respect. In such an event, because of the application of the above-mentioned formula, any such reimbursement is more likely to be a reflection (albeit indirect) of those costs incurred by the Works Contractor rather than those incurred by the Management Contractor himself.

10.4.12 The operation of the formula method of adjustment of the Management Fee serves to highlight one potential area of conflict of interest on the part of the Management Contractor. As noted at para. 10.1.7, the Management Contractor has a general duty under JCT 98 MC to co-operate with the Professional Team. More specifically, clause 8.5 of JCT 98 MC requires the Management Contractor to pass comment on Works Contractors' applications for reimbursement of direct loss and/or expense and also requires that he collaborate with the Quantity Surveyor in the ascertainment of such direct loss and/or expense. It is not difficult to envisage a situation whereby the discharging of such an obligation by the Management Contractor could require an increased input of resource by him if the payments to be made by the Employer in respect of Works Contractors' direct loss and/or expense were to be minimised. Such increased input could have the effect of reducing the likelihood that the Prime Cost would exceed the Contract Cost Plan Total by the applicable percentage and would thereby reduce the likelihood of the Management Contractor being in any way reimbursed the cost of his increased input.

10.4.13 How such a dilemma would be resolved in practice would depend upon the working relations established between the Professional Team and the Management Contractor (bearing in mind that the Conditions of Contract keep the two at arm's length) and in particular, the Management Contractor's approach to his management responsibilities as defined under JCT 98 MC.

Element 3 – sums arising out of Works Contractors' default

10.4.14 As noted at paras 10.3.11 and 10.3.12, it is possible for an entitlement to reimbursement of direct loss and/or expense to accrue to either the Management

Contractor or the Works Contractor by reason of the default of the other. It is also possible for one Works Contractor to recover from the Management Contractor direct loss and/or expense incurred as a result of the default of another Works Contractor. In much the same way that similar situations with domestic sub-contractors under JCT 98 PWQ are essentially a matter for resolution between those sub-contractors and the Contractor, direct loss and/or expense arising as a result of default between the Management Contractor and the Works Contractors is not required to be ascertained by the Quantity Surveyor but is to be agreed solely between the Management Contractor and the Works Contractor (clauses 4.49 and 4.50 of WC/2 refer).

10.4.15 The result of the application of such principles is that there may well arise a considerable quantum of alleged direct loss and/or expense, as between the Management Contractor and the various Works Contractor, which the Quantity Surveyor has played no part in ascertaining. In the ordinary course of events, this would not affect the Employer as he would not be called upon to reimburse the sums involved. However, as discussed at paras 10.1.24 to 10.1.34, it can happen that the Employer is required to provide such reimbursement by the operation of clause 3.21 of JCT 98 MC. On such an occasion, the Employer will be exposed to a degree of expenditure over which his own Quantity Surveyor is not expressly empowered by the contract to exercise control.

10.4.16 The Management Contractor is entitled to ... *any expenditure incurred by the Management Contractor for which he is entitled to reimbursement by the Employer in accordance with clauses 3.21 and 3.22* ... (clause 4.6.5 of JCT 98 MC refers). In this respect, the Employer is obliged to pay all amounts ... *properly incurred* ... by the Management Contractor in discharging claims ... *properly made* ... by Works Contractors and which the Management Contractor is unable to recover from the defaulting Works Contractor. It is suggested that, in the absence of an express power of ascertainment, any control exercised by the Quantity Surveyor on the Employer's behalf must be in terms of assessing whether or not amounts have been properly paid by the Management Contractor in respect of sums properly claimed by the Works Contractor. In carrying out this task, the Quantity Surveyor is entitled to be provided by the Management Contractor with all the necessary documentation.

10.4.17 It is possible to argue that the position in respect of such claims is not too far removed from that which pertains in respect of claims submitted and ascertained under clause 4.45 of WC/2, loss and/or expense caused by one of the matters for which the Employer is responsible and in respect of which ascertainment is carried out by the Architect/Quantity Surveyor. However, it is suggested that the terms in which the Quantity Surveyor's duty is couched in relation to sums arising by virtue of the operation of clause 3.21 of JCT 98 MC are such as to enable a less rigorous degree of control to be exercised by him in this respect. Claims properly made (and sums properly paid in respect thereof) are, according to the provisions of WC/2, simply to be ... *agreed* ... if possible between the Management Contractor and the Works Contractor.

10.4.18 Clause 4.49 of WC/2 contains no provisions (such as those contained in clause 4.45 thereof) regarding the need for any such loss and/or expense to be ascertained. In this respect, it is suggested that the Quantity Surveyor cannot subsequently apply a rigour, should such sums become payable by the Employer, which is not present at first instance in the provisions of WC/2. The Quantity Surveyor may therefore find himself constrained to allow to the Management Contractor whatever sums have been previously agreed between the Management Contractor and the various Works Contractors under clause 4.49 of WC/2.

Conclusion

10.4.19 Because of the manner in which a management contract is structured, and in particular the provisions for remunerating the Management Contractor, the manner in which loss and/or expense can arise and may be reimbursed by the Employer is somewhat more complex than is the case with a traditional lump sum contract. In particular, the Quantity Surveyor may find himself drawn into the settlement of claims as between the Management Contractor and the various Works Contractors in a way that would not occur in respect of a contract let under JCT 98 PWQ. In such situations, the normal approach to ascertainment adopted by the Quantity Surveyor may find itself at odds with the more commercial approach to such matters traditionally adopted by contracting organisations and which appears to be provided for under the terms of WC/2. In arriving at a mutually acceptable resolution to such situations it is submitted that much will depend upon the Management Contractor's approach to the co-operation ethos of this method of procurement.

REFERENCE

[1] Management Contract Documentation (1987) Commentaries on the JCT Management Contract Documentation MC/2, the Joint Contracts Tribunal for the Standard Form of Building Contract.

11

GENERAL CONDITIONS OF CONTRACT FOR BUILDING & CIVIL ENGINEERING MAJOR WORKS GC/WORKS/1 WITH QUANTITIES (1998)

11.1 INTRODUCTION

The development of GC/Works 98

11.1.1 This chapter deals with General Conditions of Contract for Building and Civil Engineering Major Works GC/Works/1 With Quantities (1998), hereinafter referred to as GC/Works 98.

11.1.2 GC/Works 98 has been produced by the Central Advice Unit of the Property Advisers to the Civil Estate (PACE), which is the agency currently responsible for advising central government on the management of its estate. For this purpose, central government comprises some 40 government departments, over 100 government agencies and 400-plus NHS trusts. GC/Works 98 is intended for use where the Employer is one of these bodies, albeit an adapted form (known as PC/Works) has been produced for use by private-sector Employers.

11.1.3 The standard form of contract for government works dates back to 1943, when it was considered necessary to draft a document which suspended some of the normally accepted commercial rights of contractors in the interests of furthering the war effort. The contract (known as CCC/Works/1) was the product of a single draughtsman, who later became a distinguished Law Lord, and was written with a clear purpose in mind – to reduce the scope for disruption to the progress of government works as a result of commercial disputes with the Contractor. The result was a form of contract with – as far as the Contractor was concerned – little opportunity to pursue claims, no right to question the validity of instructions and no right to determine the contract. The contract was perceived as harsh, but was deliberately so, and the protection of the Contractor's commercial interests was largely dependent on the fair-mindedness of the Employer and its appointed advisers.

11.1.4 CCC/Works/1 was phased out and replaced by the first edition of GC/Works/1 in 1959. This new form of contract followed much the same pattern as the old and was similarly perceived as carrying a heavy pro-Employer bias. Indeed, it was not until the publication of Edition 3 of GC/Works/1 in 1989 that there was any apparent attempt to tone down contract terms that were perceived by many as draconian. The principal aims of Edition 3 were expressed as being to improve relations between Employer and Contractor, to encourage faster settlement of accounts and claims, and to introduce the Project Manager (PM) as a central figure in the administration of the contract along lines then suggested by the British Property Federation (BPF). Whilst the new contract was still seen by many as harsh, there was a generally accepted view among those most familiar with its use that it was "tough but fair" and that it apportioned contract risk sensibly to the party best able to manage that risk. This latter perception may partly explain an increase in the use of GC/Works/1 following the publication of Edition 3. The Contracts in Use Survey, prepared for the RICS in November 1996 by Davis Langdon & Everest indicates that, between 1989 and 1995, the use of GC/Works/1 increased from under 2% by value to over 5% by value of all contracts let.

11.1.5 One aspect of the GC/Works contracts on which there is almost universal agreement is the clarity and conciseness of the language used. This can be contrasted with the JCT forms of contract, about which a distinguished construction lawyer once said that ... *the draughtsmanship adopted was a very effective means of camouflaging or* *concealing its true meaning from any but the most expert and tireless reader* ... Unlike the JCT contracts, which are drafted by a committee reflecting largely divergent interests, the GC/Works documents have been written by a single party to the contract and, with the benefit of this unilateral provenance, its objectives are clear. Its clarity of purpose is perhaps reflected in the brevity of the language.

11.1.6 Notwithstanding the above, in his 1994 report Constructing the Team [1], Sir Michael Latham had recommended the New Engineering Contract (NEC) as being that most in keeping with the set of principles which he was trying to promote. He went on to recommend that central government lead the way in implementing a proposed wholesale shift to the use of NEC, with an initial target of one-third of all government-funded schemes during the period 1994–1998 being let under that contract. However, a report entitled Construction Procurement by the Government [2], commissioned by the Department of the Environment and published in late 1995, concluded that NEC had no proven track record and that there were ... *no practical or commercial grounds for recommending the use of NEC ... under present circumstances* ... Instead, it was decided radically to revise the existing GC/Works documents so as to preserve its perceived advantages whilst bringing it into line with the principles put forward by Sir Michael Latham.

11.1.7 Accordingly, a new "Lathamised" set of documents was developed and published by PACE in 1998. Not only is this new set of contracts fully compliant with the Construction Act 1996, but it goes further than the Act itself in implementing Latham principles. The contracts are couched in language which is easy to understand, they are intended to comprise "a complete family of interlocking documents", and there is a new ... *fair dealing* ... provision aimed at encouraging teamworking and co-operation. A number of specific changes have also been made in this latest set of documents which represent a significant and deliberate shift away from the previous "tough but fair" philosophy of government contracts and towards the principles embodied in Constructing the Team [1]. In particular, gone are the "final and conclusive" decisions of the Authority (the term previously used to describe the Employer) and its PM which could not be challenged even by an arbitrator, and gone is the stipulation in relation to the adjudication of disputes that the adjudicator be either an officer of the Authority or, alternatively, someone ... *acting for the Authority* ... A further departure from previous editions is that there is now an express right on the part of the Contractor to determine the contract in the event of default by the Employer – something the Contractor was previously unable to do. Even the Employer is now referred to as such, instead of the somewhat more lofty and remote title of "the Authority".

The operation of GC/Works 98

11.1.8 Before going on to discuss in detail those particular provisions of GC/Works 98 which bear specifically upon the issue of claims, a few general points need to be made in respect of the operation of this form of contract – particularly for the benefit of those more familiar with the administration of JCT type contracts. There are a number of respects in which the two types of contract differ, those of particular significance being discussed below.

Structure

11.1.9 The structure of the new form of contract itself may be somewhat unfamiliar to those more used to dealing with the JCT family of documents.

11.1.10 GC/Works 98 comprises a total of 75 separate conditions spread over 54 pages. In terms of the volume of text, the document is approximately one-third the size of JCT 98 PWQ. Like its predecessor document, GC/Works 98 is sub-divided into a number of functional sections, each containing a set of conditions with a common theme, thus:

- Contract Documentation, Information and Staff (Conditions 1–6)
- General Obligations (Conditions 7–25)
- Security (Conditions 26–29)
- Materials and Workmanship (Conditions 30–32)
- Commencement, Programme, Delays and Completion (Conditions 33–39)
- Instructions and Payment (Conditions 40–51)
- Particular Powers and Remedies (Conditions 52–60)
- Assignment, Sub-letting, Subcontracting, Suppliers and Other Works (Conditions 61–65)
- Performance Bond, Parent Company Guarantee and Collateral Warranties (Conditions 66–68).

11.1.11 Conditions 20 and 44 are "not used" and an additional 9 suffixed conditions (e.g. Condition 1A, 8A, etc.) are inserted at various points throughout the document, bringing the number used to 75 in total. The Appendix which forms part of the JCT family of documents is also present in GC/Works 98, but is known instead as the Abstract of Particulars.

11.1.12 As a result of the sub-division of the Conditions noted in para. 11.1.10, the document is considerably easier to navigate than JCT 98 PWQ (a point recognised by the JCT, who are gradually overhauling their documents along similar lines – albeit this remains to be done in respect of JCT 98 PWQ itself). This relative ease of use is further assisted by the inclusion of a subject index and a Schedule of Time Limits. This Schedule conveniently summarises, in one place, the numerous time limits set out throughout the document (GC/Works 98 makes extensive use of these as a spur to effective management of the contract) and is fully cross-referenced to the relevant contract conditions.

11.1.13 Another distinction between the GC/Works forms and those produced by JCT is that, in the former case, it is not intended that the conditions themselves be signed by the parties to the contract. The executed agreement in the case of GC/Works 98 is a separate document known as the Contract Agreement (a model copy of which is published by PACE), which by reference incorporates the General Conditions of Contract. Any amendments to the General Conditions ought, therefore, to be accompanied by a corresponding amendment to the Contract Agreement.

Parties

11.1.14 The two parties to the contract, in the legal sense, are the Employer and the Contractor. As with the JCT family of contracts, however, several other parties are referred to and have their roles and responsibilities defined by the contract. The

Quantity Surveyor (QS) and the Planning Supervisor, for instance, are named and perform much the same functions under both types of contract. GC/Works 98 and its predecessor document differ from JCT 98 PWQ, however, in that they make no mention of the Architect. Instead, the contract administration role typically performed by the Architect under JCT 98 PWQ is performed under GC/Works 98 by the Project Manager (PM) – a position which has evolved from that of the Superintending Officer in the earlier Editions of GC/Works/1 and which, in Edition 3 (1989) was defined as ... *the "PM" means the official of the Authority, or other person employed in that capacity, named in the Abstract of Particulars and appointed by the Authority to act on his behalf (subject to the exclusions set out in the Abstract of Particulars) for the purpose of managing and superintending the Works* ... This definition is not dissimilar to that of the Client's Representative in the Manual of the BPF System [3] ... *the person or firm responsible for managing the project on behalf of and in the interests of the Client. He may be an architect, chartered surveyor, engineer or project manager. He may be an employee of the Client or a consultant* ... *The main management functions of controlling time* ... *standards and building performance are gathered together into the role of Client's Representative. He is charged with the responsibility of looking after the Client's interests* ... Whilst the definition of the PM in GC/Works 98 is not quite so concise as it was in GC/Works/1 nor, for that matter, is it as concise as the BPF's definition of the Client's Representative the duties required of him are nonetheless laid down unequivocally in the appropriate conditions.

11.1.15 Both BPF and GC/Works 98 adopt a similar philosophy insofar as the main role of the building designers (including the architect) does not necessarily extend into the construction period other than in an advisory capacity to the PM. The contract itself does not stipulate the professional discipline of the person who is to perform the PM function. Condition 1 requires that the PM be an individual named in the Abstract of Particulars whilst Condition 4(2) permits that individual to ... **expressly delegate in writing** to **named** representatives any of [his] powers and duties ... The bold text has been added here to emphasise the importance of maintaining clarity, should such delegation be considered, as to the nature and extent of any particular individual's responsibility and authority under the contract. It ought to be clear to the Contractor at all times from which source he is to accept which instructions and from whom he is to request what information. It would seem sensible to suggest that the discipline of the PM on any particular contract (and of those to whom he delegates his functions) will be dependent upon the nature of that contract and on the particular skills required. It is feasible that, in certain circumstances, most if not all of the PM's functions may indeed be carried out by the Architect. In other situations, the named PM may be an employee of the Employer's Group delegating specific functions as necessary to others better qualified to carry them out.

Programme and progress

11.1.16 Consistent with its clear identification of the management function and the separation of that function from others – such as design or cost control – GC/Works 98 also goes further than the JCT family of contracts in endeavouring to encourage sound management practice in the administration of the contract. By virtue of

Condition 1, the Programme is elevated to the status of a contract document which has the effect of making the content of the Programme – its logic, sequencing, etc. – binding on both parties. This is further reinforced by Condition 31(1)(b), which specifically requires the Contractor to ... *execute the Works ... in accordance with the Programme* ... How the Contractor delivers the project by the Date for Completion thereby becomes a contractual issue as between the parties. This can be contrasted with the position under JCT 98 PWQ, wherein the Contractor is required to provide and, upon the occurrence of certain happenings, update, a Master Programme although that programme is not accorded any contractual status. It is therefore no more than a statement by the Contractor of how he proposes to execute the work. In fact no reference is made in JCT 98 PWQ as to the format or content of the Master Programme. It is not even required properly to reflect the Dates of Possession and for Completion. (See section 2.10 – The Status of the Contractor's Programme.)

11.1.17 GC/Works 98 encourages a proactive approach to the management of the contract and to the use of the Programme in achieving this end. Condition 33 is quite specific about the format in which the Programme is to be presented. In particular, it is required to set out the proposed sequence of work, the method of working and the resources to be used and also to highlight those events which are considered critical to the completion of the works. The Programme must be achievable and be presented in a format which permits effective monitoring of progress. The intention is clear – to encourage the Contractor to give proper consideration pre-contract to the production of a document against which his progress on site will be monitored and his entitlement to extensions of time and to payment in respect of prolongation and disruption will be adjudged. As the Commentary to GC/Works 98 [4] points out ... *The programme requirements are substantial, and are intended to ensure that the Contractor enters the project fully resourced and with a properly considered and achievable method of working ...*

11.1.18 Opinions in the industry differ as to the advisability of giving the contractor's programme Contract Document status. Be that as it may, the difficulty of establishing the true effect on the Contractor's progress of the delaying events which are the subject of a dispute, in the absence of a properly produced construction programme, is all too apparent to anyone who has been involved in the resolution of construction disputes. In the authors' experience, the majority of those contracts which become mired in Adjudication, Arbitration or Legal Proceedings on the issue of delay do so in the absence – until it is too late – of any form of properly networked construction programme. It should be a matter of some concern to the industry as a whole that, in a considerable number of instances, the first appearance of any such document is as part of the Contractor's pleaded case during the course of a formal dispute.

11.1.19 Of course, any programme represents no more than its author's view as to how the completion of the contract works will be achieved, and any such view will – particularly in the early stages of a project – be based upon a certain amount of supposition rather than on hard facts. Any programme which is to be used as an effective monitoring tool must be responsive to events and must, wherever possible, be based upon facts rather than supposition. For this reason, and in order to maintain its effectiveness as a monitoring tool, Condition 33(2) permits the Contractor to submit proposals to amend the Programme at any time. It is important to note in this respect that the intention is to ensure that the Programme at all times reflects the actual progress of the Works, even if this indicates that the Works will be completed later than the Date for Completion.

11.1.20 Should the progress of the Works be delayed, amendments to the Programme will give a timely indication of the effects of any delay and ought, in the longer term, to make the task of establishing the facts behind delay (and thereby responsibility for it) that much easier. In this respect, it has to be stressed that the amendments to the Programme permitted by Condition 33(2) do not relieve the Contractor of his responsibility to complete the Works by the Date or Dates for Completion (whether original or revised). As the Commentary to GC/Works 98 [4] points out, ... *amendment of the Programme is a completely separate process from the claiming and granting of extensions of time* ... Should the Contractor's progress be delayed such that he considers himself entitled to an extension of time then the procedures set out in Condition 36 must be followed (see paras 11.2.1 to 11.2.12). In this respect, Condition 33(2) is quite specific that the submission by the Contractor of any amended Programme does not constitute the notice of delay referred to in Condition 36. The purpose of Condition 33(2) is simply to underline the fact that, at any particular point in time, the Contractor should be working in accordance with a published Programme which is reasonably reflective of the situation as it actually pertains on site.

11.1.21 The management regime of which the Programme is a part is further reinforced by a series of regular (usually monthly) progress meetings and associated reports. The requirements for these progress meetings, and for the reports which precede and follow them, are set out in detail in Condition 35. The Commentary to GC/Works 98 [4] again provides a useful summary of the intention behind these meetings, thus ... *The meetings should ensure that the parties promptly face all relevant issues, especially delay and increased cost, and should also provide valuable contemporaneous evidence in the event of disputes* ...

11.1.22 Condition 35(3) requires that, 5 days before each progress meeting, the Contractor submits a detailed written report. This report is required to set out details of the Contractor's actual progress relative to the Programme and, among other things, to explain the reasons behind any delays, to confirm details of any extensions of time requested and to set out any suggested re-programming proposals. It is the intention that all these issues be discussed at the subsequent progress meeting. Within 7 days following that meeting the PM is himself required to issue a written statement which, among other things, confirms his view as to the current state of progress and the reasons behind any delays, and also confirms any measures agreed with the Contractor to mitigate the effects of such delay. The PM is also required to set out the current situation as regards requests for, and awards of, extensions of time.

11.1.23 Of course, any management regime is only as effective as those who are responsible for implementing it. Should both the PM and the Contractor fail to give proper effect to the requirements of Conditions 33 and 35 by, for instance, continuing to work to a Programme which is patently out of date or by confining reports and progress meetings to mere platitudes, then the regime will not, have its intended effect. Properly implemented, however, such a regime (which closely links a properly networked programme with regular progress meetings and with the consideration of delays and of extension of time awards) will ensure that issues relating to progress are constructively and timeously discussed and will make the proper consideration of delays and entitlements to extensions of time that much easier to undertake.

Payments

11.1.24 Those more familiar with JCT contracts will find differences in the mechanisms for interim payments (referred to as "advances" in GC/Works 98) and in the regime for insurances, warranties and bonds which is more extensive than that adopted by JCT.

11.1.25 Since the introduction of GC/Works/1 Edition 3 in 1989, monthly advances under the GC/Works forms have been made by way of Stage Payments throughout the duration of the Works. These Stage Payments have, with the exception of the value of variations, expense attaching to prolongation and disruption and finance charges, been determined in advance by reference to a chart (traditionally an "S" curve) appended to the Contract. This system has the benefit of simplicity in that it removes the need to assess the value of the completed works at monthly intervals. However, it also has the disadvantage that the resulting assessment will be a less accurate reflection of the value of work actually carried out – albeit that this ought only to cause a particular problem in the event of the Contractor's insolvency since, with a solvent Contractor, the financial effects of any payment ahead of actual site progress can be reconciled in the final account process. Should acceleration of or delays to the progress of the Works occur, the "S" curve is simply "squeezed" or "stretched" so as to maintain the relationship between actual progress and the flow of money.

11.1.26 Recognising that such a system does have its limitations, and in response to the call in Constructing the Team [1] for more flexible payment arrangements, Condition 48 of GC/Works 98 introduces alternative means of determining the value of the Contractor's monthly entitlement. Alternative A maintains the stage payment approach described above. Alternative B varies this approach somewhat, in that payment becomes dependent upon the completion by the Contractor of certain Milestones. The Employer may define each of the relevant Milestones or, alternatively, may leave it to the Contractor to do so subject to the Employer's agreement. In either event each defined Milestone (together with the sum of money attaching to it) is set out on a Milestone Payment Chart, and payment becomes due only when all work within a particular Milestone and, unless specifically stated otherwise all preceding Milestones, has been completed. The Commentary to GC/Works 98 [4] accepts that arguments may occur where Milestones have been ill-defined by the parties, hence care needs to be taken to define Milestones where this method of payment is adopted. Alternative C is based on the valuation by the PM of completed works. However, receipt by the PM of the Contractor's timely, written application for payment of advances and valuation is a condition precedent to certification under this alternative. In the opinion of PACE, however, ... *the method is very time-consuming for both parties to the Contract* ... and is only really suited to contracts which are subject to large-scale remeasurement. It is worthy of note that unlike JCT 98 PWQ and irrespective of which alternative is adopted, the value included for varied work which is the subject of an accepted lump-sum quotation or the value of which has been otherwise agreed is not subject to retention.

11.1.27 Other payment features which will be unfamiliar to users of JCT 98 PWQ are the sharing of cost savings between the Employer and the Contractor (Condition 38(4)), the option of a bonus system for early completion (Condition 38A), the option of a retention payment bond in lieu of the withholding of money by the Employer (Condition 48A) (now similarly provided for in JCT 98 PWQ by way of optional clause 30.4A issued in January 2000) and the option for the Employer to make a mobilisation payment to the Contractor (Condition 48B). This latter measure is explained in the following terms in the Commentary to GC/Works 98 [4] ... *These optional provisions, which meet a recommendation of* Constructing the Team, *reflect the fact that for some types of work, especially mechanical and electrical engineering contracts, the Contractor may need to incur substantial expenditure early in the contract, before work commences on site* ... The mobilisation payment provision in GC/Works 98 is similar in effect to the optional advance payment now provided for by clause 30.1.1.6 of JCT 98 PWQ.

11.1.28 When the option is provided for in the Abstract of Particulars the making of a mobilisation payment is dependent upon the Contractor providing a satisfactory payment bond within 28 days of the acceptance by the Employer of his tender. This done, the Contractor is entitled to payment of the agreed sum, which will be stated in the Abstract of Particulars as a percentage of the Contract Sum, within 14 days. The sum paid is recovered by the Employer during the course of the contract by making regular deductions from the Advances on account.

11.1.29 No discussion of the payment provisions of GC/Works 98 would be complete without mention of Condition 51, which gives both parties to the contract extremely wide powers of set-off. The Commentary to GC/Works [4] explains the clause in the following terms ... *The Condition is drawn in very wide terms, allowing a global balance to be calculated between the Contractor's Group and the Government under all relevant contracts* ...

11.1.30 Condition 51 is notable not least because, with the first sentence containing over 150 words (and the associated definition of the term "Group", set out in Condition 1, running to some 80 words), it is not entirely consistent with the clear and concise approach evident in the remainder of the document. This said, it was no doubt considered unwise to deal with this issue, having regard to its nature, in anything other than strictly lawyerly terms. Put simply, any sum of money owed on a particular contract by a particular Contractor to a particular government department or agency may, if necessary, be recovered by any other government department or agency through any other contract involving any other member of the Contractor's group of companies. Lest this be considered too draconian, the system also operates in reverse, which was not the case with that of GC/Works/1 Edition 3. Such wide rights of set-off have long been considered necessary having regard to the special position of government as the single largest client of the UK construction industry (see *AB Contractors -v- Flaherty (1978) 16 BLR 8*), and to its desire to avoid a situation whereby the insolvency of one member of a contracting group can leave large sums of money owing to the Crown whilst payments continue to have to be made to other members of the same group in respect of other contracts.

Insurances and bonds

11.1.31 The provisions of GC/Works 98 which concern insurance of the Works and insurance in respect of public liability are broadly similar to those contained in the JCT family of contracts, save in the following two respects:

(i) under Alternative C of Condition 8, and where the Employer is ... *a Minister of the Crown, a government department or other Crown agency or authority* ... the Employer may accept the risk of loss or damage but is not obliged to effect insurance to cover that risk (reflecting the situation that, as a matter of policy, the Crown tends not to effect insurance); and

(ii) by virtue of Conditions 8(2) and 8(6)(b) all contract insurances, except those for professional indemnity pursuant to Condition 8A (see para. 11.1.32), are to be maintained until the expiration of the last Maintenance Period to expire. Whereas similar insurances under JCT 98 PWQ are required to be maintained only up to the date of Practical Completion or until determination, whichever is the earlier.

11.1.32 GC/Works 98 also provides for a number of other insurances, bonds and collateral agreements which do not fully feature in the JCT family of contracts. Condition 10 deals with the Contractor's liability where, he is made responsible for the design of certain elements of the Works either as set out in the Abstract of Particulars or as subsequently instructed by the PM. This issue is dealt with in a manner not dissimilar to that adopted by JCT in the Contractor's Designed Portion Supplement, save that GC/Works 98 provides for the Contractor's liability to be set either at the level of "reasonable skill and care" (Alternative A) or that of "fitness for purpose" (Alternative B). Condition 8A of GC/Works 98, which is optional, goes on to require this liability to be underpinned by an appropriate professional indemnity insurance. This has always been a curious omission from the JCT forms.

11.1.33 Conditions 66 and 67 of GC/Works 98, which only apply where so stated in the Abstract of Particulars, permit the Employer to require from the Contractor either a 10% performance bond or an ultimate parent company guarantee or (if considered necessary) both. In both instances (as with the earlier examples of retention bonds and mobilisation payment bonds), PACE have published model forms of agreement.

11.1.34 Condition 68, which is also optional, requires the Contractor to ... *use reasonable endeavours to procure* ... that any agreement between the Contractor and any of his subcontractors and suppliers contains obligations on the latter to provide collateral warranties and ultimate parent company guarantees in an approved form in favour of the Employer. Whilst the logic behind this provision may be sound, it is an onerous requirement which is likely to become quite unwieldy in practice. Should Condition 68 be implemented, it is suggested that it be so done in a discerning fashion, reflective of where the real risks lie, rather than on a blanket basis.

Acceleration, cost savings and bonuses for early completion

11.1.35 With the exception of JCT 98 MC the JCT family of contracts do not make provision for the progress of the works to be accelerated such as to achieve a Completion Date earlier than that set out in the Appendix to the Contract. Since the publication of Edition 3 in 1989, however, GC/Works has permitted such acceleration of the Works by consent. This issue is now dealt with in Condition 38 of GC/Works 98.

11.1.36 The initiative can be either with the Employer (Conditions 38(1) and (2) refer) or with the Contractor (Condition 38(3) refers). Should the Employer wish to achieve earlier completion of either the whole or any Section of the Works, he can direct the Contractor to submit priced proposals for so doing. In any such proposals the Contractor is also to set out the amendments to the Programme necessary in order to achieve the earlier completion date. By this means, the Employer is able to gauge the practical effect of the Contractor's proposals on site operations and to assess for himself (most probably on advice from the PM) whether the measures proposed have a realistic prospect of success – though it is a fact of life that the more one compresses a programme the greater the risk that it will not in the event be achieved. It is therefore necessary carefully to assess any proposed resequencing of activities both to confirm that the new sequence is indeed workable and to determine its effects on the critical path through the project. It is quite possible that the Contractor's proposals will place elements of the work on the critical path which previously contained float. This can have a consequential effect on the feasibility or otherwise of the Employer being able to supply information in respect of these newly critical elements in time to permit the programme to be achieved. Indeed, should the Contractor consider that the acceleration required by the Employer cannot feasibly be achieved he must decline to submit proposals and explain why (Condition 38(1)(b) refers).

11.1.37 If the Contractor submits proposals which are acceptable, the Employer is then to specify the new Date for Completion of the Works or of any relevant Section, any necessary amendments to the Programme, the amount by which the Contract Sum is to be adjusted and any consequential revisions to Milestone or Stage Payment Charts. Progress from that point is assessed by reference to the newly amended Programme and any subsequent entitlement to an extension of time will be similarly judged. The system operates in like fashion should the initiative behind the proposed acceleration be the Contractor's. In this event, the Employer undertakes to give proper consideration to any proposals put forward by the Contractor. It is important to note in either case that the Employer must accept the Contractor's proposals in their entirety or reject them. The Employer cannot, as is the case with Variation Instructions, insist that the Contractor accelerate the Works and subsequently require that the reasonable cost of so doing be determined by the QS.

11.1.38 The Employer may, perhaps because circumstances have changed in the interim, wish to seek acceleration of the Works in order to improve upon the Date for Completion. In such circumstances the Employer must expect to pay the reasonable costs of such acceleration. However, it may well be that the Employer turns to Condition 38 in order to mitigate the effects of delays to the progress of the Works, in other words, to attempt to recover such delays so as to complete as close as possible to the original Date for Completion. In such circumstances, the operation of Condition 38 ought to have regard to the manner in which responsibility for such delays is apportioned. In this respect, the Commentary to GC/Works 98 [4] gives the following useful guidance on the commercial realities of operating Condition 38 ... *it is not an easy option. It will often be difficult for the parties to agree on the exact causes of, and responsibility for, any delay, and therefore the financial arrangements for accelerating. If the Contractor accepts that delay is due to his own acts or omissions, he should be ready to accelerate on his own account whenever the cost of accelerating will be offset by avoiding the costs of extended time on site and liquidated damages, plus of course the effect on relations with the Employer. In other cases, the Contractor will seek some*

326

reassurance that he will recover his costs. Employers, for their part, are properly wary of paying twice for an existing delivery date ...

11.1.39 Should the delays which the Employer is seeking to recover be of his own making then invariably the Contractor will have been awarded an extension of time. In such a situation, and in the absence of any agreement under Condition 38, the Contractor's obligation is to complete by the new, extended Date for Completion. Should the Employer seek completion at an earlier date then he must again expect to pay the Contractor the reasonable costs of so doing.

11.1.40 There remains a role for the operation of Condition 38, however, even in instances where some or all of the delay which the Employer is seeking to recover is the responsibility of the Contractor. As is explained at para. 11.2.12 (and further, in respect of JCT 98 PWQ, at para. 2.7.45) there is a duty upon the Contractor to endeavour to mitigate the effects of any delay to the progress of the Works. Where such delays are the responsibility of the Contractor, the Commentary to GC/Works 98 [4] makes clear that this requires measures of the Contractor even where such measures may be ... *expensive or inconvenient* ... Case law, however, would appear to suggest that such measures would stop short of the expenditure of significant or substantial sums of money by the Contractor (see para. 2.7.45).

11.1.41 By way of illustration one may imagine a situation in which a Contractor is in culpable delay amounting to some 4 weeks, which is likely to expose him to additional preliminaries costs and liquidated and ascertained damages amounting to some £100,000. It would appear reasonable to suggest that the duty upon the Contractor to mitigate the effects of such a delay extends to the expenditure of a significant proportion of this sum which it will fall to him to pay in any event. Suppose, however, that it is of particular importance to the Employer to complete by the original Date for Completion but that it is now impossible to do so without the expenditure of an additional £150,000 – i.e. at least £50,000 more than the most that the Contractor may be obliged to expend if he did not recover the delay. It may well be that 24-hour shift working or the provision of additional and expensive craneage is required. Agreement could be reached under Condition 38 such that the Contractor will complete by the original Date for Completion in return for an additional payment of £60,000 by the Employer. The first £90,000 of the cost of acceleration is borne by the Contractor. The incentive to the Contractor is that this is some £10,000 less than would fall to him to pay were he not to accelerate the progress of the Works.

11.1.42 The above example adopts simplicity in order to illustrate the point. In reality, such situations will invariably be complex and fraught with difficulty. It is difficult, however, to avoid the impression that the Employer has, in the words of the Commentary to GC/Works 98, paid twice to achieve an existing delivery date. Further, there is a question mark over how the Employer can be said to have received that which he has paid for under Condition 38 in situations in which the Contractor initiates acceleration measures but the newly agreed Date for Completion is still not achieved because of subsequent delaying events. The scope for dispute over the respective rights and obligations of the parties in such a situation is obvious. Notwithstanding this, however, one must remember that the purpose of Condition 38 is to permit the Employer to deal in a commercial way with the practicalities of the situation with which he is faced. If the value to the Employer of achieving a particular completion date does indeed outweigh the cost to the Contractor of failing to meet that date, then Condition 38 provides a mechanism by which, given a reasonably flexible approach on both sides, the matter may be resolved to mutual benefit.

11.1.43 Condition 38 also permits the Contractor to initiate proposals which he considers . . . *will enhance the buildability of the Works, or reduce the cost of the Works, or the cost of maintenance, or increase the efficiency of the completed Works* . . . The Contractor is required to submit with his proposal an estimate of the likely cost savings. Interestingly, and unlike the situation in respect of acceleration, this is only an estimate. The actual value of the proposal, should the Employer decide to accept it, will be determined by the QS in accordance with the valuation rules in Conditions 41 to 43 (see paras 11.3.6 to 11.3.34). When the value of the proposal has been so determined, Condition 38(4) provides that the Employer and the Contractor . . . *shall share equally the relevant savings* . . . Further, should the Contractor's proposal require that work be resequenced then the Programme and the Date or Dates for Completion are to be amended accordingly and, if appropriate, an extension of time awarded.

11.1.44 A further dimension to the whole issue of completion dates, acceleration, etc. is added by the optional Condition 38A, which permits the Contractor to receive a bonus should the Works be completed ahead of the Date for Completion. The Commentary to GC/Works 98 [4] points out that this provision is aimed at complying with the recommendation in Constructing the Team [1] that contracts ought to provide incentives for exceptional performance. The mechanism is a simple one. For every calendar day that the Contractor completes the Works ahead of the Date for Completion (whether original or as adjusted by extensions of time or acceleration agreements), he receives a bonus at the rate stated in the Abstract of Particulars.

11.1.45 It does not seem unreasonable, if the Employer derives a financial benefit from early completion, that the Contractor (whose efforts may have made this possible) should also share in that benefit. However, there are practical problems connected with the use of such an incentive system which ought to be borne in mind. If, for instance, the Employer is himself unlikely to receive a financial benefit from early completion then it would be unwise to offer such a benefit to the Contractor. Early completion would then, in effect, become a financial burden upon the Employer. This is particularly relevant in the context of work carried out for government clients, for whom there is often no financial advantage in early completion. Even if there were, one would have to set the level of the bonus carefully such that some at least of the financial benefit remained with the Employer.

11.1.46 Whether Condition 38A is operative or not, it is important for the Contractor to bear in mind that should he be aiming to complete the Works by a date earlier than the Date for Completion current at the time it would be in his best interests to submit for the PM's agreement his proposals for the amendment of the Programme as provided for in Condition 33(2). At least by so doing he will determine whether or not the Employer is able to provide the requisite information and instructions in time to enable him to achieve his proposals. Should the Employer be in a position to provide this information to the Contractor in accordance with the accelerated Programme then it ought to be possible to achieve early completion to the mutual benefit of both parties. However, it should be borne in mind that, in accordance with the principle set down in the case of *Glenlion Construction Limited -v- The Guinness Trust (1988) 39 BLR 89*, the Employer is under no obligation to provide information to enable the Contractor to unilaterally accelerate the Programme in order to achieve completion before the Date for Completion.

11.2 EXTENSIONS OF TIME

Procedures

11.2.1 The extension of time provisions of GC/Works 98 are set out in Condition 36. Consistent with the approach apparent in the document as a whole, the three pages of text taken up by clause 25 of JCT 98 PWQ are reduced to a single page in GC/Works 98. Part of the reason for this is that the management regime set out elsewhere in the document, in particular the use of the Programme and of regular progress meetings, ought in theory to reduce the need for many of the procedural checks and balances contained in JCT 98 PWQ.

11.2.2 One of the more obvious differences between JCT 98 PWQ and GC/Works 98 is that the latter document has only one regime for the consideration of extensions of time, whereas the former has two – one which applies before the contractual Completion Date has passed (clause 25.3.1 refers) and one which applies after, i.e. during a period of Contractor culpable delay (clause 25.3.3 refers). A further obvious difference is that, in the case of JCT 98 PWQ, the Architect is apparently obliged to act in respect of events occurring during the original or extended contract period only in response to a notice from the Contractor except when exercising his right pursuant to clause 25.3.2 of JCT 98 PWQ to fix a Completion Date earlier than that previously fixed under clause 25. Whereas, by virtue of Condition 36(1) of GC/Works 98, the PM is free to act on his own initiative should he consider that the progress of the Works has been delayed.

11.2.3 The facility for the Authority to act unilaterally was present in a limited form in Edition 2 of GC/Works/1 and was further clarified and accorded to the PM in Edition 3 in response to the decision in *London Borough of Merton -v- Stanley Hugh Leach (1985) 32 BLR 51*. One of the many points decided in that case (which concerned a contract carried out under JCT 63) was that the absence of a notice from the Contractor did not relieve the Architect of his duty to give due consideration to the Contractor's entitlement to an extension of time. By the time of the decision in *Leach*, JCT had already amended the wording of clause 25 of JCT 80 such that the Architect arguably was not (with the exception noted in para. 11.2.2) at liberty to act unless and until receiving the necessary notice, particulars and estimate from the Contractor. In contrast, the draughtsmen of the GC/Works forms of contract gave the PM an express obligation to act on his own initiative. It is an obligation that the PM would be wise to take seriously, as any failure on his part to take account of acts of prevention on the part of the Employer, notwithstanding any corresponding failure to notify on the part of the Contractor, could well lead to loss of the Employer's right to deduct liquidated and ascertained damages in respect of delays for which the Contractor is responsible. It is intended that the management regime explained at paras 11.1.16 to 11.1.23, with its emphasis on regular reporting and discussion of matters relating to progress, will assist the PM in fulfilling this obligation.

11.2.4 GC/Works 98 therefore permits (but does not require) the Contractor to submit a notice to the PM requesting an extension of time. Other than a stipulation that this notice shall include the grounds for the Contractor's request, GC/Works 98 is silent as to the form which it must take and the point in time at which it must be submitted except that it may not be submitted after completion of the Works. It is not even clear whether the notice is required to be in writing. What is clear, however, is that should the Contractor at any time propose an amendment to the Programme pursuant to Condition 33(2) that Condition expressly provides that such a proposal ... *shall not constitute a notice from the Contractor requesting an extension of time for the Completion of the Works or of any Section* ... Further, Condition 35(3) requires regular (usually monthly) written reports from the Contractor which must include, among other things, progress by reference to the Programme, details of outstanding information requests, of any new events which have actually affected (or will potentially affect) progress, and of any recent requests for extensions of time. At some point, therefore, it will be necessary for the Contractor to reduce his request (and the facts which lie behind it) to writing by way of his regular progress report. Similarly, whilst there is no express equivalent of the requirement in JCT 98 PWQ that the Contractor's notice must be given ... *whenever it becomes reasonably apparent that the progress of the Works is being or is likely to be delayed* ... (clause 25.2.1.1 refers), the requirement of Condition 35(3) that the Contractor report on a regular basis and in a reasonable degree of detail ought to ensure that issues affecting progress are raised in a timely fashion.

11.2.5 In the absence of detailed prescription as to the serving of such notices, therefore, one is very much dependent upon the discipline provided by the Programme and progress meetings. It is, of course, in the Contractor's interests to submit timely requests for extensions of time but GC/Works 98 does not appear expressly to rule out the possibility of dubious late requests by the Contractor aimed primarily at digging himself out of a hole of his own making – which is not to say, of course, that the PM is obliged to give such requests undue consideration. Late requests which raise matters not previously aired in the Contractor's regular progress reports, or in progress meetings, may well be viewed (and not unreasonably so) with scepticism by the PM.

11.2.6 In terms of the information required from the Contractor in support of his request for an extension of time, GC/Works 98 is again less specific than JCT 98 PWQ. Clause 25.2 of JCT 98 PWQ requires the following from the Contractor in support of his application for an extension of time:

- details of the cause (or causes) of delay and identification of any which are Relevant Events;
- particulars of the effect on progress of each Relevant Event identified; and
- an estimate of the expected resulting delay to the Completion Date of each Relevant Event identified.

11.2.7 Condition 36(1) of GC/Works 98, by contrast, is no more specific than simply requiring that the Contractor's notice requesting an extension of time shall state the grounds for the request. As to the level of detail required, although the contract is not specific, it is arguable that the Contractor ought to provide sufficient in the way of detail to enable the PM to form a reasonable view as to the Contractor's entitlement. This, of course, would be supported by further details as to the extent of any delay and the reasons behind it in the Contractor's regular progress reports, and in any proposed amendments to the Programme produced by the Contractor pursuant to Condition 33(2). Condition 25 further requires the Contractor to maintain proper records in support of . . . *any claims made or to be made by the Contractor* . . . and to give the PM and QS such access to those records as they reasonably require in order to establish the Contractor's entitlement. In the context of delays and requests for extensions of time, such records would normally include site diaries, records of labour and plant usage (which is a requirement of Condition 15 in any event), details of material deliveries, weather conditions, etc. In common with the remainder of the document, the initiative lies very much with the PM in obtaining from the Contractor such supporting information as he considers reasonably necessary in order to form a view.

11.2.8 One area where GC/Works 98 is quite specific is in its use of time limits. As noted at para. 11.1.12, this form of contract makes extensive use of such time limits as a means of ensuring that matters are addressed in timely fashion and do not either lead to delays themselves or become running sores throughout the contract. In general, if the Contractor fails to comply with a time limit then he risks losing his entitlement in respect of the matter concerned, whilst a failure on the part of the PM or QS to comply with time limits can lead to the Contractor becoming entitled to the payment of finance charges under Condition 47. By virtue of Condition 1(4), any of the time limits set down in the contract can be extended by agreement between the parties even after the relevant time period has expired, with the exception of those set down in Condition 36 relative to extensions of time. It is always far easier to deal with requests for extensions of time contemporaneously with the events concerned. It would appear to be for this reason that those draughting the contract decided that the time limits would be mandatory in this respect and not subject to adjustment even by agreement.

11.2.9 The PM is given a maximum of 42 calendar days (excluding bank and public holidays) from the date of receipt of the Contractor's notice in which to notify the Contractor of his decision. All such decisions must be described either as interim or final. Interim decisions are to be kept under review until such time as the PM is satisfied that he has sufficient information available to him to make a final decision. As with JCT 98 PWQ, however, no final decision may reduce or withdraw an earlier interim extension of time unless work has subsequently been omitted from the contract. It is important, therefore, that the PM give proper consideration to the issues even before awarding an interim extension of time. It may even be advisable to err on the cautious side until the PM is satisfied that he is in receipt of the fullest possible supporting information from the Contractor.

11.2.10 No request for an extension of time may be submitted after completion of the Works. Again, this stipulation is intended to encourage the Contractor to submit such requests reasonably contemporaneously with the events to which they relate. In any event, in similar fashion to JCT 98 PWQ, the PM has a further period of 42 days following completion of the Works within which to review any outstanding requests and interim awards and to reach a final decision on the Contractor's overall entitlement.

11.2.11 Should the Contractor wish to take issue with any decision of the PM, he may do so by submitting a claim to that effect within 14 days of his having received the disputed decision. That claim must set out the grounds upon which the Contractor disputes the decision of the PM. Again the contract is silent as to the degree of detail required, but it is suggested that it is in the Contractor's firm interest to provide as much in the way of detail as possible in support of his claim. The PM is then required, within a further 28 days, to notify the Contractor of his decision on the claim. Should there remain a dispute as to the Contractor's entitlement, he is at liberty to notify the Employer, at any time, that he requires the matter to be referred to adjudication. The issue of such a notice will initiate the adjudication procedures as set out in Condition 59. By virtue of Condition 59(8), the adjudicator may subsequently vary or overrule the PM's decision if he considers it to have been incorrect. Alternatively he could seek determination of the dispute by reference to an Arbitrator under Condition 60. If he does so, however, he will have to wait until after completion, alleged completion or abandonment of the Works, or determination of the Contract, unless of course both parties agree otherwise. This is contrary to provisions in JCT 98 PWQ which do not so restrict the timing of a reference to an Arbitrator.

11.2.12 A further point of interest arises in respect of the stipulation in Condition 36(6) that no extension of time will be granted in respect of delays which are due to ... *lack of endeavour* ... on the part of the Contractor. This requirement can be read in conjunction with that in Conditions 31(1) and 34(1) that the Contractor is to proceed ... *with diligence* (and) *in accordance with the Programme* ... The combined effect is not dissimilar to the requirement in clause 25.3.4.1 of JCT 98 PWQ that ... *the Contractor shall use constantly his best endeavours to prevent delay in the progress of the Works* ... There is no real legal authority at present as to the meaning of this phrase and others like it when used specifically in connection with construction contracts, albeit, as noted at para. 2.7.45, there is a weight of authority unrelated to the construction industry. This authority would seem to suggest that the responsibility on the Contractor is an onerous one, but that it stops short of the expenditure of significant or substantial sums of money. It would appear that the Contractor is required to act as if he were in prudent and determined pursuit of his own interests and, indeed, if he is endeavouring to mitigate delays of his own making he is in fact acting in his own best interests. As to the maximum reasonable cost of such action on the part of the Contractor, one would imagine a sum approaching the level of liquidated and ascertained damages for which the Contractor would become liable were he not to act to recover the delay would be appropriate. The Commentary to GC/Works 98 [4] sheds some light on the subject by making clear that the Contractor is expected to endeavour to avoid delay even where the measures required are ... *expensive or inconvenient* ...

Matters entitling the Contractor to an extension of time

11.2.13 Condition 36(2) lists the following 7 matters which entitle the Contractor to an extension of time:

(a) *the execution of any modified or additional work;*
(b) *any act, neglect or default of the Employer, the PM or any other person for whom the Employer is responsible (not arising because of any default or neglect by the Contractor or by any employee, agent or subcontractor of his);*
(c) *any strike or industrial action which prevents or delays the execution of the*

> *Works, and which is outside the control of the Contractor or any of his subcontractors;*
>
> *(d) an Accepted Risk or Unforeseeable Ground Conditions;*
>
> *(e) any other circumstances (not arising because of any default or neglect by the Contractor or by any employee, agent or subcontractor of his, and other than weather conditions), which are outside the control of the Contractor or any of his subcontractors, and which could not have been reasonably contemplated under the Contract;*
>
> *(f) failure of the Planning Supervisor to carry out his duties under the CDM Regulations properly; or*
>
> *(g) the exercise by the Contractor of his rights under Condition 52 (Suspension for non-payment).*

1.2.14 This compares with the 18 Relevant Events currently listed in clause 25.4 of JCT 98 PWQ. This does not mean that the Contractor has a more restricted set of rights when working under GC/Works 98 but, rather, that the terms of Condition 36(2) are generally more loosely drafted than their JCT 98 PWQ equivalent and, as a result, are of wider application. For instance, Condition 36(2)(b) entitles the Contractor to an extension of time in respect of delays caused by ... *any act, neglect or default of the Employer, the PM or any other person for whom the Employer is responsible* ... This item potentially covers a multitude of sins, some (but not necessarily all) of which are dealt with on an individual basis in clause 25.4 of JCT 98 PWQ. Similarly, Condition 36(2)(e) refers to ... *any other circumstances* ... (with the exception of Contractor default and inclement weather) which are outside the control and reasonable contemplation of the Contractor.

1.2.15 GC/Works 98 is therefore somewhat less prescriptive than JCT 98 PWQ in its approach to the issue of extensions of time and, as such, is more akin to JCT 98 MW in this respect. There is a degree of scope, not present in JCT 98 PWQ, for the PM to use his discretion as to what does or does not entitle the Contractor to an extension of time. The point has already been made in relation to JCT 98 MW (see para. 7.2.14) that, in the absence of prescription, a flexible and consensual approach is required of all parties in order to ensure the smooth running of the contract. This is equally true of GC/Works 98, and is underlined by Condition 1A which states that:

> *(1) the Employer and the Contractor shall deal fairly, in good faith and in mutual co-operation, with one another ...;*
>
> *(2) both parties accept that a co-operative and open relationship is needed for success, and that teamwork will achieve this ...*

1.2.16 Whilst such clauses (which are intended in this case to reflect the philosophy of Sir Michael Latham) not unreasonably attract the criticism that they are virtually impossible to enforce and thereby constitute little more than contractual window dressing, the smooth operation of non-prescriptive contract machinery is largely dependent on an approach which avoids the adoption of entrenched positions and the vigorous pursuit of self-interest. In the operation of Condition 36, both the Contractor and the PM ought to have due regard to the spirit of Condition 1A.

1.2.17 Whilst this difference in approach between the two forms makes a detailed comparison of JCT 98 PWQ and GC/Works 98 somewhat difficult in this regard, it is possible to identify a number of significant respects in which the regime set out in the two contracts varies. These can be summarised as follows:

Force majeure

11.2.18 GC/Works 98 makes no specific reference to *force majeure* as something which will entitle the Contractor to an extension of time. However, it is possible that Condition 36(2)(e) of GC/Works 98 may in fact be of wider application than clause 25.4.1 of JCT 98 PWQ. As noted at paras 2.8.2 and 2.8.3, *force majeure* is a vague term, deliberately so according to some commentators, but case law confirms that, for an event to qualify as such, it must be one which is outside the will or control of man. Furthermore, the examples given by the Courts of events which may constitute *force majeure* suggest that, on the whole, they are required to be of somewhat seismic proportions (war, epidemics, etc.). By contrast, one can envisage a range of circumstances which, although not amounting to a plague of locusts, are nevertheless ... *outside the control of the Contractor* ... and would thereby entitle him to an extension of time under Condition 36(2)(e) of GC/Works 98.

Adverse weather

11.2.19 Whereas JCT 98 PWQ places the risk of adverse weather largely with the Contractor, in that he is entitled to an extension of time only in extreme conditions and even then he is not entitled to any loss and/or expense arising in connection therewith, GC/Works 98 places this risk entirely with the Contractor. There is no facility for the PM to award an extension of time by reason even of exceptionally adverse weather. Condition 36(2)(e) specifically excludes weather conditions from the ambit of those matters which are beyond the Contractor's control and for which he is otherwise entitled to an extension of time. One benefit of this is that it removes the need for arguments (relatively common under JCT 98 PWQ contracts) as to precisely what constitutes ... *exceptionally adverse weather conditions* ... Another argument may well be that it places the risk with the party (in this case the Contractor) best able to manage and/or price that risk. In the authors' view, whilst this may well be true of weather conditions generally it is not necessarily true of conditions which qualify as exceptional. One ought to be able to expect that an experienced Contractor will be better placed even than an experienced Employer to quantify those weather-related risks which one can normally expect to encounter on construction contracts, and to make proper allowance within his tender. The same cannot be said, however, of conditions which by their nature are outside the ambit of one's normal experience.

11.2.20 In describing the scheme of risk allocation set out in JCT contracts (in that case JCT 63) HH Judge Edgar Fay QC stated in *Henry Boot Construction Limited -v- Central Lancashire New Town Development Corporation (1980) 15 BLR 1* that ... *in cases where the fault is not that of the Contractor, the scheme clearly is that in certain cases the loss is to be shared; the loss lies where it falls* ...

11.2.21 What the learned Judge is understood to have meant by this was that neither the Employer nor the Contractor ought to be required to compensate the other where delay is the result of a neutral event – i.e. an event (such as exceptionally adverse weather) not caused by the fault of either party. In other words, the Contractor ought to be relieved of liability for liquidated and ascertained damages (by the award of an extension of time), whilst the Employer ought not to be required to pay direct loss and/ or expense to the Contractor. GC/Works 98 (and its predecessor documents) departs in this instance from the maxim that "the loss lies where it falls" in holding the Contractor liable to compensate the Employer for delays due to what are neutral and, by definition, unpredictable events. The Commentary to GC/Works 98 [4] explains the philosophy thus ... *The Contractor will need to tender in the light of the period allowed* [by the Employer] *to complete the Works and to accept the risk and or advantage that the actual weather will be better or worse than is implied by the period stipulated ...*

11.2.22 This said, this is one of the few areas in the new contract where the allocation of risk could be described as harsh toward the Contractor.

Insured risks

11.2.23 GC/Works 98 has no direct equivalent of clause 25.4.3 of JCT 98 PWQ, which entitles the Contractor to an extension of time in respect of delays caused by certain insured risks (fire, lightning, storm, flood, etc.). Arguably, any such event would give rise to an entitlement under Condition 36(2)(e) of GC/Works 98. This permits the PM to award an extension of time for events which are outside the Contractor's control and which are not caused by his default or neglect, which would be the case with most, if not all, of the events covered by clause 25.4.3 of JCT 98 PWQ. However, the Contract also requires that the events in question ... *could not have been reasonably contemplated under the Contract ...* This requirement creates a potential difficulty in this instance in that, by virtue of the fact that such events are expressly required to be the subject of insurance under Condition 8, they quite clearly were within the contemplation of the parties to the Contract. Furthermore, Condition 36(2)(d) goes as far as to make express provision with regard to Accepted Risks (so called because they are "accepted" by the Employer as being risks which are generally uninsurable such as sonic booms, radiation, etc.) whilst not doing the same in respect of insurable risks. This potentially raises the argument that, had it been the intention that the Employer would share the risk of delay arising from insurable events, then the Contract would expressly have stated so. Incidentally, it is worth noting that the equivalent term to ... *Accepted Risks ...* to be found in JCT 98 PWQ is ... *Excepted Risks ...* (because they are excepted from the Contractor's range of risks); the words "Accepted" and "Excepted" are quite different yet they are used to equal effect as between GC/ Works 98 and JCT 98 PWQ by virtue of describing the same situation from the different standpoints of the Employer and Contractor respectively.

11.2.24 Whilst the rather loose wording of Condition 36(2)(e) does afford the PM a measure of discretion and flexibility in approaching such matters, it is suggested that any such discretion ought to be exercised very much with the spirit of Condition 1A (fair dealing and teamworking) in mind. One has to ask oneself whether it is really the intention of either party to this Contract that, for instance, a 6-month delay caused by a fire which was the result of a strike of lightning (which, whilst not expressly included in Condition 36(2)(e), is not the result of the Contractor's default and is therefore not expressly excluded by it) ought reasonably to expose the Contractor to 6 months' worth of liquidated and ascertained damages.

PM's instructions

11.2.25 As explained in more detail at para. 11.3.2 below, the PM's power to issue instructions to the Contractor is rather less tightly defined under GC/Works 98 than is the Architect's under JCT 98 PWQ. With this in mind, it is interesting to compare the way in which the two forms of contract deal with the Contractor's entitlement to an extension of time when this power is exercised.

11.2.26 Clause 25.4.6 of JCT 98 PWQ grants the Contractor an entitlement where any instruction has been issued late. Where an instruction is not late, however, the Contractor is entitled to an extension of time only in those situations listed in clause 25.4.5 of JCT 98 PWQ. GC/Works 98 arguably entitles the Contractor to an extension of time in respect of delays caused by any instruction of the PM, whether issued late or not. Condition 36(2)(a), for instance, entitles the Contractor to an extension of time in respect of ... *the execution of any modified or additional work* ... thereby expressly making allowance in respect of Variation Instructions (VI's) as defined in Condition 1(1). Condition 36(2)(b) then goes on to cover delays caused by ... *any act, neglect or default of the Employer, the PM or any other person for whom the Employer is responsible* ... A reading of this text suggests that any instruction issued by the PM will entitle the Contractor to an extension of time if it can be demonstrated to have caused a delay, even in the absence of any default or neglect on the part of the PM.

11.2.27 This said, the distinction here is rather more of academic than practical interest. Clause 25.4.5 of JCT 98 PWQ, even though not exhaustive, arguably covers the majority of instances whereby the Contractor's progress is likely to be materially affected by the issue of Architect's Instructions. In practice, therefore, there are unlikely to be many situations in which the Contractor would be entitled to an extension of time under GC/Works 98 but not so under JCT 98 PWQ.

Delay by Nominated Sub-Contractors

11.2.28 Clause 25.4.7 of JCT 98 PWQ gives the Contractor an express entitlement to an extension of time for delays on the part of Nominated Sub-Contractors. There is no equivalent clause in GC/Works 98. With the exception of costs generated as a result of insolvency, which are reimbursable by the Employer, Nominated Sub-Contractors under GC/Works 98 are entirely the risk and responsibility of the Contractor.

11.2.29 With the exception of the items discussed above, the balance of the risk attaching to time in GC/Works 98 is not substantially different from that in JCT 98 PWQ, albeit that the approach in the former document is somewhat less prescriptive and leaves correspondingly more room for the PM to use his discretion as to the Contractor's entitlement. The only real guidance given in the Commentary to GC/Works 98 [4] as to how the PM ought to interpret and apply Condition 36(2) is in the following terms ... *the PM should only award extensions of time if the relevant delay is "due to" one of the permitted causes of delay. Particularly difficult problems may arise in relation to concurrently operating permitted and non-permitted causes of delay. The solution to this problem has been intentionally left to the PM, who must decide, as a matter of fact, whether or not the relevant delay is "due to" one of the permitted causes of delay* ...

.2.30 No guidance is given as to how to interpret the somewhat loosely defined causes of delay set out in Condition 36(2). As noted above, the best guidance available is probably that set out in Condition 1A. The PM is free to, and ought to, approach the matter fair-mindedly and with a view to maintaining the generally sensible balance of risk between the parties which the contract as a whole sets out to achieve.

11.3   VARIATIONS AND DISRUPTION

Variations

1.3.1 All matters connected with the issuing of Instructions by the PM, with the exception of how they are valued, are dealt with under Condition 40 of GC/Works 98. Indeed, the definition of the term "Instruction" given in Condition 1(1) is ... *any instruction given in accordance with Condition 40* ... Like JCT 98 PWQ, the PM is empowered to give instructions to the Contractor only in respect of those matters expressly laid down in the contract. Unlike JCT 98 PWQ, these matters are all listed together for ease of reference under Condition 40(2). The matters listed under Condition 40(2) are as follows:

(a) *the variation or modification of all or any of the Specification, Drawings or Bills of Quantities, or the design, quality or quantity of the Works;*

(b) *any discrepancy in or between the Specification, Drawings and Bills of Quantities;*

(c) *the removal from the Site of any Things for incorporation and their substitution with any other Things;*

(d) *the removal and/or re-execution of any work executed by the Contractor;*

(e) *the order of execution of the Works or any part of them;*

(f) *the hours of working and the extent of overtime or night work to be adopted;*

(g) *the suspension of the execution of the Works or any part of them;*

(h) *the replacement of any person employed in connection with the Contract;*

(i) *the opening up for inspection of any work covered up;*

(j) *the amending and making good of any defects under Condition 21 (Defects in Maintenance Periods);*

(k) *cost savings under Condition 38 (Acceleration and cost savings);*

(l) *the execution of any emergency work as mentioned in Condition 54 (Emergency work);*

(m) *the use or disposal of material obtained from excavations, demolition or dismantling on the Site;*

(n) *the actions to be taken following discovery of fossils, antiquities or objects of interest or value;*

(o) *measures to avoid nuisance or pollution;*

(p) *quality-control accreditation of the Contractor as mentioned in Condition 31 (Quality); and*

(q) *any other matter which the PM considers necessary or expedient.*

1.3.2 It will be apparent that the last of these serves to widen the scope of the PM's powers quite considerably, albeit it is suggested that any discretion which is granted to the PM by virtue of this provision would require to be exercised reasonably. In light of this considerable discretion, it appears not to have been considered necessary by the authors of GC/Works 98 to include a provision (such as that at clause 4.2 of JCT 98 PWQ) expressly permitting the Contractor to challenge the PM's right to issue any particular instruction. However, the position is clarified in the Commentary to GC/Works 98 [4] in the following terms ... *The PM's powers under paras (1) and (2) are very wide, especially in view of the sweeping words of sub-para. (2)(q). However, it should be noted that all Instructions will need to satisfy Condition 1A (Fair dealing and teamworking), and that no Instruction, without the Contractor's consent, may unilaterally alter the terms of the Contract itself ...*

11.3.3 All Instructions must be complied with by the Contractor "forthwith" except when the PM, pursuant to Condition 40(5) makes a VI (i.e. a Variation instruction as defined in condition 1(1)) ... *conditional upon agreement of such a lump-sum price, pending which agreement the Contractor is not to begin complying with the VI* ... and all are to be in writing, except those issued under sub-paras (2)(b), (d), (g) and (l) which may be given orally but which must be confirmed in writing within 7 calendar days of issue. It is perhaps sensible where, for instance, confusion arises as a result of discrepancies between the various contract documents and where this is causing work on site to be held up, that immediate oral instructions can be issued under Condition 40(2)(b) with a view to restarting progress. It is also reasonable that emergency work can be instructed in such a manner. This sense of urgency would appear to explain the provision in Condition 40(3) that oral Instructions may be given to ... *such employee or agent of the Contractor as the PM thinks fit* ... rather than to the Contractor's agent as is the case with written instructions. Written confirmation of oral Instructions, however, must be directed to the Contractor's agent. This, of course, potentially gives rise to a situation whereby the Contractor's agent may be temporarily unaware of Instructions issued by the PM. This is, unfortunately, unavoidable if the Contractor's agent is not personally in attendance when the Instruction is first issued. It is suggested, however, that care should be taken by the PM to confirm any such Instruction to the Contractor's agent sooner rather than later in the interests of sound management of the project. GC/Works 98 provides a further useful tool in this respect by requiring the Contractor's agent to confirm receipt of all Instructions, including written confirmation of oral Instructions. It ought therefore quickly to become apparent if the Contractor's agent is unaware of any oral Instructions which have been issued to others in his absence.

11.3.4 One provision which could potentially be a source of difficulty is the requirement in Condition 40(3) that oral Instructions are ... *immediately effective in accordance with their terms* ... The Contractor must comply with such Instructions immediately, despite the fact that they may not be confirmed in writing for as much as 7 days. It is possible (if unlikely) to envisage a situation whereby such confirmation fails to materialise. This may happen, for instance, in situations where the PM has delegated authority to more than one person but their respective areas of responsibility have not properly been defined. In such circumstances, the Contractor's entitlement to be paid in respect of an oral instruction which he has put into effect may be questionable. Should this be the case it is suggested that the PM, in approaching the matter, ought to have proper regard for the fair dealing provisions set out in Condition 1A.

11.3.5 The position in circumstances where the Contractor fails to comply with an Instruction is set out in Condition 53 and is not dissimilar from the position under

JCT 98 PWQ. In this event, the PM is to issue a notice to the Contractor requiring compliance, within a specified time period, with the Instruction concerned. The period for compliance is left to the discretion of the PM, but would generally be, one would suggest, of the order of 7 to 14 days. Should the Contractor fail to comply with the Instruction within the period specified, the Employer may pay others to undertake whatever is necessary in order to give effect to the Instruction and subsequently recover the additional costs of so doing from the Contractor. The Employer may also use non-compliance with Instructions as grounds for determining the Contract under Condition 56. In such cases, the Employer may issue a notice of determination should the Contractor fail to comply with an Instruction ... *within a reasonable period of its issue* ... (Condition 56(6)(a) refers). The Employer's right to determine arises only should the Contractor's non-compliance continue for 14 days after the date of issue of the notice under Condition 56(1)(a). As to which remedy the Employer would be best advised to pursue in such instances, the authors would refer to the sound advice given in the Commentary to GC/Works 98 [4] that ... *specific legal advice should be taken before determining any significant Contract, as determination is full of pitfalls, and the source of many legal disputes* ... In the great majority of cases, action under clause 53 ought to be more than sufficient.

11.3.6 Variation Instructions (VI's) are defined in the following terms in Condition 1(1) of GC/Works 98:

> ... *any Instruction which makes any alteration or addition to, or omission from, the Works or any change in the design, quality or quantity of the Works; and "the Works" means the works described or shown in the Specification, Bills of Quantities and Drawings, including all modified or additional works to be executed under the Contract* ...

11.3.7 This definition is somewhat more truncated than that set out at clause 13.1 of JCT 98 PWQ. In particular, it is specifically concerned with changes to the scope and nature of the Works as designed. Instructions which do not alter the "quality or quantity" of the Works but which do alter the conditions under which they are carried out (working hours, access to the site, etc.) do not qualify as VI's under GC/Works 98. There is a significance to this distinction, in that different valuation rules apply, on the one hand, to VI's and, on the other, to all other Instructions. The general principles which underpin the valuation of all Instructions are set out at Condition 41. Specific rules as to the valuation of VI's are set out at Condition 42, whilst Condition 43 deals with the valuation of all other Instructions.

11.3.8 One striking contrast with the position under JCT 98 PWQ is that, by virtue of Condition 41(2) of GC/Works 98, the value ascribed to any Instruction ... *shall include any disruption to or prolongation of both varied and unvaried work* ... Clause 13.5 of JCT 98 PWQ, of course, states that ... *no allowance shall be made under clause 13.5 for any effect upon the regular progress of the Works or for any other direct loss and/or expense* ... such matters being dealt with according to the somewhat different set of rules contained in clause 26.

11.3.9 One consequence of the different approach adopted by GC/Works 98 is that – certainly as regards the effect of *variations* – the strictures which JCT 98 PWQ has traditionally applied to the provenance of direct loss and/or expense (as set out in clause 26 thereof) are conspicuous by their absence. The JCT contracts have, on the whole, treated direct loss and/or expense as a form of quasi-legal loss which requires a level of proof approaching that required by the Courts when pursuing a claim for common law damages. Hence the requirement in clause 26.1 of JCT 98 PWQ that the QS … *ascertain* … (i.e. determine as a matter of fact) the Contractor's entitlement having been provided with such documentary evidence as he may reasonably require. In effect, the QS must be satisfied, on the basis of the information provided to him, that the sums claimed have indeed been expended and as a direct result of the matters complained of. Inasmuch as this constitutes a more forensically based methodology than is normally apparent in the valuation of varied works, it is not an approach which is expressly required by Condition 42 in respect of the prolongation and disruption element of VI's.

11.3.10 A further point to be made in this respect is that the costs which flow from prolongation and disruption tend not to be of a kind which lend themselves to the sort of neat allocation as between individual Instructions that GC/Works 98 envisages. This is particularly the case with costs attaching to disruption, the more so in complex situations where there may be a considerable number of contributory causes of, for instance, uneconomic use of labour and plant. It is suggested that the approach adopted by GC/Works 98, where the QS, in theory at any rate, commits himself incrementally to reimbursing the Contractor in respect of prolongation and disruption every time he values a VI, may result in a less accurate overall assessment and could potentially lead to a technical "overpayment" to the Contractor in the final analysis. Nevertheless, this appears to be an approach which GC/Works 98 wishes to encourage in the valuation of VI's generally (refer paras 11.3.19 and 11.3.20) and which it considers, when operated in a sensible commercial fashion, will lead to fewer disputes without necessarily sacrificing the interests of the Employer. It is the authors' view that the avoidance of disputes in this area will depend very much on the attitude displayed by the QS and the Contractor. A hardened (or "claims conscious") commercial approach on the part of the Contractor, when combined with an overly officious approach by the QS, will serve to create friction whatever the contract form. In apparently moving away (in respect of VI's at least) from treating prolongation and disruption costs as a quasi-legal form of loss, GC/Works 98 appears to permit of a sensible commercial approach by both sides, which carries scope for a degree of compromise, and which may act to defuse the potential for friction. One is again minded, in this respect, to refer to the fair-dealing provisions contained in Condition 1A.

11.3.11 Condition 42(1) provides two alternative means for arriving at the value of a VI. Either by the acceptance of a lump-sum quotation from the Contractor or by valuation by the QS. The facility for agreeing a lump sum with the Contractor has been present since the publication of Edition 3 of GC/Works/1 in 1989. However, as the Commentary to GC/Works 98 [4] points out … *The pre-pricing of variations was endorsed in general terms by "Constructing the Team". Edition 3 procedures have been retained and enhanced …*

11.3.12 The new Condition 42(4) appears to suggest that the QS ought only to be required to value VI's in situations where it has not been possible, for whatever reason, to agree a price with the Contractor. In other words, the Employer's first priority ought to be to agree the pre-pricing of VI's and only if this proves impossible should the QS be required to provide a valuation. In similar vein to JCT 98 PWQ clause 13A, Condition

40(5) provides for the PM to make compliance with a VI conditional upon agreement by the Employer of the Contractor's quotation for the additional/varied work. In other words the PM can stipulate that unless/until there is agreement as to the value of a VI the work comprising that VI is not to be executed. It is clear that the authors of GC/Works 98 are keen to encourage a consensual approach to the valuation of VI's (which – together with the related issues of extensions of time and direct loss and/or expense – are traditionally the more contentious areas of the construction process) and that the emphasis upon agreed lump sums and upon negotiation which is apparent in Condition 42 is regarded as a key element of that approach.

1.3.13 The procedure by which lump sum quotations are obtained from the Contractor and considered by the Employer is set out in sub-paras (2) and (3) of Condition 42. As to how the procedure ought to be operated, the Commentary to GC/Works 98 [4] contains the following guidance ... *The procedure is predicated on the basis that both sides will approach matters in a reasonable way. Contractors should not load quotes, nor is it expected of QS's acting for the Employer consistently to reject or recommend against lump sums because they cannot be absolutely sure of their overall price compared to evaluation against bill rates. The lump-sum system is inherently less certain and embodies an element of commercial negotiation in order to realise the advantages of pre-pricing ...*

1.3.14 According to Condition 40(5), upon issuing a VI the PM may require the Contractor to submit a lump sum quotation of the cost of complying with the VI. The quotation is to be submitted within 21 days of the date upon which the VI is issued. It is important to note that the Contractor is only to provide such a quotation on instruction from the PM. Unsolicited quotations from the Contractor have no contractual status and fall outside the procedure set down in Conditions 40 and 42.

1.3.15 The Contractor's quotation is required to set out how the sum has been calculated by distinguishing between, on the one hand, the direct cost of complying with the Instruction and, on the other, the cost of any prolongation or disruption attaching to it. The Contractor is also to include with the quotation ... *such other information as will enable the QS to evaluate that quotation* ... (Condition 42(2) refers). It would appear that the nature and extent of such supporting information is, initially at any rate, within the Contractor's discretion. There is no express power for the QS to call for further information in support of a quotation should he consider that the information submitted by the Contractor is inadequate (Condition 41(4) gives this power to the QS only in relation to his own valuation of VI's). Of course, if the Contractor wants his quotation to be accepted then it is in his interests to provide the QS with such information as he may reasonably require. Indeed, subparas (3) and (4)(b) of Condition 42 expressly envisage a negotiating process arising out of the lump-sum system. It is not unreasonable to expect that the provision of supporting information will form a part of such a negotiating process, which the authors of GC/Works 98 require to be conducted within the spirit of the guidance quoted above and of the fair dealing provisions in Condition 1A. Unreasonable requests for information from the QS or an unco-operative approach from the Contractor are likely to lead to a breakdown of the system.

1.3.16 The PM has 21 days from receipt of the Contractor's quotation in which either to accept it or, alternatively, to reject it and to state in lieu a lump sum which he would be prepared to accept. Should it be possible to agree upon a figure (whether it be that initially quoted or otherwise) then it is this figure which will be added to the Contract Sum and the matter is concluded. Should it not be possible to reach agreement, however, the PM must then instruct the QS to value the VI.

11.3.17 Clause 13.4.1.1 of what was then JCT 80 provided that ... *all Variations ... shall be valued by the Quantity Surveyor*... unless either the Contractor and the Employer had already agreed a sum between themselves or unless the Variation was to be the subject of a 13A Quotation. Valuation of Variations by the QS was therefore very much the default position, and will be the situation with which the majority of QS's are still most familiar. However, as noted at Section 3.3, this is no longer the situation under JCT 98 PWQ. Nor is it the situation under GC/Works 98 where, as noted above, it appears to be the intention of the authors of the document that, as is now also the case with JCT 98 PWQ, pre-pricing by the Contractor of VI's becomes the rule rather than the exception. By virtue of Condition 42(5), the QS is empowered to value a VI only when required to do so, either because the PM has elected not to seek a quotation from the Contractor or because it has not been possible to agree a lump sum. In the latter instance, some 6 weeks or more may well have passed since the issue of the VI.

11.3.18 The basic rules in accordance with which the QS is required to value the VI are set out at Condition 42(5) and, in principle, are not inconsistent with those in clause 13.5 of JCT 98 PWQ – that is, wherever possible VI's are to be valued by reference to rates contained in the Bills of Quantities and, where this is not possible, by reference to ... *fair rates and prices, having regard to current market prices* ... Condition 42(5)(d) also provides for work to be valued by reference to daywork charges ... *if it is not possible to value by any of the preceding methods of measurement and valuation* ... again, an approach consistent with that set out at clause 13.5.4 of JCT 98 PWQ. In common with its approach to matters generally, GC/Works 98 manages to dispose of these rules in a little over 8 lines of text rather than the full page occupied by clauses 13.5.1. to 13.5.4 of JCT 98 PWQ which deal with essentially the same subject matter. Much of the detail which does not find its way into the text of GC/Works 98 is intended to be dealt with by reference to good professional practice and to the fair-dealing provisions in Condition 1A.

11.3.19 By virtue of Condition 41(2), the QS is (as noted at para. 11.3.8) required to include within the value of the VI an allowance in respect of any prolongation and/or disruption to both varied and unvaried work which he considers is consequent upon the VI. Condition 42(6) further requires that the QS allow for prolongation and/or disruption to "work not within the direct scope of the VI" by adjusting the rates and prices for such work. There is no guidance, either within the contract itself or in its accompanying Commentary, as to the means by which the QS is to make due allowance for such prolongation and disruption. Condition 42(6) appears to suggest that it ought to be by means of adjustments to the rates and prices included in the Bills of Quantities, an approach which, as noted at para. 11.3.9, is somewhat at odds with the more forensic method normally adopted by QS's in ascertaining direct loss and/or expense.

11.3.20 GC/Works 98 appears to envisage an element of judgment and discretion being used insofar as the prolongation and disruption attaching to VI's is concerned (the approach to prolongation and disruption arising otherwise than by reason of VI's is dealt with in a manner much more consistent with that of JCT 98 PWQ – see paras

11.3.24 to 11.3.34). The fact that some of the prolongation and/or disruption costs attaching to a particular VI may not yet have been incurred at the point in time when the QS ascribes a value to that VI also serves to highlight the element of estimation inherent in the process – an element which is not entirely consistent with the forensic approach to costs actually incurred which is required by JCT 98 PWQ.

11.3.21 The Contractor is required to provide the QS, within 14 days of a request, with any information which the QS reasonably requires in order to value an Instruction (Conditions 41(4) and 42(7) refer). Notably, in terms of time limits, the clock begins to run only when the QS makes such a request. The contract does not stipulate a time period within which a request must be made. In this instance, as distinct from that where a quotation is submitted, it is the QS (rather than the Contractor) who is the arbiter of that which is reasonably required. Should the QS be satisfied that he has the information he requires, he then has 28 days in which to notify the Contractor of his valuation. The contract is not specific about what happens should the Contractor fail to provide the information requested within the time stipulated or, alternatively, should there be a dispute as to the sufficiency of the particulars. It is certainly open to the PM, by virtue of Condition 1(4)(a), to agree to extend the 14 days. However, to acquiesce too often in such a situation would operate to undermine the discipline which the various contractual time limits in GC/Works 98 are intended to instil. It is suggested that such action ought to be considered only where the Contractor has good reason for having failed to provide the information by the due date or where it would be patently unreasonable for the PM to withhold his agreement. Where the 14-day period is not extended, it is further suggested that, in accordance with normal practice under JCT 98 PWQ (which admittedly contains no express requirement, other than at clause 30.6.1.1 which specifically relates to the preparation of the final account), the QS should nevertheless endeavour to provide a valuation. Such a valuation would be the QS's best estimate in the absence of the information he reasonably requires from the Contractor.

11.3.22 Once the QS has notified the Contractor of his valuation, the Contractor has a further 14 days within which to dispute the QS's figures. Should he do so, Condition 42(9) requires that he provide reasons for his disagreement and also his own valuation of the VI. Should the Contractor fail to provide such notification within the stipulated time period, he will be deemed to have accepted the QS's valuation and is not permitted to make any further claim in respect of that particular VI. Again, this time period may be extended by agreement and it is suggested that this is done only in situations where not to do so would be patently unreasonable. Having received such notification of the Contractor's disagreement under Condition 42(9), the contract is then oddly silent as to how the QS ought to respond. It is suggested that in such circumstances he ought, as soon as is reasonably practicable, to revise his valuation to the extent that the Contractor has satisfied him that this is warranted. Alternatively, the QS must inform the Contractor that his original valuation stands. Inasmuch as any disagreement remains at this point, it is then open to the Contractor to refer the matter to Adjudication or Arbitration.

11.3.23 There are no time limits in clause 13 of JCT 98 PWQ which apply to the exercise by the QS of his power to value variations (albeit that there are time limits set down in clause 30.6.1 thereof in the context of the preparation of the final account). Perhaps unsurprisingly in light of this, it is often the case that disagreements concerning the value of variations have a tendency to persist well into the final account discussions and serve in the long term to sour relations between the Contractor and the QS, what is commonly known as a "running sore". Part of the rationale behind the system of time limits in GC/Works 98 would appear to be to avoid just such a situation developing. However, it is worthy of note that – in circumstances where a quotation is initially sought from the Contractor but is found unacceptable – at least 12 weeks may elapse following the issue of an Instruction before the QS is required to produce a valuation. The Contractor then has a further 2 weeks within which to notify the QS of any disagreement. It remains highly probable that arguments concerning the value of a VI may continue between the Contractor and the QS many weeks after the work required by the VI has been carried out, bearing in mind that unless specifically noted otherwise by the PM pursuant to Condition 40(5), all Instructions are required to be complied with "forthwith" (see para. 11.3.3). Should there be a large number of VI's on any particular contract, therefore, it is likely that the fair-dealing provisions of Condition 1A will be tested to the limit in order for the "running sore" scenario to be successfully avoided.

Procedures for the determination of prolongation and disruption costs (other than by reason of VI's)

11.3.24 The valuation of Instructions which do not constitute VI's is dealt with under Condition 43. As compliance with these Instructions does not necessarily result in any physical output, they do not attract a value of their own in the sense that VI's do. The Contractor is, however, entitled to recover what the Commentary to GC/Works 98 [4] refers to as ... *out-of-pocket expense* ... This is defined in the following terms in Condition 43(4) ... *In this Condition "expense" shall mean money expended by the Contractor, but shall not include any sum expended, or loss incurred, by him by way of interest or finance charges however described* ...

11.3.25 Condition 47 (discussed in detail at paras 11.4.4 to 11.4.6) deals expressly with the treatment of finance charges. These are recoverable only in circumstances either where the PM, the QS or the Employer have failed to comply with a contractual time limit (e.g. as regards certification or payment) or where the QS varies any decision of his. In any other circumstance, finance charges are not an allowable head of claim under GC/Works 98. Condition 47 is the means by which GC/Works 98 provides the "substantial contractual remedy" which is necessary if the provisions of the Late Payment of Commercial Debts (Interest) Act 1998 are to be avoided (see para. 4.8.24). The extent to which Condition 47 does provide such a "substantial" remedy, and thereby constitutes an effective bar to the Late Payment Act, is dependent upon the applicable rate of interest entered in the Abstract of Particulars. In this respect it is suggested that anything below Base Rate plus 5% may run the risk of being considered insufficiently substantial as a remedy (because that is the rate set by the JCT in its own contracts and the JCT is a consensual body whose opinions can be regarded as acceptable on an industry-wide basis), thereby potentially exposing the Employer to the level of Base Rate plus 8% enshrined in the Act. JCT 98 PWQ now makes similar provision in clause 30.1.1.1, albeit this clause is only applicable in situations where the Architect has certified amounts as being due and payable but the Employer has failed to discharge the full amount by the final date for payment set down in the Contract. The applicable rate of interest is fixed by the JCT at Base Rate plus 5%. Interestingly, GC/Works 98 goes further than either the Late Payment Act or JCT 98 PWQ in that it entitles the Contractor to quarterly-compounded interest. The other two documents restrict the Contractor's entitlement to simple interest. Other than the provision in clause 30.1.1.1 of JCT 98 PWQ, the JCT family of contracts makes no express provision in respect of interest payments. However, following the decision of the Court of Appeal in *F. G. Minter Limited -v- Welsh Health Technical Services Organisation (1981) 11 BLR 1*, such charges are in fact accepted as an allowable head of claim under JCT 98 PWQ, even though (save in the very limited instances provided for by Condition 47) they have been expressly excluded from the ambit of GC/Works 98.

11.3.26 A further obvious casualty of the approach of GC/Works 98 to the issue of "expense" (as distinct from "*loss* and/or expense") is any element of profit. The terms of Conditions 43 and 46 (which deals with prolongation and disruption) effectively rule out any claim from the Contractor for loss of profit, consequential loss or indeed in respect of recovery of overheads where specific items of additional expenditure cannot be demonstrated. This said, there is a legal argument to the effect that the word "expense" should not necessarily be taken to exclude the entitlement to claim loss of profit as the jury is still out on this one (see *Stratton and Others -v- Inland Revenue Commissioners [1957] 3 WLR*).

11.3.27 The approach taken by GC/Works 98 to the reimbursement of ... *expense* ... under Conditions 43 and 46 is much more analogous to the treatment of direct loss and/or expense under clause 26 of JCT 98 PWQ than is the case with the valuation of prolongation and disruption arising from VI's under Condition 42. Condition 43(1)(a), for instance, requires that the Contractor must have ... *properly and directly* ... incurred the expense claimed and that it must be ... *beyond that provided for in, or reasonably contemplated by, the Contract* ... The first of these requirements echoes that in clause 26.1 of JCT 98 PWQ that the Contractor's entitlement be limited to ... *direct loss and/or expense* ... i.e. that the Contractor must have succeeded in establishing a causal nexus between the matters complained of and the sums claimed. The second is analogous to the requirement of clause 26.1 of JCT 98 PWQ that any sums to be paid to the Contractor are those ... *for which he would not be reimbursed by a payment under any other provision in this Contract* ... Furthermore, Condition 43(1) of GC/Works 98 refers to the Contractor's entitlement to sums ... *determined by the QS* ... The language used is very much reminiscent of the use of the word ... *ascertain* ... in clause 26.1 of JCT 98 PWQ. Indeed, the current edition of the *Shorter Oxford English Dictionary* carries the following definitions of the two terms:

> Ascertain – *to establish as a certainty*
> Determine – *to ascertain definitely*

11.3.28 It is apparent, therefore, that the rigours to be applied in establishing the Contractor's entitlement under clause 26 of JCT 98 PWQ (see para. 11.3.9) apply equally to Conditions 43 and 46 of GC/Works 98. Prolongation and disruption costs are here (as distinct from their treatment under Condition 42) being treated by GC/Works 98 as a quasi-legal form of loss requiring a standard of proof approaching that which would normally be expected by a Court in relation to common-law damages.

11.3.29 Under Condition 43, the clock begins to run from the date on which the Contractor complies with the Instruction. Within 28 days of this date, the Contractor is to submit to the QS such information as is reasonably required to enable the QS to determine the Contractor's entitlement. The QS then has a further 28 days in which to notify the Contractor of his determination. In this instance, and in contrast with the position with regard to the valuation of VI's (see para. 11.3.21), it is suggested that, should the Contractor fail to provide the QS with the information he reasonably requires, there is no overriding obligation upon the QS nonetheless to arrive at a figure. As with clause 26 of JCT 98 PWQ, should the Contractor have failed to establish to the QS's reasonable satisfaction his entitlement to sums claimed (subject to the practical considerations discussed at paras 11.3.9 and 11.3.19) then the QS is himself entitled, even obliged, to determine the Contractor's entitlement as "nil". In any event, should the Contractor dispute the QS's determination he has a further 14 days to notify the QS of his disagreement and of his estimate of the correct amount, failing which he will be deemed to have accepted the QS's determination. Again, the contract provides no guidance as to what the QS ought to do in the event that the Contractor disputes his determination but it is suggested that the QS ought, within a reasonable time, to revisit his determination on the basis of any information subsequently provided by the Contractor. Should a disagreement remain following the conclusion of such an exercise, the only available means of resolution would then appear to be a reference to Adjudication or Arbitration.

11.3.30 Condition 46 deals with prolongation and disruption costs which are incurred by the Contractor otherwise than by reason of PM's Instructions. The scope of application of Condition 46, therefore, is somewhat more restricted than that of clause 26 of JCT 98 PWQ. The principles applied, however, are not dissimilar. As with Condition 43, the Contractor must have ... *properly and directly* ... incurred the expense concerned. Further, the term ... *expense* ... is defined as in Condition 43 (see para. 11.3.24). Unlike Condition 43, which is specifically concerned with the consequences of PM's Instructions, Condition 46 requires that the expense must be incurred by reason of a number of specified matters. These will be dealt with in more detail at paras 11.3.35 to 11.3.43 below, but such matters are not dissimilar in principle to those set out at clause 26.2 of JCT 98 PWQ. Condition 46(1) also stipulates that the Contractor will only be reimbursed by reason of that which ... *unavoidably results in the regular progress of the Works or any part of them being materially disrupted or prolonged and which is beyond that reasonably contemplated by the Contract* ...

11.3.31 This, again, is analogous to the requirement in clause 26.1 of JCT 98 PWQ that the Contractor's entitlement arises only where ... *he has incurred* ... *direct loss and/or expense* ... *for which he would not be reimbursed by a payment under any other provision in this Contract* ... *because the regular progress of the Works or of any part thereof has been or is likely to be materially affected* ...

11.3.32 The similarity with clause 26.1 continues with the requirement at Condition 46(3)(a) that the Contractor is to notify the PM ... *immediately upon becoming aware that the regular progress of the Works or any part of them has been or is likely to be disrupted or prolonged* ... In that notice, the Contractor is required to give particulars of the circumstances which have given rise to the prolongation or disruption and is also to notify the PM that he considers that there will be a financial consequence. It is notable that the requirements of GC/Works 98 in this respect are somewhat more stringent than is the case with extensions of time, in respect of which there is no specific requirement for the Contractor to submit a written notice (see paras 11.2.3 and 11.2.4).

11.3.33 Condition 46(3)(b) further requires the Contractor, at the latest within 56 days of having incurred the expense concerned, to provide the QS with ... *full details of all expenses incurred and evidence that the expenses directly result from the occurrence of one of the events described in para. (1)* ... Again, this is consistent with the requirement in clause 26.1.3 of JCT 98 PWQ that the Contractor is to provide the QS with ... *such details ... as are reasonably necessary* ... The only difference in approach, apart from the time limit imposed by GC/Works 98, is that the Contractor is required as a matter of course to provide such information whereas, under JCT 98 PWQ, he is required to do so only "upon request". Following receipt of this information, the QS is required to "determine" (see para. 11.3.27) the Contractor's entitlement and to notify the Contractor of his decision within 28 days. According to Condition 1(4)(a), either of these time limits may be extended by agreement, even after the relevant deadline has passed. Should no such agreement be made it is arguable that, where the Contractor exceeds the 56-day period set down in Condition 46(3)(b), he will lose his entitlement completely whereas should the QS exceed the 28-day period set down in Condition 46(5) then the Contractor would be entitled to interest on any sums withheld or delayed as a result by virtue of Condition 47(1)(a). It is suggested that, in such situations, it ought to be an implied term of the contract that neither the PM nor the Contractor unreasonably withhold their agreement to an extension of the specified time period.

11.3.34 It is apparent that the treatment of prolongation and disruption costs under both Conditions 43 and 46 (more particularly the latter) is much more consistent with the treatment of direct loss and/or expense under clause 26 of JCT 98 PWQ than it is with the approach adopted in respect of VI's by Condition 42. All of the elements in clause 26 of JCT 98 PWQ which set the treatment of such costs in a quasi-legal context (see para. 11.3.27) are present in Condition 46 of GC/Works 98, i.e. that the expense concerned must arise as a ... *direct* ... consequence of a number of ... *matters* ... having ... *materially affected* ... the ... *regular progress* ... of the Works, that the Contractor is required to provide ... *immediate* ... notice and ... *full particulars* ..., and that the Contractor's entitlement is to be ... *determined* ... by the QS. Interestingly, and unlike the situation in respect of extensions of time, and the valuation of VI's and other Instructions, Condition 46 gives no express right to the Contractor to dispute the QS's determination. Presumably, should the Contractor consider the QS's determination to be inadequate, he would have recourse to Adjudication or Arbitration.

List of matters

11.3.35 As noted at para. 11.3.8 above, the Contractor is entitled to recovery of any prolongation and disruption costs consequent upon compliance with VI's and other Instructions. The Contractor's entitlements in this respect are dealt with under Conditions 42 and 43. Condition 46 sets out the remaining matters in respect of which the Contractor is entitled to recover prolongation and disruption costs. Condition 46 is therefore, of itself, more limited in scope than is clause 26 of JCT

98 PWQ. Taken together with Conditions 42 and 43, however, the matters in respect of which the Contractor's entitlement arises under GC/Works 98 are broadly consistent with those set out in the JCT form.

11.3.36 The matters which give rise to an entitlement under Condition 46 (i.e. in addition to VI's and other Instructions) are as follows:

The execution of "other works" pursuant to Condition 65
(Condition 46(1)(a) refers)

11.3.37 This concerns works carried out on the site, whether or not in connection with the Works, by others directly engaged by the Employer. By virtue of Condition 65(1), the Contractor is required to provide ... *reasonable* ... site facilities for such works. To the extent that the Contractor is required to provide facilities beyond that which is "reasonable", presumably by having to buy in additional facilities (scaffolding, storage, site accommodation, etc.) rather than simply make available that which is already there, and also in respect of any further disturbance caused to the progress of the Works, the Contractor is entitled to be reimbursed. This matter is broadly analogous to that set out at clause 26.2.4.1 of JCT 98 PWQ.

Delay in being given possession of the Site or any part of it
(Condition 46(1)(b) refers)

11.3.38 This is a similar entitlement to that set out at clause 26.1 of JCT 98 PWQ. The two contracts differ, however, in that delayed possession of the site under JCT 98 PWQ is permitted up to a maximum of 6 weeks (optional clause 23.1.2 applies) whereas there is no such limit provided for by GC/Works 98.

Delay in respect of decisions, information etc. to be provided by the PM
(Condition 46(2)(a) refers)

11.3.39 Again, this matter is broadly similar to that set out at clause 26.2.1 of JCT 98 PWQ. However, Condition 46(4) further provides that the delayed receipt of such information from the PM is to be measured from a date either agreed beforehand with the Contractor or, alternatively, from a date upon which the Contractor gave reasonable notice of his need for the information in question. The second of these provisos is present, in a fashion, in clause 26.2.1.2 of JCT 98 PWQ, whilst the first is, in effect, dealt with by clause 26.2.1.1 of JCT 98 PWQ. It is now an option for the Employer under JCT 98 PWQ to provide the Contractor with an Information Release Schedule (the Sixth Recital thereto refers) which sets out ... *what information the Architect will release and the time of that release* ... Except to the extent that the dates on the Schedule may be varied by agreement (clause 5.4.1 refers) the Contractor's entitlement arises should any of those dates be missed and loss and/or expense be incurred as a direct result. To the extent that information is not included on the Schedule, the position is governed by clause 5.4.2. Under this clause the Architect is to provide information at times when it is reasonably necessary for the Contractor to receive it having regard either to the Contractor's actual progress or, should such progress indicate completion ahead of the Completion Date, then at such times only as would permit the Contractor to achieve the Completion Date itself (rather than any accelerated version of it) – a position which is consistent with the decision of the Court in *Glenlion Construction Ltd. -v- The Guinness Trust (1987) 11 ConLR 126*. The position as between JCT 98 PWQ and GC/Works 98 in respect of delayed release of information is, therefore, not materially dissimilar. Not untypically, however, the approach taken by the JCT document involves considerably more verbiage than is apparent in GC/Works 98.

Delay in respect of the execution of work or supply of goods and materials direct from the Employer (Condition 46(2)(b) refers)

11.3.40 This stipulation is different from that set out at Condition 46(1)(a) and referred to above (see para. 11.3.37) in that it relates to the execution of work or the supply of materials and goods by the Employer in connection with the Works (rather than simply the carrying out of works on site, whether or not in connection with the Works). The Condition goes on to circumscribe the Contractor's entitlement by making clear in Condition 46(4) that there must have been some delay on the Employer's part in supplying materials or in carrying out work. The Contractor's entitlement under this head does not arise merely by dint of the fact that the Employer is supplying such material or carrying out such work. In this way, the Contractor's rights are not necessarily as extensive as those set out at clauses 26.2.4.1 and 26.2.4.2 of JCT 98 PWQ.

Delay in respect of directions or instructions from the Employer or the PM with regard to security passes and the nomination of sub-contractors and suppliers (Condition 46(2)(c) refers)

11.3.41 There is no equivalent in JCT 98 PWQ of the first of these, in that JCT 98 PWQ has no equivalent of Condition 27 of GC/Works 98 which deals with the security passes. Delayed instructions with regard to nomination would appear to be covered by clause 26.2.1 of JCT 98 PWQ. In respect of both of these matters, the provisos set out at Condition 46(4) apply to the Contractor's entitlement (see para. 11.3.39). A further important proviso is added, to the effect that both the Employer and the PM are entitled to a reasonable period for consideration of the matters covered by this paragraph and, further, remain entitled to use their proper discretion in such matters. The Contractor's rights in respect of delayed nomination under GC/Works 98 would appear to be co-extensive with those under JCT 98 PWQ.

Advice from the Planning Supervisor outside the scope of the CDM Regulations (Condition 46(1)(c) refers)

11.3.42 Under clause 26.2.9 of JCT 98 PWQ, the Contractor is entitled to loss and/or expense only if the Employer fails to ensure that the Planning Supervisor and the Principal Contractor (if this is not the Contractor himself) carry out their proper duties under the CDM Regulations. There is no equivalent of this in GC/Works 98. Neither is there an equivalent in JCT 98 PWQ of the above entitlement on the part of the Contractor where the Planning Supervisor provides advice outside his remit.

11.3.43 The following matters set out at clause 26.2 of JCT 98 PWQ likewise have no equivalent in Condition 46 of GC/Works 98:

- failure by the Employer to give proper access to, from and over the site (clause 26.2.6 refers) – albeit that an initial failure by the Employer to give the Contractor possession of any part of the site (as opposed to access thereto once possession has been granted) does, as noted at para. 11.3.38 above, give rise to an entitlement under Condition 46(1)(b) of GC/Works 98;
- disturbance caused by the inclusion of Approximate Quantities which are not a reasonable forecast of the quantity of work in fact required (clause 26.2.8 refers); and
- proper suspension of the Works for non-payment (clause 26.2.10 refers).

11.4 DETERMINING THE EXPENSE OF PROLONGATION AND DISRUPTION

11.4.1 Having regard to the fact, as noted at para. 11.3.27 above, that the terms "ascertain" (as used in JCT 98 PWQ) and "determine" (as used in GC/Works 98) are essentially interchangeable, the discussion in Chapter 4 as regards the heads under which claims for prolongation and disruption expense may be submitted and the principles which must be brought to bear in assessing the Contractor's entitlement, is largely of equal relevance in the context of GC/Works 98. These matters, therefore, will not be discussed further at length here. However, there are a number of issues relating to the determination of the Contractor's entitlement which are peculiar to GC/Works 98 and are therefore worthy of some further comment.

11.4.2 The first of these concerns the fact that GC/Works uses the term "expense" rather than the "direct loss and/or expense" referred to by JCT 98 PWQ. The word "loss" is conspicuous by its absence. "Expense" is defined in Condition 46(6) in the following terms ... *money expended by the Contractor, but shall not include any sum expended, or loss incurred, by him by way of interest charges however described* ...

11.4.3 The Commentary to GC/Works 98 [4] also makes clear that the term is restricted to "out-of-pocket" expenses, i.e. money demonstrably flowing out of the Contractor's coffers. The *Shorter Oxford English Dictionary*, on the other hand, defines the term "loss" as ... *detriment or disadvantage resulting from deprivation or change of conditions* ... One can, of course, be deprived of something that one is entitled to but which one has not yet received. Certain elements of loss, therefore, would not fall within the definition of "expense" set out in GC/Works 98 as they do not constitute an actual flow of funds from the Contractor. The most obvious example of a head of claim which, whilst allowable under JCT 98 PWQ does not qualify under GC/Works 98, is loss of profit. Further, it is suggested that the recovery of head-office overheads by reference to a formula such as Hudson's or Emden's would likewise be disbarred as there has been no demonstration of an actual outflow of funds. Interest foregone on the Contractor's own capital (as distinct from interest paid on borrowed money) was also, until relatively recently, disallowed for the same reason, the Contractor had simply been denied the opportunity to earn interest rather than having been put to the expense of paying it. However, following advice received on the implications of the decision in *FG Minter* (see para. 11.3.25), the PSA (as it then was) amended its official guidance so that loss of interest which the Contractor could have earned on his capital is allowable within the confines of what is now Condition 47.

11.4.4 Chapter 4 deals at some length with the question of interest, both in relation to the position as laid down in the *Minter* decision, which gave a general right to recovery of finance charges as part of a claim for direct loss and/or expense, and in respect of the Late Payment of Commercial Debts (Interest) Act 1998 (hereinafter referred to as the Late Payment Act). The position under GC/Works 98 is somewhat different in that the Contractor's right to recovery of finance charges has been expressly circumscribed by Condition 47. This Condition provides the "substantial" contractual remedy now required in order for the Employer to avoid the provisions of the statute. However, when combined with Conditions 43(4) and 46(6), Condition 47 also serves effectively to remove the Contractor's right (in all but the most restricted of circumstances) to what could be termed *Minter* interest.

11.4.5 This approach to the treatment of finance charges (which is explained in more detail at para. 11.4.6) was first introduced with the publication in 1989 of Edition 3 of GC/Works 1. The applicable rate of interest was fixed at Base Rate plus 1% simple but, under GC/Works 98, it has been left to the Employer's discretion and is to be entered by him into the Abstract of Particulars but it is to be compounded quarterly. As noted at para. 11.3.25 above, it is suggested that a rate of the order of Base Rate plus 5% be selected in order to avoid the Employer potentially being required to pay interest at the higher level of Base Rate plus 8% simple enshrined in statute. JCT 98 PWQ likewise incorporates a provision aimed at relieving the Employer of the full implications of the Late Payment Act; clause 30.1.1.1 now provides for the recovery of interest by the Contractor on monies certified by the Architect but not properly paid by the Employer. The applicable rate of interest is fixed by JCT at Base Rate plus 5% simple. This provision was first introduced in April 1998 as part of Amendment 18 to what was then JCT 80. Prior to that date, the Contractor's only entitlement under JCT 80 was to *Minter* interest. Unlike GC/Works 98, the Contractor under JCT 98 PWQ remains entitled to *Minter* interest (in respect of loss and/or expense) in addition to that provided for by clause 30.1.1.1 (in respect of late payment on certified amounts).

11.4.6 Under Condition 47 of GC/Works 98, interest can be recovered only in situations where either the Employer, the PM or the QS fail to meet a deadline specified in the Contract or, alternatively, where the QS varies a decision (e.g. as to the valuation of a VI or as to his determination of prolongation and disruption expenses) previously notified to the Contractor. This latter provision is particularly important to practising QSs, as neglect on their part to provide timely and accurate valuations of VI's (or determination of prolongation and disruption expenses) can lead to the Employer incurring interest charges at an uncomfortably high rate. Of course, if the QS varies a decision in such a way as a result of the belated production of information by the Contractor then Condition 47(5)(b) serves to remove the Contractor's right to interest. In the situations set out above, and only in such situations, quarterly-compounded interest, at the rate stated in the Abstract of Particulars, is payable from the date upon which monies would have been certified but for the default of the Employer, the PM or the QS until the date upon which the monies in question are in fact certified. Whilst this provision is somewhat wider than that either provided for by the Late Payment Act or under clause 30.1.1.1 of JCT 98 PWQ (neither of which, for instance, would provide a remedy to the Contractor should the QS simply increase his valuation of a Variation over a period of time), all other interest payments which would normally form part of the Contractor's entitlement under clause 26 of JCT 98 PWQ are expressly disallowed by Conditions 43(4), 46(6) and 47(6) of GC/Works 98.

11.4.7 It remains only to return briefly to the subject of VIs and, more specifically, to the way in which prolongation and disruption consequent upon the issue of a VI is to be evaluated. Reference has already been made (at paras 11.3.9, 11.3.19 and 11.3.20 above) to the absence from Condition 42 of the strictures normally applicable to the assessment of such costs and, also, to the problems that may arise from the fact that the QS is required to approach the matter incrementally in respect of each VI as it arises. Particular care must be taken by the QS, when evaluating the prolongation and disruption element of VI's, to bring the requisite degree of forensic rigour to the somewhat subjective exercise required of him in this instance. Furthermore, it is particularly important that the QS takes care to ensure that there is no element of duplication between those amounts evaluated under Condition 42 and those determined under Conditions 43 and 46. Given the fragmented and subjective nature of the exercise required by Condition 42, such a task will be less easy than it may otherwise have been.

REFERENCES

[1] Sir Michael Latham (1994) *Constructing the Team (the Latham Report)*.
[2] Sir Peter Levene (1995) *Construction Procurement by Government – An Efficiency Unit Scrutiny (The Levene Report)*.
[3] British Property Federation (1983) *Manual of the BPF System – The British Property Federation System for Building Design and Construction*.
[4] PACE (1998) *GC/Works/1 (1998) – Model Forms and Commentary*.

12

NEC ENGINEERING AND CONSTRUCTION CONTRACT

12.1 INTRODUCTION

Preface

12.1.1 This section of the book deals with the Second Edition of the New Engineering Contract, renamed the Engineering and Construction Contract, published in 1995 and incorporating addendum Y(UK)2 dated April 1998, hereinafter referred to as NEC/2.

12.1.2 When the First Edition of the New Engineering Contract (hereinafter referred to as NEC/1) was published in 1993, it marked a radical departure from the approach adopted by more traditional forms of contract such as those published by the JCT. The structure and content of NEC/2 is somewhat different to the other contracts addressed in this book, as is the philosophy underlying it. Because of this, and because many readers may be unfamiliar with this form of contract, it is considered to be deserving of a fuller introduction. This introduction is divided into four parts, thus:

   (i)   the background to NEC/1;
  (ii) the Latham Report and the manner in which it impacted on NEC/1;
 (iii) the structure and content of NEC/2;
 (iv) the level of acceptance of NEC by the industry.

The background to NEC/1

12.1.3 In September 1985, the Council of the Institution of Civil Engineers approved a recommendation from its Legal Affairs Committee to ... *lead to a fundamental review of alternative contract strategies for civil engineering design and construction with the objective of identifying the needs for good practice* ... In July 1985, Dr Martin Barnes and Dr John Perry were commissioned to prepare the specification for the new style of contract which would meet the Legal Affairs Committee's "challenge" – the specification was submitted to the Legal Affairs Committee in December 1986 which, in turn, led to a consultative process amongst a limited group invited to comment by the Institution. Following this initial consultative process, the Council decided to develop a draft of the new style of contract as described in Barnes' and Perry's specification which draft was to be developed by a working group of the Institution's members comprising employers, consulting engineers and contractors – on the one hand, a working group so composed would bring a broad spectrum of views to the drafting table (contrast this arrangement with GC/Works which is drafted by the employer – see Chapter 11 of this book) but, on the other hand, the "spectrum" would be taken largely from the membership of the Institution of Civil Engineers.

12.1.4 A draft of the new style of contract was prepared by the New Engineering Contract Working Group, led by Dr Martin Barnes, with a consultative version of the contract being published in January 1991; the consultative version was field tested on a number of schemes and, based upon feedback from the end users, the contract was completely revised (albeit retaining the original tenets) and published as NEC/1 in 1993.

12.1.5 The New Engineering Contract Working Group was set three main objectives to be achieved in the drafting of NEC/1 (which objectives apply equally to NEC/2), namely flexibility, clarity and simplicity and stimulus to good management. Each of these objectives are examined in turn:

(i) *Flexibility* was desired so that NEC/2 could be used for engineering or building works containing any or all of the traditional disciplines, including civils, electrical, mechanical and building works. Further aspects of its flexibility are in its options for different price structures and payment mechanisms and those for design responsibility. NEC/2 provides this flexibility by incorporating such things as "main option clauses" and "secondary option clauses" around the "core clauses" thereby allowing the procurer to adopt its chosen method of procurement in each case without having to use a wholly different standard form of contract.

(ii) *Clarity and simplicity* is sought by the use of ordinary language, as opposed to "legalese", so that NEC/2 can be understood by those whose first language is not English and so that it can be easily translated into other languages. The structure of the NEC/2 documentation and relatively precise definition of the duties of the parties further aids clarity as does the absence of cross-referencing between clauses. Whilst NEC/2 is drafted in English, it provides for the language of the contract to be expressly stated otherwise (in part one of the contract data) and for the contract to be governed by whatever law is stated in the contract to apply (see also part one of the contract data).

(iii) *Stimulus to good management* – this is the area in which NEC/2 seems to be making its mark in the construction industry albeit that during these early stages of its use this has to be treated with caution.

The Latham Report

12.1.6 In July 1993, The Joint Review of Procurement and Contractual Arrangements in the United Kingdom was announced to the House of Commons. The review was financed jointly by the Department of the Environment and the construction industry and was undertaken by Sir Michael Latham. An interim report entitled Trust and Money was published as a discussion document in December 1993 and the final report, Constructing the Team, was published by HMSO in July 1994. The final report is widely referred to as The Latham Report and that title is used in this book.

12.1.7 The main elements of the terms of reference for Sir Michael Latham were to consider:

(i) current procurement and contractual arrangements; and
(ii) current roles, responsibilities and performance of the participants, including the client with particular regard to:
- the processes by which clients' requirements are established and presented;
- methods of procurement;
- responsibility for the production, management and development of design;
- organisation and management of the construction processes; and
- contractual issues and methods of dispute resolution.

12.1.8 So far as contractual issues are concerned, a widely held view in the construction industry is that relations are conducted in too adversarial a fashion – and there are those (but not the authors of this book) who hold the view that this may actually be encouraged by the manner in which contracts are drafted. The Latham Report, at para. 5.13, suggested that there were a number of ways to approach the concerns expressed about contracts, namely:

1. *To do nothing.*
2. *To amend existing Standard Forms to meet some of the concerns.*
3. *To try to define what a modern construction contract ought to contain ... (and having achieved this) ... change existing contract forms to take account of such requirements, and/or to introduce a new contract which will deliver them ...*

12.1.9 Option 1 was discarded since ... *it is no longer possible to do nothing* ... (para. 5.14 of The Latham Report refers) whilst Option 2 fared little better when viewed in the context of Point 9 to The Executive Summary of The Latham Report which states:

> *Endlessly refining existing conditions of contract will not solve adversarial problems. A set of basic principles is required on which modern contracts can be based. A complete family of interlocking documents is also required. The New Engineering Contract fulfils many of these principles and requirements, but changes to it are desirable and the matrix is not yet complete. If clients wish, it would also be possible to amend the Standard JCT and ICE Forms to take account of the principles ...*

12.1.10 At para. 5.17.5 of the Latham Report, the approach of NEC/1 was described as ... *extremely attractive* ... with point 11 of the Executive Summary going on to state:

> *Public and private sectors clients should begin to use the NEC, and phase out 'bespoke' documents. A target should be set of one third of Government funded projects started over the next 4 years to use the NEC ...*

12.1.11 From the above, it is clear that Sir Michael Latham favoured Option 3 (... *to define what a modern construction contract ought to contain* ...) and, in this connection, para. 5.18 of The Latham Report sets out the features which should be included in ... *the most effective form of contract in modern conditions* ... In summary form, such features include the following;

  (i) a specific duty for all of the parties to treat one another fairly;
  (ii) firm duties of teamwork with financial motives to achieve same;
  (iii) a wholly interrelated package of documents, e.g. to include all methods of procurement;
  (iv) easily followed language together with guidance notes;
  (v) separation of the roles between contract administrators and adjudicators;
  (vi) a choice of risk allocation;
  (vii) pre-agreement of variations;
  (viii) milestone related payments in lieu of monthly valuations;
  (ix) clear time limits for honouring payments due;
  (x) secure trust fund routes for payment;
  (xi) speedy dispute resolution by a predetermined adjudicator;
  (xii) incentives for exceptional performance;
  (xiii) advance mobilisation payments where appropriate.

12.1.12 Sir Michael Latham was of the view that NEC/1 contained virtually all of the above features (para. 5.19 of his report refers) albeit that he proposed some alterations – his proposal was influential in bringing about the publication of NEC/2 as evidenced by the Guidance Notes thereto which, in the Foreword, state:

> ... *Following the publication of the Latham report, the Institution of Civil Engineers decided to bring forward the publication of a second edition of the New Engineering Contract. This second edition includes a large number of small refinements to the first edition prompted by further comment on the first edition and by feedback from projects on which the first edition has been used. Extensive changes have been made to the insurance and adjudication provisions. It also includes the changes recommended in the Latham report in order that the New Engineering Contract should comply with the principles for a modern contract set out in the report and that it should be entirely appropriate for wide use.*
>
> *One of the recommendations in the Latham report was that the name of the document should be changed. This is why, in the second edition, the name of the main contract has been changed to Engineering and Construction Contract. This now forms part of the NEC family of contracts which includes the Professional Services Contract, the Engineering and Construction Subcontract and the Adjudicator's Contract ...*

12.1.13 Albeit that NEC/2 is little known within the construction industry, the obvious enthusiasm for its predecessor, NEC/1, in the Latham Report is likely to make it an increasingly well known – and well used – contract in the years to come (see paras 12.1.20 to 12.1.25 below).

The structure and content of NEC/2

12.1.14 NEC/2 is published in 10 volumes, comprised thus:

(i) the complete NEC/2;
(ii) six versions of NEC/2, one for each of the main options;
(iii) the NEC Engineering and Construction Subcontract;
(iv) flow charts; and
(v) guidance notes.

12.1.15 The content of each of these volumes is as follows:

(i) The Complete NEC/2

This volume contains all the clauses and schedules constituting NEC/2 including:

- the core clauses which are common to all of the contract options;
- clauses for each of the main options;
- clauses for each of the secondary options;
- schedule of cost components, applicable to main options A to E;
- schedule of cost components using the short method, applicable to main options A to E; and

- contract data. The contract data are set out in two parts, thus:
  - (a) Data Provided by the Employer – this section includes information as to the identity of the Employer, the Project Manager, the Supervisor and the Adjudicator. It also states which of the main and secondary options apply to the contact. So far as reference to the core clauses is concerned, data are inserted concerning the Works Information, the starting date, the defects correction period, the currency of contract, levels of insurance cover and so on.
  - (b) Data Provided by the Contractor – this section includes information as to the identity of key people, Works Information in connection with Contractor's design, hourly rates for various categories of employee, overheads percentages and so on.

(ii) Six versions of NEC/2, one for each of the main options

There are six types of payment mechanism available in the main options and these are as follows:

- option A – priced contract with activity schedule;
- option B – priced contract with bill of quantities;
- option C – target contract with activity schedule;
- option D – target contract with bill of quantities;
- option E – cost-reimbursable contract;
- option F – management contract.

As previously noted, the complete NEC/2 contains all of the core clauses together with the clauses relevant to each of the main options; the aforementioned main option documents incorporate both the core clauses and the main option specific clauses into what are termed "merged versions" thus permitting the conditions for each of the main options to be read as a whole. For ease of identification, the main option specific clauses are printed in bold type.

The merged versions also include:

- those secondary option clauses which are relevant to each main option;
- the Contract Data, adapted for each main option; and
- the two schedules of cost components (except for option F where they are not required).

The procurement principles underlying each of the main options are set out below:

Option A: priced contract with activity schedule

An *activity schedule* lists the activities which the Contractor expects to carry out in providing the works. Each of the activities are priced by the Contractor and the total of these Prices is the Contractor's price for providing the whole of the *works*.

Option B: priced contract with bill of quantities

Procurement in this manner is directly analogous to that pertaining with JCT 98 PWQ insofar as *bills of quantities* are prepared on the Employer's behalf for pricing by the Contractor.

Options C and D: target contracts (with activity schedule or bill of quantities)

Target contracts are sometimes used where the extent of work to be done may not be fully defined or where anticipated risks are greater which, if transferred to the Contractor, may be priced at a disproportionately high amount. In such circumstances, it can be preferable for the Employer and the Contractor to share the financial risk. Options C and D of NEC/2 provide for financial risk to be shared in the following way:

- The Contractor tenders a target price in the form of a priced *activity schedule* or *bill of quantities*. The target price includes the Contractor's estimate of Actual Cost plus other costs, overheads and profit to be covered by his Fee.
- The Contractor tenders his Fee in terms of a *fee percentage* to be applied to Actual Cost. During the course of the contract, the Contractor is paid the Actual Cost (being the amount net of any Disallowed Costs) plus the Fee. This is defined as the Price for Work Done to Date (PWDD). At the end of the contract, the Contractor is paid his share of the difference between the final total of the Prices and the final PWDD according to a formula stated in the Contract Data. If the final PWDD is greater than the final total of the Prices, the Contractor pays his share of the difference to the Employer. If, on the other hand, the final PWDD is lower than the final total of the Prices, the Employer pays his share of the difference to the Contractor which share is paid provisionally at completion and is corrected in the final account.

One means by which a Contractor may be "protected" from excessive cost overruns is for the Contractor's Fee to be reduced proportionately to the cost overrun but for the Fee to be nonetheless the subject of a minimum level. NEC/2 addresses this matter by stating (in the Contract Data) the appropriate range over which the Fee may be reduced. Another means is for the target price to be regularly reviewed so as to be adjusted in the light of compensation events, changes in the Works Information and so on – the need for such adjustments is to be anticipated where the design is not fully formulated at the time the contract is placed.

Option E: cost-reimbursable contract

A cost-reimbursable contract may be used where the definition of the work to be done is inadequate even as a basis for a target price and an early start to construction is nonetheless required. In such circumstances, the Contractor cannot be expected to take cost risks (or at least not at levels commercially attractive to the Employer) other than those which the control of his employees and other resources entails. The Contractor carries minimal risk and is paid Actual Cost plus his tendered Fee.

Option F: management contract

The Contractor's responsibilities for construction work carried out under a management contract are the same as those taken by the Contractor working under the other main options although he does not carry out any construction himself. Although the Contractor's services apply mainly to the construction phase he would usually be appointed before construction starts. If substantial pre-construction services are required, and if the Employer wishes to have the option to change the Contractor before construction starts, the Guidance Notes to NEC/2 recommend that a separate contract should be awarded for a pre-construction service contract.

The Contractor's Fee will increase as Subcontractors' prices (Actual Cost to the Contractor) increase due to compensation events. However, the Contractor will not receive separate payment for his work in dealing with compensation events nor will he receive any additional Fee for work on compensation events which does not lead to an increase in the Subcontractors' prices.

(iii) The NEC Engineering and Construction Subcontract

This volume contains all the clauses (core and options) and schedules constituting the Subcontract. Since it is intended to be used in connection with work to be subcontracted under NEC/2, it has been drafted in a manner which permits it to be used "back to back" with any of the main options.

(iv) Flow Charts

As is explained in paras 12.2.28 to 12.2.33 below, NEC/2 is a contract which places a considerable premium on effective management systems. The mechanics of these systems are illustrated by a separate document containing a total of 71 Flow Charts. These Flow Charts show the procedural logic on which the NEC/2 is based and takes the user through the decision making process associated with operation of most of the NEC/2 clauses. Both this document and the Guidance Notes which accompany NEC/2 (see below) are described as being intended for guidance purposes only. They have no legal effect and users are requested not to use them as an aid in legally interpreting the contract.

(v) Guidance Notes

The Guidance Notes explain the background to NEC/2, its structure and the means of using it – the notes are both comprehensive and informative, which is essential given that NEC/2 is still relatively new to the construction industry. As with the Flow Charts discussed above, however, they are described as being intended for guidance purposes only. They have no legal effect and users are requested not to use them as an aid in legally interpreting the contract. This said, and largely as a result of the economic use of language in the contract itself, the operation of many of the systems used in NEC/2 becomes clear only from a reading of the Guidance Notes and the Flow Charts. The fact that these ancillary documents are expressly stated as having no legal effect could well raise interesting questions should a Court ever be required to determine the rights and responsibilities of the parties to the Contract in situations where these are not readily apparent from the sparse wording of the Contract itself.

12.1.16 NEC/2 (be it the complete version, one of the merged versions or the sub-contract) is arranged in nine sections, thus:

1. General;
2. The Contractor's main responsibilities;
3. Time;
4. Testing and Defects;
5. Payment;
6. Compensation events;
7. Title;
8. Risks and insurance;
9. Disputes and termination.

12.1.17 The first digit of a clause number denotes the section to which it belongs, e.g. clause 25 is to be found in section 2.

12.1.18 After deciding upon the main option, the user may choose any of the secondary options; the secondary options clauses are numbered separately and are prefixed by the option letter, e.g. clause L1 relates to Sectional Completion. The secondary options are as follows:

- option G – performance bond;
- option H – parent-company guarantee;
- option J – advanced payment to the Contractor;
- option K – multiple currencies *(used only with options A and B)*;
- option L – sectional completion;
- option M – limitation of the Contractor's liability for his design to reasonable skill and care;
- option N – price adjustment for inflation *(used only with options A, B, C and D)*;
- option P – retention *(not to be used with option F)*;
- option Q – bonus for early Completion;
- option R – delay damages;
- option S – low-performance damages;
- option T – changes in the law;
- option U – The Construction (Design & Management) Regulations 1994 *(only for contracts in the UK)*;
- option V – trust fund;
- option Z – additional conditions of contract.

There is no requirement for any of the secondary options to be used albeit where secondary option R is **not** used the Employer will be left to rely on general common law damages in the event of delay. Where a decision is made to use the secondary options, they may be used in any combination save where indicated above.

12.1.19 The procurement principles underlying each of the secondary options are outlined below:

Option G: performance bond and option H: parent-company guarantee

These options should be included where the Employer requires security for performance of the Contractor in accordance with the Contract. Unlike GC/Works 98, NEC/2 does not include any standard forms of words for either the performance bond or the parent company guarantee.

Option J: advanced payment to the Contractor

The option of making an advanced payment is intended for contracts in which the Contractor has to make a heavy investment right at the beginning of the contract – this could include the purchase of major items of plant which are of significant expense and might not have been purchased but for the contractor securing the contract in question. The start time for repayment of the advanced payment and the repayment amounts are to be stated in Part One of the Contract Data.

Whereas such a provision may appear similar to payment for materials off site, it is in fact quite different inasmuch that, with the advanced payment, the Contractor pays back the advanced monies to the Employer, whilst a payment for materials off site represents a payment to the Contractor that he was in any event entitled to receive albeit that he is paid prior to the materials being incorporated into the *works*. NEC/2's accommodation of advanced payments is likely to relate more to the civil engineering industry than to the construction industry.

Recognising the likelihood of an Employer wishing to obtain a bond as security against any advanced payment, option J requires the Contractor to furnish a bond. Again, unlike JCT 98 PWQ and GC/Works 98, NEC/2 does not include any standard form of wording in respect of the bond.

Option K: multiple currencies

This option is applicable to main options A and B only and is used when it is intended that payment to the Contractor should be made in more than one currency and the risk of exchange-rate changes should be carried by the Employer. This option is based on the procedure used by the World Bank the effect of which is that the Contractor is protected from the currency exchange-rate changes which may take place after a fixed date as they affect designated parts of the work.

The Contract Data sets out which items of work are to be paid for in currencies other than the currency of the contract, what those currencies are, the maximum amounts payable in each currency and the exchange rates to be used in calculating the payments.

Where certain items in the contract are to be paid for by the Employer in a given currency yet the Contractor chooses to procure those items in a different currency, then it is the Contractor who takes the risk (or benefit) of any movement in exchange rates. If, however, there are certain items in the contract to be paid for by the Employer in multiple currencies, then it is the Employer who takes the risk (or benefit) of any movement in exchange rates on the occasion of making his payment to the Contractor in the *currency of the contract.*

Option L: sectional completion

The provision for Sectional Completion is broadly analogous to that pertaining with the Sectional Completion Supplement published by the JCT.

Each *section* will have a Completion Date, either stated by the Employer or tendered by the Contractor. Bonus for early Completion and delay damages can be related to the sectional Completion Dates by using options Q and R respectively. Option R does not deal with proportionally reducing damages in the event of sectional completion/partial take over, albeit that the Guidance Notes recommend that damages ought to be proportionally reduced in such situations (see the earlier comments on the legal status of material contained in the Guidance Notes).

Option M: limitation of the Contractor's liability for his design to reasonable skill and care

Where a Contractor is contractually obliged to undertake the design (or parts thereof) but in the absence of clauses which state the nature of his design liability, an implied term is likely to exist to the effect that the Contractor will have a strict liability which is to say that the design must be fit for its purpose and it will be no defence, in the event of design failure, to say that all reasonable skill and care had been exercised. To avoid the possibility of a term being implied in this fashion, option M provides for the Contractor's liability for design to be assessed on a threshold referable to the exercise of ... *reasonable skill and care to ensure that* [the design complies] *with the works information* ...

The whole issue of whether design liability should operate on a strict liability or reasonable skill and care basis is both significant and complex.

Option N: price adjustment for inflation

Without option N, the contract is firm price and the Contractor carries the risk of inflationary increases in the costs of labour, plant, materials, etc., in a manner analogous to the selection of clause 38 of JCT 98 PWQ. If the Employer's preference is for a fluctuating price contract, then option N may be used (albeit that this option may only be used in conjunction with main options A, B, C and D).

For cost reimbursable and management contracts (main options E and F respectively), the Employer already carries the risk of inflationary increases since payments are of Actual Cost which are "current costs" and thereby automatically include for price increases occurring after the contract is signed.

Option P: retention

The Guidance Notes to NEC/2 describe this option in the following way:

> ... *The purpose of retention is to enable the Employer to retain a proportion of the price for work done to date as security and an additional motivation*

for the contractor to complete the works. The procedure used ... has no effect on the contractor's cash flow in the early part of the contract period ...

We would agree that retention provides ... *security* ... for the Employer but not with the view that it creates an ... *additional motivation* ... for the Contractor to complete the *works* since sums of money held in retention represent the amounts which would have been paid to the Contractor (but for the withholding provisions in the contract) and which are retained in anticipation of the emergence of Defects. In consequence, retention provides an incentive for the Contractor to achieve certain quality thresholds rather than to provide him with an incentive as to timeous Completion of the works – the incentive to complete the works is more likely to stem from a wish to earn a bonus (see Option Q) or to avoid delay damages (see Option R).

The retention option is not to be used in management contracts (option F) on the premise that the Employer is adequately protected by the retention in the sub-contracts.

Option Q: Bonus for early Completion

Where Completion of the *works* is achieved, or the Employer takes over the *works*, ahead of the Completion Date, this option provides for the Contractor to be paid a bonus at a predetermined daily rate for each day that the earlier of these two conditions is achieved ahead of the Completion Date.

Option R: delay damages

Delay damages are directly analogous to liquidated and ascertained damages with which people in the construction industry are more familiar; the whole matter of liquidated and ascertained damages is addressed in section 2.13 of this book. It is recommended that this option is used since without it the Employer's recourse to damages will instead be at large (i.e. on a proven loss basis) thereby making the recovery of damages both a costly and time consuming task.

Option S: low-performance damages

This option enables the Employer to recover damages in the event that the perform-ance of the *works* in use does not reach a specified level due to a design or other fault of the Contractor.

The performance of the *works* will be measured against criteria stated in the Works Information, e.g. the air-handling capacity of air-conditioning plant, the U-values of the building's components and such like. Insofar as the actual performance of the *works* falls below that laid down in the Contract Data, then the resultant shortfall in performance will represent a diminution in the building's worth which diminution option S seeks to calculate and express in the form of liquidated damages.

Option T: changes in the law

Inclusion of this option has the effect of transferring the risk of changes in law which occur after the contract date to the Employer. Although there might be many changes in law, only those changes which affect the Contractor's costs are relevant. Examples of such law are import duties, customs payments and building regulations.

Option U: The Construction (Design and Management) Regulations 1994

This option is only to be used for contracts in the UK. Since the majority of construction work carried out in the UK is subject to the CDM Regulations the Guidance Notes to NEC/2 recommend that this option should be used.

If the option is used, any delay to the work or changed work caused by the application of the CDM Regulations will constitute a compensation event provided that ... *an experienced contractor could not reasonably be expected to have foreseen it* ...

Option V: trust fund

The Trust Fund has been designed to protect the Contractor, and any other firm at each tier of the supply chain, against the Employer's insolvency. The protection is only in respect of covering payments due for work carried out but not paid for at the time of the insolvency.

Option Z: additional conditions of contract

The Guidance Notes to NEC/2 state that:

> ... *This option should be used where the Employer wishes to include additional conditions. These should be carefully drafted in the same style as the core and optional clauses, using the same defined terms and other terminology. They should be checked for consistency with the other conditions.*
>
> *Additional conditions should be used only when absolutely necessary to accommodate special needs such as those peculiar to the country in which the work is to be done. The flexibility of the (NEC/2) main and secondary options minimises the need for additional conditions. Additional conditions should never be used to limit how the Contractor is to do the work in the contract as this is part of the function of the Works Information* ...

Notwithstanding the above, some lawyers and construction professionals have observed that option Z presents an opportunity to introduce, on behalf of the Employer, clauses which counteract drafting weaknesses/contractor biases elsewhere in NEC/2.

The level of acceptance of NEC by the industry

12.1.20 The publication of NEC/1 in 1993 represented a radical departure from the approach traditionally adopted by the more established forms of contract. Perhaps unsurprisingly in light of this, the response of a traditionally conservative industry to the arrival of NEC/1 was (and in the case of NEC/2 remains) mixed. From its very earliest days the document has attracted supporters. As noted above, Sir Michael Latham concluded that NEC/1 fulfilled many of the principles embodied in his Report and accordingly recommended that ... *public and private sector clients should begin to use the NEC* ... Further, he went so far as to recommend that Government set an example of what he considered best client practice by letting one-third of all Government contracts over the following 4 years on NEC.

12.1.21 Sir Peter Levene's Efficiency Unit scrutiny of 1995, however, came down in favour of Government developing the GC/Works forms of contract so as to embody the principles set down in The Latham Report rather than moving over to NEC. The Departments consulted by Sir Peter's team were broadly supportive of GC/Works/1 Edition 3, with many describing themselves as "strongly opposed" to NEC. Two Departments in particular had commissioned comparative studies of the various contract forms and both studies had favoured GC/Works over NEC. Accordingly, The Levene Report recommended that GC/Works/1 ought to be developed ... *into a family of contracts meeting the key Latham principles, taking account of reports from Departments on their trialling of the New Engineering and Construction Contract and other forms of contract* ... [1]. As noted in Chapter 11, the result of this exercise was the eventual publication of GC/Works 98.

12.1.22 As noted above, however, NEC did attract support – both during the consultative phase and following publication – from a number of influential clients of the industry. Chief among these early supporters were BAA (formerly the British Airports Authority) and a number of the privatised utility companies. Following the publication of The Latham Report in 1994 they were joined by other big players such as London Underground, Sainsbury, Mercury Communications and the Highways Agency. The form also gained popularity overseas. This was particularly the case in South Africa where it was adopted enthusiastically by the world's fifth-largest electrical utility, and in the Far East where it was used on the construction of the 40-storey Peninsula Hotel in Bangkok.

12.1.23 After an initial silence, the reaction from the legal fraternity was generally hostile. Criticism was principally aimed at the language used in the contract, which was considered by some to be oversimplified, stilted and cumbersome. In particular it was considered a mistake on the part of the draughtsmen of NEC to have abandoned accepted (if sometimes convoluted) terminology which, for all its apparent faults, at least had the benefit of familiarity and a substantial body of legal authority which clarified its meaning. The Guidance Notes to NEC/2 explain that ... *A fundamental objective of [NEC/2] is that its use should minimise the incidence of disputes. Thus words like 'fair', 'reasonable' and 'opinion' have been used as little as possible* ... The response of one construction lawyer to this was that ... *the draughtsmen of the NEC seem to have had a pathological aversion to the use of the word 'reasonable' when the accrued wisdom of centuries of legal experience has concluded that for many situations it is the best word* ...

12.1.24 A number of legal commentators applauded NEC's emphasis on management systems aimed at ensuring the smooth running of the contract but criticised its lack of detailed provisions in the event of serious difficulties being encountered. Supporters of NEC pointed out that, should the parties approach the contract in the manner required, then any such provisions would be superfluous and in any event would be inconsistent with NEC's stated objectives of stimulating good management and encouraging collaborative working. The view of one legal commentator, in contrast, was that ...

it may be a lot safer for all concerned in a contract to assume everyone will be nasty to each other than to pretend they won't … [2]. The broad view of the legal fraternity was, and remains, that NEC has sacrificed clarity in the interests of brevity.

12.1.25 The most recent RICS Survey of Contracts in Use (covering contract usage during 1995) caught only 2 projects which had been let on NEC/1 and only 1 project which had been let on NEC/2. Together they represented only 0.15% by value of all projects caught by the survey. Applied to the total value of construction output in the UK, however, the survey would suggest that contracts worth some £80 million in total may have been let on one or other of the NEC forms during the course of 1995. A survey carried out in early 1996 and reported in the research journal *Engineering, Construction and Architectural Management* found that NEC had already been used on over 1,500 contracts worldwide, albeit only 40 of these were in the UK and the bulk of the remainder were let by a single company well known for its enthusiastic support for NEC – South African electricity utility Eskom [3]. Anecdotal evidence, compiled by the authors more recently suggests that this form of contract has yet to gain widespread acceptance in the UK but that it has certainly taken root firmly, is still being used by its earlier supporters (which suggests that their experience of it has been favourable) and that its use generally is growing albeit slowly.

12.2   PECULIARITIES OF NEC/2

The terminology of NEC/2

12.2.1 One of the more obvious features of NEC/2 to those who are using it for the first time is the strikingly unfamiliar use of language and terminology. The Guidance Notes to NEC/2 explain that one of the stated objectives of NEC/2 is to achieve … *clarity and simplicity* … The Notes go on to explain that … *it uses only words which are in common use so that it is easily understood* … *It has few sentences which contain more than 40 words* … *It is arranged and organised in a structure which helps the user to gain familiarity with its contents* … *The quantity of text used is much less than existing standard forms and the amount of text needed to give effect to the options is small* … It is this feature of NEC/2 which, as noted at para. 12.1.23, has attracted criticism from a number of lawyers who feel that clarity has in fact been sacrificed for the sake of brevity.

12.2.2 One particular linguistic oddity within NEC/2 that has attracted particular comment from lawyers and others is the use throughout of the present tense. In some instances this is used to denote a statement of fact whilst in others it is intended to impose a legal obligation. As an example, clause 11.2(3) states that … *The Contract Date is the date when this contract came into existence* … This is a simple statement of fact. Clause 18.1, on the other hand, states that … *The Contractor acts in accordance with the health and safety requirements stated in the Works Information* … This means that the Contractor *shall* act in accordance with health and safety requirements – i.e. a legal obligation is being imposed on the Contractor. A number of lawyers have pointed to the confusion that may be caused by this approach in the event that a Court is required to interpret the contract. They also point out that NEC/2's constant use of the present tense renders the language stilted and often difficult to follow. Whether one agrees with this sentiment or not, it certainly helps to give the document an unfamiliar feel.

12.2.3 The unfamiliar feel of the document is further emphasised by the decision of its authors to depart from commonly accepted construction-industry terminology. There are, for instance, no references to such familiar terms as "extensions of time", "variations" or "direct loss and/or expense". Again, NEC/2 has been criticised by lawyers for departing in such a radical fashion from terms which are legally understood in that they have been clarified by years of accumulated judicial precedent. The more immediate effect for the user of NEC/2, however, is the need to become acquainted in a practical sense with a wealth of new and unfamiliar terms so as to be able properly to understand and operate the document.

12.2.4 In order to do this, it is first necessary to understand NEC/2's system of **identified** and **defined** terms. Identified terms are terms which are specific to the project in question and will be stated in the Contract Data (which is the equivalent in NEC/2 of the Appendix in JCT contracts and the Abstract of Particulars in GC/Works contracts). Identified terms are denoted by italics, examples being the *Contractor*, the *starting date* and the *completion date*. All are specific to the project. Defined terms are terms which have the same meaning wherever the contract is used. These terms are denoted by the use of capital letters and are all defined in clause 11.2. Examples are the Works Information, the Accepted Programme and (again and somewhat confusingly) the Completion Date. The use of this system does mean, unfortunately, that the unfamiliar user will find himself having constantly to cross-refer between the particular clause with which he is concerned, clause 11.2 and the Contract Data (not to mention the Guidance Notes and the Flow Charts).

12.2.5 A brief examination of the main elements of NEC/2 is sufficient to highlight the unfamiliarity of the terminology. The Contractor's principal obligation, for instance, is stated as being to "Provide the Works". This is defined at clause 11.2(4) as being ... *to do the work necessary to complete the* works *in accordance with this contract and all incidental work, services and actions which this contract requires* ... What are traditionally referred to as drawings, specifications, etc. in JCT contracts are referred to in NEC/2 as the "Works Information" and the "Site Information". The Contract Data identifies the particular documents in which they are contained. The Contract Sum, with which users of JCT will be familiar, becomes the "Prices" in NEC/2.

12.2.6 The terminology used in NEC/2 to denote the various dates for commencement and completion of the work can be particularly difficult to understand for the first time user. For instance, the contract refers to a *starting date* and a *possession date*. The former of these denotes the date upon which the Contractor is to start work on the project. As this may involve pre-planning, design and manufacturing work it is possible that the *starting date* may be somewhat earlier than the date on which the Contractor is actually given possession of the site – i.e. the *possession date*. The Contractor is not permitted to commence work on site until the *possession date*, regardless of the *starting date*. There can only ever be one *starting date*, although there may be a number of *possession dates* if the Site is divided into parts. These *possession dates* are set out in the Contract Data. The situation can be further complicated by the fact that clause 31.2 requires the Contractor to state in his programme the dates when he **needs** possession of parts of the Site, albeit he will only be given possession on those dates if they are later than the relevant *possession dates* in the Contract Data.

12.2.7 Unless option L is used there can only be one *completion date*. This can either be stated by the Employer in Part 1 of the Contract Data or, alternatively, left for the Contractor to insert in Part 2 of the Contract Data. If option L is used, separate *completion dates* (albeit not *possession dates*) for each *section* can be entered by the Employer in Part 1 of the Contract Data. On his programme, however, clause 31.2 requires the Contractor to indicate the date on which he **plans** to achieve completion. This date, the significance of which is explained in greater detail at paras 12.2.37 and 12.2.38 below, may be earlier than the *completion date* stated in the Contract Data.

12.2.8 As to completion of the works themselves, NEC/2 divides this into two distinct elements – "Completion" and "taking over". Completion is defined in clause 11.2(13) as being when the Contractor has ... *done all the work which the Works Information states he is to do by the Completion Date and ... corrected notified Defects which would have prevented the Employer from using the* works ... There is doubt as to whether Completion under NEC/2 would be interpreted by a Court differently to practical completion under JCT forms or substantial completion under ICE forms. It is for the Project Manager to decide when Completion has occurred and to certify the same within 1 week of its occurrence. The Employer is then required to take over the relevant part within a further 2 weeks, at which point possession of the relevant part of the Site returns to the Employer. Only at the point of take over does responsibility and risk in respect of the relevant part pass from the Contractor back to the Employer. Until this point, therefore, the Contractor carries the risk of damage to the *works* notwithstanding the fact that Completion has occurred. If the Employer occupies or uses any part of the site prior to Completion, he is generally deemed to have taken it over. Take over, which is also to be certified by the Project Manager in accordance with clause 35.4, is a conceptually useful tool which does not appear anywhere in the JCT family of contracts.

12.2.9 A further area where NEC/2's terminology may appear particularly obscure on first reading is that of variations and claims, the subject matter of this book. Whilst this area is dealt with in detail at sections 12.3 to 12.7 below, it is considered useful at this point to clarify some of the terminology used as an aid to understanding. NEC/2, of course, uses none of the terminology traditionally associated with claims. Indeed, it is the intention of NEC/2 to replace the familiar claim scenario with an ongoing process of dispute resolution by agreement covering both time and money. This system is set out in clause 6 of the contract, which deals with what are known as compensation events. At their simplest, these are events which are not the fault of the Contractor and which, should they occur, will entitle the Contractor to additional time and money. Whilst the compensation events are listed in clause 60.1, in the same way that JCT 98 PWQ lists Relevant Events at clause 25.4 and matters giving rise to an entitlement to direct loss and/or expense at clause 26.2, there the similarity ends. There are no references to extensions of time, variations (which are referred to in NEC/2 as changes to the Works Information), ascertainment or direct loss and/or expense. Instead, individual compensation events are simply notified either by the Project Manager or the Contractor and are valued on the basis either of quotations submitted by the Contractor or of assessments by the Project Manager, and are implemented by the Project Manager. Implementation of a compensation event merely involves changing the Prices and the *completion date* in order to take account of the agreed effect. No other mechanism exists within NEC/2 for changing the Contractor's entitlement to time and money.

2.2.10 There are, of course, further presentational aspects of NEC/2 which distinguish it from the more widely used forms of contract. Chief among these are the clause numbering system and the deliberate lack of any cross-referencing within the document. Clause 60, rather confusingly, is a sub-clause of clause 6. Additionally, clauses which contain several different provisions do not denote these in the traditional fashion by means of separate numbered sub-clauses but instead by the use of unnumbered bullet points (and even indented sub-bullet points). Whilst this does not necessarily cause a problem within the document itself, because of the lack of cross-referencing, it can cause a problem to the user who wishes to refer others to a particular provision. The lack of cross-referencing itself, whilst leaving the text of the form relatively uncluttered and thereby more readable, can leave the user unsure as to whether he is fully aware of all the separate contract provisions which may bear on a particular situation (albeit that this is where the Guidance Notes and Flow Charts are intended to be of assistance). The usage of certain words and phrases within the document can also lead to confusion. Examples of this are the terms "Others" (clause 11.2(2) refers), "others" (clause 28.1 refers), "other people" (clause 45.1 refers), "people" (clause 44 refers) and – finally – "People" (the Schedule of Cost Components refers). Perhaps in anticipation of some of the above comments the Guidance Notes to NEC/2 do point out, somewhat ironically, that ... *the initial impact of reading* [NEC/2] *may not convey its full simplicity* ...

The parties to NEC/2

General

2.2.11 Like the other forms of contract with which this book deals, the parties to the contract in the strict legal sense are the Employer and the Contractor. NEC/2 gives the Employer relatively little in the way of responsibilities as these are generally undertaken on his behalf by the Project Manager. The Employer's own responsibilities are limited largely to giving possession of the site to the Contractor and maintaining access to it, providing facilities as stated in the Works Information, taking over the works as provided for in the contract and making due payment to the Contractor.

2.2.12 The Contractor's responsibilities are, of course, extensive. Duties in respect of specific issues (such as dealing with compensation events) are set down in the appropriate clause of the contract and, where relevant, will be dealt with in detail in the sections which follow. The main responsibilities of the Contractor, however, are collected together in clause 2. Not surprisingly, chief among these is that the Contractor ... *Provides the Works in accordance with the Works Information* ... (clause 20.1 refers). Clause 11.2 confirms that ... *To Provide the Works means to do the work necessary to complete the* works *in accordance with this contract and all incidental work, services and actions which this contract requires* ... Where so stated in the Works Information, the Contractor is to design part or all of the Works. Where secondary option M is employed, the Contractor's liability for such design is limited to one of reasonable skill and care, otherwise it is the much more onerous liability of fitness for purpose.

12.2.13 There is a specific duty on all the parties named in the contract, including the Contractor, to ... *act as stated in this contract and in a spirit of mutual trust and co-operation* ... (clause 10.1 refers). In the case of the Contractor, this is supplemented by a further duty to co-operate and share the site with Others (defined at clause 11.2(2) as being all those parties not specifically named in the contract but nevertheless involved in the project). The Contractor is responsible for all sub-contractors and, in the absence of agreement to the contrary with the Project Manager, is required to sub-let all work on either the NEC Engineering and Construction Subcontract or the NEC Professional Services Contract. Clause 29.1 further provides that the Contractor ... *obeys* ... instructions given ... *in accordance with this contract* ... either by the Supervisor or the Project Manager.

12.2.14 In addition to the legal parties to the contract (defined in clause 11.2(1) as The Parties) a number of other parties, with which those more accustomed to JCT forms may be unfamiliar, have roles assigned to them by NEC/2. Essentially, these are all roles which, under engineering forms of contract, were traditionally performed by the Engineer. In NEC/2 the roles have been separated and defined.

The Project Manager

12.2.15 The Project Manager is appointed directly by the Employer, ideally at an early stage in the project, and is a central figure in the NEC/2 matrix of roles and responsibilities. As the Guidance Notes point out, NEC/2 ... *places considerable authority in the hands of the Project Manager* ... The list of powers and duties of the Project Manager are extensive and include inter alia the giving of instructions to the Contractor, the acceptance or otherwise of programmes and other information provided by the Contractor, the assessment of payments due to the Contractor, the consideration and assessment of compensation events and the certification of Completion. Duties in respect of specific issues which are relevant to this book (such as dealing with compensation events) are dealt with in detail in the sections which follow.

12.2.16 In undertaking these responsibilities the Guidance Notes make it clear that the Project Manager's objective is to safeguard the Employer's interests in achieving the completed project. In so doing, the Guidance Notes further stress that the Project Manager ought not to be constrained unduly by the terms of his own appointment, as it is considered that this will hamper effective management of the project and may make disputes more difficult to resolve. Having said that, the Project Manager has (or ought to have) a considerable degree of autonomy to act in the Employer's interests and it is important to bear in mind that, notwithstanding the absence of any express requirement in NEC/2 for the Project Manager to act impartially, it is nevertheless likely that when required to use his discretion (such as when deciding that Completion has been achieved or when assessing compensation events) he is under a common law duty to act fairly. The Guidance Notes point out that the Project Manager ... *is constrained from acting unreasonably in this role by statements* [within the contract] *of the basis on which he is to make each type of decision but not what decisions he is to make* ... Whilst opinions on this may differ, it is certainly the case that clause 10.1 requires the Project Manager to act ... *in a spirit of mutual trust and co-operation* ... , whilst the Guidance Notes again encourage him at various points to work collaboratively with others and to engage with the Contractor in ... *positive manage-ment* ... of the project.

2.2.17 The Project Manager may be an employee of the Employer or, alternatively, a retained consultant. In any event, the Project Manager is to be named in Part 1 of the Contract Data and is not to be replaced by the Employer without prior notification to the Contractor. There is no express requirement, unlike in the case of the Engineer under the ICE forms, that the Project Manager be a named individual. The Project Manager referred to in the Contract Data may be a firm. Indeed, on any project of reasonable size or complexity, it is likely that the functions of the Project Manager will need to be carried out by a team and that it will therefore make more sense to name the firm by which they are employed as Project Manager in the Contract Data. Clause 14.2 of NEC/2 gives the Project Manager the power, after notifying the Contractor accord-ingly, to delegate any of his actions. The contract itself gives no further instruction as to how such delegation ought to be managed, albeit that the Guidance Notes do point out that delegation ought to be of ... *specific authorities and duties under the contract to particular members of staff or others* [and] ... *notification of the Contractor should include details of the actions delegated and the person to whom they are delegated* ... It is suggested that, in the absence of such clarity in communications, it will be difficult for the Contractor to remain aware of who has authority for which actions under the contract. As noted at para. 12.1.15, however, the Guidance Notes which contain this advice are not themselves contractually binding and the publishers of NEC/2 state that ... *they should not be used for legal interpretation of the meaning* ... of the contract itself.

The Supervisor

2.2.18 Whilst the Project Manager's duties and powers are concerned with the management and administration of the contract, the Supervisor holds specific responsibility for controlling and approving the quality of the constructed works. The role is similar to that of a clerk of works or resident engineer and generally comprises the carrying out of tests and inspections, notification of Defects found and the issuing of the Defects Certificate. It is important to note that the roles and responsibilities of the Project Manager and the Supervisor are quite separate and distinct and that the Supervisor has authority under the contract to act independently of the Project Manager in respect of the matters which are within his remit (subject, of course, to his duty to act co-operatively). The Project Manager has no authority to countermand or otherwise influence the Supervisor in respect of his notifying Defects to the Contractor, instructing searches or issuing the Defects Certificate – notwithstanding that the actions and decisions of the Supervisor in these respects may well affect the progress of the project and may give rise to a compensation event under clauses 60.1(8), (10) or (11). Interestingly, however, it is the Project Manager rather than the Supervisor who is empowered by clause 44 to accept a Defect rather than require the Contractor to amend it.

The Adjudicator

2.2.19 Any dispute which arises under NEC/2 may be referred to the Adjudicator for a decision which will be binding upon the Parties either until agreement is reached on the matter or until the decision is eventually settled in proceedings before the tribunal (which, dependent on the entry made in the Contract Data, will either be an Arbitrator or a Court). NEC/1's inclusion of adjudication as a means of dispute resolution pre-dated the support expressed for it in The Latham Report. After publication of the Latham Report, however, the adjudication provisions in NEC/2 were revised and expanded in scope and were then further substantially restructured in 1998 in order to secure compliance with Section 108 of the Housing Grants, Construction and Regeneration Act 1996.

12.2.20 The Adjudicator is appointed jointly by the Employer and the Contractor and ought ideally to be appointed in accordance with The Adjudicator's Contract which is part of the NEC family of documents. Initially, the Employer's proposed Adjudicator ought to be named in Part 1 of the Contract Data. However, the Employer cannot unilaterally impose an Adjudicator without the agreement of the Contractor. In this regard the Guidance Notes state that, should the Contractor object to the individual named in Part 1 of the Contract Data ... *a suitable person will be the subject of discussion and agreement before the Contract Date* ... The Adjudicator is required by clause 10.1 of NEC/2 to act ... *in a spirit of independence* ... Furthermore, and in accordance with Sections 108(2)(e) and (f) of the Housing Grants, Construction and Regeneration Act 1996, the Adjudicator is required to act impartially and may himself take the initiative in ascertaining the facts and the law (clause 90.8 of NEC/2 refers). He is to arrive at a decision within the timetable laid down by Section 108(2)(c) and (d) of the Housing Grants, Construction and Regeneration Act 1996 and is to give reasons for his decision.

12.2.21 It was noted earlier that Addendum Y(UK)2 to NEC/2 was issued in April 1998 and that, amongst other things, it substantially altered clause 90 in order to make the adjudication provisions of the contract compliant with the Housing Grants, Construction and Regeneration Act 1996. It is true that certain of the timetables incorporated in NEC/2's original provisions were inconsistent with those required by the Construction Act 1996 and that this is no longer the case with the amended provisions. However, other changes were introduced by reason of Addendum Y(UK)2 which, in the opinion of the authors of this book and of other commentators, potentially operate to render clause 90 non-compliant with Sections 108(1) and (2)(a) of the Housing Grants, Construction and Regeneration Act 1996.

12.2.22 Clauses 90.1 to 90.4 of NEC/2 interpose a new stage in the dispute resolution process which could well fall foul of the requirements of the Housing Grants, Construction and Regeneration Act 1996. Prior to being recognised as a dispute, any difference of opinion between the parties is referred to simply as a ... *dissatisfaction* ... which must be notified by the Contractor (or, in some instances, by the Employer) to the Project Manager. The parties are then required (clause 90.2 refers) to attend a meeting in order to attempt to resolve the disagreement. Only following this, and if the matter remains unresolved for 4 weeks following notification, does NEC/2 recognise that a dispute exists. Only then can notice be given to refer the dispute to the Adjudicator. It seems reasonably apparent that this procedure is in contravention of the Housing Grants, Construction and Regeneration Act 1996, which clearly states that a party may **at any time** give notice of his intention to refer **any** difference to adjudication and that an adjudicator ought then to be appointed within 7 days. By preventing such referral for at least 4 weeks, NEC/2 risks non-compliance with the Housing Grants,

Construction and Regeneration Act 1996 – in which case the provisions of the Scheme for Construction Contracts would apply instead. The stated intention of the authors of NEC/2 in introducing this provision is that it affords the parties the opportunity to resolve differences between them before resorting to formal dispute resolution procedures and thereby aids collaborative working and the smooth running of the project. The response of one legal commentator, when discussing an almost identical provision introduced into the ICE sixth Edition at about the same time, was that its effect was to ... *drive a coach and horses through the legislation* ... [4]. It remains to be seen whose point of view will prevail.

Designers

2.2.23 NEC/2, as pointed out in the Guidance Notes which accompany it, is predicated on the fact that designers will be retained by the Employer in order to prepare a design for the works. The Guidance Notes even anticipate that a number of separate disciplines may be involved and, in such circumstances, a lead designer is recommended to perform a co-ordinating role. Despite the obvious importance of the design role in securing a successful project, however, designers merit no mention whatsoever in the contract itself. This has led to a degree of frustration amongst designers, who consider themselves to have been sidelined somewhat. As one past-president of the RIBA with experience of NEC/1 commented in 1995, neglecting the importance of the design team ... *endangers the quality of the product* [and] *a management method that gives the process a higher value than the product reverses priorities* ... [5]. The authors of NEC are nonetheless content to remove the role of the designers from the scope of the contract, but do recommend in the Guidance Notes that the Project Manager's brief include management of the input of the designers.

Communication between the Parties

2.2.24 NEC/2 is unique in terms of the detail which it provides as to the manner, form and timing of the various communications between the parties. This is perhaps unsurprising in a document which has a stated aim of promoting and encouraging good collaborative management. One immediate effect for those endeavouring to understand the contract, however, is that it requires a degree of cross-referring between clause 13 (which deals specifically with the subject of communications) and the remainder of the contract (including the Contract Data), many of the provisions of which involve the types of communication which are regulated by clause 13. The unfamiliar user is not necessarily assisted in this task by the deliberate avoidance of any cross-referencing within the document itself.

2.2.25 Clause 13.1 requires that ... *each instruction, certificate, submission, proposal, record, acceptance, notification and reply which this contract requires is communicated in a form which can be read, copied and recorded* ... It is apparent from this that the contract envisages a considerable number of different types of communication between the parties and that they are all required to be in writing. Where it is stated in the contract that a reply is required to a communication then, unless stated otherwise, such reply is to be given within the *period for reply* set down in Part 1 of the Contract Data. This period will vary from project to project and it is therefore particularly important that the parties ensure that they are cognisant with the particular period applicable to the project in question. Failure on the part of the Project Manager or the Supervisor to reply within the stated period is a compensation event (clause 60.1(6) refers), albeit that the Project Manager can extend the *period for reply* provided this can be agreed with the Contractor before the reply is due. As such agreement is likely to deprive the Contractor of a compensation event, however, it is difficult to see what incentive he has to agree. Strangely, no sanction can be applied to the Contractor should he fail to comply.

12.2.26 Where a communication from the Contractor requires acceptance from the Project Manager, the Project Manager is free either to accept or not to accept the Contractor's submission. Should the Project Manager not accept the submission he is to state his reasons for so doing. The Contractor is then required to resubmit within the *period for reply*. If the Project Manager's reason for withholding his acceptance is not one which is stated in the contract then a compensation event arises (clause 60.1(9) refers). The Guidance notes explain that this is aimed at reducing the Contractor's risk should the Project Manager's withholding of acceptance be in any way vexatious or unreasonable. By virtue of clause 14.1, no acceptance of a communication from the Contractor can operate to lessen the Contractor's responsibility to Provide the Works.

12.2.27 Notifications have also been singled out for special treatment by virtue of clause 13.7, presumably because they deal with matters of particular note or have the effect of initiating periods within which other parties are required to respond. Were they to be overlooked because, for instance, they were buried within correspondence relating to other matters then the system of communication upon which much of the contract relies would be somewhat undermined. Accordingly, and in a move that has been criticised in some quarters as potentially giving rise to an unnecessary paperchase, clause 13.7 requires that notifications are always communicated separately from other forms of communication.

Management and teamwork

12.2.28 One of the more important points to bear in mind about NEC/2 is that it is primarily intended as a management tool rather than simply as a statement of legal liabilities. It is not possible properly to understand the document and how to put it to use without an appreciation of the philosophy behind it and where this places the document within the management matrix of the project as a whole. Accordingly, the Guidance Notes which accompany the contract go to some lengths to explain the management philosophy behind NEC/2. The objective of stimulating good management practices is described in the Guidance Notes as being ... *perhaps the most important characteristic of* [NEC/2] ... The Notes go on to explain that ... *every procedure has been designed so that its implementation should contribute to rather than detract from the effectiveness of management of the work* ... Furthermore, the good management practices which the contract seeks to encourage are intended to be based upon a system of teamworking aimed at reducing the scope for tension and disagreement between the parties and, as a result, lessening the incidence of disputes. The Guidance Notes explain that ... *foresighted, co-operative management of the interactions between the parties can shrink the risks inherent in construction work* ... The contract aims to provide the management tools to encourage the parties to work collaboratively rather than contentiously, with the objective of reducing the scope for and incidence of difficulties and disputes. In this it found in Sir Michael Latham its most prominent supporter to date.

12.2.29 Opinions differ over the extent to which it is possible, within the confines of a document, to encourage such collaborative working on sometimes complex construction schemes involving many parties often with divergent interests and where considerable sums may be at stake. The Guidance Notes to the contract adopt the position that ... *people will be motivated to play their part in collaborative management if it is in their commercial and professional interest to do so* ... An obvious response to this is to ask how it is possible by means of a contract to tie the commercial and professional interests of parties to their ability successfully to collaborate. The Guidance Notes respond to this by giving as an example the manner in which NEC/2 deals with the compensation events which form the main subject matter of this chapter. It is stated that NEC/2 endeavours to remove uncertainty as to the manner in which unexpected events which arise during the course of the contract will affect the parties' interests. It is claimed that such an approach will ... *improve the outcome of projects generally for parties whose interests might seem to be opposed* ... Whilst the system of compensation events itself is dealt with in considerably more detail at sections 12.3 to 12.7 it will be instructive at this stage, in endeavouring to understand the intended operation of the system, to look at how it is regarded by its authors as fitting into the collaborative philosophy of the contract as a whole.

12.2.30 The Guidance Notes state that it is the intention of NEC/2 that, in any given situation, the actions of the Project Manager in pursuing the interests of the Employer ought not to affect adversely the position of the Contractor. This is taken as a reference to the not unfamiliar situation whereby a Contractor may be faced with enforced changes to his planned method of working, together with numerous variations to his works, and only an uncertain prospect (after protracted and maybe acrimonious debate) of recovering his costs let alone making the required return on his investment. The result is that the Contractor adopts a defensive and even antagonistic position which conditions his approach to the project as a whole and which is harmful to its smooth running. NEC/2 seeks to avoid this by endeavouring to base the valuation of compensation events as far as possible on the Contractor's own forecast of his actual cost, on top of which he also recovers a margin. The Contractor takes the risk that his forecast may be too low, but this is a lower risk than were he to have been required to assess and price the same issues as part of his tender. Because the forecast is made reasonably contemporaneously with the events to which it relates it is that much more likely to be accurate. The Employer, of course, takes the risk that the forecast will be too high, but in return he receives a firm commitment from the Contractor and certainty as to the cost and time implications of the event concerned. Furthermore, the timetable set down in NEC/2 for the handling of compensation events (which is dealt with in more detail at paras 12.7.1 to 12.7.18) ensures that matters are disposed of as they arise – thus removing long-term uncertainty as to the project's out-turn costs and also reducing the scope for "running sores".

12.2.31 In concluding their comments on the system of compensation events, the Guidance Notes explain that ... *this arrangement is intended to stimulate foresight, to enable the Employer to make rational decisions about changes to the work with reasonable certainty of their cost and time implications, and to put a risk on the Contractor which*

is tolerable and which motivates him to manage the new situation efficiently ...
Having committed himself to a fixed price and timescale for the compensation event in
question, it is in the Contractor's interest not to exceed that sum or the time allowed.
The system thereby, in theory, motivates him toward efficiency in dealing with the
event in question.

12.2.32 Having said all this, the director-general of the ICE (the publishers of NEC/2) was
himself quoted in 1995 as saying that ... *it is naive to think that a form of contract will
prevent a dispute if the parties involved are determined to have one* ... [6]. An architect,
having used NEC/1, was quoted at about the same time as saying that ...
*if there is an adversarial attitude, you can't get over it just with a well-written
contract* ... [7]. Indeed, and as noted at paras 12.1.23 and 12.1.24, there is much
legal opinion to the effect that the use of radical language in NEC/2 and its emphasis on
brevity may, if anything, encourage such disputes. There does appear to be a measure
of agreement, therefore, that what is required if NEC/2 is to achieve its stated objectives
is a change in attitude and approach on the part of those involved in the planning,
design and construction of projects – something, indeed, not dissimilar to the industry-
wide change of culture espoused in The Latham Report and, more recently, by Sir John
Egan. The systems set down in NEC/2 can certainly provide an environment within
which such a reformed culture is able to function, but those systems cannot on their
own be expected to succeed in bringing about the required change.

12.2.33 The minimising of conflict and the promotion of good management undoubtedly
requires proper resourcing by the parties involved and a degree of foresight. Whilst
some of the procedures and systems in NEC/2 will no doubt encourage rather than
discourage this, it is suggested that the required foresight must extend to well before
those systems come into use – to the design and planning stage and to the compilation
of the Contractor's tender submission. At this stage all parties must endeavour to
assess the level of resources they will be required to deploy in order to make the
systems work. One of the most consistent pieces of anecdotal evidence gained from
studying the usage of NEC/1 and NEC/2 is the degree of surprise on the part of those
working with the contracts for the first time at the level of management resourcing
required in order to permit them to comply with their contractual responsibilities. The
promoters of NEC/2 suggest that this additional resource applied to resolving
problems on an ongoing basis during the course of the project is more than offset
by the consequent saving in resource which, under more traditional forms of contract,
tends to be directed toward the pursuit of claims and other such disputes long after the
project itself is finished. Anecdotal evidence (albeit limited) collected by the authors of
this book suggests that the management systems laid down in the contract can indeed
encourage commitment of the necessary level of management resource, whether
anticipated or not, rather than the unpleasant alternative of allowing a disaster to
unfold and become the subject of protracted debate, dispute and even legal
proceedings over a number of years. In this, NEC/2 may indeed be having its desired
effect of encouraging good management and effective teamworking.

The programme

12.2.34 In a contract which places so much emphasis on management systems, it is perhaps
not surprising that the Contractor's programme receives such prominence. The

Guidance Notes explain that ... *the Programme is an important document for administering the contract. It enables the Project Manager and Contractor to monitor progress and to assess the time effects of compensation events including changes to the Completion Date* ... Unlike JCT 98 PWQ, which simply requires the Contractor to produce a programme, NEC/2 contains detailed provisions as to what the programme is to contain and how it is to be used. Oddly, NEC/2 contains no requirement either for the Contractor to work in accordance with his programme (in which respect it differs from GC/Works 98) nor to progress the works regularly and diligently. Nonetheless, a proper understanding of the practical operation of the system of compensation events will be difficult to achieve without also being aware of the function of the Contractor's programme within NEC/2.

12.2.35 There ought at all times to be an Accepted Programme in existence. This is defined in clause 11.2(14) as ... *the programme identified in the Contract Data or is the latest programme accepted by the Project Manager* ... The first Accepted Programme ought ideally to be that identified by the Contractor in Part 2 of the Contract Data. If no such programme has been identified, the first programme is to be submitted by the Contractor within the period specified by the Employer in Part 1 of the Contract Data. Having been submitted, this programme may – within 2 weeks – either be accepted or not accepted by the Project Manager. The Project Manager may only refuse to accept the Contractor's submitted programme for the reasons stated at clause 31.3 of NEC/2, thus:

- it is not practicable;
- it does not carry the information required by the contract;
- it is not a realistic representation of the Contractor's intentions; or
- it fails to comply with the Works Information.

12.2.36 Whilst one can envisage some interesting discussions as to whether or not the submitted programme is practicable or faithfully represents the Contractor's intentions, should the Project Manager decline to accept it then (failing a referral to the Adjudicator) the Contractor is required to resubmit his programme to the Project Manager within the *period for reply* stated in the Contract Data (clause 13.4 refers). It is foreseeable that, in situations where the Contractor has failed to anticipate in advance the rigours which will be applied to his programme by NEC/2, some time can elapse and several submissions can be required before his first programme is finally accepted by the Project Manager. If this process extends into the period for carrying out the works on site it may well cause a problem with the operation of those management systems within the contract which are dependent upon there being an Accepted Programme in place. When the Contractor's first programme is finally accepted by the Project Manager it becomes the first Accepted Programme.

12.2.37 Clause 31.2 of NEC/2 goes into some considerable detail as to what is to be shown on any programme submitted by the Contractor. Failure to show all of the information required entitles the Project Manager to decline to accept the programme, and any delay or additional cost thereby caused will **not** be a compensation event. The Guidance Notes to NEC/2 summarise the information to be shown on the programme in the following terms ...

- *dates which are stated in the Contract Data or the Works Information* [e.g. *possession dates* and the *completion date*];
- *dates decided by the Contractor* [e.g. his planned completion date and dates on which he requires information from the Employer];
- *method statements* [including resources];
- *order and timing* [including operations to be carried out by the Employer and Others];
- *float and, separately, time risk allowances*;
- *health and safety requirements; and*
- *other information required in the Works Information* . . .

12.2.38 The programme is therefore a very substantial document and the resources required in order to produce and update it ought not to be underestimated. Particularly important in terms of assessing compensation events are the contract provisions concerning float and "time risk allowances". Time risk allowances are described in the Guidance Notes as . . . *allowances attached to the duration of each activity or to the duration of parts of the* works. *These allowances are owned by the Contractor as part of his realistic planning to cover his risks* . . . In other words, these allowances (which ought to be realistic) cannot be used in order to absorb delays consequent upon compensation events. By contrast, **float** contained within the programme (that is, between the *starting date* and the Contractor's planned Completion) is available to accommodate the effects of compensation events. In other words, float within the Contractor's programme belongs to the Employer. Float between the Contractor's planned Completion and the Completion Date, however, does not. This is known as "terminal float" and is treated in the same way as time risk allowances.

12.2.39 One effect of this is that, should a Contractor place all of the float in his programme at the end and thereby bring forward his planned Completion to a date earlier than the Completion Date, the Contractor will own **all** of the float in the programme and the Employer will have the benefit of none of it. Delays to activities, which in the absence of such tactics by the Contractor would not be critical, will then become so by virtue of the compressed programme. As the time effects of compensation events are judged by reference to the Contractor's **planned** date for Completion (clause 63.3 refers), compensation events would thereby occur in circumstances where otherwise – because of the presence within the programme of float – they would not. The Project Manager would be required to set the Completion Date at a later point in time than would otherwise be the case. It is important that Project Managers remain alert to such "programmesmanship" on the part of Contractors, albeit their options for rejecting a programme are constrained by the provisions of clause 31.3.

12.2.40 Having had his first programme accepted by the Project Manager, clause 32.2 then requires the Contractor to issue revised programmes in response to particular situations. This is the case whenever he is instructed to do so by the Project Manager (which will usually be in response to a compensation event), whenever the Contractor himself considers it necessary (perhaps in conjunction with an early warning notice issued in accordance with clause 16.1) and, in any event, throughout the course of the contract at no greater interval than that stated in the Contract Data. This latter provision is intended to ensure that, at all times, there is a programme in existence which accurately reflects progress to date, incorporates the expected effects of known compensation events and sets out the Contractor's intentions regarding progress to Completion. In this way, the programme remains an active management tool throughout the course of the contract. Once a revised programme has been accepted by the Project Manager it then becomes the Accepted Programme. In this event, as is made clear by clause 11.2(14), ... *the latest programme accepted by the Project Manager supersedes previous Accepted Programmes* ... Assessments of compensation events are always carried out by reference to the then current Accepted Programme.

12.2.41 Should the Project Manager accept a Contractor's programme (whether first or revised) outside the 2-week period stipulated in clause 31.3 then his acceptance is likely to be valid but will constitute a compensation event (clause 60.1(6) refers). Should he fail to accept the programme at all then it is likely that, in the absence of a positive rejection, his acceptance will nevertheless be deemed to have been given but a compensation event will again have occurred. If the Project Manager cannot get what he considers to be an acceptable first programme from the Contractor his options are less than clear. If the reason for the Project Manager's rejection is that the programme does not show the information required by the contract then he is empowered to withhold one-quarter of the Price for Work Done to Date from any payments due to the Contractor (clause 50.3 refers). If the Project Manager's rejection is for one of the other prescribed reasons, however, this option is not available. Other than to instruct the Contractor to stop work until an acceptable programme is provided (clause 34.1 refers) it is difficult to see what other options are available to the Project Manager. Clause 61.4 provides that any such instruction to stop in such circumstances would not be a compensation event.

12.2.42 A final warning note for Project Managers on the subject of revised programmes submitted by the Contractor, particularly in accordance with the second bullet point of clause 32.2 (i.e. ... *when the* Contractor *chooses to* ...). There is the possibility of the Contractor retaining a full-time planner based on site who continually bombards the Project Manager with revised programmes on an almost daily basis and in response to the most trifling of matters. On this subject one commentator who has written an authoritative work on NEC/2 points out that ... *the burden this throws onto the Project Manager who is obliged to respond to each revision is immense. But failure to respond is not only a breach of contract but also gives advantages to the contractor in the assessment of compensation events* ... [8]. It is suggested that, in order to enable what is generally a laudable system to operate as intended, all parties need to give proper consideration in advance to the resources they are going to require and to have in mind their duty to co-operate as set down in clause 10.1.

12.3   COMPENSATION EVENTS – THE GENERAL CONCEPT

12.3.1 Section 6 of NEC/2's core clauses, entitled ... *Compensation events* ..., addresses both those events which are to be the subject of compensation and the means of that compensation being brought into effect. As described at page 57 to NEC/2's Guidance Notes ... *compensation events are events which, if they occur, and do not arise from the Contractor's fault, entitle the Contractor to be compensated for any effect the* *event has on the Prices and Completion Date. The assessment of a compensation event is always of its effect on both Prices and the Completion Date . . .*

12.3.2 Compensation events represent a novel cocktail of the evaluation of variations, extensions of time and direct loss and/or expense (thereby consolidating clauses 13, 25 and 26 respectively of JCT 98 PWQ) and its precise mix will depend upon the compensation event under consideration and its particular effect on time and cost. By way of illustration, consider the compensation event under clause 60.1(1) of NEC/2 where . . . *the Project Manager gives an instruction changing the Works Information* . . . (in effect, a variation), the assessment of which will address the following:

- a forecast of the Actual Cost of carrying out the varied work;
- a forecast of the Actual Cost of any other costs which the contractor will incur which costs are analogous to loss and/or expense under JCT 98 PWQ;
- such extension to the date for Completion as is adjudged appropriate.

The carrying into effect of the above components is referred to as . . . *implementing compensation events* . . . in clause 65.

12.3.3 In the following section 12.4, each of the Compensation events within NEC/2 is addressed in turn. Whilst many of the compensation events may appear similar to JCT 98 PWQ's Relevant Events, addressed in detail in section 2.8 hereof, the terminology within NEC/2 is different and inasmuch that it is, as yet, untested by the Courts one cannot be certain as to their proper interpretation following the decision by the draughtsmen to abandon much of industry recognised terminology. However, those who may desire to challenge NEC/2 in the future must take note of clause 10.1 which provides as follows:

> . . . *The Employer, the Contractor, the Project Manager and the Supervisor shall act as stated in this contract and in a spirit of mutual trust and co-operation* . . .

12.3.4 The intention behind clause 10.1 is presumably to discourage the parties from poring over the contract with a view to exploiting any potential for alternative interpretation that there might be. Setting aside the issue of the enforceability or otherwise of clause 10.1 (which, as laymen, we regard as setting out a position of worthy intent rather than something capable of legal definition), it will be to no avail where there is genuine, as distinct from contrived, uncertainty as to a clause's application.

12.4 COMPENSATION EVENTS – CORE CLAUSES

Clause 60.1(1) The Project Manager gives an instruction changing the Works Information except
- **a change made in order to accept a Defect or**
- **a change to the Works Information provided by the Contractor for his design which is made at his request or to comply with other Works Information provided by the Employer**

12.4.1 This clause is broadly equivalent to clause 25.4.5.1 of JCT 98 PWQ (see paras 2.8.8–2.8.27) inasmuch as it addresses the consequences of a variation being made to the

Works Information. The Works Information itself will be set out in the Contract Data (see para. 12.1.15), Part 1 of which is provided by the Employer and will include a general description of the *works*, drawings (for location, general arrangement and detail), specification and any constraints on access or sequences of construction. Part 2 of the Contract Data (see para. 12.1.15) which is provided by the Contractor will include information concerning those parts of the works which the Contractor is to design.

12.4.2 It is to be noted that even clarifications to previously issued drawings may only be made by changing the Works Information such that they too become compensation events. Further changes to the Works Information may also be necessary in respect of:

- resolving an ambiguity or inconsistency in or between the contract documents (to which clause 17.1 refers); or
- eliminating anything within the Works Information requiring the Contractor to perform something which is "illegal or impossible" (to which clause 19.1 refers).

12.4.3 Clause 60.1(1) sets out two exceptions to changes in the Works Information giving rise to a compensation event. Each such exception is considered in turn:

- ... *a change made in order to accept a Defect* ... will not qualify as a compensation event. The mechanism for accepting a Defect is set out in clause 44.2 which provides that if the Contractor and Project Manager are prepared to consider a change to the Works Information such that a Defect does not have to be corrected then the Contractor is obliged to furnish the Project Manager with a quotation for ... *reduced Prices or an earlier Completion Date or both* ... which will come in to effect if accepted by the Project Manager.
- ... *a change to the Works Information provided by the Contractor for his design which is made at his request or to comply with other Works Information provided by the Employer* ... will not qualify as a Compensation event. Ostensibly, this provides a sensible safeguard in terms of preventing the Contractor from changing his design solutions and resting any additional costs in respect thereof at the Employer's door. However, consider the alternative possibility in which a change to the design will result in a saving – should not the Prices be reduced in such a circumstance and as would be provided for in clause 63.2 but for this embargo?

Clause 60.1(2) The Employer does not give possession of a part of the Site by the later of its *possession date* and the date required by the Accepted Programme

12.4.4 Consideration must first be given to clause 33 which provides both that the Employer is to give possession of each part of the Site to the Contractor on or before the later of the *possession date* or the date required by the Contractor and set out in his Accepted Programme (clause 33.1 refers) and that while the Contractor has possession of a part of the Site the Employer gives him access to and use of it (clause 33.2 refers). Whilst "possession" is not one of the defined terms in clause 11, it is reasonable to assume in this context that it equates with giving the Contractor a licence to occupy the Site up to the date of Completion and for the purpose of carrying out the *works* – and this is consistent with the provision in clause 33.2 that the Contractor has ... *access to and use of* ... the Site.

12.4.5 Since clause 33.1 recognises the possibility that the Contractor may programme to start the *works* later than the *possession date* set out in Part 1 of the Contract Data (thereby allowing the Employer to defer in giving possession of the Site by that same margin), the Employer may choose to retain possession of the Site to the later date indicated by the Contractor. If the Employer makes plans to this end and the Contractor revises his programme to indicate an earlier start (though *not* earlier than indicated in the Contract Data) and the Employer cannot accommodate that then this may be the subject of a compensation event. Clause 31.3 sets out those grounds in respect of which the Project Manager may reject the Contractor's programme – and the setting of an earlier programmed date for possession is not one of them. Whilst a Project Manager might be minded to withhold acceptance of the Contractor's revised programme in this sort of circumstance, it must be recognised that such a withholding of acceptance would in itself give rise to a compensation event by virtue of clause 60.1(9) which clause is considered in para. 12.4.28 to 12.4.30. In the circumstances where the Employer has indicated a *possession date* in the Contract Data, it would be advisable for him to maintain that date (or at least have the ability to do so) irrespective of subsequent indications from the Contractor.

Clause 60.1(3) The Employer does not provide something which he is to provide by the date for providing it required by the Accepted Programme

12.4.6 This clause is broadly equivalent to clause 25.4.6 of JCT 98 PWQ (see paras 2.8.29–2.8.44) inasmuch as it addresses the consequences of the Employer failing to provide "something" to the Contractor by the date set out in the Accepted Programme. However, the use of the word "something" goes beyond the narrower definition within JCT 98 PWQ which deals specifically with the provision of, amongst other things, instructions and drawings. This said, should the Employer fail to provide anything which it is within his control or obligation to do so then it is reasonable that the Contractor has a means of redress.

12.4.7 Whereas clause 25.4.6 of JCT 98 PWQ (via clause 5.4) puts the onus on the Architect to provide the Contractor with relevant information timeously, under NEC/2 the onus is on the Project Manager to observe the Contractor's requirements as set out in the Accepted Programme and it is not immediately clear as to how the Project Manager will satisfy this obligation. This is because clause 31.2 lists that which the Contractor is to include in any programme put forward for acceptance and, despite the list being comprehensive, it does not specifically include the dates on which any outstanding information is required. Clause 31.2 imposes many requirements on the Contractor to indicate on his programme factors such as when he will complete the *works*, by what means, the risk allowances he has made and so on. Less is said as to what must be indicated on the programme if the Project Manager is to deduce by what dates information and the like must be provided by the Employer.

Clause 60.1(4) The Project Manager gives an instruction to stop or not to start any work

12.4.8 An instruction by the Project Manager either to stop or not to start any work is a compensation event and this is a prudent inclusion if, as the NEC/2's Guidance Notes indicate, such an instruction is necessary for reasons of safety. Since the compensation event arises out of any instruction being issued, clause 61.4 goes on to add that the Prices and Completion Date will not be changed where such an instruction has been rendered necessary due to ... *a fault of the Contractor* ..., e.g. a failure to comply with the CDM Regulations.

12.4.9 Because NEC/2 does not provide cross-references between clauses, the proviso within clause 61.4 will not be evident when considering clause 60.1(4) in its own right. It is interesting to note that in the case of clause 60.1(1) (see paras 12.3.1 to 12.3.3) and clause 60.1(9) (see paras 12.4.28 to 12.4.30) the provisos as to their range of application are contained within the clauses themselves.

Clause 60.1(5) The Employer or Others do not work within the times shown on the Accepted Programme or do not work within the conditions stated in the Works Information

12.4.10 The part of this clause referring to the Employer has similarities with clause 25.4.8 of JCT 98 PWQ (see paras 2.8.51 to 2.8.55) inasmuch that the Employer will be responsible to the Contractor if work he has directly commissioned causes delay and/or disruption to the Contractor's progress. The part of this clause referring to Others is understood by reference to clause 11.2(2) which categorises them as ... *people or organisations who are not the Employer, the Project Manager, the Supervisor, the Adjudicator, the Contractor, or any employee, Subcontractor* [sic] *or supplier of the Contractor* ... On this basis, "Others" could include statutory undertakers (to the extent that they are required to carry out work pursuant to their statutory obligations) and might therefore be regarded as being similar to clause 25.4.11 of JCT 98 PWQ (see para. 2.8.61), however, this compensation event is of wider reaching effect in two respects. First, NEC/2 also permits there to be an adjustment to the Prices and, secondly, the term "Others" is of such broad meaning as to apply to any party not excluded by the precise definition within clause 11.2(2).

12.4.11 Clause 31.2 envisages that the Works Information will detail the order and timing of the work to be undertaken by the Employer and Others and obliges the Contractor to accommodate such work within his programme. To the extent that the requisite details are not available for inclusion in the Works Information, clause 31.2 also allows for subsequent agreement on them with the Contractor. It is obviously important that the Employer notes carefully the information set out in the Contractor's programme in order that he may plan his own work in a manner which will not give rise to a compensation event. This said, the Employer must be prepared to be flexible since having made plans to align with the Contractor's programme, the Contractor may revise his programme in a manner which is consistent with the Works Information but which the Employer can no longer accommodate thus giving rise to a compensation event. As observed in para. 12.4.5, there are limited grounds for the Project Manager being able to reject a Contractor's programme.

Clause 60.1(6) The Project Manager or the Supervisor does not reply to a communication from the Contractor within the period required by this contract

12.4.12 Clause 13.3 states that ... *if this contract requires the Project Manager, the Supervisor or the Contractor to reply to a communication, unless otherwise stated in this contract, he replies within the period for reply* ... The period for reply referred to will be set out in Part 1 of the Contract Data and, consistent with the intention of NEC/2 to encourage good management, the period should not be excessive otherwise this will militate against the ability of the parties to respond to any problems which may emerge. The sanction against the Employer in the event that his Project Manager or Supervisor fails to respond to a Contractor's communication within the agreed timescale is provided for by way of clause 60.1(6). It is unclear what sanction the Employer would have against the Contractor should the Contractor fail to respond within the agreed timescale, other than – it is suggested – for the Project Manager to discount any delay thereby caused in any subsequent assessment of compensation events.

12.4.13 Some commentators have criticised clause 60.1(6) as imposing time limits which may not always be appropriate to the given circumstances, e.g. either (i) where a party requires an immediate response in the face of an emergency or (ii) where a party is compelled to respond to a communication against the backdrop of competing priorities and the knowledge that the response could be made at a later date without affecting programme or cost. With regard to (i), there is nothing to prevent the parties from responding within the stated time limits if they are able to do so and it is in the interests of the project. With regard to (ii), the following should be noted:

- clause 13.5 permits the Project Manager to extend the *period for reply* in circumstances where he has the Contractor's agreement to this; and
- should the Project Manager be late in making a response but also in circumstances which will not affect programme or cost then even should the Contractor notify the occurrence of a compensation event the ... *Prices and Completion Date are not changed if the Project Manager decides that an event ... has no effect upon Actual Cost or Completion* ... (clause 61.4 refers).

Clause 60.1(7) The Project Manager gives an instruction for dealing with an object of value or of historical or other interest found within the Site

12.4.14 This clause is broadly equivalent to clause 25.4.5.1 of JCT 98 PWQ (see paras 2.8.8 to 2.8.27) to the extent of it referring to clause 34 thereof (see para. 2.8.25) dealing with the discovery of antiquities and the like on site. So far as NEC/2 is concerned, this compensation event must be read in conjunction with clause 73.1 which provides as follows:

> ... *The Contractor has no title to an object of value or of historical or other interest within the Site. The Contractor notifies the Project Manager when such an object is found and the Project Manager instructs the Contractor*

how to deal with it. The Contractor may not move the object without instructions ...

12.4.15 Clauses 60.1(7) and 73.1 of NEC/2 differ from the JCT 98 PWQ approach in two fundamental respects, thus:

- the care to be taken by the Contractor with objects of value prior to the receipt of instructions is not provided for; and
- the Contractor's entitlement to compensation runs not from the date of the objects of value being discovered but, rather, from the receipt by the Contractor of instructions from the Project Manager.

Insofar as the Contractor will have to bear his own costs for the period of time between discovering objects of value and receiving the Project Manager's instructions in respect thereof (which period of time may be for as long as the *period for reply* stated in the Contract Data), the drafting of this clause might be regarded as favourable to the Employer when compared with the corresponding position under JCT 98 PWQ. The drawback of this is that the Contractor may, quite understandably, endeavour to progress the *works* as best he can pending the receipt of the Project Manager's instructions and potentially upset the exact location and condition of objects of value prior to their being inspected by experts.

Clause 60.1(8) The Project Manager or the Supervisor changes a decision which he has previously communicated to the Contractor

12.4.16 Clause 13.1 governs what is communicated between the parties and it provides as follows:

> ... *Each instruction, certificate, submission, proposal, record, acceptance, notification and reply which this contract requires is communicated in a form which can be read, copied and recorded* ...

Inasmuch that clause 60.1(8) refers to the communication of a "decision", it cannot be understood by reference to clause 13.1. Consideration therefore needs to be given to the meaning of the word decision (or any of its derivatives) having regard to the fact that:

- it is not a communication within the meaning of clause 13.1;
- it is not a defined term in clause 11.2 (because it does not begin with a capital letter); and
- it is not defined in the Contract Data (because it is not written in italics).

The question to arise is whether the word decision is to be construed as having its ordinary meaning (a judgment; conclusion; resolution reached or given; or making up one's mind) or whether it can be construed by reference to the use of the word in NEC/2. With the latter in mind, there follows a consideration of those clauses within NEC/2 which use the word decision or one of its derivatives.

12.4.17 Words connoting decisions are to be found in clauses 16, 61, 63 and 64 – and there may be other instances but they are not readily apparent given the policy in NEC/2 of dispensing with cross-references between clauses. The clauses in question are considered in turn:

12.4.18 *Clause 16* – As addressed at para. 12.7.8, early-warning meetings are held to seek solutions to problems impacting on the cost, time and quality objectives of the project. Arising out of those meetings, the decisions taken are recorded by the Project Manager and copied to the Contractor. In this situation, a Contractor could propose a solution which the Project Manager may rely upon and thereby confirm back to the Contractor as a "decision". If the Contractor's advice were subsequently to emerge to be flawed such that the Project Manager had then to instruct that an alternative and corrective course of action be taken then it would appear that the Contractor is entitled to treat the instruction as a compensation event.

12.4.19 *Clauses 61.1 and 61.2* – These clauses deal with the actions to be taken in the event of there being a changed decision. The clauses, being of a procedural nature, do not assist in the understanding of how the word decision (or one of its derivatives) is to be interpreted.

12.4.20 *Clause 61.4* – This clause provides that the Prices and Completion Date are not changed if the Project Manager decides an event notified by the Contractor (a) arises from the Contractor's fault; or (b) neither has happened nor is expected to; or (c) does not affect cost or completion; or (d) is not one of the compensation events set out in NEC/2. It may be observed, therefore, that if the Project Manager decides that any of conditions (a)–(d) are in play then the Prices and Completion Date are not changed whilst the converse is that *none* of the conditions may be in play if the Price and Completion Date are to be changed. Should the Project Manager change an earlier decision as to any of conditions (a)–(d) being in play then this would bring about a Compensation event but it is hard to see what practical benefit this would bring to the Contractor as set out in Examples 12.1–12.5 below:

Example 12.1   The Project Manager decides that only condition (a) is in play but later changes that decision

If the Contractor notifies an event but the Project Manager decides that it has arisen through the Contractor's own fault then neither the Prices nor Completion Date are changed. If the Project Manager later changes this decision, a compensation event will have occurred but this will not in itself give rise to time or cost consequences. Instead, regard will be had to the compensation event first notified by the Contractor as being that around which the assessment of time and cost would have to be made.

Example 12.2   The Project Manager decides that only condition (b) is in play but later changes that decision

If the Contractor notifies an event but the Project Manager decides that it has neither arisen nor is likely to then neither the Prices nor Completion Date are changed. If the Project Manager later changes this decision (say in consequence of an event which was considered unforeseeable by the Project Manager but which nonetheless came about as forecast by the Contractor), a compensation event will have occurred but this will not in itself give rise to time or cost consequences. Instead, regard will be had to the compensation event first notified by the Contractor as being that around which the assessment of time and cost would have to be made.

Example 12.3   The Project Manager decides that only condition (c) is in play but later changes that decision

If the Contractor notifies an event but the Project Manager decides that it will not affect cost or completion then it follows that neither the Prices nor Completion Date will be changed. If the Project Manager later changes this decision, a compensation event will have occurred but this will not in itself give rise to time or cost consequences. Instead, regard will be had to the compensation event first notified by the Contractor as being that around which the assessment of time and cost would have to be made.

Example 12.4   The Project Manager decides that only condition (d) is in play but later changes that decision

If the Contractor notifies an event but the Project Manager decides that it is not one of the compensation events set out in NEC/2 then it follows that neither the Prices nor Completion Date will be changed. If the Project Manager later changes this decision, a compensation event will have occurred but this will not in itself give rise to time or cost consequences. Instead, regard will be had to either (i) the compensation event first put forward by the Contractor as being that around which the assessment of time and cost consequences would have to be made or (ii) such other compensation event as the Project Manager then decides is appropriate in the circumstances.

Example 12.5   The Project Manager decides that any two of conditions (a)–(d) is in play but later changes a decision in respect of one condition only

If the Contractor notifies an event but the Project Manager decides that it arises from the Contractor's fault *and* that it is not one of the compensation events set out in NEC/2 then it follows that neither the Prices nor Completion Date will be changed. If the Project Manager later changes *one* of his decisions, say to the effect that the event notified by the Contractor is one of the compensation events set out in NEC/2, then a compensation event will have occurred but it remains the case that neither the Prices nor Completion Date are to be changed since one of the *other* conditions (in this example, the Project Manager's opinion that the Contractor is at fault) is still in play.

12.4.21 Clause 61.4 goes on to add that if the Project Manager decides none of conditions (a)–(d) is in play then he shall instruct the Contractor to submit a quotation in respect of the notified event. If the Project Manager later changes his decision to the effect that one or more of conditions (a)–(d) is in play then a compensation event will have occurred and the effect of the Project Manager's changed decision is to prevent the Contractor from having any contractual entitlement to the Prices and Completion Date being changed.

12.4.22 *Clause 61.5* – Should the Project Manager decide that the Contractor has not given an early warning of an event … *which an experienced Contractor could have given* … then clause 63.4 goes on to provide that the assessment of the Compensation event takes place … *as if the Contractor had given early warning* … since such a warning may have allowed some costs to be reduced and time to be saved. If the Project Manager later changes his decision, a Compensation event will have occurred but this will not itself figure in the subsequent assessment of time and cost and, instead, regard will be had to the position that would have pertained had the Project Manager originally accepted that early warning had been given.

12.4.23 *Clause 61.6* – If the Project Manager decides that the effects of a Compensation event are … *too uncertain to be forecast reasonably, he states assumptions about the event in his instruction to the Contractor to submit quotations* … Insofar as the assessment of the Compensation event will be based on the assumptions given by the Project Manager and should those assumptions be shown to be wrong by reference to what actually happens then the Project Manager notifies a correction to the assumption previously given and an appropriate adjustment to the earlier assessment of the Compensation event is made. Such an adjustment is made by reference to the Compensation event described in clause 60.1(17) (see para. 12.4.54) and not to the Compensation event in clause 60.1(8) under consideration here.

12.4.24 *Clause 63.4* – See para. 12.4.22.

12.4.25 *Clause 64.1* – Should the Project Manager decide … *that the Contractor has not assessed the compensation event correctly in a quotation and he does not instruct the Contractor to submit a revised quotation* … then he assesses the compensation event himself. If the Project Manager later changes this decision, a compensation event will have occurred but this will not itself figure in the subsequent assessment of time and cost and, instead, regard will be had to the quotation initially advanced by the Contractor which the Project Manager erroneously regarded as incorrect.

12.4.26 The analysis in paras 12.4.16 to 12.4.25 as to the effect of clause 60.1(8) is summarised in the Table 12.1.

12.4.27 With the exception of the position deduced from clause 16, the compensation event described under clause 60.1(8) is either of no effect or the Contractor's remedy is by reference to another compensation event. As to clause 16, the compensation event could produce a result which is unfair to the Employer and this prompts the thought that the word decision (or one of its derivatives) is not necessarily intended to be construed by reference to how it is used in NEC/2 but, rather, to its ordinary meaning. If this is correct then *any* change of decision by the Project Manager will give rise to a compensation event.

Table 12.1. Matters relating to clause 60.1(8)

Clause	Paragraph(s) in this book	Comments
16	12.4.18	The operation of the compensation event could produce a result which is unfair to the Employer.
61.1 and 61.2	12.4.19	No effect.
61.4 (Examples 12.1–12.4)	12.4.20–12.4.21	The operation of the compensation event is of no effect per se since matters are instead settled by reference to the compensation event alleged by the Contractor prior to the Project Manager's erroneous rejection of it.
61.4 (Example 12.5)	12.4.20–12.4.21	The operation of the compensation event is of no effect.
61.5	12.4.22	The operation of the compensation event is of no effect per se since matters are instead settled by reference to the facts as they pertained prior to the Project Manager's erroneous decision.
61.6	12.4.23	The operation of the compensation event is of no effect per se since matters are instead settled by reference to the compensation event described in clause 60.1(17).
63.4	12.4.24	The operation of the compensation event is of no effect per se since matters are instead settled by reference to the facts as they pertained prior to the Project Manager's erroneous decision.
64.1	12.4.25	The operation of the compensation event is of no effect per se since matters are instead settled by reference to the facts as they pertained prior to the Project Manager's erroneous decision.

Clause 60.1(9) The Project Manager withholds an acceptance (other than acceptance of a quotation for acceleration or for not correcting a Defect) for a reason not stated in this contract

12.4.28 Acceptance is one of the things that can be communicated between the parties as referred to in clause 13.1 (see para. 12.4.16). Whilst clause 13.8 provides that the Project Manager may withhold acceptance of a Contractor's submission, the clause goes on to add that … *withholding acceptance for a reason stated in* [the] *contract is not a compensation event* …; the corollary of this is set out in clause 60.1(9).

12.4.29 Although the Project Manager has total power to withhold acceptance of any of the Contractor's submissions, it is a power that is to be exercised with some care if a compensation event is to be avoided. In the circumstances, it is important to note those clauses which make reference to submissions being made to the Project Manager for his acceptance and the reasons he may give (if any) for withholding that acceptance.

Table 12.2 sets out those clauses referring to the Project Manager's acceptances – and there may be other instances but they are not readily apparent given the policy in NEC/2 of dispensing with cross-references between clauses.

12.4.30 As can be seen from Table 12.2, the Project Manager's grounds for withholding acceptances are, with the exception of clause 13.4, defined with reasonable precision. If it were the intention of the drafters of NEC/2 to be precise about the reasons for withholding acceptances then that intention is largely undone by clause 13.4 – and the potential for a Project Manager to seek refuge behind it when unable to respond within the period for reply is considerable, itself a Compensation event by reason of clause 60.1(6) (see paras 12.4.12 to 12.4.13).

Table 12.2. References to Project Manager's acceptances

Core clause	Matter for acceptance	Reason(s) for witholding acceptance
13.4	Communication submitted or resubmitted by the Contractor.	More information needed in order to be able to assess the Contractor's submission fully.
13.8	Contractor's submission.	See para. 12.4.28.
15.1	Contractor's proposal for adding to the Working Areas.	The proposed addition is not necessary for Providing the Works or the proposed area will not be used for work under the contract.
21.2	Particulars relating to the Contractor's design.	The Contractor's design does not comply with the Works Information or it does not comply with the applicable law.
21.3	Particulars relating to the Contractor's design being submitted in parts if each part is capable of full assessment.	None given.
23.1	Particulars relating to the design by the Contractor of an item of Equipment.	The design of the item of Equipment will not allow the Contractor to Provide the Works in accordance with the Works Information or the Contractor's design as accepted by the Project Manager (see clause 21.2) or the applicable law.
24.1	Replacement of Contractors' personnel named in the Contract Data.	The proposed replacement's qualifications and experience are not as good as those of the person who is to be replaced.
26.2	Proposed Subcontractor.	The proposed Subcontractor's appointment will not allow the Contractor to Provide the Works.
26.3	Proposed subcontract conditions where the corresponding NEC form is not being used.	The conditions will not allow the Contractor to Provide the Works or the conditions fail to include a statement that the parties shall act in a spirit of mutual trust and co-operation.

Core clause	Matter for acceptance	Reason(s) for withholding acceptance
31.3	Contractor's first programme.	The Contractor's plans which it shows are not practicable or it does not show the information which the contract requires (see clause 31.2) or it does not represent the Contractor's plans realistically or it does not comply with the Works Information.
32.2	Contractor's revised programme.	Though not specifically stated in clause 32, the embargoes stated in clause 31.3 will equally apply here.
44.2	Acceptance of Defects.	This is subject to the mechanics for accepting or rejecting Contractor's quotations. See also Compensation event 60.1(1) as addressed at paras 12.4.1 to 12.4.3.
50.3	Contractor's first programme.	The grounds for withholding acceptance are given in clause 31.3.
62.3	Quotations for Compensation events.	The Project Manager notifies the Contractor that (i) the proposed instruction/proposed changed decision will not be given or (ii) he will be making his own assessment or (iii) a revised quotation is required. In the latter event, clause 62.4 requires the Project Manager to explain his reasons for requesting a revised quotation.
85.1	Contractor's submission of policies and certificates of insurance.	The policies and certificates do not comply with the contract.

Main option clauses	Matter for acceptance	Reason(s) for withholding acceptance
31.4 (options A and C)	Contractor's programme details relating to each activity on the activity schedule.	Though not specifically stated in clause 31.4 (option A), the embargoes stated in clause 31.3 (Core Clause) will equally apply here.
54.2 (options A and C)	Contractor's revision to the activity schedule.	Clause 54.3 (option A) gives reasons for witholding Acceptance as being the revision not complying with the Accepted Programme or the pricing of the activity schedule not being reasonably distributed or the total of the prices being changed.
26.4 (options C, D, E and F)	Contractor's submission of contract data concerning each subcontract.	The use of the contract data will not allow the Contractor to Provide the Works.

(continued)

Table 12.2. (cont.)

Main option clauses	Matter for acceptance	Reason(s) for withholding acceptance
36.5 (options C, D, E and F)	Contractor's submission of a Subcontractor's proposal to accelerate.	Though not addressed within the main option clauses, the presumption is that the Project Manager has the right to withold acceptance of the proposal by virtue of Compensation event 60.1(9) which excludes from such an event the matter of the Project Manager witholding acceptance to an acceleration propasal.

Secondary option clauses	Matter for acceptance	Reason(s) for withholding acceptance
G1.1	Contractor's submission of a Performance Bond.	The commercial position of the bank/insurer is not strong enough to carry the bond.
J1.2	Contractor's submission of an Advance Payment Bond.	The commercial position of the bank/insurer is not strong enough to carry the bond.

Notes
(i) A number of the main option clauses refer to the word acceptance (or one of its derivatives) as to be given by the Project Manager but do so in terms of the consequence of acceptance (whether given or properly withheld); see clauses 36.3 (options A, B, C and D), 36.4 (options E and F), 65.4 (options A, B, C, D), 11.2(30) (options C, D, and E), 53.5 (option C) and 11.2(29) (option F).
(ii) A number of core clauses refer to the word acceptance (or one of its derivatives) as to be given by the Contractor but without always providing an express remedy to the Employer in the event of that acceptance not being forthcoming; see clauses 64.2, 80.1, 87.1, 87.2 and 97.1. This is also the case with the following main option clauses; 31.4 (option A), 36.5 (options E and F) and 11.2 (22) (option F).

Clause 60.1(10) The Supervisor instructs the Contractor to search and no Defect is found unless the search is needed only because the Contractor gave insufficient notice of doing work obstructing a required test or inspection

12.4.31 Clause 42.1 empowers the Supervisor, provided he gives his reasons, to instruct the Contractor to make searches. Such searches may include the carrying out of tests and inspections other than those already required either by law or by the Works Information, and may be instructed with the aim of identifying Defects. Should a search be instructed in accordance with clause 42.1 and no such Defects be found then, subject to the proviso in clause 60.1(10), a compensation event will have occurred. In this way clause 60.1(10) provides an obvious source of protection to the Contractor in that it compensates him should he be instructed to open up the *works* on the suspicion of there being a Defect and yet no Defect is found. If, however, the Supervisor instructs a search only because he has been denied the opportunity to carry out a test or inspection required either by the law or by the Works Information (clause 40.1 refers) as a result of the Contractor having covered up the relevant portion of the *works* without giving the notice required by clause 40.3, then the aforementioned proviso to clause 60.1(10) removes the Contractor's rights to compensation.

2.4.32 The situation under NEC/2 is somewhat different to that under JCT 98 PWQ. Under NEC/2, if a search is ordered and no Defect is uncovered as a result then, except where the Contractor has previously failed to give a notice required by clause 40.3, a compensation event will have occurred and the Contractor will be entitled to a change in both the Prices and the Completion Date. This is in contrast to the situation pertaining under JCT 98 PWQ whereby, provided the Architect in instructing the Contractor to open up work has had due regard to the Code of Practice referred to in clause 8.4.4, the Contractor will not receive any additional payment (albeit he will remain entitled to an extension of time) even where no defects are revealed as a result of the opening up.

Clause 60.1(11) A test or inspection done by the Supervisor causes unnecessary delay

2.4.33 Clause 40.5 requires the Supervisor to carry out his ... *tests and inspections without causing unnecessary delay* ..., the corollary of which is that the Contractor should be compensated in the event that the Supervisor does not act with due despatch. Clause 60.1(11) provides the means for that compensation to be carried into effect.

2.4.34 Under clause 40.5 ... *unnecessary delay* ... can be gauged by reference to that which the Contractor ought in any event to have anticipated by reference to the Works Information or the applicable law. In the wider sense, and with regard to clause 42.1, ... *unnecessary delay* ... could be gauged by reference to the extent to which a test or inspection causes a delay which is greater than would have occurred had the Supervisor acted reasonably and timeously.

Clause 60.1(12) The Contractor encounters physical conditions which
- are within the Site,
- are not weather conditions and
- which an experienced contractor would have judged at the Contract Date to have such a small chance of occurring that it would have been unreasonable for him to have allowed for them

2.4.35 There is both advantage and disadvantage to the Employer in reserving to himself the risk of unforeseen ground conditions. The advantage is that the Employer will only pay for conditions which are *actually* encountered rather than meet the allowance included in the Contractor's tender for conditions which *might* be encountered – assuming, of course, that the Contractor would price the risk. By way of a corollary to this, the disadvantage to the Employer is that he is less likely than the Contractor to be able to average out the impact of difficult ground conditions over numerous projects. NEC/2 is a contract which imposes upon the Employer the risk of, inter alia, ground conditions and its provisions require closer inspection to see if they go beyond this particular theme which pertains in most construction and engineering contracts.

2.4.36 The first thing to note is that the clause refers to *physical* rather than *ground* conditions, the effect of which is to produce a very wide meaning following the finding in *Humber Oil Terminals Trustee -v- Harbour and General (1991) 59 BLR 1 (CA)* which addressed a similar provision in the ICE Form of Contract, 5th Edition. In that case, the term "physical conditions" was construed as applying to a combination of soil conditions and applied stresses; whilst the soil conditions were themselves foreseeable, the action of stresses on the soil conditions leading to a collapse was not.

12.4.37 The second bullet point of the clause excludes the Contractor's right to compensation in the event that the physical conditions are ... *weather conditions* ... and this is largely due to the fact that weather conditions are addressed in clause 60.1(13) (see paras 12.4.43 to 12.4.45). However, an interesting comparison may be made with clause 12(1) of the ICE Form of Contract, 6th Edition which refers also to ... *conditions due to weather conditions* ... the effect of which is to exclude the *consequences* of weather conditions, e.g. the site being flooded following heavy rainfall. This being so, it would appear likely under NEC/2 that the Contractor would be compensated under clause 60.1(12) in the event of his suffering the consequences of adverse conditions caused by weather conditions.

12.4.38 The third bullet point of the clause poses a novel test as to the extent to which the Contractor should allow in his tender for the occurrence of adverse physical conditions. In contrast to the foreseeability test which is ordinarily imposed (see, for example, clause 12(1) of the ICE Form of Contract, 6th Edition which takes as a reference point conditions which could be ... *foreseen by an experienced contractor* ...), NEC/2 instead imposes a probability test. The Guidance Notes to NEC/2 observe that the foreseeability test has been ... *the source of many disputes* ...; the text of the notes then runs out (perhaps due to a printing error) at the point where the Guidance Notes to NEC/1 would have gone on to add ... *mainly because judgment is involved. The NEC includes an overhauled but not radically different procedure. The test to be applied is a much less subjective one, namely, what an experienced contractor would have judged* ...

12.4.39 In determining whether or not to make allowances for adverse physical conditions, the Contractor will also have regard to clauses 60.2 and 60.3. Clause 60.2 requires that the Contractor, in judging the physical conditions, takes into account the Site Information (including any publicly available information referred to therein), information available from a visual inspection of the Site and ... *other information which an experienced contractor could reasonably be expected to have or to obtain* ... Ironically, this latter criterion is not dissimilar to the foreseeability test of the type criticised in the Guidance Notes to NEC/2 (see para. 12.4.38). Clause 60.3 provides that if there is any inconsistency in the Site Information (including information referred to in it), the Contractor is assumed to have taken into account ... *the physical conditions more favourable to doing the work* ... and this follows the contra proferentem rule.

12.4.40 It is suggested that the NEC/2 approach is likely to be as productive of disputes as the foreseeability approach which it seeks to replace. This is primarily because it is unclear from whose perspective (i.e. the Employer or Contractor) the word "unreasonable" is to be viewed. If there is a particularly remote chance of adverse physical conditions being encountered, it may be unreasonable for the Contractor to make allowance in his pricing therefore because the Employer will then have to accept a tender which is inflated because it has contingencies within it relating to the circumstances which may never arise. In sharp contrast, the Contractor may be compelled for commercial reasons to keep his tender as low as possible by excluding the costs of circumstances which are foreseeable but which he can later argue had so remote a chance of occurring that had he included allowances in respect thereof it might have proved *unreasonable* to his prospects of winning the work in competitive tender!

12.4.41 The test set out in the third bullet point of clause 60.1(12) is more likely than not to produce a result which operates to the Employer's detriment, as demonstrated in Examples 12.6 and 12.7 below:

Example 12.6 The Contractor makes no allowance for physical conditions which, though foreseeable, are remote

Having studied the site investigation report and other relevant items of tender documentation, the Contractor foresees that some physical conditions may arise causing him to question the need to include an allowance in his tender in respect thereof. Having considered the risk to be remote (and in circumstances where some allowance would have to be made were clause 12(1) of the ICE Form of Contract, 6th Edition to have been in force), the Contractor decides that it would be unreasonable for him to make allowances in his tender in respect thereof; the allowance would be unreasonable both for the Employer in respect of pricing for something which may never occur and for the Contractor in terms of making his tender less competitive. Should the remote risk of adverse physical conditions not materialise then the Contractor's pricing policy will be of no concern to the Employer. If, however, adverse physical conditions do materialise then the Employer will be disadvantaged in the following respects:

(i) even though the Contractor might be better able than the Employer to average out the effect of adverse physical conditions over numerous projects, he will be entitled to full reimbursement in respect of matters which he foresaw but was nonetheless permitted to exclude from his reckoning when tendering; and

(ii) the Contractor's right to compensation will not be pegged at the level that could have been included in the tender (which level would have regard to both the need to be competitive and the likelihood of the risk eventuating) but, rather, he will be reimbursed the costs actually incurred in coping with the adverse physical conditions.

Example 12.7 The Contractor makes allowance for physical conditions which are foreseeable albeit that he is not required to do so in light of the remoteness of those conditions eventuating

Having studied the site investigation report and other relevant items of tender documentation, the Contractor foresees that some physical conditions may arise in respect of which he decides to make an allowance in his tender albeit that he does not (nor need not) disclose this. Having secured the contract, the Contractor later encounters the physical conditions which he foresaw but which he by then contends had such a small chance of occurring that he was not obliged to allow for them – the Project Manager gives careful consideration to the circumstances and agrees with the Contractor's view. In the subsequent assessment of the compensation event, the Contractor will be permitted to recover the actual cost of dealing with the particular adverse physical conditions without the Employer being given credit for any allowance that may already have been included in the tender.

12.4.42 Examples 12.6 and 12.7 reveal an interesting relationship between clauses 60.2 and 60.1(12); the former invites the Contractor to consider the possibility of his encountering adverse physical conditions whilst the latter permits him, in certain circumstances, to take the conscious decision of ignoring the consequences of his deliberations.

Clause 60.1(13) A weather measurement is recorded
- **within a calendar month,**
- **before the Completion Date for the whole of the works and**
- **at the place stated in the Contract Data the value of which, by comparison with the weather data, is shown to occur on average less frequently than once in 10 years**

12.4.43 Whereas a number of construction contracts refer to ... *exceptionally adverse weather conditions* ... (e.g. clause 25.4.2 of JCT 98 PWQ as considered at para. 2.8.4) the subjective judgment that this calls for can be productive of disputes. NEC/2 seeks to overcome this by including an objective and measurable approach to the consideration of weather conditions and this is achieved by comparing *weather measurements* as they are experienced with *weather data* being the baseline for comparison purposes and should the weather conditions in the former be more adverse than those in the latter then the Contractor is entitled to compensation.

12.4.44 According to Part 1 of the Contract Data, the *weather measurements* to be recorded for each calendar month are:

- the cumulative rainfall (in millimetres);
- the number of days with rainfall more than 5 mm;
- the number of days with minimum air temperature less than zero degrees Celsius;
- the number of days with snow lying at a declared time (in the UK, this will be 0900 hours GMT as this is when the Met Office takes its readings); and
- any other measurements regarded as appropriate to the circumstances, e.g. the number of working hours during which wind speeds exceed a given level since this will be relevant to the operation of tower cranes.

As to the weather data, these will comprise the records of past weather measurements for each calendar month as taken from an independent authority such as the United Kingdom Meteorological Office (the Met Office) weather station nearest to the site.

12.4.45 In the course of setting out an objective and measurable approach to the consideration of weather conditions, the necessarily prescriptive details which are needed to replace the sort of subjective test to be found in JCT 98 PWQ (see para. 2.8.4) will produce a certain result albeit that on occasions it may prove unjust. This can be so where some adverse conditions may not qualify for compensation purposes even where less adverse conditions do. This possibility is demonstrated in Examples 12.8 to 12.10 below which share the following *weather data* characteristics (i.e. the 10-year average) for the month of November:

- cumulative rainfall: 75 mm;
- 5 days with rainfall more than 5 mm;
- 5 days with minimum air temperature less than zero degrees Celsius;
- 25 working hours with wind speeds exceeding 60 km/hour.

Example 12.8   Adverse weather conditions, confined to rainfall, evenly spaced throughout the month

On 6 days evenly spaced throughout the month the rainfall is 6 mm each together with sporadic rainfall on 8 other days of 2 mm each.

Albeit that the total rainfall of 52 mm (i.e. 6 days × 6 mm plus 8 days × 2 mm) is less than the monthly cumulative rainfall of 75 mm indicated in the *weather data*, there are more than 5 days with rainfall of more than 5 mm making this a compensation event. Because the rainfall has been evenly spread across the month, the inconvenience to the Contractor may not be that great and this would be reflected in the assessment of the compensation event.

Example 12.9   Adverse weather conditions, multiple in nature, occurring in quick succession

In a 4-day period, there is evenly falling rain of 60 mm the suddenness of which causes trenches to collapse and leads to the Contractor being involved in extensive de-watering. The Contractor embarks upon planking and strutting the trenches in the following days and just as he is ready to start upon pouring the concrete foundations there is a 4-day period in which the air temperature does not rise above zero degrees Celsius resulting in concreting operations being deferred. As soon as the temperature rises on the 5th day the Contractor is ready to pour the concrete foundations – but the confines of the site are such that the concrete has to be transported in skips by crane. However, the crane is rendered inoperable when, in a 24-hour working hour period over 3 days (i.e. 8 hours per day), there are wind speeds exceeding 60 km/hour.

In the above circumstances, the Contractor would not be entitled to compensation because each of the adverse weather conditions he experiences falls narrowly within the threshold in the *weather data* albeit that the effect of their acting in succession causes grave disruption to the Contractor. Were the subjective test of ... *exceptionally adverse weather conditions* ... instead to have applied then surely the Contractor would be entitled to compensation.

Note: Although the extent of rainfall will not qualify as a compensation event under clause 60.1(13), it is possible that its *consequences* may instead qualify under clause 60.1(12) and this is for the reasons given in para. 12.4.37.

Example 12.10 The weather conditions are as in Example 12.9 but later in the month further adverse weather conditions are encountered

Following the conditions encountered in Example 12.9 which have occurred from, say, the beginning of the month, there is a 4-day period at the end of the month in which there is rainfall of 18 mm (falling evenly over those days). In that same 4-day period, 2 days have temperatures which do not rise above zero degrees Celsius and there are 3 working hours where wind speeds exceed 60 km/hour.

At this juncture, all of the weather measurements exceed that which was to be anticipated from the *weather data*, thus:

(i) *Rainfall* – The Contractor took on the risk of rainfall up to 75 mm with no more than 5 days producing 5 mm. In the event, the rainfall was 78 mm (i.e. 60 mm as per Example 12.9 plus 18 mm as per this Example 12.10) and the Contractor is therefore entitled to be compensated for coping with the additional 3 mm of rainfall. This alone is enough to carry the Contractor's right to compensation even though there were only 4 days when rainfall exceeded 5 mm (i.e. 60 mm in 4 days as per Example 12.9 is an average of 15 mm per day whilst 18 mm in 4 days as per this Example 12.10 is an average of only $4\frac{1}{2}$ mm per day).

(ii) *Temperature* – The Contractor took on the risk of there being 5 days when the temperature would not rise above zero degrees Celsius. In the event, there were 6 such days (i.e. 4 days as per Example 12.9 plus 2 days as per this Example 12.10) and the Contractor is therefore entitled to be compensated for coping with the additional day of freezing temperatures.

(iii) *Wind speed* – The Contractor took on the risk of there being 25 working hours when the wind speed would exceed 60 km/hour. In the event, there were 27 such hours (i.e. 24 hours as per Example 12.9 plus 3 hours as per this Example 12.10) and the Contractor is therefore entitled to be compensated for coping with the additional 2 hours of high wind speeds.

As may be seen from the calculations in (i)–(iii) above, although the Contractor is now in the position of being entitled to compensation, such entitlement only begins to run once the thresholds in the *weather data* have been exceeded. Accordingly, the Contractor will have had to absorb the effects of what has been described in Example 12.9 thereby confining his compensation to the weather conditions experienced at the end of the month. It may prove that the level of this compensation is not that great if, during the reasonable weather in the middle of the month, the Contractor were able to complete the weather sensitive laying of foundations.

Clause 60.1(14) An Employer's risk event occurs

12.4.46 The Guidance Notes to NEC/2 confirm that the Employer's risk events referred to in clause 60.1(14) are those set out at clause 80.1. Unlike the majority of standard forms of contract, NEC/2 endeavours to prescribe and define the risks (in a commercial, legal and general sense) which the Employer is to carry. The Contractor's risks, by contrast, are simply described in clause 81.1 as ... *the risks which are not carried by the Employer* ... Should any of the Employer's risk events listed at clause 80.1 occur then this will constitute a compensation event.

12.4.47 The risks in question can be summarised as follows:

- Those general risks which arise unavoidably out of the fact that the *works* are taking place on the site, which are the result of negligence or breach of duty on the part of the Employer or those for whom he is legally responsible, or which arise by reason of some other fault on the part of the Employer or in his design (or in designs prepared by his retained consultants). Such risks, which may include damage to the *works* themselves, to property other than the *works* and personal injury to third parties, would normally be covered by the Employer's public liability insurance or (in the case of faulty design) by professional indemnity insurance held by the Employer's designers.
- The risk of damage to Plant and Materials (defined at clause 11.2(10) as ... *items intended to be included in the works* ...) which are supplied to the Contractor either by the Employer or by Others (clause 11.2(2) refers) on the Employer's behalf. The risk of such damage becomes a Contractor's risk once he has received and accepted such Plant and materials.
- The risk of loss or damage to the *works* or to Plant and Materials caused by such things as war and rebellion, riot and civil commotion, or radioactive contamination. These are described by the Guidance Notes as being ... *caused by outside influences beyond the control of the Parties* ... They are generally uninsurable risks which almost invariably are left for the Employer to carry. Interestingly, most other standard forms of contract tend also to include damage caused by sonic boom within this category. NEC/2, however, leaves this as a Contractor's risk despite the fact that it is generally not possible for him to insure against it.
- Risks arising once the Employer has taken over part of the *works* (as defined at clause 35). These risks arise out of what the Guidance Notes call ... *the consequences of normal ownership* ... The only exception to this rule is loss or damage consequent upon either a Defect or a Contractor's risk event which existed at the time of take over or, alternatively, loss or damage caused by the Contractor following take over. These limited risks remain with the Contractor until the issue of the Defects Certificate.
- Risk of loss or damage to the *works* or to any Equipment (clause 11.2(1) refers) or Plant and Materials (clause 11.2(10) refers) which are retained on site by the Employer following termination in accordance with clauses 94 to 97. Items which may have been provided by the Contractor, therefore, become the Employer's risk should they be retained on site after termination. Again, the only exception to this is loss or damage caused by the Contractor himself following termination.

- The Employer is free to state within Part 1 of the Contract Data any additional risks which he is prepared to assume. One such example may be the risk of loss or damage to any existing buildings within which the *works* are taking place and any associated contents. These are traditionally risks which remain with the Employer.

12.4.48 The risks defined above give rise to several points of particular note for those practitioners more familiar with other standard forms of contract such as JCT or GC/Works. A number of those risks which are carried by the Employer under NEC/2 do not necessarily arise out of any default on the part of the Employer. Indeed, they may well be completely outside the control of the Employer. Examples of such events may be incidents of theft or vandalism to Plant and Equipment provided by the Employer under the second bullet point above, or damage caused during rioting under the third bullet point above. The approach adopted by more familiar contract forms in such instances is to entitle the Contractor to additional time (such that he does not become exposed to liquidated and ascertained damages) but not to reimbursement of any loss or damage which he suffers as a result of the event. Only in the case of matters within the control of the Employer, such as those in the List of Matters contained in clause 26.2 of JCT 98 PWQ, does the Contractor become entitled to such additional reimbursement. NEC/2, by contrast, raises the possibility of the Contractor becoming entitled to additional payment in respect of matters which are genuinely outside the control of the Employer but which are nevertheless at the Employer's risk. In this respect, the Employer is placed in a less favourable position than that which pertains under JCT or GC/Works forms.

12.4.49 In addition, and again unlike the situation under more familiar contract forms, both the fourth and fifth bullet points above give rise to a limited number of instances in which the Contractor may become entitled to additional payment even after Completion, take over or termination of the *works*. This in turn must lead to an uncomfortable realisation on the part of the Employer that, notwithstanding the fact that the *works* have been completed, he may nevertheless continue (albeit in a limited number of respects) to incur additional financial liabilities towards the Contractor. Under JCT 98 PWQ, by contrast, the Contractor's entitlement to reimbursement of direct loss and/or expense under clause 26 is restricted to that arising out of events which affect the progress of the Works prior to Practical Completion.

12.4.50 One final point to be made in respect of clause 60.1(14) concerns its interaction with clause 83.1. This latter provision requires the Employer to indemnify the Contractor in respect of ... *claims, proceedings, compensation and costs* ... arising from an event which is at the Employer's risk. This covers precisely the same set of circumstances as clause 60.1(14) but arguably provides a wider financial right of recovery for the Contractor in that he will be entitled to the reimbursement of all his costs which reasonably flow from the event in question, including third-party claims made against him and the associated costs of such claims. As is explained more fully at para. 12.7.23, clause 63.1 restricts the Contractor's financial right of recovery to those costs (which may even be forecasts) as listed in the Schedule of Cost Components, together with the appropriate Fee. It is not unlikely that assessments produced using this approach will differ somewhat from those produced by the indemnity approach of clause 83.1. Assessments under clause 63.1 will certainly not provide for the Contractor to recover any losses flowing from third-party proceedings in which he becomes involved. Assessments under clause 83.1, on the other hand, cannot lead to changes to the Completion Date. It would therefore appear that, although there are certain remedies exclusive to each of clauses 63.1 and 83.1, there is also a degree of overlap which needs to be borne in mind when making assessments under either or both systems.

Clause 60.1(15) The Project Manager certifies take over of a part of the works before both Completion and the Completion Date

2.4.51 Clause 35 of NEC/2 deals with take over of the *works* by the Employer. This concept, which has no counterpart in the JCT family of contracts, is discussed and explained in greater detail at para. 12.2.8. Essentially, however, take over marks the point at which the risk and responsibility attaching to the *works* (or the relevant part) passes back from the Contractor to the Employer. Clause 35.3 stipulates that, with a limited number of exceptions, take over occurs by default should the Employer begin to use the *works* (or any part of it) before Completion has been certified. Clause 35.4 requires the Project Manager to certify take over within a further week of the Employer having commenced using parts of the *works*. Such certification by the Project Manager constitutes a compensation event under clause 60.1(15) should it occur both before Completion **and** before the Completion Date. Should take over occur before Completion but **after** the Completion Date (i.e. during a period of culpable delay on the part of the Contractor) then it would appear that the Employer is permitted to use parts of the *works* without giving rise to a compensation event. Furthermore, he will be permitted to use parts of the building whilst simultaneously (where secondary option R has been selected) levying the full amount of delay damages in respect of the whole of the *works* against the Contractor. In such situations, where parts of the *works* have been taken over by the Employer, NEC/2 contains no facility for proportionally reducing the delay damages as is the case with partial possession under JCT 98 PWQ. The Guidance Notes do recommend that, where parts of the *works* are taken over by the Employer, ... *any delay damages should be reduced to represent the proportion of the cost of delay which is obviated by the release of that part of the works* ... However, as pointed out in the Guidance Notes themselves, they have no contractual effect and ... *should not be used for legal interpretation of the meaning of* ... NEC/2. The Employer appears therefore, at the risk of ignoring his fair-dealing obligations under clause 10.1, to be at liberty to overlook this recommendation.

2.4.52 It is important to note that the compensation event occurs only when the Project Manager certifies take over and not when the Employer begins to use the relevant part of the *works*. The significance of this is that if the Employer has begun to use parts of the *works*, either for a reason stated in the Works Information or in order to suit the Contractor's work methods, then take over is not deemed to have occurred and no compensation event can arise (clause 35.3 refers). It is obviously important that, should the Employer be aware of any intended early use on his part of areas of the building (e.g. for the purpose of storage of furniture or equipment), he is advised to make this clear in the Works Information. Should he do this then any such early use will not constitute a compensation event and the Contractor will be deemed already to have accommodated its effects in his Prices and in his planned date for Completion. The remaining exception noted in clause 35.3 simply ensures, not unreasonably, that the Contractor does not become entitled to additional time and money where the Employer agrees to early take over of parts of the *works* simply to suit the Contractor's work methods.

Clause 60.1(16) The Employer does not provide materials, facilities and samples for tests as stated in the Works Information

12.4.53 Clause 40.2 requires both the Contractor and the Employer to provide ... *materials, facilities and samples for tests and inspections as stated in the Works Information ...* This obligation covers only those tests and inspections which are specifically required by the Works Information or by law. It does not apply to tests or inspections consequent upon searches instructed by the Supervisor in accordance with clause 42.1. Examples of the type of materials, facilities or samples which come within the purview of clause 40.2 are given in the Guidance Notes and may include testing apparatus and instruments, laboratory facilities, fuel, etc. Should the Employer fail to provide the necessary facilities, etc. as stated in the Works Information then such failure constitutes a compensation event. Should the Employer provide the necessary facilities, etc. but not at the times shown on the Accepted Programme, then a compensation event would also have arisen but under clause 60.1(3) rather than clause 60.1(16). Should the Contractor fail to provide those facilities, etc. required of him by the Works information then, of course, a compensation event does not arise.

Clause 60.1(17) The Project Manager notifies a correction to an assumption about the nature of a compensation event

12.4.54 As explained in more detail at paras 12.7.9 and 12.7.10, should the Project Manager decide that the effects of a compensation event are too uncertain to be forecast then clause 61.6 permits him to state certain assumptions about the event in his instruction to the Contractor to submit a quotation. Should any of those assumptions later be discovered to have been incorrect, the Project Manager is required to notify a correction and this in itself will constitute a compensation event. The assessment of this later event will essentially comprise an adjustment to the earlier, and provisional, event. Clause 60.1(17) is necessary because clause 65.2 otherwise provides that ... *the assessment of a compensation event is not revised if a forecast upon which it is based is shown by later recorded information to be wrong* ... Clauses 60.1(17) and 61.6 provide the only opportunity for the subsequent revisiting of earlier assessment of compensation events. The Guidance Notes describe the intention behind this as being to emphasise ... *the finality of the assessment of compensation events* ... and to encourage the ongoing resolution of such issues during – rather than long after – the progress of the project.

Clause 60.1(18) A breach of contract by the Employer which is not one of the other compensation events in this contract

12.4.55 This is described in the Guidance Notes to NEC/2 as ... *an 'umbrella' clause* ... Its intention is to bring within the scope of the compensation event system any breaches of contract by the Employer which may involve the Contractor in additional time or cost but which receive no mention elsewhere in clause 60. This particular compensation event was introduced only in NEC/2 and was not present in NEC/1. It was introduced in order to avoid the possibility of time becoming "at large" were the Employer to do something (or omit to do something) which affected the Contractor's progress but in respect of which there was no contractual mechanism for compensating the Contractor. The ability to extend the period for completion of the works in response to the occurrence of certain events is intended primarily to preserve the Employer's right to liquidated and ascertained damages should the Contractor be in culpable delay. In the absence of any such mechanism, time would become "at large" should the Employer do anything to hamper the Contractor's progress. In other words, the Contractor would not be bound to complete by a certain date but would instead have a reasonable time in which to complete the works. Should he fail to complete within a reasonable time the Employer's only remedy would be to go to law to seek to prove general (i.e. unliquidated) damages. It was to avoid this eventuality that clause 60.1(18) was introduced, by bringing all possible breaches on the part of the Employer within the scope of the compensation event system.

12.4.56 Another point of note concerning clause 60.1(18) is its likely effect on the common law rights of the Parties, particularly those of the Contractor. The system of compensation events in NEC/2, like its counterparts in other forms of contract, is intended to provide a mechanism within the contract by which the Contractor can be recompensed in the event of defaults by the Employer. The intention is to remove the need for the Contractor to go to law in order to obtain such recompense. His entitlement will be calculated by reference to the rules contained in the contract. In the case of clause 26 of JCT 98 PWQ, these rules are intended to produce a measure of the Contractor's entitlement which is based upon similar principles to those employed by the Courts in assessing general damages for breach of contract, i.e. the reasonable losses which flow directly from the breach. JCT 98 PWQ, however, is careful to note (clause 26.6 thereof refers) that the mechanism set out in clause 26 is without prejudice to any rights or remedies the Contractor may possess. In other words, the Contractor is not precluded from going to law should he consider that in a particular instance the contractual mechanism affords an unacceptable remedy. There is no such express reservation of the Parties' common law rights in NEC/2. Indeed, it is arguable that clause 60.1(18) of NEC/2 endeavours instead to bring the Contractor's rights in respect of all breaches by the Employer within the scope of the compensation-event system. In some instances it may well be that this provides the Contractor with a less favourable remedy, as the assessment rules for compensation events set out in clause 63.1 will generally act to preclude any right of recovery of consequential losses. However, following the decision of the Court of Appeal in *Strachan & Henshaw Ltd -v- Stein Industrie (UK) Ltd and Another (1998) 87 BLR 52*, it appears unlikely that the Parties' common law rights will be held to have been excluded in the absence of an express and clearly worded term to that effect. There is no such term within NEC/2.

Clause 60.7 Suspension of performance

12.4.57 Addendum Y(UK)2 was introduced in April 1998 in an effort to ensure that NEC/2 was compliant with the Housing Grants, Construction and Regeneration Act 1996. Among other things Y(UK)2 made a number of changes to the payment provisions of NEC/2 in order to comply with the timetable for payment set down in Section 110 of the Housing Grants, Construction and Regeneration Act 1996. Clause 50.1 of NEC/2 provides for assessment dates at the intervals stated in the Contract Data. Clause 56.1 provides that payment becomes due 7 days after the assessment date, and the Project Manager must issue a payment certificate during that 7-day period. The final date for payment is either 21 days after the date on which payment becomes due or such alternative period as is stated in the Contract Data. Should the Employer intend to withhold any or all of the payment due then clause 56.2 provides that he must notify the Contractor to that effect in writing at least 7 days before the final date for payment. Clause 13.2 makes clear that this notice must actually be received by the Contractor by this date in order to be effective. The notice is required to state the amount to be withheld and the grounds for doing so.

12.4.58 Should no notice be given but the Employer nevertheless withholds payment after the final date for payment then Section 112 of the Housing Grants, Construction and Regeneration Act 1996 gives the Contractor a right to suspend his performance of the contract until payment is made. Unlike many contract forms, NEC/2 makes no express provision for this right of suspension. It is therefore left to be implied in terms as set out within the statute. Clause 60.7, also introduced by way of Addendum Y(UK)2, provides that any such suspension of work by the Contractor will qualify as a compensation event. This will only be the case, of course, where the Contractor exercises his right properly and in accordance with the statute. Section 112(1) of the statute, for instance, requires the Contractor to give at least 7 days' notice of his intention to suspend together with his reasons for so doing, whilst Section 112(3) confirms that the right of suspension ends upon proper payment being made. Interestingly, as these provisions are implied into the contract rather than being express terms, it is arguable whether the notice required by Section 112(2) is subject to the requirement in clause 13.1 of NEC/2 that all such communications be in writing. This requirement applies only to communications ... *which* [the] *contract requires* ... Arguably, this particular notice is not required by the contract but, instead, is required by statute. Section 112(4) of the statute already provides for the Completion Date to be changed to reflect the period of suspension. By making such suspension a compensation event, NEC/2 goes further than the statute in also allowing the Contractor to recover any additional costs brought about as a result.

12.5 COMPENSATION EVENTS – MAIN OPTION CLAUSES

Options A and C: priced contract and target contract with activity schedule

12.5.1 There are no additional compensation events set out in these options.

Options B and D: priced contract and target contract with bill of quantities

12.5.2 Both main options B and D assume that the Prices are based upon *bills of quantities* which will have been priced by the Contractor. Only under these main options is the use of *bills of quantities* anticipated. Accordingly, three additional compensation events which relate specifically to the *bills of quantities* apply only to main options B and D. These additional events are set out at clauses 60.4, 60.5 and 60.6 and can be summarised as follows:

- a difference between the final total quantity for an item and the quantity in the *bill of quantities*, which difference affects the unit cost of the item (clause 60.4 refers);
- a difference between the final total quantity for an item and the quantity in the *bill of quantities*, which difference delays Completion (clause 60.5 refers); and
- a correction by the Project Manager of mistakes in the *bill of quantities* (clause 60.6 refers).

12.5.3 Despite the fact that the payment provisions of main options B and D operate quite differently, the wording of the above clauses as applied to both options is identical. The practical effect of this may surprise some users of NEC/2 and is explained in more detail at paras 12.7.49 to 12.7.63. Another notable factor is that clauses 60.4 and 60.6 are among the few compensation events which can give rise to a reduction in the Prices. This issue is addressed more fully at para. 12.5.10. Each of clauses 60.4, 60.5 and 60.6 is now discussed in turn.

Clause 60.4 A difference between the final total quantity of work done and the quantity stated for an item in the *bill of quantities* at the Contract Date is a compensation event if
- **the difference causes the Actual Cost per unit of quantity to change and**
- **the rate in the bill of quantities for the item at the Contract Date multiplied by the final total quantity of work done is more than 0.1% of the total of the Prices at the Contract Date.**
If the Actual Cost per unit of quantity is reduced, the affected rate is reduced

12.5.4 Clause 60.4 provides, in a limited number of circumstances, that changes in quantity from those stated in the *bill of quantities* may constitute a compensation event. For this to be the case, two tests must be satisfied. The first of these is that the change in quantity must have a measurable effect on the Actual Cost per unit of the item. The fact that the Actual Cost may be different from the rate in the *bill of quantities* will not suffice to trigger a compensation event if the Actual Cost itself has not been affected by the change in quantity (the subject of assessing compensation events by reference to their effect on the Actual Cost is discussed in more detail at paras 12.7.20 to 12.7.34). The second test which must be satisfied is essentially one of magnitude. Changes in quantity will only rank as compensation events if their financial effect is sufficiently significant. The original tender rate, when multiplied by the final total quantity, must comprise at least 0.1% of the Prices at the Contract Date.

12.5.5 This provision is best illustrated by way of an example. A project has Prices totalling £500,000 at the Contract Date (i.e. the JCT 98 PWQ equivalent of the original Contract Sum). 0.1% of the Prices therefore amounts to only £500. Any item in the *bill of quantities* which amounts in the final analysis to more than £500 may therefore give rise to a compensation event, even where the change in quantity concerned is relatively minor. Of course, the requirement that **both** parameters must be satisfied in order to trigger a compensation event ought in theory to limit the number of occasions on which this occurs.

12.5.6 Notwithstanding this, however, the test is unlikely to be as restrictive as its equivalent in JCT 98 PWQ. The JCT form envisages changes in quantity in only two circumstances – either consequent upon a Variation or, alternatively, where an Approximate Quantity included in the Bills of Quantities subsequently changes on remeasurement. In the first of these situations clause 13.5.1.2 of JCT 98 PWQ requires that the change in quantity must be "significant" in order to trigger a change in the rate whilst, in the second, clause 13.5.1.5 of JCT 98 PWQ gives rise to a change in the rate only where the originally included Approximate Quantity is "not a reasonably accurate forecast" of the final quantity. As can be seen from the example given above, NEC/2 does not necessarily require the change in quantity to be significant in order to trigger a compensation event. In addition, both of the tests in JCT 98 involve a measure of subjectivity which the more strictly defined parameters in NEC/2 endeavour to avoid. The application of subjective judgment under JCT 98 PWQ is likely in practice to yield fewer instances of re-rating than the application of the objective tests under NEC/2 – particularly in view of the low threshold set down in clause 60.4 of NEC/2. Similarly, whilst the cost effect of changes in quantity are again reflected subjectively in a "fair allowance" under JCT 98 PWQ they are intended to be assessed objectively under NEC/2 by reference to changes in the Actual Cost (albeit that, as explained in paras 12.7.19 to 12.7.29, the approach of NEC/2 is in fact somewhat less objective in practice than in theory).

12.5.7 Clause 61.3 (which is discussed in more detail at para. 12.7.4) requires the Contractor to notify the Project Manager within 2 weeks of becoming aware of something that he **believes** may be a compensation event. In the case of clause 60.4, until it is known whether the alleged change in quantity satisfies the parameters set out in that clause then it will not be clear as to whether or not a compensation event has indeed occurred. If the Project Manager decides that the parameters in clause 60.4 are not likely to be satisfied then, in accordance with clause 61.4, the Prices will not be adjusted on the grounds that the event notified by the Contractor . . . *is not one of the compensation events stated in this contract* . . . Alternatively, the Project Manager may instruct the Contractor to submit a quotation based upon the stated assumption (clause 61.6 refers) that the criteria in clause 60.4 will be met. Should the criteria not subsequently be met then the Project Manager notifies a correction to his earlier assumption. This in itself will give rise to a further compensation event under clause 60.1(17) which will operate to cancel out the effect of the earlier, and incorrect, compensation event.

Clause 60.5 A difference between the final total quantity of work done and the quantity for an item stated in the *bill of quantities* at the Contract Date which delays Completion is a compensation event

12.5.8 Whilst clause 60.4 deals with situations whereby a change in quantity affects the Actual Cost of an item, clause 60.5 deals with the situation whereby the effect of a change in quantity is to delay Completion. It will again generally be the case that this compensation event falls to be notified by the Contractor in accordance with clause 61.3 rather than by the Project Manager. Having received notification it may well be prudent, should there be any uncertainty as to what the final quantity is likely to be, for the Project Manager initially to instruct the Contractor to submit a quotation on the basis of a stated assumption as to the likely final quantity. Should the final measured quantity be different to that originally assumed, the Project Manager can notify a correction in accordance with clause 61.6 and a compensating adjustment can then be made. In such situations the Actual Cost of the item concerned, as well as the Contractor's progress, may be affected should additional time-related costs be incurred as a result of the additional work. Such additional costs would form part of the assessment of the compensation event and the Prices, as well as the Completion Date, would be adjusted accordingly.

12.5.9 The requirements of clauses 31 and 32 concerning the Contractor's programme ought to facilitate the task of identifying and assessing such compensation events. The operations affected by the change in quantity ought to be readily identifiable on the Accepted Programme, and the requirement of clause 31.2 that each operation have a resourced method statement ought also to assist in assessing the effect of the change on the Actual Cost. In accordance with clause 62.2, the Contractor's quotation is to include a revised programme demonstrating the effect of the change in quantity on the remaining work. This ought to assist the Project Manager in giving proper considera-tion to the Contractor's quotation. It is worthy of note in respect of clause 60.5 that **all** changes of quantity, of whatever magnitude, fall to be considered provided their effect is to delay Completion. Having said this, it is unlikely that negligible changes in quantity will have an effect on Completion and therefore – by default – it is likely that only changes in quantity of a reasonable magnitude will fall to be considered under clause 60.5. An obvious exception to this, of course, is where a small but unavoidable change in quantity causes delivery problems for the Contractor leading to extended procurement periods or to premium payments.

Clause 60.6 The Project Manager corrects mistakes in the bill of quantities which are departures from the method of measurement or are due to ambiguities or inconsistencies. Each such correction is a compensation event which may lead to reduced Prices

12.5.10 Clause 60.6 permits the Project Manager to correct mistakes in the *bill of quantities*. Such corrections will entitle the Contractor to additional payment should their effect be to increase his Actual Cost, and to additional time should they affect his planned completion. However, clause 60.6 is also one of the few compensation events which can lead to a reduction in the Prices. This is a sensible measure as it may well be that an item has been erroneously included in the *bill of quantities* and an anomalous result would be produced were the Contractor to remain entitled to be paid in respect of such an item notwithstanding that he was not required to carry out the work required by it. Other than in respect of an acceleration agreement pursuant to clause 36, no such correction of a mistake under clause 60.6 may give rise to an earlier Completion Date than that stated in the Contract Data.

12.5.11 The mistakes which fall to be considered under clause 60.6 are restricted to those ... *which are departures from the method of measurement or are due to ambiguities or inconsistencies* ... No guidance is given as to how the phrase "ambiguities or inconsistencies" ought to be interpreted and it therefore falls to be given its literal meaning in practice. "Ambiguity" is defined in *The Shorter Oxford English Dictionary* as of ... *double or dubious meaning* ... whilst "inconsistency" is defined as ... *want of agreement between two things or parts of a thing* ... It would appear, therefore, that this element of the clause serves as a form of modified contra proferentem rule relative specifically to *bills of quantities* – in that anything which is unclear or of dubious meaning may be construed in a manner which can lead to a reduction in the Prices and thereby benefit the Employer. In such situations clause 60.6 would appear to be inconsistent with clause 63.7, which states that instructions ... *to change the Works Information* [provided by the Employer] *in order to resolve an ambiguity or inconsistency* ... will be assessed ... *as if the Prices and the Completion Date were for the interpretation most favourable to the Contractor* ... Having said this, clause 55.1 confirms that ... *information in the bill of quantities is not Works Information* ... It would therefore appear that there are two potentially contradictory rules relating to ambiguities and inconsistencies – one relating to the Works Information and one (potentially more favourable to the Employer) relating to the *bill of quantities*.

Option E: cost-reimbursable contract

12.5.12 There are no additional compensation events set out in this option.

Option F: management contract

12.5.13 There are no additional compensation events set out in this option.

12.6 COMPENSATION EVENTS – SECONDARY OPTION CLAUSES

Option J: advanced payment to the contractor

Clause J1.2 ... Delay in making the advanced payment is a compensation event

12.6.1 Clause J1.2 requires the Employer to provide the advanced payment within a specified time period. Where no payment bond is required from the Contractor, the Employer must forward the advanced payment within 4 weeks of the Contract Date. Where a payment bond is required from the Contractor and the Contractor is dilatory in providing it in an acceptable form, the Employer has 4 weeks following receipt of an acceptable bond within which to forward the advanced payment. Whilst late payment on the Employer's part would ordinarily entitle the Contractor to recover interest in accordance with clause 51.2, the Guidance Notes to NEC/2 point out that ... *if the Employer is late in making the advanced payment the financial consequences for the Contractor may be significant* ... Such consequences may not be adequately catered for simply by a right to recover interest. For instance, the price agreed by the

Contractor with a supplier (and on which the Contractor's Prices are based) may be contingent upon the supplier receiving payment by a particular date. Should this not happen the Contractor may be required to pay a premium which will not be catered for in his own Prices. Similarly, delay in payment may lead to the loss of a production slot and potentially serious delays. Failure on the part of the Employer to make the advanced payment within the period specified in clause J1.2 has therefore been made a compensation event.

Option T: changes in the law

Clause T1.1 ... A change in the law of the country in which the Site is located is a compensation event if it occurs after the Contract Date. The Project Manager may notify the Contractor of a compensation event for a change in the law and instruct him to submit quotations. If the effect of a compensation event which is a change in the law is to reduce the total Actual Cost, the Prices are reduced

12.6.2 Generally speaking, this option serves to place the risk of any changes in the law of the country in which the Site is situated onto the Employer, provided the change takes place after the Contract Date. This said, if the effect of such a change is to effect a reduction in the Prices then the Employer is permitted to take the benefit of this. The Guidance Notes to NEC/2 point out that it is only those changes which affect the Contractor's own costs and/or progress which will qualify as compensation events. Therefore changes in the level of income tax payable by the Contractor's employees would not qualify as a compensation event, whereas changes in the Contractor's National Insurance contributions payable by him on behalf of his employees would qualify as a compensation event. The Guidance Notes further explain that ... *for the purposes of this clause, law would include a national or state statute, Ordinance, decree, regulation (including building or safety regulations) and a by-law of a local or other duly constituted authority or other delegated legislation* ... The clause itself states that the Project Manager **may** notify the Contractor of a change in the law and instruct the Contractor to submit a quotation. However, the Guidance Notes make clear that the Contractor is also free to provide such notification, in accordance with clause 61.3, should the Project Manager himself be unaware that a change has occurred which may qualify as a compensation event.

Option U: The CDM Regulations 1994 (rules for contracts in the UK)

Clause U1.1 ... A delay to the work or additional or changed work caused by application of The Construction (Design and Management) Regulations 1994 is a compensation event if an experienced contractor could not reasonably be expected to have foreseen it

12.6.3 The CDM Regulations apply to most construction work undertaken in the UK and the Guidance Notes to NEC/2 recommend that option U is used wherever the CDM Regulations apply. The various duties and powers arising out of the application of the Regulations are not set out in the contract itself, albeit that the provisions of the

Regulations are nevertheless binding upon the parties. Should the application of those provisions cause a delay to the works or, alternatively, give rise to varied or additional work then this may – in certain circumstances – give rise to a compensation event. This happens in circumstances where an experienced contractor could not reasonably have been expected to have foreseen the effect which the application of the Regulations has produced. The Guidance Notes explain that ... *the financial risk is shared between the Employer and the Contractor, the dividing line being based on foreseeability at the time of tender* ...

12.6.4 It is worthy of note that this is not the same type of test as that set out in clause 60.1(12) in respect of physical conditions encountered on the Site (discussed at paras 12.4.38 to 12.4.42). The test in clause 60.1(12) incorporates an element of commercial judgment in that an experienced contractor may have foreseen certain conditions as a possibility but, having considered that there was a small chance of their actually occurring, it was reasonable for him to have made no allowance in his tender. Should the conditions be encountered, a compensation event will have occurred notwithstanding that the Contractor foresaw the event and opted to disregard it. The test in clause U1.1 is a simpler test and does not rely on any element of commercial judgment. If an experienced Contractor could not have foreseen the matter then its occurrence will qualify as a compensation event. The reverse position is that, if an experienced Contractor could have foreseen the matter, its occurrence will not rank as a compensation event.

12.7 PROCEDURES ATTACHING TO COMPENSATION EVENTS

General

12.7.1 Clauses 61 to 65 set out the procedures to be followed by the parties in connection with the occurrence of a compensation event. These procedures can be summarised as follows:

- notification;
- quotation and assessment; and
- implementation.

12.7.2 The contract sets down a challenging timetable (too challenging in the opinion of some commentators) within which all of the above stages are to be completed. The intention of this timetable is, as explained in the Guidance Notes, ... *to promote efficient management of the contract procedures* ... and to ensure, as far as possible, that matters are resolved on an ongoing basis during the currency of the contract rather than remaining as a source of contention until long after its completion. These and other management procedures in NEC/2 can indeed shorten considerably the period within which the final account can be resolved – sometimes within a matter of days according to one supporter. Others, however, have expressed the opinion that the system set out in clauses 61 to 65 is based on the premiss that compensation events will be relatively rare occurrences and that large numbers of them on a regular basis will put a serious strain on resources and lead to a breakdown of the system.

Anecdotal evidence shows that there is also some merit in this view, with some users experiencing a backlog of compensation events and an enforced relaxation of the contract timetable. Again, it is particularly important that both Contractor and Project Manager are properly resourced to deal with a flow of compensation events that can run into hundreds – even on the ostensibly more straightforward of schemes.

Notification

12.7.3 If the compensation event is an instruction or a changed decision from either the Project Manager or the Supervisor, the Project Manager is required by clause 61.1 to notify the Contractor at the time of the event. According to the Guidance Notes, this applies in the case of compensation events (1), (4), (7), (8), (10), (15) and (17) as listed in clause 60.1. Further, consequent upon clause 13.7, the required notification is to be given separately from the instruction or changed decision which prompted it. It is easy to see from this the paper-generating potential of these procedures, particularly in the event that a large number of instructions are given. At the same time that the Project Manager gives notification he is also required to request a quotation for the compensation event from the Contractor, unless a quotation has already been submitted or unless the event has been caused by the default of the Contractor (in which case no adjustment is made either to the Prices or to the Completion Date). The Contractor must begin complying with the instruction or changed decision which forms the subject matter of the compensation event as and when it is received by him. Clause 61.2 also enables the Project Manager to request quotations in respect of **proposed** instructions or changed decisions, the purpose presumably being to permit him to assess the effect on time and cost of his proposal before deciding whether to proceed with it. In such situations, of course, the Contractor refrains from complying with the instruction until it has been confirmed by the Project Manager.

12.7.4 Should the Project Manager fail to notify the Contractor as required above, this may well constitute a breach by the Employer and itself be a compensation event under clause 60.1(18). It is important, therefore, that the Project Manager take care to provide the required notification each time an instruction is issued either by him or (which may be more of an administrative problem for him) by the Supervisor. In any event, if the Contractor considers that a compensation event has occurred (whether one of those covered by clause 61.1 or otherwise) and no notification has been received to that effect from the Project Manager, then clause 61.3 entitles the Contractor himself to notify the Project Manager. The further qualification is provided that the Contractor's notification must be submitted within 2 weeks of his having become aware of the event. It is worthy of note that this is a subjective rather than an objective test. There is no question here of imposing upon the Contractor a date when he **ought** to have become aware of the event. Whilst this may come as a relief to contractors generally, it does potentially have the unfortunate side effect of leaving the door open for the raising of issues long after the event on the grounds simply that they have been hitherto overlooked by all concerned. This is an eventuality which the entire compensation-event system is otherwise designed to avoid.

12.7.5 Clause 61.4 sets out how the Project Manager is to respond should the Contractor notify an event in accordance with clause 61.3. There will be no adjustment either of the Prices or of the Completion Date if the Project Manager decides that the matter raised by the Contractor fails one of the following tests:

- the event has been caused by some default of the Contractor himself;
- the event either has not happened or is not expected to happen;
- there will be no effect on Actual Cost or on Completion; or
- the event is not a compensation event as defined in clause 60.1 of the contract.

12.7.6 If the matter passes these tests then the Project Manager instructs the Contractor to submit a quotation for the compensation event. Either way, and unless some longer period is agreed with the Contractor, the Project Manager is to arrive at his decision within 1 week of the Contractor's notification. The decision is at the complete discretion of the Project Manager, subject only to the constraint in clause 10.1 to act in a spirit of mutual trust and in accordance with any implied common law duty to act fairly.

12.7.7 Should no extension to the 1-week period for making a decision be sought or agreed, the resolution of what may be complex matters unilaterally by the Project Manager and in such a short timescale may seem rather arbitrary when seen from the Contractor's viewpoint – particularly if subsequent events prove the Project Manager's decision to have been wrong. In this regard, and should the Contractor wait until he is proved right, he may well find himself out of time for referring the matter to the Adjudicator. As a prerequisite to referral to the Adjudicator, clause 90.2 requires the Contractor first to notify his dissatisfaction to the Project Manager within 4 weeks of the offending decision. In such circumstances, particularly where the matter is one of any complexity, it is recommended that the Contractor agree to an extension of the 1-week period within which the Project Manager must arrive at his decision. The contract appears to envisage that the Project Manager will ordinarily take the initiative in suggesting such an extension, albeit there is nothing to prevent the Contractor from doing so, and to notify the Project Manager immediately of his dissatisfaction should he not agree with the eventual decision. Should the Project Manager at any point subsequently change his decision this in itself will constitute a compensation event under clause 60.1(8). All the more reason for the Project Manager to give careful consideration to the Contractor's notification under clause 61.3 and to suggest to the Contractor an extension of the 1-week period if appropriate.

12.7.8 Clause 16.1 requires both the Contractor and the Project Manager to give early warning to the other, immediately upon becoming aware, of any matter which could affect the use of the completed project or lead to an increase in the Prices or a delay to Completion. The remainder of clause 16 then sets out how the parties are required to act in order to neutralise as far as possible the effect of the anticipated problem. Compensation events will fall within the scope of the matters anticipated by clause 16, and clauses 61.5 and 63.4 set out the Project Manager's powers should he consider that an event has been notified of which he considers that the Contractor ought to have given early warning but did not. Clause 63.4 (the Project Manager's assessment of the compensation event) is discussed in detail at paras 12.7.45 to 12.7.48. In order to exercise the powers contained in clause 63.4, however, the Project Manager must first give the notification required by clause 61.5. Should the Project Manager decide, again in his complete discretion, that an experienced contractor could have given early warning of the matter concerned (and one can imagine some interesting arguments on this subject) he is required at this stage to do no more than notify the Contractor to that effect. No reasons for his decision need be given.

12.7.9 Of course, it is not always possible to forecast accurately the effects of a compensation event particularly if its precise extent is uncertain (which can often be the case in works of alteration and repair) or if it is not expected to happen for some time. NEC/2 recognises this in clause 61.6 by permitting the Project Manager (but not the Contractor) to decide whether the effects of an event are too uncertain to be reasonably forecast. If the Project Manager does so decide he can then state certain assumptions about the event in his instruction to the Contractor to submit a quotation. The subsequent assessment of the event will be based on these assumptions. Should any of these assumptions subsequently prove to have been wrong then the Project manager is to notify a correction. This in itself will constitute a compensation event under clause 60.1(17) and the assessment of that event will essentially comprise an adjustment to the earlier one.

12.7.10 The importance of clause 61.6 is apparent from a reading of clause 65.2 which states that ... *the assessment of a compensation event is not revised if a forecast upon which it is based is shown by later recorded information to have been wrong* ... Assessments based on assumptions notified by the Project manager in accordance with clause 61.6 are therefore the only assessments which can subsequently be amended. The Guidance Notes explain that ... *the reason for this strict procedure is to motivate the parties to decide the effects of each compensation event either before or soon after it occurs. Since each quotation can include due allowance for risk (Clause 63.5), and the early warning procedure should minimise the effects of unexpected problems, the need for later review is minimal* ... One commentator on NEC/2, however, has raised the point that where any sort of uncertainty is perceived by the Contractor he may decide to load his quotations as a reflection of the commensurate risk, with the resulting possibility of overpayment by the Employer. This commentator goes on to state that, as a reaction to this, ... *some project managers using the NEC have a confirmed policy of automatically stating assumptions. They do this as a precautionary measure to retain some element of cost control* ... [8]. It is suggested that such an approach would be inconsistent with the philosophy set down in the Guidance Notes to NEC/2 and could well operate to unravel the ongoing cost and time certainty which the system of compensation events is designed to encourage.

12.7.11 Clause 61.7 confirms that the compensation-event system ceases to operate after the *defects date*. The *defects date* is not a fixed date in NEC/2 but, instead, is set at the end of a period following Completion of the whole of the *works* which is stated in the Contract Data. Its equivalent in JCT 98 PWQ is the end of the Defects Liability Period. Accordingly, either party can still notify the other of a compensation event until the end of the NEC/2 equivalent of the Defects Liability Period. This can be contrasted with the position in JCT 98 PWQ whereby all matters concerning extensions of time are required to be resolved within 12 weeks from Practical Completion.

Quotations

12.7.12 Clause 62 regulates the submission of quotations by the Contractor which have been instructed by the Project Manager in accordance with clause 61. Recognising that there may be more than one way of accommodating the effects of a compensation event, clause 62.1 permits the Project Manager to instruct the Contractor to submit ... *alternative quotations based upon different ways of dealing with the compensation event which are practicable* ... In the first instance, it is anticipated that the Project Manager will instruct the Contractor as to what the suggested alternatives are. It may be possible, for instance, to propose solutions employing different methods which alternatively prioritise time or cost. The Employer can then, through the Project Manager, select the option which best suits his priorities. Clause 62.1, however, also permits the Contractor to propose practicable alternatives of his own should he consider any to be available in addition to those identified by the Project Manager. In the case of changes to the Works Information this provision will permit the Contractor to bring his own expertise to bear on, for instance, questions of buildability. It does not appear, however, that the Contractor can act on his own initiative in this respect in the absence of a prior instruction issued by the Project Manager under clause 62.1. This is not to say, of course, that the Contractor is prevented from taking a similar initiative in respect of an early-warning matter. In such situations, clause 16.3 allows all parties to submit proposals for dealing with the matter. This said, once a matter has been notified as a compensation event, it would appear that the initiative in making proposals as to how best to accommodate it passes to the Project Manager.

12.7.13 It has been explained in section 12.3 that assessments of compensation events cover both the time and cost effects of the matter in question. Quotations for compensation events are to set out the Contractor's proposed changes to the Prices and to the Completion Date consequent upon the event. Other than in assessing his risk allowance in accordance with clause 63.5, however, the Contractor has only limited discretion as to how he arrives at his quotation. He is in fact required to operate within a set of rules when preparing his quotation. These rules, which are intended to encourage transparency and trust between the parties, determine how the Contractor is to assess the time and cost effects of the compensation event. They are set out at clause 63 and are discussed in detail at paras 12.7.19 to 12.7.44. In order to assist in maintaining transparency, and to permit the Project Manager to reach his own appraisal of the Contractor's quotation, the Contractor is also required to submit details of his assessment. The amount of such detail would, it is suggested, be dependent upon the significance and complexity of the event in question. Where the programme for the remainder of the work is affected by the event, the details to be provided by the Contractor are to include a revised programme setting out the effects.

12.7.14 The Contractor's quotation is to be submitted within 3 weeks of the Project Manager's instruction, whilst the Project Manager is required to respond to it within a further 2 weeks following submission. Both of these periods may be extended by the Project Manager provided agreement on this is reached with the Contractor before either the submission or the reply respectively are due (clause 62.5 refers). Should such agreement be reached, it is important that the Project manager bear in mind that a notification is required from him to that effect. It is suggested that such extensions ought to be considered either where the compensation event in question is of a particular significance or complexity or, alternatively, where a large number of events are awaiting consideration. This is of particular importance when one considers that, in the absence of any stated assumptions in his instruction to the Contractor to submit a quotation, it will not be possible subsequently to revise the assessment of the event should it prove to be inaccurate (clause 65.2 refers). The Guidance Notes suggest that extensions ought also to be considered in situations where, for instance, ... *a weather compensation event ... occurs very early in a calendar month. The effect of the event cannot be assessed until the end of the month when the extent of the weather exceeding the 10-year return weather data becomes known* ... Otherwise the Guidance Notes suggest that an extension of the time periods be considered the exception rather than the rule, as too liberal a use of this mechanism will operate to undermine the stated intention of the compensation event system which is to resolve matters quickly and on an ongoing basis.

12.7.15 Upon receipt of the Contractor's quotation, clause 62.3 limits the Project Manager's response to one of four alternatives as follows:

- he instructs the Contractor to submit a revised quotation;
- he accepts the quotation as it stands;
- he notifies the Contractor that the proposed instruction or changed decision which would form the subject matter of the compensation event (clause 61.2 refers) will not now be given; or
- he notifies the Contractor that he will be making his own assessment in accordance with clause 64.

12.7.16 The contract is not specific as to the reasons why the Project Manager may request the Contractor to submit a revised quotation. Presumably the Project Manager is empowered, at his complete discretion, to decide that a quotation is unacceptable and either to request a revised quotation from the Contractor (first bullet point above) or, alternatively, to carry out his own assessment of the compensation event (fourth bullet point above). One reason why the Project Manager may request a revised quotation is that the Contractor's original quotation has not been properly prepared in accordance with the provisions of clause 63. As to whether the Project Manager opts to implement either the first or the fourth bullet point above, it is suggested that this is likely to be conditioned in practice by the administrative burden it will place upon the Project Manager (and his capacity to cope with it) and also by his confidence in the Contractor's ability to produce an acceptable assessment. The philosophy underlying NEC/2, that compensation events should wherever possible be implemented by agreement between the Project Manager and the Contractor, would appear to suggest that assessment by the Project Manager ought to be the exception rather than the rule.

12.7.17 One further point worthy of mention in respect of quotations is defrayment of the Contractor's costs of preparing them. Under main options C, D, E and F the Contractor's overall entitlement is based upon the Actual Cost of the *works* as defined in the Schedule of Cost Components. As the Schedule of Cost Components is likely to include those costs which the Contractor incurs in preparing quotations, it follows that, under these main options, the Contractor will recover those costs. This is not the situation, however, in the case of main options A and B (the traditional lump-sum options) where the Contractor's entitlement is based on the Prices as adjusted by the implementation of compensation events. As explained at paras 12.7.19 et seq. below, the assessment of compensation events is based upon Actual Cost as defined in the Contract. In the case of main options A and B, clause 11.2(28) expressly excludes the cost of preparing quotations from this definition. This cost is therefore left to be carried by the Contractor. Recognising this, the Guidance Notes recommend that

Contractors employed under main options A and B make an allowance in their tendered *fee percentage* (which is applied to the Actual Cost) to cover the likely level of costs attaching to the preparation of quotations.

12.7.18 The Guidance Notes point out that ... *this policy has been adopted in options A and B in order to retain the certainty of the Prices relative to the work done ...* The Guidance Notes do go on to suggest however, where the effect on the Contractor's costs of preparing a quotation is likely to be unusually large or where the Employer may be asking ... *for multiple quotations for significant design changes ...,* that the Employer consider making special arrangements to reimburse the Contractor. In approaching such situations, it is presumably envisaged that the Employer will have regard to his duty to act co-operatively as set down in clause 10.1.

The rules for assessment

12.7.19 In many ways, clause 63, which sets out the rules in accordance with which compensation events are assessed, represents the essence of NEC/2 in that it is the hub of its intended system of dispute avoidance. It provides the mechanism by which the Contractor and the Project Manager can agree the cost and time effects of compensation events in a manner which is timeous, which is reasonably transparent, which gives the Employer a significant degree of cost and time certainty and which places a risk on the Contractor which is tolerable. The rules set down in clause 63 are, however, somewhat complex and may well be difficult to apply in practice. In particular, the proper application of the rules is likely to require a level of resource from both the Contractor and the Project Manager somewhat in excess of that required for the administration of more traditional forms of contract. The rules set out in clause 63 are primarily intended to be used to enable the Contractor to assess compensation events for the purpose of submitting quotations in accordance with clause 62. Should the Project Manager be required to make his own assessment consequent upon the provisions of clause 64, this will be done in accordance with the same rules.

12.7.20 Clause 63.1 stipulates that the Prices are changed to reflect the effect which a compensation event has on **Actual Cost**. In simple terms, this is done by calculating the Actual Cost both with, and without, the effect of the compensation event. The Contractor's tendered rates and prices are not relevant to the exercise. Unlike JCT 98 PWQ, therefore, the Contractor will not be tied to his tender rates (or even the familiar concept of "fair rates and prices") in the event of varied work. He will theoretically be reimbursed on the basis of the actual effect on his costs of the compensation event, albeit that there are complications to this as explained at paras 12.7.23 to 12.7.29.

12.7.21 The Guidance Notes explain that ... *the reason for this policy is that no compensation event for which a quotation is required is due to the fault of the Contractor or relates to a matter which is at his risk under the contract. It is therefore appropriate to reimburse the Contractor his forecast additional costs or actual additional costs if work has already been done arising from the compensation event. Disputes arising from the applicability of contract rates are avoided ...* The only situation where this may not apply is under main options B and D, whereby clause 63.9 permits the Project Manager and the Contractor to agree that ... *rates and lump sums in the bill of* *quantities may be used as a basis for assessment instead of Actual Cost and the resulting Fee* ... The phrasing of this clause appears to envisage assessments using both the rates and prices themselves and rates and prices derived from them where the nature of the work necessitates this. It is suggested that those more familiar with the valuation of variations under JCT forms of contract may well find themselves opting for this method of assessment in preference to the more unfamiliar procedure otherwise required by NEC/2. The Guidance Notes recommend the use of this procedure ... *particularly for small items, where calculations using the Schedule of Cost Components* (explained at para. 12.7.23) *may be unduly lengthy in relation to the value of the compensation event* ... The Notes also suggest that this method is used only where the parties are reasonably confident that it will produce results broadly in line with those which would be produced were the assessment to be based on Actual Cost.

12.7.22 It is to be noted that assessment is **not** based on a comparison between, on the one hand, the Contractor's Actual Cost of incorporating the effect of the compensation event and, on the other, the Contractor's **tendered price** for the work as it was prior to the compensation event. Such an assessment would have the effect of additionally reimbursing the Contractor in respect of any errors contained in his tendered price should that have been too low. In order to avoid this eventuality, NEC/2 introduces a potentially problematic concept – the Contractor's **supposed** Actual Cost had it not been for the impact of the compensation event. It is this figure which is used as a basis for comparison with the Actual Cost as affected by the compensation event. The picture is further complicated by the fact that assessment can often take place before the compensation event has actually occurred and therefore before any of the costs in question have themselves been incurred. In such circumstances, **both** figures used in the assessment (i.e. with and without the effect of the compensation event) will, to a greater or lesser degree, be theoretical. In this sense, it is apparent that the term Actual Cost as used in clause 63 can be something of a misnomer.

12.7.23 Additionally, Actual Cost is defined in NEC/2 as being only those costs attaching to the items listed in the Schedule of Cost Components attached to the contract (there is also an alternative Shorter Schedule of Cost Components which, although not appreciably shorter, is recommended for use where there are a large number of compensation events). It is, of course, quite possible that the Contractor may incur costs which are not listed in the Schedule of Cost Components. Should this be the case, albeit that this ought to be relatively rare, then these costs will not be included in the Contractor's reimbursement. Also, clause 52.1 stipulates that ... *amounts included in Actual Cost are at open market or competitively tendered prices with all discounts, rebates and taxes which can be recovered deducted* ... Should the Contractor find himself paying premium rates and prices for reasons which do not entitle him to be reimbursed similarly (e.g. if he could have, but did not, provide an early warning of the event) then the amount of the premium will not form part of his Actual Cost notwithstanding that the sums in question will have been paid by him. Clause 63.6 introduces a further discipline by additionally requiring that ... *assessments are based upon the assumptions that the Contractor reacts competently and promptly to the compensation event* [and] *that the additional Actual Cost and time due to the event are reasonably incurred* ... Again, costs incurred as a result of inefficiency on the part of the Contractor will not be reimbursed. In the context of clause 63, therefore, it is apparent that the Contractor's reimbursement is not in fact based on his actual costs in the commonly accepted sense.

12.7.24 Should all or part of the work consequent upon the compensation event already have been undertaken then it ought to be possible for the Actual Cost of the work carried out to be determined by reference to the Contractor's records, albeit that it will not necessarily be the case that such records will be maintained in a format consistent with the Schedule of Cost Components. This latter problem may be a particular issue where Sub-contractors are involved in carrying out the affected work. The Schedule of Cost Components (where main options A and B are used) appears to be based on the premise that the Contractor will have access to his Sub-contractors' payroll and other such records – clause 11.2(28) confirms that the Schedule of Cost Components applies ... *whether work is subcontracted or not* ... Under main options A and B there is no facility within the Schedule of Cost Components for the Contractor simply to pass on lump sum figures quoted by his Sub-contractors. One would imagine that the reason for this is to maintain as far as possible the transparency of the system.

12.7.25 Certainly, if the Subcontractor has been appointed on the basis of the NEC Engineering and Construction Subcontract, as he should, then there is an obligation on him to provide information in accordance with the Schedule of Cost Components in like fashion to the Contractor's obligation to do the same under NEC/2. However, whilst use of the NEC form of subcontract is encouraged by clause 26.3 of NEC/2 it is not mandatory and the evidence to date is that it is not in particularly wide use. In situations where subcontracts do not provide for information to be supplied in the format, at the times and for the purposes envisaged in NEC/2, the Contractor may well face difficulties in providing the required details of his Actual Cost. This is likely to be particularly the case where one is dealing with **forecast** Actual Cost, i.e. in situations where the work in question has not yet been carried out.

12.7.26 Under options C, D and E, by contrast, the Schedule of Cost Components relates only to the Contractor's own costs and not to those of his Subcontractors. The remainder of the Actual Cost is made up of ... *the amount of payments due to Subcontractors for work which is subcontracted* ... This would appear to give the Contractor a greater degree of flexibility, when producing assessments of compensation events, simply to pass on as part of his own quotation lump sum figures provided by Subcontractors – albeit that clause 52.2 requires the Contractor to keep, among other things ... *records of communications and calculations relating to assessment of compensation events for Subcontractors* ... and to permit the Project Manager to inspect them. Any costs which the Project Manager decides ... *should not have been paid to a Subcontractor in accordance with his subcontract* ... or ... *results from paying a Subcontractor more for a compensation event than is included in the accepted quotation or assessment* ... form part of the Disallowed Cost as defined in clause 11.2(30) and will not be reimbursed to the Contractor.

12.7.27 It is worthy of note that only under main options C, D, E and F is the Contractor required to maintain records in a stated format and to provide access for the Project Manager to inspect those records. One would presume that this is because the Actual Cost forms the basis of the Contractor's whole entitlement under those options whilst, under main options A and B, it is only used to assess compensation events and an "open book" policy is therefore considered unnecessarily cumbersome. In forming a view on the Contractor's quotations under main options A and B, therefore, the Project Manager will have access to no more than the "details" which the Contractor is required to submit in accordance with clause 62.2. Whilst the contract is silent on the form which these "details" ought to take it is suggested that, at the very least, they are presented in a format consistent with the Schedule of Cost Components and capable of being referenced to the activities on the Contractor's resourced programme.

12.7.28 In respect of work which has already been carried out at the time of the assessment, the Guidance Notes suggest that ... *Actual Cost should be readily accessible from records* ... In light of the above discussion it is suggested that, in all but the simplest of situations, this may in fact be a far-from-straightforward task. In instances where the work of a number of Subcontractors is affected by the compensation event particularly, the administrative burden on both the Contractor and the Project Manager may be considerable. In the case of work which has yet to be carried out, the Guidance Notes themselves concede that ... *forecasting future Actual Cost is less straightforward* ... Records relating to the work itself will not be available, of course, because it has not yet been carried out. What is required, therefore, is an estimate of the likely cost of the work following the impact of the compensation event. This in itself ought not to be too taxing an exercise for the Contractor, as he ought to be relatively familiar with the resources likely to be required and their probable levels of output. The Contractor ought also to be assisted by the fact that he will have available to him a detailed and reasonably accurate resourced programme.

12.7.29 However, it is not only the forecast Actual Cost with the impact of the compensation event which is required, but also the forecast Actual Cost of the work as originally planned. There will, of course, be original tendered rates within the Prices which cover this work, but it is not these which are to be used for the assessment. This raises the prospect of some potentially interesting discussions between the Contractor and the Project Manager should the forecast Actual Cost of the originally planned work bear little or no relation to the tendered price. This will be the case particularly should the compensation event involve the deletion of work and the Contractor's forecast Actual Cost for the deleted work be somewhat less than his originally tendered price. In respect of assessing the effect on the Prices particularly of deleted work, the Guidance Notes suggest that ... *there is ... no obstruction to the Project Manager and the Contractor agreeing to delete the price if both are satisfied that it adequately represents the reduction in forecast Actual Cost* ... Where there appears to be a disparity between the original price and the forecast Actual Cost it is suggested that a sufficiently robust explanation of this would be required from the Contractor, in the absence of which the Project Manager may be persuaded to arrive at his own assessment of the event in accordance with clause 64 (albeit see the comments paras 12.7.47 and 12.7.48 as to potential limits on the Project Manager's capacity to implement clause 64). Where there is likely to be no particular disparity between the forecast Actual Cost and the Prices, it is difficult to see why assessment of the deleted work on the basis of the original price should not become the rule rather than the exception. It is certainly likely that this may happen in practice, if for no other reason than to reduce the administrative burden on the Contractor and the Project Manager where there are a large number of compensation events.

12.7.30 It is worthy of note in this respect that, consequent upon clause 63.2, the Prices will only be **reduced** in response to a compensation event where the event is either a change to the Works Information (clause 60.1(1) refers) or where it involves the correction of an assumption by the Project Manager in respect of an earlier compensation event (clause 60.1(17) refers). Where a *bill of quantities* is used under main options B and D, if the *bill of quantities* contains either ambiguities or departures from the *method of measurement* requiring correction by the Project Manager then such corrections constitute a compensation event which can also lead to a reduction in the Prices (clause 60.6 refers). Additionally, where secondary option T is employed, if there is a change in the law (e.g. relating to levels of tax or other duties) then this can also lead to a reduction in the Prices. In the case of all other compensation events the Prices are not reduced even if the effect of the event is to reduce the Actual Cost. Should the Employer decide to introduce additional compensation events by means of an entry in the Contract Data and it is required that these may give rise to a reduction in the Prices, then it is important that an additional condition to this effect is introduced by way of secondary option Z.

12.7.31 The intention of the process described above is, of course, to produce an assessment of the effect of the compensation event on the Contractor's Actual Cost which is more or less accurate despite the fact that it may have been based largely upon estimates and notional cost bases. In addition to the Actual Cost, clause 63.1 also permits the Contractor to recover his Fee. In practice, this is likely to be achieved by applying the Contractor's tendered *fee percentage* to the forecast Actual Cost with and without the effect of the compensation event. The assessment of the compensation event will thereby automatically incorporate the consequent effect upon the Fee. Where work is deleted from the Contract the assessment must also include an adjustment (effectively a reduction) to the Fee.

12.7.32 In either event, where main options A and B are employed, the Prices are only ever subject to a single adjustment in line with the Contractor's tendered *fee percentage*, which is deemed to include for any fees payable to Subcontractors affected by the compensation event (clause 63.10 refers). In this regard, the Guidance Notes recognise that the Contractor may himself incur administration costs consequent upon co-ordinating the Subcontractor's input in connection with the compensation event which will have to be met from his own Fee. Accordingly, the Guidance Notes recommend that the Contractor anticipate this by agreeing with his Subcontractors tendered *fee percentages* which are lower than his own – thereby leaving him a margin from which he can fund his administrative costs. It is unlikely that this will always be possible, however, and the Contractor may well experience problems where compensation events affect the work of Subcontractors with *fee percentages* equal to or higher than his own. He will be required to reimburse those Subcontractors at levels higher than his own reimbursement. Under main options C, D, E and F, of course, the Contractor's *fee percentage* will only be applied to his own directly incurred costs. Fees paid to Subcontractors will, provided they are deemed reasonable and do not fall to be part of the Disallowed Cost, be reimbursed to the Contractor as part of his Actual Cost.

12.7.33 One further feature of the assessment of compensation events by reference to Actual Cost is worthy of mention. Where secondary option N is not used then the Contractor, under main options A to D, carries the risk of price inflation and is expected to make due allowance for this risk in the Prices. Where secondary option N is used (and it may only be used with main options A to D) it places the risk of price inflation on the

Employer and also provides a mechanism for adjusting the Prices based upon the use of price indices selected by the Employer. It is beyond the scope of this book to provide a detailed explanation of the operation of NEC/2's price-adjustment mechanism, but there are issues concerning its effect on the assessment of compensation events of which the user of NEC/2 should be aware.

12.7.34 The Guidance Notes to NEC/2 point out that any assessment of the Actual Cost of compensation events will include some items which are priced at levels pertaining at the time of the assessment and some which are priced at levels pertaining at the *base date* (which the Guidance Notes recommend should normally be 4 to 6 weeks before the tender submission date). This is because the Actual Cost is based upon the Schedule of Cost Components. Some of the items contained in the Schedule, such as the hire of equipment and the purchase of materials, will be paid at levels pertaining at the date of the assessment. Other items, such as hourly rates for employees of the Contractor, will be paid at the rates set in Part 2 of the Contract Data and which will generally be consistent with rates pertaining at the *base date*. In order to ensure consistency in operating the price-adjustment mechanism, clause N3.1 requires that any elements of the Actual Cost of a compensation event which are priced at levels current with the assessment date are first adjusted back to the *base date* index. The whole of the Actual Cost of the compensation event is then adjusted in accordance with either clause N4.1 (where main option A or B applies) or clause N4.2 (where main option C or D applies). Following this adjustment, the whole of the Actual Cost of the compensation event will then be set at a level current with the assessment date. It is worthy of note in this respect that, even where secondary option N does not apply, then the Employer nonetheless carries the risk of price inflation in respect of compensation events – specifically, those elements of the Actual Cost which are not fixed at the *base date* by virtue of Part 2 of the Contract Data.

12.7.35 One additional point which arises out of secondary option N ought to be borne in mind. The risk of price inflation remains with the Employer up to the Completion Date for the whole of the *works*, after which the price-adjustment index is frozen (clause N2.2 refers). Should secondary option L be employed and the Contractor be in culpable delay in respect of any *section* of the *works*, he will nevertheless continue to be indemnified against increases in cost up until the Completion Date for the whole of the *works*. If he remains in culpable delay beyond the Completion Date for the whole of the *works* then he will no longer be thus indemnified. Whilst this may seem unreasonably generous toward the Contractor, the Guidance Notes to NEC/2 explain the reason for it as being to avoid making the price-adjustment calculations unnecessarily complicated.

12.7.36 Clause 63.3 deals with the assessment of delays consequent upon the compensation event. In this regard, it is important to note that the delay to the Completion Date consequent upon the compensation event will actually be assessed by reference to its effect on the Contractor's **planned** Completion. As explained at paras 12.2.37 and 12.2.38, this may be some time before the Completion Date. In this event, any terminal float, that is the period between planned Completion and the Completion Date, will not be available to absorb the effects of the compensation event. This can best be explained by reference to an example whereby the Contractor's planned Completion is, say, 1 March 2000 and the Completion Date is 31 May 2000. The Contractor's terminal float is, in this example, 13 weeks. If the effect of a compensation event is to delay planned Completion by 2 weeks to 15 March 2000, then the Completion Date is also deferred by 2 weeks to 14 June 2000. The Contractor's terminal float is maintained at 13 weeks. In this way the Completion Date can be deferred from 31 May to 14 June notwithstanding that Completion is still expected during March. It is to be noted that no compensation event can ever result in a reduction of the time available to complete the contract. This can only be done by securing the Contractor's agreement to an acceleration in accordance with the provisions of clause 36.

12.7.37 By virtue of the fact that an Accepted Programme will be in place, and that clause 62.2 requires the Contractor to submit a revised programme with his quotation ... *if the programme for the remaining work is affected by the compensation event* ..., it ought to be possible to assess with reasonable accuracy the effect of the compensation event on the remainder of the work and on the Contractor's planned Completion. In this regard, clause 63.3 confirms that the time effect of any compensation event must always be assessed by reference to the current Accepted Programme. When carrying out an assessment of a compensation event it is important to bear in mind, as noted at para. 12.2.37, that float within the programme is available to absorb the effects of the event whereas the Contractor's time-risk allowances are not and are to be preserved in any revised programme unless the Contractor himself chooses to adjust them. Should the Project Manager accept the Contractor's quotation for the compensation event then the revised programme submitted with it will become the Accepted Programme for all the purposes of the contract. The Contractor will be bound by the Completion Date (or, where secondary option L is employed, Completion Dates) shown on the revised programme and the Employer will also be bound by any revised dates for the provision by him of information and other things.

12.7.38 In respect of both the time and cost attaching to a compensation event, clause 63.5 requires the Contractor to include ... *risk allowances for matters which have a significant chance of occurring and are at the Contractor's risk* ... There is little difference in principle between this requirement and the position traditionally taken in engineering contracts that, for instance, the Contractor is to bear the risk of ground conditions which ought to have been anticipated by an experienced engineering contractor. It is also consistent with the philosophy of NEC/2 that contract risk should be placed with the party best able to manage it.

12.7.39 The contract is not clear as to whether the Contractor is required to identify his risk allowances as such within his quotation. If he does so it will add to the transparency of the system, which NEC/2 seeks to encourage as a means of avoiding disputes, and may also assist the Project Manager in properly evaluating the quotation. If the Contractor's risk allowances are not identified they will nevertheless be deemed to be included. Further, they will not subsequently be subject to adjustment should the assumptions on which they were based be shown to have been incorrect. As noted above, except where the Project manager makes stated assumptions in accordance with clause 61.6, no assessment of a compensation event can subsequently be revised (clause 65.2 refers). Contractors should therefore approach the inclusion of such risk allowances with particular care. The Project Manager should also bear in mind, when assessing the Contractor's quotation, that risk allowances are likely to be greater where uncertainty attaches to the work and that the Contractor is permitted (even obliged) to make a proper commercial allowance in respect of such uncertainty. Where the Project Manager is uncomfortable with the level of risk allowance apparent within a quotation it is suggested that he either takes measures to reduce the uncertainty so as to permit the Contractor to reduce his risk allowance or, alternatively, invite a quotation which is based upon stated assumptions which remove the uncertainty in the short term but which will leave the assessment subject to possible future revision.

12.7.40 Two further rules aimed at encouraging a disciplined approach on the part of the Contractor are included at clauses 63.4 and 63.6. These rules are intended to ensure that any costs which were caused by inefficiency on the part of the Contractor remain to be borne by the Contractor and will not form part of the assessment of the compensation event. As noted at para. 12.7.8, clause 16.1 requires the Contractor to give early warning of likely compensation events so as to enable the most effective means of dealing with them to be formulated in good time. If the Project Manager decides that such early warning could have been given by an experienced contractor but in this instance was not, he must notify the Contractor to this effect (clause 61.5 refers) and the Contractor is required to take account of his own failure in assessing the compensation event. The compensation event is to be assessed as if the Contractor had given the required early warning.

12.7.41 In practice, this will mean that the assessment of the Contractor's forecast Actual Cost will be that much more notional than would otherwise be the case. This is because the Contractor will be required to remove from his forecast Actual Cost those costs which he has incurred but which he theoretically would not have incurred had he provided the required early warning. One example of such costs would be premiums attaching to shortened order periods which would not have been necessary had orders been placed in good time. Other costs attaching to a failure to provide early warning are, however, likely to be somewhat less readily identified and could well lead to disputes. The requirement for the Contractor to provide the Project Manager with details of the cost effects of his own alleged failings (bearing in mind that the Contractor may not necessarily agree that he ought to have given an early warning) ought additionally to provide scope for some interesting arguments.

12.7.42 Clause 63.6 provides the added discipline that the additional costs attaching to a compensation event must have been reasonably incurred and must be ... *based on the assumption that the Contractor reacts competently and promptly to the compensation event* ... The Contractor will find it difficult to price his own anticipated inefficiencies into a quotation for a compensation event and, once the assessment of the event has been fixed, will be unable to recover such costs after the event.

12.7.43 Clause 17.1 provides for the Project Manager to issue instructions to resolve any ambiguity or inconsistency within or between any of the documents involved in the contract, whether provided by the Employer or the Contractor. Where such an instruction involves a change in the Works Information this will potentially give rise to a compensation event. Clause 63.7 sets down how such compensation events are to be assessed depending upon whether the original ambiguity or inconsistency arose in respect of information provided by the Employer or by the Contractor. If the information was provided by the Employer then the assessment is made ... *as if the Prices and the Completion Date were for the interpretation most favourable to the Contractor* ... The opposite is, of course, the case if the information were originally provided by the Contractor. This is essentially a reformulation of the rule of law known as contra proferentem, which provides that no party to a contract may gain advantage from ambiguities in contract terms drafted by itself. Such ambiguities will be resolved in favour of the other party.

12.7.44 Once the cost effect of the compensation event has been assessed and a figure arrived at, the Prices are then adjusted accordingly. In the case of main options A and C this takes the form of changes to the *activity schedule* while, in the case of main options B and D, it takes the form of changes to the *bill of quantities*. In the case of main options A and B the Contractor will be reimbursed in respect of the compensation event once the work has been done by virtue of the fact that the Price for Work Done to Date will be based on the revised *activity schedule* or *bill of quantities* which incorporate the assessment of the event. In the case of main options C, D, E and F the Contractor will be similarly reimbursed by virtue of the fact that the Price for Work Done to Date is the Actual Cost, which will likewise include the effect of compensation events.

Assessment by the Project Manager

12.7.45 It is envisaged by NEC/2 that the assessment of compensation events will primarily be carried out by the Contractor and subsequently be agreed with the Project Manager. Clause 64 lists a limited set of circumstances in which the Project Manager may carry out the assessment himself. It is intended, however, that this be the exception rather than the rule as excessive use of this power on the part of the Project Manager will tend to cut across the consensual approach towards dealing with compensation events which NEC/2 endeavours to encourage. The Guidance Notes point out that the four circumstances listed in clause 64.1 ... *are all derived from some failure of the Contractor* ... Only in the event of such failure is the Project Manager empowered to set aside the Contractor's assessment and substitute one of his own. The four circumstances set out in clause 64.1 can be summarised as follows:

- the Contractor has failed to submit a quotation, together with the required details, in the time allowed;
- the Project Manager has decided that the Contractor has not assessed the compensation event correctly (i.e. not in accordance with the rules for assessment set down in clause 63) but decides not to request a revised quotation from the Contractor;
- the Contractor has not submitted a revised programme with his quotation; or
- at the time the Contractor submits his quotation the Project Manager has not accepted the Contractor's latest programme for one of the reasons set out in clause 31.3.

12.7.46 In addition, the Project Manager is to base his assessment of the compensation event on his own assessment of the programme for the remainder of the work where either there is no Accepted Programme or where the Contractor has failed to submit a revised programme for acceptance as required by clause 32. This particular provision is described in the Guidance Notes as ... *a major incentive on the Contractor to keep his programme up to date* ... Where the Project manager does opt to make his own assessment he is to notify the Contractor to that effect and is to provide details of his assessment to the Contractor within the same period provided within the contract for the Contractor to submit his quotations (usually 3 weeks). This time period starts to run from the date when it became apparent (presumably to the Project Manager) that the Project Manager's assessment would be required.

12.7.47 It appears from the wording of clause 64.1 (which refers to the Contractor not having ... *assessed the compensation event correctly ...*) that it is only in the event of some error on the part of the Contractor as regards the mechanics of the assessment that the Project Manager can intervene to impose his own assessment. It is arguable, in light of such an interpretation, whether the Project Manager can substitute his own view for that of the Contractor simply because he disagrees with the assumptions which underpin it. This may be a potentially significant distinction. What it would mean in practice is that, provided the Contractor has properly applied the rules in clause 63 and has also complied with the various other provisions concerning timescales and the submission of programmes, the Project Manager would be unable effectively to challenge the Contractor's assessment no matter how unpalatable he may find it. This would be the case with changes to the Works Information as with any other compensation event and would mark a significant shift from the position under the JCT 98 PWQ and other similar contracts, whereby the final decision in respect of such matters lies invariably with the Employer's QS rather than the Contractor.

12.7.48 An alternative view, based on a less narrow interpretation of the wording of the second bullet point of clause 64.1, is that the Contractor's assessment could be construed as incorrect (and thereby subject to replacement by the Project Manager's own assessment) should the Project Manager disagree with it in any respect. This situation would be little removed from that pertaining under JCT and similar contracts in that the Project Manager would always, in the final analysis, have the option of assessing the compensation event himself. The Guidance Notes to NEC/2 would appear, however, to lend support to the position as set out in para. 12.7.47, in that they explain that the Contractor not having assessed the compensation event correctly *... means that the changes to the Prices have not been correctly assessed in accordance with Clause 63.1 and/or the change to the Completion Date has not been correctly assessed in accordance with Clause 63.3 ...* Notwithstanding this guidance, it is suggested that the wording of clause 64.1 remains ambiguous.

Implementation

12.7.49 A compensation event takes effect when it is "implemented". Until that point in time it would appear that the Contractor is entitled neither to a change in the Prices nor to a change in the Completion Date. Clause 65 sets out to confirm the means by which a compensation event is implemented and to define the point in time at which this happens. According to clause 65.1, the Project Manager implements a compensation event by way of notification to the Contractor. Under main options A, B, C and D, the Project Manager includes in his notification the changes to the Prices and the Completion Date from either the accepted quotation or from his own assessment (clause 65.4 refers). Under main options E and F, the Project Manager includes in his notification the forecast of such changes, presumably to aid the Contractor in meeting his obligations under clause 20.4 to provide regular forecasts of the total Actual Cost. In the absence of notification by the Project Manager, it would appear that the compensation event will not have been implemented and the Contractor will not be entitled either to a change in the Prices or to a change in the Completion Date even though the effect of clause 61.1 may be to require the Contractor to put the instruction or changed decision giving rise to the compensation event into effect before the event is implemented. The Project Manager is required to notify implementation of the compensation event on the later of his acceptance of the Contractor's quotation, the completion of his own assessment or the occurrence of the compensation event. If such occurrence is not for some time following the assessment then the contract evidently envisages the prospect of some delay between the assessment and the event itself being implemented.

12.7.50 The mechanics of putting into effect a change in the Completion Date is the same whichever main option is being used under NEC/2. The Project Manager notifies a change in the Completion Date and a revised programme is issued which incorporates the effects of the compensation event in question. When it comes to putting into effect a change in the Prices, however, the practical consequences – in terms of determining the Contractor's overall entitlement – differ markedly according to which main option is used. The reason for this is the distinction which NEC/2 draws between, on the one hand, the Prices and, on the other hand, the Price for Work Done to Date (PWDD). It is the PWDD, and **not** the Prices, which primarily determines the Contractor's entitlement. Furthermore, the composition of the PWDD differs according to which main option is used. Under the lump-sum options A and B, for instance, the PWDD is simply the Prices adjusted to take account of matters such as changes in quantity, compensation events, etc. In this sense the final PWDD is not dissimilar to the final account under JCT 98 PWQ, whereby the Contract Sum is adjusted to take account of variations, direct loss and/or expense and other such matters.

12.7.51 Under the target cost options C and D, however, the PWDD comprises the Actual Cost paid or committed by the Contractor together with the Fee. This sum need bear little relation to the Prices, which – in the final analysis – will simply be the original Prices as adjusted by the effects of compensation events. The Prices under main options C and D do little more than to set the level of the target cost against which the Contractor's performance is measured. In addition to his being paid the PWDD, the Contractor under main options C and D is also entitled to his share (in the proportions stated in Part 1 of the Contract Data) of the difference between the PWDD and the target cost as manifested in the adjusted Prices. Should the Contractor's share be a negative amount (i.e. the PWDD exceeds the adjusted Prices) then this will obviously serve to reduce the contractor's entitlement to a level below the PWDD. In this way main options C and D represent a hybrid form of cost-reimbursement contract in that the Contractor's final entitlement is made up of those costs he has actually incurred (plus his Fee) together with an adjustment reflective of his share of the difference between his actual costs and his adjusted tender price. He is thereby rewarded for achieving savings against the target cost but, should the PWDD exceed the Prices, must also take his share of the burden of any overspend.

12.7.52 Main options E and F are what most practitioners would recognise as true cost-reimbursement contracts. In the final analysis the Actual Cost (plus Fee), the PWDD and the Prices are one and the same. Under option E, the Actual Cost (and, thereby, the PWDD and the Prices) comprises both the amount of payments due to Subcontractors and those of the Contractor's directly incurred costs set out in the Schedule of Cost Components. Under option F, the Actual Cost comprises only those amounts due to Subcontractors. Because option F is a management contract, it is anticipated that all of the Contractor's directly incurred costs are covered by his Fee. The Schedule of Cost Components does not therefore apply.

12.7.53 The means by which practical effect is given to a compensation event differs according to which main option is employed. It is possible to identify three quite distinct approaches, in line with the payment options discussed above. These can be summarised as follows and are discussed in detail at paras 12.7.54 to 12.7.63:

- the "lump-sum" approach (main options A and B);
- the "target-cost" approach (main options C and D); and
- the "cost-reimbursement" approach (main options E and F).

The "lump-sum" approach

12.7.54 The Guidance Notes to NEC/2 explain, in respect of both main options A and B, that ... *the Prices are the basis of the PWDD* ... and that ... *in the final account, the PWDD is the total of the Prices of the completed work* [under main option A and] ... *of the final quantity of work completed* [under main option B]. The original priced document upon which the Prices are based (the *activity schedule* in the case of main option A and the *bill of quantities* in the case of main option B) becomes a working document which is used throughout the course of the contract and into which the effects of compensation events can be incorporated. The Guidance Notes also point out in respect of main options A and B that ... *the only use of Actual Cost ... is as the basis of the assessment of compensation events* ... The Contractor's final entitlement will be the total produced by the priced document adjusted to take account of changes in quantity, compensation events, etc.

12.7.55 Clauses 63.8 (main option A) and 63.9 (main option B) explain that assessments of changes to the Prices consequent upon compensation events take the form of changes either to the *activity schedule* or to the *bill of quantities*. It seems apparent that the system of assessing and implementing compensation events set out in clauses 63, 64 and 65 envisages that each assessment will generally produce a single figure (or set of figures) which is used to make a one-off adjustment to the Prices. This figure will reflect the effect of the compensation event on the Contractor's Actual Cost but, for payment purposes, will then be incorporated into the priced document. In the case of main option A, the adjustment can be made to the price for the activity (or group of activities) which are affected by the compensation event. On completion of that activity (or group of activities) the Contractor is then paid the adjusted amount incorporating the effect of the compensation event. In the case of main option B, the contract envisages that the *bill of quantities* may be made up of a combination of, on the one hand, rates applied to quantities and, on the other, lump sums. The assessment of a compensation event could take the form of one or more lump sums inserted into the *bill of quantities*. The Contractor would then be paid the lump sums as the work was completed as part of the PWDD. Alternatively, the assessment may take the form of an adjustment to the rate (or rates) for the affected work. The Contractor would then be paid for the quantity of work completed at the adjusted rate incorporating the effect of the compensation event.

The "target-cost" approach

12.7.56 Determining the Contractor's entitlement is somewhat more complicated when using main options C and D because it is dependent, in the final analysis, on the precise relationship between the Contractor's Actual Cost and the Prices. The above comments regarding the mechanics of adjusting the Prices to incorporate the effects of compensation events would appear to hold equally good relative to main options C and D, in that assessments take the form of changes to the priced document. Under main option C, therefore, prices for activities (or groups of activities) would be adjusted in the same way as with main option A. Similarly, under main option D lump sums may be inserted into the *bill of quantities* or changes may be made to rates in the same way as with main option B. The difference is that, under main options A and B, these changes to the priced documents will feed directly through into the PWDD and will therefore directly affect the amount which the Contractor is to be paid. Under main options C and D, however, all that is changed is the target cost. The PWDD is not based upon the priced document but is instead based upon the Contractor's Actual Cost plus his Fee. Thus the elaborate process of assessing compensation events set out in clause 63 of NEC/2 will, under main options C and D, have no direct effect on the PWDD and therefore only an indirect effect on what the Contractor is paid. The whole system serves simply to adjust the target cost, whilst the Contractor is being paid on an entirely different basis.

12.7.57 The practical implications of this situation are probably best explained by way of an example. A contract has original Prices totalling £500,000. Throughout the course of the contract there are no compensation events and the Prices therefore remain at £500,000. The final PWDD, however, is £450,000. The Contractor is entitled to this sum **plus** his share of the difference between the Prices and the PWDD. Clause 53 of NEC/2 (which applies to both main options C and D) contains a somewhat elaborate system for assessing the value of the Contractor's share of this saving. This system is based on value bands which extend either side of the Prices and which are referred to in the contract as *share ranges*. These *share ranges* (and the contract appears to assume that there will usually be four of them with the Prices set somewhere in the middle) are identified in Part 1 of the Contract Data. Next to each *share range* is also identified a percentage referred to as the *Contractor's share percentage*. The Contractor's share of any saving or overspend which falls within a given *share range* will be the total of the saving or overspend multiplied by the *Contractor's share percentage* for the relevant *share range*.

12.7.58 For the purposes of this example, we can assume that the *Contractor's share percentage* for the *share range* £400,000 to £600,000 is 50%. In other words, the Contractor is entitled to 50% of any saving up to the value of £100,000 – i.e. the range between the lower band limit of £400,000 and the Prices. Similarly he must bear 50% of the cost of any overspend up to the value of £100,000 – i.e. the range between the Prices and the upper-band limit of £600,000. In our example, the PWDD represents a saving of £50,000 relative to the Prices. In addition to the PWDD of £450,000, therefore, the Contractor will also be entitled to a 50% share of this saving – i.e. £25,000. The Contractor's overall entitlement, therefore, will be £450,00 + £25,000 = **£475,000.**

12.7.59 Let us take the same contract again but this time let us assume that the Prices have been affected by compensation events. The assessments of these compensation events, carried out in accordance with the rules in clause 63, total £80,000. These assessments are incorporated into the priced document and the Prices now become £580,000. Of course, at the time when the assessments were carried out they were based largely on **forecasts** of the Actual Cost plus Fee (paras 12.7.28 and 12.7.29 refer). In the event, a number of these forecasts have proved to be incorrect and the Actual Cost which the Contractor has incurred plus his Fee amounts to £100,000. In accordance with clause 65.2, however, the earlier assessments of the compensation events cannot be revised as a result of this and the Prices therefore remain at £580,000. Because the PWDD is not based on the Prices however, but is instead based upon the Contractor's Actual Cost plus Fee, it will have increased by £100,000 to £550,000. The Contractor's entitlement in this case therefore is to the PWDD of £550,000 plus 50% of the difference of £30,000 between this figure and the adjusted Prices. The Contractor's overall entitlement, therefore, will be £550,000 + £15,000 = **£565,000**. It will be noted that this figure is actually £90,000 higher than the figure of £475,000 produced in the absence of any compensation events (para. 12.7.58 refers). In other words, the Contractor has received £90,000 in respect of compensation events which had been assessed as having a value of only £80,000 but which actually cost him £100,000 to put into effect.

12.7.60 Under main options A and B, when compensation events occur the amounts assessed in accordance with clause 63 are the amounts which are ultimately paid to the Contractor. This only happens under main options C and D, however, when the amounts assessed in accordance with clause 63 are identical to the Actual Cost which the Contractor in the event incurs. Where the cost of a compensation event differs from an earlier assessment the Contractor will in fact be paid neither the earlier assessed figure nor his Actual Cost but, instead, a figure somewhere between the two. It is unfortunate that this is not clear to the user of NEC/2 either from a reading of the form itself or from any of the guidance documents published with it.

The "cost-reimbursement" approach

12.7.61 Under main options E and F both the Prices and the PWDD are based upon the Contractor's Actual Cost. There is no priced document which forms the basis for the Prices as there is with main options A to D. Instead, clause 20.4 requires the Contractor to provide regular forecasts to the Project Manager of the total Actual Cost for the whole of the *works*. It would appear from the wording of clause 65.3 that assessments of compensation events prepared in accordance with clause 63 are incorporated into these forecasts as and when implementation of the event occurs. Meanwhile the Contractor will actually be paid on the basis of his Actual Cost plus Fee. Should the assessment of the compensation event carried out under clause 63 be incorrect, therefore, this will not affect the Contractor's entitlement nevertheless to be paid the Actual Cost he incurs. It will simply mean that the Contractor's forecast of the Actual Cost will have been incorrect.

12.7.62 In the example given at paras 12.7.57 to 12.7.59 a Contractor employed under main option E or F would receive £100,000 in respect of compensation events if that were the value of his Actual Cost plus Fee, regardless of the fact that the assessments carried out in accordance with clause 63 amounted only to £80,000. It would therefore appear that, apart from permitting the Completion Date to be changed, the only function of the compensation-event system under main options E and F is to assist in providing reasonably accurate forecasts of the final cost. The system has no other effect on the Contractor's entitlement to payment. It is for this reason that some commentators have questioned the need for a system of compensation events (particularly one as complex and elaborate as that in NEC/2) in the context of a cost reimbursement contract.

12.7.63 It will be apparent from the above discussion that, in terms at any rate of changes to the Prices, the application of the compensation-event system can give rise to different results according to which of the main options is employed. Under main options A and B, the Contractor will be paid the amounts assessed under clause 63 regardless of what his Actual Cost is. Under main options C and D, the Contractor's entitlement may be neither the assessed amount nor his Actual Cost but, instead, a figure somewhere in between. Under main options E and F, the Contractor will be paid his Actual Cost regardless of the amounts assessed under clause 63. In the example quoted above, the same set of circumstances would give rise to a payment to the Contractor of £80,000 under main options A and B, £90,000 under main options C and D and £100,000 under options E and F. This provides a graphic illustration of the manner in which NEC/2 allocates price risk differently as between the various main options. Unfortunately for the user of NEC/2, at no point in any of the documentation is attention specifically drawn either to this distinction or to its practical effects. The unfamiliar user must rely either on a careful reading of the documents or, alternatively (and probably more likely), on a degree of trial and error.

REFERENCES

[1] Sir Peter Levene (1995) *Construction Procurement by Government – An Efficiency Unit Scrutiny (The Levene Report)*.
[2] Mike Gibson of Solicitors Berwin Leighton quoted in *Contract Journal*, 7 September 1995.
[3] Research published in *Engineering Construction & Architectural Management* and quoted in *Architect's Journal*, 22 February 1996.
[4] Barrister Rudi Klein quoted in *Building*, 19 June 1998.
[5] Architect Michael Manser quoted in *Building*, 27 January 1995.
[6] Roger Dobson (Director-General and Secretary of the Institute of Civil Engineers) quoted in *Building*, 27 January 1995.
[7] Architect Jonathan Manser quoted in *Building*, 9 December 1994.
[8] Brian Eggleston (1996) *The New Engineering Contract – A Commentary*, Blackwell Science, Oxford.

Part 4

WORKED EXAMPLE OF THE ASCERTAINMENT OF DIRECT LOSS AND/OR EXPENSE

A WORKED EXAMPLE OF THE ASCERTAINMENT OF DIRECT LOSS AND/OR EXPENSE

W.1   INTRODUCTION

w.1.1 The following worked example is intended to demonstrate how the principles put forward in the text of the book would be applied in practice to one theoretical project.

w.1.2 Inevitably the background data – that is to say, the detail of the programme, the correspondence, the instructions and the cost data kept by the Contractor – would be much more extensive on a live project than it is practicable to record here.

w.1.3 The Form of Contract used is the JCT Standard Form of Building Contract 1998 Edition Private with Quantities. Where appropriate, references are given to the relevant text in the book.

w.1.4 It is assumed that the detailed claim in this example is submitted by the Contractor shortly after Practical Completion. Following the advice given at paras 5.3.3 and 5.3.4, an initial reaction to the claim is formulated by the Quantity Surveyor (Appendix D hereto). The Quantity Surveyor has also written to the Contractor requesting specific data to assist him in his assessment.

w.1.5 This example comprises the following sections:

- Outline of the Project.
- Contractor's programme prepared prior to the signing of the Contract.
- Anticipated use of preliminary items throughout the contract period.
- Principal observations on Contractor's programme.
- Contractor's claim for Extension of Time and Reimbursement of Direct Loss and/ or Expense and Commentary thereon by the Architect.
- Annotated chart of progress achieved.
- Quantity Surveyor's ascertainment of Direct Loss and/or Expense.
- Some Practical Advice!
- Appendices
  A   Analysis of the Contract Sum
  B   Analysis of Preliminaries section of the Contract Bills
  C   Analysis of Final Account
  D   Initial assessment of Contractor's claim submission.

W.2   OUTLINE OF THE PROJECT

w.2.1 The size and type of project is in line with that upon which the measured work prices section of *Spon's Architects' and Builders' Price Book 1999* is based but allowances for overheads and profit have been included both to make the figures more representative of the market generally and to reflect some transfer of values from the rates to the Preliminaries.

w.2.2 The project comprises two similar, two storey blocks of flats, Block A and Block B, where Block B is slightly larger than Block A. The contract period is 80 weeks with one Date for Completion for the whole project. The Contract Sum is £1,688,897 including £284,972 for preliminaries; details of these figures, as provided by the Contractor prior to the signing of the Contract (paras 3.3.8 to 3.3.11), are given in Appendices A and B, respectively. The Form of Contract is the JCT Standard Form of Building Contract 1998 Edition Private with Quantities.

W.3 CONTRACTOR'S PROGRAMME PREPARED PRIOR TO THE SIGNING OF THE CONTRACT

WEEK NO.

DATE FOR COMPLETION

NO.	DESCRIPTION
1	SET UP HUTS, ETC.
2	DIG FOUNDATIONS AND INTERNAL DRAINS
3	CONCRETE AND BRICKWORK IN FOUNDATIONS
4	HARDCORE, CONCRETE BED AND MEMBRANE
5	BRICKWORK UP TO 1ST-FLOOR LEVEL
6	CONCRETE FRAME
7	BRICKWORK TO ROOF LEVEL
8	ROOF AND RAINWATER GOODS
9	INTERNAL PARTITIONS
10	CARCASS SERVICES*; 1ST FIX CARPENTER
11	PLASTERBOARD AND PLASTER†
12	FLOOR SCREED†
13	PLUMBING AND SERVICES*
14	WINDOWS AND GLAZING†
15	2ND FIXINGS*
16	WALL TILING AND DECORATION†
17	FLOOR FINISHES
18	EXTERNAL WORKS
19	CLEAN AND SNAG
20	CLEAR THE SITE

ACTIVITY

BLOCK A
BLOCK B

* PART OR ALL NOMINATED SUB-CONTRACTORS
† PART OR ALL DOMESTIC SUB-CONTRACTORS

(a)   (b1)   (b2)   (c1)   (c2)

W.4 ANTICIPATED USE OF PRELIMINARY ITEMS THROUGHOUT THE CONTRACT PERIOD

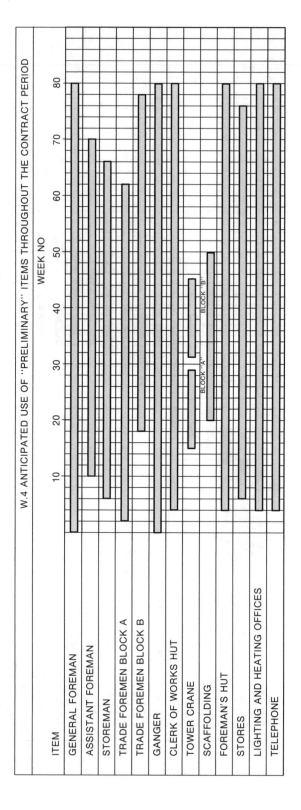

W.4 ANTICIPATED USE OF "PRELIMINARY" ITEMS THROUGHOUT THE CONTRACT PERIOD

w.2.3 The Contractor produces a master programme at the outset of the project and this is represented in Section W.3.

W.5 PRINCIPAL OBSERVATIONS ON THE PROGRAMME

(NB Every reference to a week number is to the end of the week in question.)

w.5.1 Although there is only one Date for Completion, the Contractor plans to finish Block A 16 weeks in advance of Block B. Thus a delay in Block A will not necessarily affect the Date for Completion.

w.5.2 No specific contingency is shown in the programme either in individual elements or the job as a whole; the Contractor must therefore be presumed to have allowed for such contingencies – i.e. to cover delays which do not give rise to an entitlement to an extension of time – in the periods allotted in the programme (paras 2.7.16 to 2.7.19).

w.5.3 The Contractor plans for most of the major trades to follow on from Block A to Block B rather than to work in parallel on both blocks. This will give his men the benefit of familiarity with the work when they come to Block B with a resultant improved performance upon which the rates in the Contract Bills are based. Conversely, any change in this programme strategy, resulting in any trade in both blocks having to be executed concurrently, will result in the loss of this expected improvement in output. If this programme change is the result of causes which come within the scope of clause 26.2 then any loss of output must be included in any ascertainment of loss and/or expense.

w.5.4 The Contractor plans to commence the windows and glazing 1 week before the end of the external brickwork on each block (see items (b1) and (b2) on the programme in section W.3). Moreover, he will not start wall tiling or decorating until the services-installation engineer and the plumber have finished (see items (c1) and (c2) on the programme in section W.3). The execution of external works (including external drains) is not critical in the early part of the project; it is thus undertaken as and when suitable labour from other elements becomes available. However, it becomes critical during the last few weeks before completion.

w.5.5 The Contractor's proposed use of preliminary items (Section W.4) should be noted. For example, the two trades foreman are not on site for the same periods; the scaffolding is on site from just after the commencement of external brickwork and roof on Block A until part-way through the erection of windows and glazing on Block B, it being transferred from one block to the other part way through this period.

W.6 CONTRACTOR'S CLAIM FOR EXTENSION OF TIME AND REIMBURSEMENT OF DIRECT LOSS AND/OR EXPENSE AND COMMENTARY THEREON BY THE ARCHITECT

Contractor's claim	*Commentary*
1 Introduction and details of the project	Although clause 26.1 permits the Contractor to give his quantification, the Contract does not require him to make a detailed application such as is submitted here (see para. 4.10.5).
1.1 We hereby claim reimbursement of direct loss and/or expense under clause 26 of our contract totalling £155,080 all as detailed below and an extension of time for completion of the works of 16 weeks, all as detailed below.	He is only required to:
	a) give a written notice under clause 25 when progress is delayed;
1.2 The Contract was in the JCT Standard Form of Building Contract 1998 Edition, Private with Quantities. The Contract Sum totalled £1,688,897 and the contract period was 80 weeks.	b) make written application under clause 26 if he believes he has incurred direct loss and/or expense which he is entitled to recover; and
	c) supply data necessary for the Architect to form his opinion under clause 26.1.2 and, if so requested, data necessary for the Architect or the Quantity Surveyor to ascertain the amount of loss and/or expense to be reimbursed as stated in clause 26.1.3.
	Thus omnibus claims of this sort, submitted at the end of the day with no earlier notifications will probably be out of time so far as reimbursement of loss and/or expense is concerned (paras 3.5.20 and 3.5.21). Nevertheless, for the purpose of this commentary it is assumed that proper notifications from the Contractor had been made in good time.
	In any event, absence of the proper written notice of delay will only deprive the Contractor of the right to an extension of time during the progress of the work; the Architect must review all events after Practical Completion whether notified or not, when considering fixing a new completion date (para. 2.7.34).

Moreover, the Contractor will probably still retain his right to claim reimbursement of his direct loss and/or expense even if he failed to give the proper notice because under clause 26.6 he retains his Common Law rights (see para. 1.2.8). However, in that event, he will not be able to rely upon the terms of the Contract and obtain payment through interim certificates but will have to go to Adjudication, Arbitration or the Courts for his remedy; and he should lose his entitlement to recover costs that could have been avoided had he given such notices.

Wherever possible, direct loss and/or expense in relation to each item should be separately identified and the Contractor should have kept separate records of the effects of each event provided that he was requested to do so under clause 26.1.3. This highlights the need for the Architect or Quantity Surveyor to make the proper request for data and to be precise in that request.

We do not agree that costs cannot be isolated. Thus each event should be considered on its merit (para. 4.2.9).

1.3 The individual events recorded below are in chronological order but the effect upon progress, and therefore upon our costs, cannot be isolated in a similar fashion. Thus the calculation of our entitlement is given at the end of this submission.

2 **History of the project**
(NB The numbers against each of the following events relate to the annotation on the bar chart appearing in section W.7.)

W.6 CONTRACTOR'S CLAIM FOR EXTENSION OF TIME AND REIMBURSEMENT OF DIRECT LOSS AND/OR EXPENSE AND COMMENTARY THEREON BY THE ARCHITECT (*continued*)

Contractor's claim	Commentary
1. *Week ending No. 4.* Soft spots in the ground were encountered which had to be excavated and back filled to the satisfaction of the Local Authority. The effect of undertaking this extra work was to extend the contract period by 1 week. We notified the Architect of this by our letter dated week 6. Clause 25.4.5.1 Clause 26.2.7	A Provisional Sum for Defined Work of £19,000 was included in the Contract Bills for this work and in consequence the Contractor is deemed to have made due allowance both in his pricing of the preliminaries and in his programme for executing this work. The value of this work in the final account is of the order of £16,990 and the nature of it was as described in the Contract Bills. Therefore, this matter does not come within either clause 25 or 26. In any event, the additional week was due to lack of organisation on the Contractor's part in this initial stage.
2. *Week Ending No. 6.* The invert level of the existing sewer was found not to be as shown on the drawings issued to us requiring a further instruction from the Architect varying the drain levels and gradients. This delayed work on the drains by one week as recorded at the following site meeting. Clause 25.4.5.1 Clause 26.2.7	The facts are agreed. However, this activity was not critical to achieving Completion Date (there being substantial float in the programme for this element). Thus no extension of time is due but any loss and/or expense directly caused will be reimbursed (paras 2.5.5 and 4.4.13). Records of site meetings probably do not qualify as a written application under the terms of the Contract and therefore this item would probably not qualify for reimbursement under clause 26.1. However, the Contractor would be entitled to recover damages at common law under clause 26.6 – and thus the Architect might well obtain the Employer's agreement to permit this matter to be dealt with under the Contract and this is assumed to be the case here (para. 1.2.8).
3. *Weeks ending Nos 15 and 16.* Architect's instruction No. 13 contained significant variations to the frame of Block A,	The facts are agreed. An extension of time will be granted. It is judged that this variation delayed progress on the frame and

introducing a number of subsidiary beams not shown on the contract documents. This delayed work on this critical element for 2 weeks and changed the conditions under which the frame and subsequent trades were executed.
Clause 25.4.5.1
Clause 26.2.7

4. *Week ending No. 23.* The compounding effect of the aforementioned delays has resulted in loss of a further week's progress.

5. *Week ending No. 27.* The carcassing of the services was delayed due to late instructions received from the Architect on the services nomination. In order to get this sub-contractor on site by the end of week 22, it would have been necessary for us to receive the nomination by the end of week 18. Mindful of the wording of the last sentence in clause 5.4.2, we advised the Architect of this in week 14. The nomination instruction was received at the end of week 23 – a delay of 5 weeks.
Clause 25.4.6.2
Clause 26.2.1.2

changed the conditions under which both the frame and the external brickwork were executed. The changed conditions of the execution of the frame has been valued under clause 13.5.1.2 together with the consequent extra use of preliminaries directly related to that work under clause 13.5.3.3; the extra cost arising from the changed conditions of the execution of the external brickwork has been dealt with under clause 13.5.5. It is the opinion of the Architect that no disruption to the regular progress of the project as a whole resulted from implementing this Variation.

This is judged to be due to the Contractor's lack of management and not the direct effect of the earlier delays to which he refers; this lack of management resulted in a delay of one week on Block A.

The facts are agreed but their effect is challenged. The nomination instruction was issued at the end of week 23 and after allowing 4 weeks to get the sub-contractor on site he commenced his work at the end of week 27. Although this was 5 weeks later than shown on the original programme, progress was already 4 weeks behind. Thus the effective delay on this critical element was only 1 week.

Contractor's claim	Commentary
6. *Week ending No. 31.* During the erection of the internal partitions a discrepancy between the Architect's and Structural Engineer's drawings was noticed. Work came to a virtual standstill on this critical element on Block A for a week whilst this matter was resolved (see Event No. 11 for the effect upon Block B). Clause 25.4.5.1 Clause 26.2.3	The facts are agreed. However, this discrepancy, once resolved for Block A should not have caused a similar problem on Block B (paras 2.8.16 and 2.8.17) – see comments under Event No. 11. There was no variation to the work included in the Contract Bills.
7. *Weeks ending Nos 37 and 38.* Access to the site was severely restricted due to the Employer permitting a fair-ground to occupy the adjoining car park (which he owned) over which we were granted the only access to the site. As this is a breach of the undertakings given by the Employer there was no need for us to give notice, so none was given at the time. A delay to progress of 2 weeks was experienced. Clause 25.4.12 Clause 26.2.6	The facts are not disputed. Clause 25.4.12 recognises this as a relevant event qualifying for consideration for an extension of time. The lack of proper notice will not deprive the Contractor of his entitlement for such an extension to be granted after Practical Completion but will deny him this right during the progress of the work (para. 2.7.11). Clause 26.2.6 also recognises this event as a matter that qualifies for consideration in ascertaining the reimbursement of any direct loss and/or expense. However, proper written application is a condition precedent to payment and, as such application was not made in this instance, the Architect is not empowered to ascertain the loss and/or expense (para. 1.2.7).
8. *Week ending No. 41.* A statutory undertaker did not undertake its work on programme which delayed Block A by 1 week.	
9. Moreover, at the **same time** there was a strike by the glaziers and bricklayers on the site so no work could be undertaken on Block A in any case.	The facts are agreed. Additionally, it is noted that in this instance the Employer has a Contract with the statutory undertaker so it is not executing this work pursuant to its statutory obligations; it

Clause 25.4.8.1 Clause 26.2.4.1	is therefore ... *work ... by the Employer himself or by persons employed or otherwise engaged by the Employer ...* as defined in clause 29 (paras 2.8.51 to 2.8.53). In principle, any direct loss and/or expense arising from this delay should be reimbursed to the Contractor under clause 26.2.4.1. However, as there was a strike concurrently in progress, the loss and/or expense due to the delay by the statutory authority (in this instance "a person employed or engaged by the Employer") would not be reimbursable (para. 2.9.2) but an extension of time of 1 week will be granted under clause 25.4.4.
10. *Week ending No. 43.* Apparently the Architect needed to obtain the approval of the Employer to parts of the scheme involving second fixings and accordingly these were postponed under Clause 23.2 for *1 week.* Clause 25.4.5.1 Clause 26.2.5	The facts are agreed but this element is not critical to progress to completion. There is therefore no extension of time, but any loss and/or expense directly occasioned will be ascertained under clause 26.2.5.
11. *Week ending No. 48.* The discrepancy referred to at Event No. 6 above applied to Block B also, causing a further delay of *1 week.* Clause 25.4.5.1 Clause 26.2.3	This discrepancy was rectified on Block A 17 weeks earlier; the Contractor could and should have anticipated this problem on Block B and sought confirmation that the same solution developed for Block A applied also to Block B. No extension of time nor reimbursement of loss and/or expense will therefore be recognised (paras 2.8.16 and 2.8.17).
12. *Week ending No. 50.* We received an instruction not to undertake the decoration to the kitchens in both blocks for the time being awaiting the views of the prospective tenants. Clause 25.4.5.1 Clause 26.2.5	Agreed.

W.6 CONTRACTOR'S CLAIM FOR EXTENSION OF TIME AND REIMBURSEMENT OF DIRECT
LOSS AND/OR EXPENSE AND COMMENTARY THEREON BY THE ARCHITECT (*continued*)

Contractor's claim	Commentary
13. *Week ending No. 56.* Drawings for the electrical installation and the builder's work in connection therewith were the responsibility of the relevant nominated sub-contractor. For this reason he had a Warranty Agreement with the Employer and we have no responsibility for the timely provision or sufficiency of such drawings. The drawings were received 5 weeks later than necessary, we having reminded the Architect that we required them at week 50. Clause 25.4.6.2 Clause 26.2.1.2	The facts are agreed. The nominated sub-contractor was dilatory in preparing the drawings and delayed the main contractor for 5 weeks as claimed. This will rank for extension of time under clause 25.4.6.2 and not clause 25.4.7 as the main contractor is entitled to the drawings in good time, particularly as no reference was made in the main Contract Documents to design being part of the sub-contract works (paras 2.8.13 and 2.8.14). By the same token, any direct loss and/or expense is reimbursable to the main contractor under clause 26.2.1.2. The Employer must be advised to recover direct from the sub-contractor under the Warranty the amount of this loss and/or expense and the Liquidated and Ascertained Damages denied him due to the extension of time. The Architect is not empowered to deduct this from amounts otherwise due under the Contract.
14. *Week ending No. 62.* The decorations to the kitchens – posponed under Event No. 12 above – was omitted by an Instruction from the Architect. This was a delay of 12 weeks since the postponement. Clause 25.4.5.1 Clause 26.2.7	Agreed but no delay to progress nor to the Completion Date occurred (para. 2.7.15). When part of the work is postponed in this way the Contractor is not entitled to go slow awaiting further instructions; he must still proceed regularly and diligently with the work (para. 2.3.8). Had this instruction occurred after the Architect had either fixed a new Completion Date or confirmed the existing then it could have been grounds for reducing some of the extensions of time as it is an omission of work (paras 2.7.36 to 2.7.41) but, in any event, it transpires not to have been an omission of work that was time critical.
15. *Week ending No. 66.* Following the delay by the electrical sub-contractor in providing drawings, work began to slow down after week 66 and some 4 weeks' progress was lost in all; this is the "knock on" effect of the earlier delay.	There was no evidence that the late drawings caused any such further delay. Our records and the Clerk of Works' reports indicate a lack of management on site. No extensions of time will be granted nor will any extra costs be reimbursed.

Contractor's claim	Amount £	Commentary

3. Summary of claim

3.1 *Extensions of time.* Practical completion took place at the end of week 96, a delay of 16 weeks in all. The separate delays noted above total 37 weeks and although some of these might have occurred concurrently there can be little doubt and they would have affected the completion date by at least 16 weeks. Indeed, it is likely that more than 16 weeks' extension of time is due indicating that, other things being equal, we accelerated the works. Accordingly we claim an extension of time of the 16 weeks.

Many of the delays claimed by the Contractor either did not in fact mean a delay (e.g. Events No. 1 and No. 14), were not on critical elements (e.g. Events No. 2 and No. 10) or were not due to the cause claimed (e.g. Event No. 4).

The following is the list of extensions that were critical to the completion of each block.

Event		Block A (weeks)	Block B (weeks)
3	Frame	2	2
5	Late instructions	1	1
6	Block A discrepancy	1	–
7	Restricted access	2	2
8/9	Strike by workmen	1	–
13	Late electrical drawings	–	5
		7	10

carried forward 106,494

Contractor's claim	*Amount £*	*Commentary*
		However, as there is only one Date for Completion, we are only concerned with the final completion of the work, in this case on Block B (para. W.2.2) and we therefore, in accordance with clause 25.3.3, grant a final total extension of time of 10 weeks in respect of the events listed above (of which we have already granted 6 weeks), thus:

		Clause 25.4.5.1 2 weeks
		Clause 25.4.6.2 6 weeks
		Clause 25.4.12 2 weeks
		                  _____
		                  10 weeks

		Thus a new Completion Date is fixed at the end of week 90. The 2 weeks granted under clause 25.4.12 for Event No. 7 will not give rise to an entitlement to recover loss and/or expense. No allowance can be made for accelerating the works; no instruction was given, nor could be given under the Contract, for this (paras 4.10.2 and 4.10.3).
3.2 *Loss and/or expense.* Under Clause 26, we claim total loss and/or expense of £155,080, which is detailed below.		The Contractor purports to have made this submission pursuant to his right under clause 26.1 to give his quantification. However, as is clear from our response tabulated above, it is of little or no use in determining his contractual entitlement. Accordingly we will make our observations on the Contractor's submission known to the Quantity Surveyor and

instruct him to ascertain the loss and/or expense on the basis of those observations (para. 3.5.27). In doing so he will write to the Contractor asking for specific cost records to be made available (paras 4.4.3 to 4.4.6, 4.10.5 and 4.10.6).

Extended preliminaries must be evaluated in detail (paras 3.3.10 and 4.6.2). In any event, in ascertaining the loss and/or expense, it is the Contractor's costs rather than rates in the Contract Bills with which we are concerned although it will be necessary to satisfy ourselves on any inordinate difference between these two sets of figures (para. 3.5.30).

The value of loss of output on labour must be assessed on an individual basis; overall calculations are in any event misleading (para. 4.4.21). On each occasion comparison must be made between the costs properly incurred and those that would have been incurred had the variation or disruption not taken place (para. 4.2.7).

The reduced output due to the frame variation has been evaluated in the final account under clause 13.5.1.2.

3.3 *Extended preliminaries.* 16 weeks' delay at the average rate per week for preliminaries in the Contract Bills. That is:

$$\frac{£284,972}{80 \text{ weeks}} \text{ per week} = £3,562.15 \times 16 \qquad 56,994$$

3.4 *Loss of output on labour.* The net labour value in the final account we calculate to be £495,000 based upon the proportions of labour and material in the Contract Bills. The contract took 20% longer than planned (i.e. 16 weeks in excess of the original 80 weeks); 16 weeks of the delays noted above are reimbursable. Our labour costs from our records total some £625,000 but we restrict this element of the claim to 10% of the final account value (i.e. half our delay as we were able to redistribute our resources).

10% of £495,000   49,500

carried forward   106,494

(continued)

W.6 CONTRACTOR'S CLAIM FOR EXTENSION OF TIME AND REIMBURSEMENT OF DIRECT LOSS AND/OR EXPENSE AND COMMENTARY THEREON BY THE ARCHITECT (*continued*)

Contractor's Claim	Amount £	Commentary
brought forward	106,494	
3.5 *Loss and/or expense on concrete frame.* The rates for formwork in our Contract Bills contain an error in that they were £6.00/m² too low. These rates cannot be rectified where used to build up the tender amount but this error should be made good as direct loss and/or expense for any variations – clause 26.		This is not a correct view of the meaning of clause 26 (paras 3.3.17 to 3.3.20).
130 m² at £6.00/m²	780	
3.6 *Inflation.* This project was let under a contract incorporating Clause 40 (NEDO Price Adjustment) and with a special provision that 10% of such fluctuations would not be reimbursable, i.e. it contained a 10% non-adjustable element*. At tender stage inflation was anticipated at 3% per annum and we allowed in our tender prices generally for the non-adjustable element of this. We based this on the contract sum (less p.c. sums) and inflation by the end of the contract of 4.6% (i.e. $3 \times \frac{80}{52}$)		This was a contract embodying the NEDO Formula with a 10% non-adjustable element (similar to that included in the JCT Standard Form of Building Contract 1998 Edition Local Authorities with Quantities). But the only increase we can recognise is that which arises directly from those matters that come with clause 26 – i.e. Event Nos. 2, 3, 5, 6, 10 and 13. Once again, the comparison has to be made between what the impact of the non-adjustable element was with what it would have been had these disruptions not taken place.
$\dfrac{£1,450,897 \times 4.6\% \times 10\%}{2} = £3,337$		
carried forward	107,274	

* This is a standard provision in the JCT Standard Form of Contract 1998 Edition Local Authorities With Quantities which, for the purposes of this example, has been imported by way of a special provision into JCT 98 PWQ

In practice inflation ran at an average of just over 6% per annum, i.e. the increase at the end of week 96 was 11% (i.e. $6 \times \frac{96}{52}$) it therefore cost us:

$$\frac{£1,450,897 \times 11\% \times 10\%}{2} = £7,980$$

an increase of $£7,980 - £3,337$ 4,643

3.7 *Overheads and profit.* We have applied the "Hudson formula" to this project as follows:

$$\frac{£1,688,897 \times 16 \times 10\%}{80}$$
    33,778

The caveat on the use of this formula has been overlooked (para. 4.8.61); moreover, no allowance has been made for mitigating the loss, nor for any recovery in the final account (section 4.8).

Interest is allowable as loss and/or expense (para. 4.9.1) but its calculation must be undertaken having regard to the detailed costs involved and the timing of the relevant written applications for reimbursement (paras 4.9.4 to 4.9.6).

3.8 *Interest charges.* We only allowed for financing this project until week 80. At this time some £155,000 at contract rates was still outstanding and was progressively paid over the remaining 16 weeks. The rate of interest on our overdraft was 10%. Thus our extra financing on this balance was

$$\frac{£155,000}{2} \times 10\% \times \frac{16}{52}$$
    2,385

(*continued*)

W.6 CONTRACTOR'S CLAIM FOR EXTENSION OF TIME AND REIMBURSEMENT OF DIRECT LOSS AND/OR EXPENSE AND COMMENTARY THEREON BY THE ARCHITECT (continued)

Contractor's claim	Amount £	Commentary
brought forward	148,080	
3.9 Cost of preparing this claim. Our consultant's charges for preparing this claim £7,000	7,000	Whilst the Contractor may choose to submit his quantification, he is not required to undertake this task; moreover, none of the quantification data submitted by him will be of assistance in determining his entitlement. Consequently these fees are inadmissible. Any costs directly incurred in providing the specific data requested by the Quantity Surveyor will be considered when looking at overhead reimbursement generally.
Total claim for direct loss and/or expense	£155,080	

W.7 ANNOTATED CHART OF PROGRESS ACHIEVED

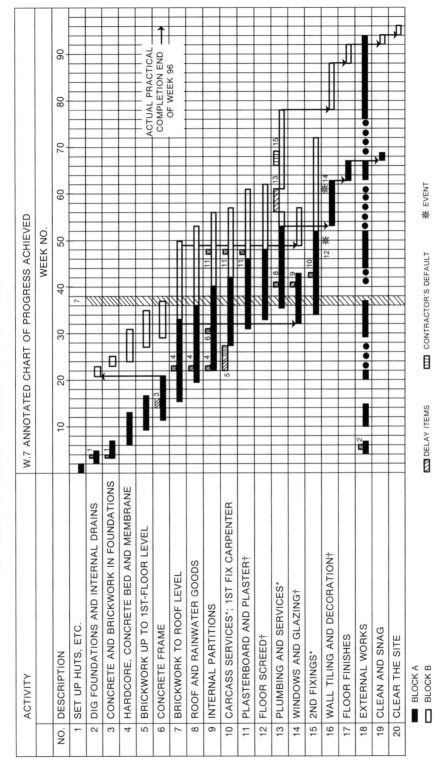

W.7 ANNOTATED CHART OF PROGRESS ACHIEVED

W.8   QUANTITY SURVEYOR'S ASCERTAINMENT OF DIRECT LOSS AND/OR EXPENSE

Introduction

W.8.1 Within a week of the receipt of the Contractor's submission we wrote, on behalf of the Architect, to the Employer with our initial reaction and assessment (paras 5.3.3 and 5.3.4) – see Appendix D.

W.8.2 We have the advantage of the annotation by the Architect of the details in the Contractor's claim (section W.6). As advised by the Architect (para. 3.2 of section W.6), we have obtained from the Contractor such particulars of his preliminaries costs and/or labour costs at the following events as are available and our detailed ascertainment of his entitlement under the contract is set out below:

Event 2
Event 3
Event 5
Event 6
Event 10
Event 13

W.8.3 The events complained of by the Contractor and the effect of the Architect's reaction thereto is summarised in Table W.1.

Loss of output on labour

£

W.8.4 Event 2
No specific records were kept highlighting the disruption on drains due to the change in invert levels (the breaking out of earlier work was valued on Daywork under clause 13.5.4). Some delay was experienced in the execution of the rest of the drainage. It is estimated that 5 men lost $\frac{1}{2}$ week's productive work as a result:

5 weeks at £235.00 per week × 50%   588

W.8.5 Event 3
Although the Clerk of Works has provided records of labour involved on the brickwork during the frame variation, these have been valued by us in the final account under clause 13.5.5.   –

W.8.6 Event 5
In our view, and that of the Architect, the late instructions on the services did not affect the productivity of other trades.   –

W.8.7 Event 6
The discrepancy discovered on the partitions in Block A resulted in lost output on both the partitions and the first fixings. There were 14 men involved who over the period concerned are judged to have each lost 70% of one week's output resulting in loss totalling:

14 men at £238.00 per week × 70%   2,332

Table W.1. Architect's response to Contractor's claim

Event	Principal reaction of Architect	Recognised delay	Extension of time for completion	Disruption of labour	Extended preliminaries
1. Soft spots	No variation involved	–	–	–	–
2. Variation to drains	Agreed but not critical	–	–	Yes	No
3. Variations to frame	Agreed	2 weeks	2 weeks	Effect on this and other work, such as it is, to be covered in valuation of variation	Covered in valuation of variation
4. Compound effect of earlier delays	This was lack of management	–	–	–	–
5. Late instruction on services nomination	Agreed	1 week	1 week	No	Yes
6. Discrepancy on Block A	Agreed	1 week but not critical to Block B	–	Yes	Yes – Block A only
7. Restricted access	Notice not given therefore loss and/or expense cannot be dealt with by Architect but a new Completion Date will be fixed at the end of the Contract – clause 25.3.3.1 (see para. 2.7.34)	2 weeks	2 weeks	–	–
8. Delay by statutory undertaker (in this instance, directly engaged by Employer)	Agreed	1 week but not critical to Block B	–	Concurrent with strike therefore no loss experienced	–
9. Strike	Agreed	Concurrent with last item	–	–	–
10. Postponement of second fixings	Agreed but not critical to progress	–	–	Yes	No

(continued)

Table W.1. (cont.)

Event	Principal reaction of Architect	Recognised delay	Extension of time for completion	Disruption of labour	Extended preliminaries
11. Effect of discrepancy on Block B	Contractor's responsibility	–	–	–	–
12. Postponement of kitchen decorations	Agreed but no effect on progress	–	–	–	–
13. Delay in issue of drawings by electrical sub-contractor	Agreed	5 weeks	5 weeks	Yes	Yes
14. Kitchen decorations omitted	Agreed but no effect on progress	–	–	–	–
15. General accumulation of delays	This was lack of management in the latter part of the job	–	–	–	–

w.8.8 Event 8

The delay by the statutory authority (engaged, in this instance, directly by the Employer rather than in a "statutory" capacity) caused no extra expense as there was a strike at the time and the Contractor is deemed to have allowed in his prices for such an eventuality (paras 2.9.2 and 3.6.1).   –

w.8.9 Event 10

The postponement of some of the second fixings was not critical and did not affect other trades. However, it is judged that the five carpenters involved at the time lost a total of some 25 days effective production over the period which is calculated at:

25 man days at £58.00 per day   1,450

w.8.10 Event 13

The delay in the electrical installation due to the late receipt of drawings from the electrical sub-contractor also delayed and disrupted the second fixings and decoration which was closely tied to the progress of the electrical work. Once it became clear that a significant delay would occur the Contractor had a duty to mitigate his losses by proper management. Of the 5 weeks' delay we judge that $2\frac{1}{2}$ weeks effective production was lost to these other trades comprising eight men, costing:

20 weeks (i.e. 8 men at $2\frac{1}{2}$ weeks each) @ £238.00 per week   4,760

Total loss of output on labour – carried to Summary at para. W.8.22   **£9,130**

Extended preliminaries

w.8.11 The effect on the works as a whole of extra involvement of Preliminaries items is given in Table W.2. This is arrived at by applying the pattern of use established in section W.4 to the delays noted in section W.6.

Table W.2. Summary of the effect on preliminary items of Extensions of Time granted

Element of Preliminaries affected by Relevant Events	(Event No. 5) Late instruction on services nomination (weeks)	(Event No. 6) Discrepancy on Block A (weeks)	(Event No. 13) Late drawings from electrical sub-contractor (weeks)	Totals (weeks)
General Foreman	1	–	5	6
Assistant Foreman	1	–	5	6
Storeman	1	–	5	6
Trade Foreman A	1	1	–	2
Trade Foreman B	1	–	5	6
Ganger	1	–	5	6
C of W hut	1	–	5	6
Crane hire	1 (Block A)	1 (Block A)	–	2 (Block A)
Crane use	–	1 (Block A)	–	1 (Block A)
Scaffolding	1 (Block A)	1 (Block A)	–	2 (Block A)

Table W.2 (cont.)

Element of Preliminaries affected by Relevant Events	(Event No. 5) Late instruction on services nomination (weeks)	(Event No. 6) Discrepancy on Block A (weeks)	(Event No. 13) Late drawings from electrical sub-contractor (weeks)	Totals (weeks)
Foreman's hut	1	–	5	6
Stores	1	–	5	6
Lighting and heating	1	–	5	6
Telephone	1	–	5	6

NB The extra preliminaries resulting from the variation to the frame (Event 3) were dealt with in valuing the variation (para. 3.3.10). The delay from Event No. 7 would qualify for an extension of time (i.e. for relief from Liquidated and Ascertained Damages) but not for recovery of loss and/or expense (but see para. W.8.24).

W.8.12 So the extra cost of time-based preliminaries, based on Table W.2, would be:

		£
General Foreman:	6 weeks at £570 per week	3,420
Assistant Foreman:	6 weeks at £275 per week	1,650
Storeman:	6 weeks at £200 per week	1,200
Trade Foreman: Block A,	2 weeks at £245 per week	490
Trade Foreman: Block B,	6 weeks at £245 per week	1,470
Ganger:	6 weeks at £210 per week	1,260
Clerk of Works hut including heating and lighting and rates	6 weeks at £60 per week	360
Crane:	2 weeks' hire at £625 per week	1,250
	1 week's use at £575 per week including operator, fuel and oil	575
Scaffolding:	2 weeks' hire at £73 per week	146
Foreman's hut and storage sheds including lighting, heating, attendance and rates:	6 weeks at £145 per week	870
Telephone rental and calls:	6 weeks at £23 per week	138
		12,829

Note that the costs are actual as distinct from the pricing assumptions made by the Contractor when pricing his Preliminaries.

No allowance has been made for inflation on the removal costs as these are negligible.

Extra cost of non-time based preliminaries:

(i) Value based preliminaries (e.g. transport) are not affected as the overall value of measured work is virtually unchanged. —

(ii) Allow for an extra maintenance of hoarding. 475

Extended Preliminaries – Carried to Summary at para. W.8.22   **£13,304**

Inflation

W.8.13 Part of the works have been delayed by the matters referred to in section W.6. This has exacerbated the effect of the non-adjustable element in the fluctuations formula. The proper comparison is between the adjustment for inflation that would have occurred had the project finished in accordance with the original contract period of 80 weeks and that which would have occurred had the project finished by the end of week 88, i.e. after allowing for an 8-week delay for reasons which also carry an entitlement to reimbursement of direct loss and/or expense (section 4.7). In both instances, an element of conjecture is involved as neither course represents what actually happened (para. 4.7.13). The results of computer printouts of both hypotheses, following curves of expenditure in line with those experienced in practice, produce a total adjustment under NEDO (i.e. 90% of the whole adjustment) as set out below and it is $\frac{1}{9}$ of the difference that is reimbursable, i.e. the 10% non-adjustable element.

	£
On an 88 week programme	68,850
On an 80 week programme	(56,719)
Extra adjustment	12,131

This represents 90% of the total, thus the 10% non-adjustable element amounts to £1,348.

Inflation – carried to Summary at para. W.8.22 **£1,348**

It is worth noting here that JCT has published *Formula Rules* (dated October 1987) which explain the application of NEDO; rule 4 provides that the formula shall not apply, amongst other things, to direct loss and/or expense. This is because direct loss and/or expense is based upon actual costs as they are incurred thus they do not require updating for inflation – if, however, there is an interval of time between those costs being incurred and reimbursement in respect thereof then the situation will be made good by way of the payment of interest (para. W.8.21).

Head office overheads and profit

W.8.14 The Contractor planned to recover the cost of head-office overheads and profit as follows (see Appendices A and B hereto). "Net" refers to amounts before any addition for head office overheads and profit:

		£
10%	on £860,000 net of measured work	86,000
5%	on £120,500 net of domestic sub-contractors' work	6,025
5%	on £238,000 net of Nominated Sub-Contractors' work	11,900
10%	on £259,064 net of Preliminaries (the number indicated is fractionally larger than 10% to which end see Appendix B hereto which shows the precise make up of this amount)	25,908
Total		129,833

459

This is an average of 8.72% on the total net amount of the Contract Sum (i.e. less profit, overheads and provisional sums) of £1,488,064.

W.8.15 Notwithstanding the cautionary notes concerning the use of the formula (para. 4.8.1 et seq.), for the sake of simplicity, a formula is used in this example. On the basis of the Hudson formula (para. 4.8.58) the Contractor would be entitled to recover head-office overheads and profit over the 8 weeks reimbursable extension of time (i.e. the extension of time for causes which also give an entitlement to reimbursement of direct loss and/or expense) at the rate stated in the Contract Sum, namely:

$$8.72\% \times £1,688,897 \frac{8 \text{ weeks}}{80 \text{ weeks}} = £14,727$$

W.8.16 But simply because the Contractor used 10% and 5% as percentages for head-office overheads and profit recovery in his tender it does not follow that these are the percentages to apply when ascertaining the amount of his loss. Moreover, by following this route, profit and overheads percentages are added to figures which already include profit and overheads; and no allowance is made for the recovery of head-office overheads and profit contained in the valuation of any Variations. The test for the relevant percentage must be that percentage recovery which was prevalent in the market place when the extension of time prevented the Contractor earning a return on other work (paras 4.8.37 and 4.8.38). Assume for this purpose that since the tender was submitted the market for construction work has hardened. From an analysis of the Contractor's accounts and tenders submitted successfully on other projects during the period of extension of time (the time when this alternative work might have been undertaken) it was clear that he could not afford to add more than 6.5% on his own work and 3.25% on sub-contract work for the recovery of head-office overheads and profit and remain competitive. On the basis of the above subdivision in the Contract Sum this equates to an average of 5.67%. Accordingly, this is the percentage to apply.

W.8.17 Furthermore, the Contractor will already have recovered some contribution to the cost of his head-office overheads and profit by virtue of the fact that Variations to the frame (Event No. 3) have been valued using the rates and prices in the Contract Bills which will already include the percentages for head-office overheads and profit. Such costs thus recovered must be set against the recovery of head-office overheads and profit arising from the extension of time for that particular event (para. 4.8.40 et seq.).

W.8.18 By reference to the formulae set out at para. 4.8.56 the Contractor's entitlement under this heading is as follows:

1. For each extension of time for causes which also carry an entitlement to recover direct loss and/or expense, taken individually, in respect of a matter which also entitles the Contractor to an adjustment to the Contract Sum by reference to the rates and prices in the Contract Bills (i.e. the Extension of time in respect of Event No. 3 – Variations to frame):

£

$$\left(\frac{2 \text{ weeks} \times £1,448,064 \times 5.67\%}{80} \right) - (£5,507 \times 10\%) = \qquad 1,559$$

2. For all other extensions of time for causes which also carry an entitlement to recover loss and/or expense, taken in aggregate:

$$\frac{6 \text{ weeks} \times £1,488,064 \times 5.67\%}{80} = \qquad\qquad 6,328$$

7,887

The need to distinguish between the above two components is not clear from this example but had the result of the application of the first formula above been a negative amount, as it could well have been, then this negative amount would have had to be ignored and not set against any of the amounts accruing as a result of applying the second formula above. To the above should be added the cost to the Contractor of collating and submitting data to support the above ascertainment which was a cost incurred after the period considered in the above calculation of head-office overheads. It is therefore to be calculated separately and amounts to say: 650

Finance charges formed part of the head-office overheads and profit and are therefore not separately calculated (para. 4.8.22).

Head office overheads and profit – carried to Summary at para. W.8.22   **£8,537**

Interest payments

W.8.19 Interest forms part of loss and/or expense (para. 4.9.1). It is therefore necessary to establish how much loss and/or expense is chargeable to each occurrence and calculate interest on each amount from the time the expenditure was incurred until the date that it was paid.

W.8.20 The various rates of interest payable by the Contractor on a weekly basis are recorded in Table W.3.

W.8.21 The interest payable on the above loss, given that settlement was made in week 104 (8 weeks after Practical Completion) would be as set out in Table W.4; this totals £3,324.

Interest – carried to Summary at para. W.8.22   **£3,324**

Summary of ascertainment of loss and/or expense

		£
W.8.22	Loss of output on labour	9,130
	Extended preliminaries	13,304
	Inflation	1,348
	Head-office overheads and profit	8,537
	Interest	3,324
	Total	**£35,643**

Table W.3. Interest rates payable by the Contractor during the Contract

A Week No.	B Prevailing interest rate %	C Accumulator	A Week No.	B Prevailing interest rate %	C Accumulator	A Week No.	B Prevailing interest rate %	C Accumulator	A Week No.	B Prevailing interest rate %	C Accumulator
1	8.50	8.50	27	9.25	248.25	53	9.25	485.25	79	9.00	722.50
2	8.50	17.00	28	9.25	257.50	54	9.25	494.50	80	9.00	731.50
3	8.50	25.50	29	9.25	266.75	55	9.25	503.75	81	9.00	740.50
4	8.50	34.00	30	9.25	276.00	56	9.25	513.00	82	9.00	749.50
5	8.50	42.50	31	9.25	285.25	57	9.25	522.25	83	9.50	759.00
6	8.50	51.00	32	9.25	294.50	58	9.25	531.50	84	9.50	768.50
7	8.50	59.50	33	9.25	303.75	59	9.25	540.75	85	9.50	778.00
8	8.50	68.00	34	9.25	313.00	60	9.25	550.00	86	9.50	787.50
9	9.25	77.25	35	9.25	322.25	61	9.25	559.25	87	9.50	797.00
10	9.25	86.50	36	9.25	331.50	62	9.25	568.50	88	9.50	806.50
11	9.25	95.75	37	9.25	340.75	63	9.25	577.75	89	9.50	816.00
12	9.25	105.00	38	9.25	350.00	64	9.25	587.00	90	9.50	825.50
13	9.25	114.25	39	9.00	359.00	65	9.25	596.25	91	9.50	835.00
14	9.25	123.50	40	9.00	368.00	66	9.00	605.50	92	9.50	844.50
15	9.25	132.75	41	9.00	377.00	67	9.00	614.50	93	9.50	854.00
16	9.25	142.00	42	9.00	386.00	68	9.00	623.50	94	9.50	863.50
17	9.25	151.25	43	9.00	395.00	69	9.00	632.50	95	9.50	873.00
18	9.25	160.50	44	9.00	404.00	70	9.00	641.50	96	9.75	882.75
19	9.25	169.75	45	9.00	413.00	71	9.00	650.50	97	9.75	892.50
20	9.25	179.00	46	9.00	422.00	72	9.00	659.50	98	9.75	902.25
21	10.00	189.00	47	9.00	431.00	73	9.00	668.50	99	9.75	912.00
22	10.00	199.00	48	9.00	440.00	74	9.00	677.50	100	9.75	921.75
23	10.00	209.00	49	9.00	449.00	75	9.00	686.50	101	9.75	931.50
24	10.00	219.00	50	9.00	458.00	76	9.00	695.50	102	9.75	941.25
25	10.00	229.00	51	9.00	467.00	77	9.00	704.50	103	9.75	951.00
26	10.00	239.00	52	9.00	476.00	78	9.00	713.50	104	9.75	960.75

To calculate interest payable from any week to the date of payment take the accumulator for that week, deduct it from 960.75 (the accumulator at the date of payment) and divide by 52. For week 34 the calculation would be:

$$\frac{960.75 - 313.00}{52} = 12.46$$

Or divide the relevant accumulator by 52 (e.g. 6.02 in week 34) and take this from the accumulator at the date of payment divided by 52 (e.g. 18.48 in week 104), i.e. 18.48 − 6.02 = 12.46.

Table W.4. Calculation of interest on the various cost components

	Drains Event 2 (section W.6)	Frame variations Event 3 (section W.6)	Late instructions Event 5 (section W.6)	Discrepency on Block A Event 6 (section W.6)	Postponement – second fixings Event 10 (section W.6)	Late drawings from Nominated Sub-Contractor Event 13 (section W.6)	Totals
1. Week(s) covered by the event	6	16–18	27–28	31	43	56–61	
2. Accumulator (Table W.3)	51.00	151.25	252.88	285.25	395.00	536.13	
3. Line 2 divided by 52	0.98	2.91	4.86	5.49	7.60	10.31	
4. Interest payable as a %: 18.48 minus line 3	17.50	15.57	13.62	12.99	10.88	8.17	
	£	£	£	£	£	£	£
5. Net labour loss	588			2,332	1,450	4,760	9,130
6. Preliminaries			2,770	1,574		8,960	13,304
7. Inflation		337	168			843	1,348
8. Overheads and profit		2,134	1,067			5,336	8,537
9. Total cost	588	2,471	4,005	3,906	1,450	19,899	32,319
10. Interest: line 4 × line 9	103	385	545	507	158	1,626	3,324
11. Grand totals	691	2,856	4,550	4,413	1,608	21,525	35,643

NB Where the delay is in excess of a week, the accumulator is based on the average accumulator over the period of delay.
The preliminaries are apportioned in relation to the various periods set out in Table W.2.
Inflation, overheads and profit are apportioned in relation to the respective extensions of time.

W.8.23 The Employer, however, has two counterclaims to pursue:

(a) **Liquidated and ascertained damages.** These are entered into the Appendix at £3,000 per week. Practical completion took place at week 96; and the extended completion date was week 90 (week 88 plus the 2 extra weeks for Event No. 7), a delay in completion of 6 weeks. Provided that the Architect has properly certified as required by clause 24 (para. 2.11.4), the Employer may deduct $6 \times £3,000 = £18,000$ from any amount due or to become due to the Contractor; the Architect cannot make the deduction from payment certificates under the Contract. Moreover, the Employer is required to register his claim to damages in accordance with clause 24.2.1 (para. 2.11.5).

(b) **Default by the nominated sub-contractor in issuing drawings for electrical work to be claimed under the Warranty Agreement.** The amount of loss and/or expense resulting from this delay totals £21,525 (see Table W.4). In addition, the main contractor was granted a 5-week extension of time for this delay, thus denying the Employer $5 \times £3,000$ or £15,000 of Liquidated and Ascertained Damages. Thus the total loss suffered by the Employer is $£21,525 + £15,000 = £36,525$. This amount must be recovered direct by the Employer under the Warranty and cannot be deducted from payment certificates. However, the Employer is likely to look to the Architect or Quantity Surveyor for advice on the amount involved.

W.8.24 Finally, it is noted that the Contractor may have cause to proceed against the Employer for breach of contract in restricting access under Event No. 7 (paras 3.5.20 and 3.5.21) as although he was ultimately granted an extension of time he did not recover any associated loss and/or expense. It would be pragmatic in such an event for the Architect to seek the Employer's consent to deal with any such claim as part of the Final Account rather than bear the costs of settling what is likely to be a relatively minor amount in the context of the Contractor pursuing a remedy at common law.

W.9 SOME PRACTICAL ADVICE!

W.9.1 The Worked Example is intended to provide a practical application of the subject matter addressed in this book rather than amount to an exhaustive listing of all of the heads of claim that a Contractor could pursue. Also, there are respects in which the Worked Example could have gone into further detail but to do so may have obscured the message. For example:

(a) The cost of the hoarding within the extended preliminaries (para. W.8.12) could have been ascribed to certain, rather than all, events giving rise to an extension of time rather than be applied on a pro-rata basis to the additional cost of time based preliminaries (Table W.4).

(b) Similarly, the calculation for inflation, overheads and profit (paras W.8.13 and W.8.14 et seq. and Table W.4) could have been calculated against each individual matter giving rise to an extension of time rather than be calculated as an overall figure subsequently spread back through the respective extensions of time. It is possible, for example, that an individual calculation of inflation for Event No. 5

(Table W.4) may show that there was no inflation between weeks 27 and 28 such that the allowance in respect thereof ought instead to be allocated to Event Nos. 3 or 13.

(c) Where a delay lasts for more than 1 week, the interest charge accumulator (Tables W.3 and W.4) is based on the average accumulator in that delay period. However, closer scrutiny may reveal that the Contractor's outlay of monies in the delay period under consideration was not on a straight-line basis but, rather, more towards the end of the delay period. To account for this, it would be necessary to calculate the interest in each individual week having first ascribed the principal sum to the applicable week.

W.9.2 Whilst this book seeks to encourage a proper and detailed analysis of Contractor's claims in a manner which is also consistent with the Contract, ascertainment is not an exercise which should be taken to extremes which are incommensurate with the sums of money being addressed. The Architect ordinarily delegates the function of ascertaining of direct loss and/or expense to the Quantity Surveyor whilst he, in turn, is likely to be engaged by the Employer on a basis which entitles him to be paid additional fees where such work is carried out. This being so, the Quantity Surveyor should use his professional judgement to arrive at an ascertainment of direct loss and/or expense of such accuracy as is necessary to do justice to the Contractor whilst avoiding embarkation on an exercise which is so detailed that such accuracy is negligibly improved but at considerable cost, in additional fees, to the Employer. Technically though, the Quantity Surveyor is not granted any degree of latitude under the Contract and he should, therefore, obtain the consent and/or approval for this course of action from the Employer and the Contractor.

Appendix A

ANALYSIS OF THE CONTRACT SUM

1. Main Contractor

		£	£
1.1	Measured work	860,000	
	NB This total represents £465,000 labour and £395,000 materials		
1.2	Profit (4%) and overheads (6%) = 10%	86,000	946,000

2. Sub-contractors

2.1 Domestic sub-contractors

	£
Finishes	78,000
Glazing	10,000
Painting	32,500
	120,500

			£	£
2.2	Profit and overheads	5%	6,025	126,525
2.3	Nominated sub-contractors			

	£
Plumbing and engineering	150,000
Electrical	88,000
	238,000

			£	£
2.4	Profit and overheads	5%	11,900	
2.5	General and special attendance		10,500	260,400

3. Preliminaries, contingencies and provisional Sums

		£	£
3.1	Preliminaries (see Appendix B)	284,972	
3.2	Contingencies	32,000	
3.3	Dayworks (including percentages)	20,000	
3.4	Provisional Sum for soft spots and additional work in foundations	19,000	355,972

		£
4.	Contract Sum	£1,688,897

Note: The foregoing figures, net of preliminaries, profit and overheads add up to £1,300,000 which ties in with the example at page 97 of *Spon's Architects' and Builders' Price Book 1999*.

Appendix B

ANALYSIS OF PRELIMINARIES
SECTION OF THE CONTRACT BILLS

Based upon the preliminaries section of *Spon's Architect's and Builders' Price Book* (1999) pages 97 to 111 with additions for profit and overheads. Specific adjustments have been made to suit this worked example, e.g. the Site Administration is resourced to a greater extent than is set out in the *Spon's Price Book*.

	Quantity unit	Rate	Fixed charge £	Value charge £	Time charge £	Total £
Site Administration						
General foreman for contract period	80 wks	£530			42,400	
Holiday relief	4 wks	£530			2,120	
Assistant foreman (weeks 10 to 70)	60 wks	£265			15,900	
Storeman/checker (weeks 6 to 66)	60 wks	£195			11,700	
Trades foreman:						
Block A (weeks 2 to 62)	60 wks	£235			14,100	
Block B (weeks 18 to 78)	60 wks	£235			14,100	
Ganger for contract period	80 wks	£220			17,600	
					117,920	
Profit and Overheads		10%			11,792	
					129,712	129,712
Defects after completion						
0.25% of estimated value of contract before preliminaries, profit and overheads are added	£1,300,000	0.25%		3,250		
				3,250		
Profit and Overheads		10%		325		
				3,575		3,575

carried forward 133,287

467

	Quantity unit	Rate	Fixed charge £	Value charge £	Time charge £	Total £
					brought forward	133,287
Insurance of Works against Specified Perils						
Estimated value of contract including preliminaries – say	1,689,000					
Estimated increase in price during contract period 6%	101,340					
	1,790,340					
Estimated increase in cost during reinstatement period 3%	53,710					
	1,844,050					
Professional fees 16%	295,048					
Premium on	£2,139,098	0.1%		2,139		
				2,139		
Profit and Overheads		10%		214		
				2,353		2,353
Office for Clerk of Works						
Haulage to and from the site	item	£150	150			
Hire charge for 15 m^2 (weeks 4 to 80)	76 wks	£2/m^2			2,280	
Lighting, heating and attendance	76 wks	£25			1,900	
Rates for 15 m^2	76 wks	£14/m^2 pa			307	
			150		4,487	
Profit and Overheads		10%	15		449	
			165		4,936	5,101
					carried forward	140,741

	Quantity unit	Rate	Fixed charge £	Value charge £	Time charge £	Total £
					brought forward	140,741
Administrative charges						
1% of estimated value of contract before preliminaries, profit and overheads are added	£1,300,000	1%			13,000	
					13,000	
Profit and Overheads		10%			1,300	
					14,300	14,300
Site accommodation						
Haulage to and from site:						
Foreman's office	item	£150	150			
Sheds	item	£300	300			
Hire charge:						
Foreman's office 15 m² (weeks 4 to 80)	76 wks	£1.70/m²			1,938	
Sheds 40 m² (weeks 6 to 76)	70 wks	£1.70/m²			4,760	
Lighting, heating and attendance on office	76 wks	£25			1,900	
Rates for 15 m²	76 wks	£14/m² pa			307	
Rates for 40 m²	70 wks	£14/m² pa			754	
			450		9,659	
Profits and Overheads		10%	45		966	
			495		10,625	11,120
Lighting and power for the Works						
1% of estimated value of contract before preliminaries, profit and overheads are added	£1,300,000	0.03%	390			
	£1,300,000	0.97%			12,610	
			390		12,610	
Profit and Overheads		10%	39		1,261	
			429		13,871	14,300
					carried forward	180,461

	Quantity unit	Rate	Fixed charge £	Value charge £	Time charge £	Total £
					brought forward	180,461

Water for the Works
0.33% of estimated
value of contract before
preliminaries, profit
and overheads are

	Quantity unit	Rate	Fixed charge	Value charge	Time charge	Total
added	£1,300,000	0.33%			4,290	
Connection	item	£130	130			
Standpipe	item	£50	50			
Piping	44 m	£5/m	220			
			400		4,290	
Profit and Overheads		10%	40		429	
			440		4,719	5,159

Temporary telephones for use of the Contractor
Connection charge

	Quantity unit	Rate	Fixed charge	Value charge	Time charge	Total
for line	item	£99	99			
Connection charge for external ringing device	item	£10	10			
Line rental (6 quarters)	1 month	£28.67			29	
	17 months	£14.33			244	
External ringing device rental	6 quarters	£2.50			15	
Calls	76 wks	£22			1,672	
			109		1,960	
Profit and Overheads		10%	11		196	
			120		2,156	2,276
					carried forward	187,896

	Quantity unit	Rate	Fixed charge	Value charge	Time charge	Total
			£	£	£	£
					brought forward	187,896

Safety and the like
0.66% of estimated
value of contract before
preliminaries, profit
and overheads are

	Quantity unit	Rate	Fixed charge	Value charge	Time charge	Total
added	£1,300,000	0.03%	390			
	£1,300,000	0.63%			8,190	
			390		8,190	
Profit and Overheads		10%	39		819	
			429		9,009	9,438

Removing rubbish
0.2% of estimated
value of contract before
preliminaries, profit
and overheads are

	Quantity unit	Rate	Fixed charge	Value charge	Time charge	Total
added	£1,300,000	0.05%	650			
	£1,300,000	0.15%			1,950	
			650		1,950	
Profit and Overheads		10%	65		195	
			715		2,145	2,860

Drying the Works
0.1% of estimated
value of contract before
preliminaries, profit
and overheads are

	Quantity unit	Rate	Fixed charge	Value charge	Time charge	Total
added	£1,300,000	0.1%	1,300			
			1,300			
Profit and Overheads		10%	130			
			1,430			1,430
					carried forward	201,624

	Quantity unit	Rate	Fixed charge	Value charge	Time charge	Total
			£	£	£	£
					brought forward	201,624
Small tools						
1% of estimated value of labour	£465,000	1%			4,650	
					4,650	
Profit and Overheads		10%			465	
					5,115	5,115
Mechanical plant						
Tower crane static 30 m radius 4/5 tonne maximum load: Haulage to and from the site commissioning and dismantling	item	£5,000	5,000			
Hire (weeks 15 to 45)	30 wks	£600			18,000	
Electricity, fuel and oil (weeks 15 to 29 and 31 to 45)	28 wks	£170			4,760	
Operator at 40 hours per week (weeks 15 to 29 and 31 to 45)	28 wks	£10/hr			11,200	
			5,000		33,960	
Profit and Overheads		10%	500		3,396	
			5,500		37,356	42,856
Transport for personnel						
Labour content is assessed at 2,050 man weeks and each man receives £1.54 per day or £7.70 per 5-day week	2,050 man/ wks	£7.70			15,785	
					15,785	
Profit and Overheads		10%			1,579	
					17,364	17,364
					carried forward	266,959

472

	Quantity unit	Rate	Fixed charge	Value charge	Time charge	Total
			£	£	£	£
				brought forward		266,959
Scaffolding						
Delivery and erection:						
Block A	item	£2,650	2,650			
Hire for 14 weeks	14 wks	£70			980	
Dismantle and remove	item	£1,300	1,300			
Delivery and erection:						
Block B	item	£2,650	2,650			
Hire for 16 weeks	16 wks	£70			1,120	
Dismantle and remove	item	£1,300	1,300			
			7,900		2,100	
Profit and Overheads		10%	790		210	
			8,690		2,310	11,000
Temporary hoardings						
100 m of plywood decorated hoarding	100 m	£61.50	6,150			
Extra for pair of gates	item	£225	225			
			6,375			
Profit and Overheads		10%	638			
			7,013			7,013
TOTAL OF PRELIMINARIES						£284,972

SUMMARY OF ANALYSIS OF PRELIMINARIES

	Fixed Preliminaries	Value Based Preliminaries	Time Based Preliminaries	Profit and Overheads	Totals
	£	£	£	£	£
Site administration			117,920	11,792	129,712
Defects after completion		3,250		325	3,575
Insurance of Works		2,139		214	2,353
Office for Clerk of Works	150		4,487	464	5,101
Administrative charges			13,000	1,300	14,300
Site accommodation	450		9,659	1,011	11,120
Lighting and power for the Works	390		12,610	1,300	14,300
Water for the Works	400		4,290	469	5,159
Temporary telephones for Contractor	109		1,960	207	2,276
Safety and the like	390		8,190	858	9,438
Removing rubbish	650		1,950	260	2,860
Drying the Works	1,300			130	1,430
Small tools			4,650	465	5,115
Mechanical plant	5,000		33,960	3,896	42,856
Transport for personnel			15,785	1,579	17,364
Scaffolding	7,900		2,100	1,000	11,000
Temporary hoardings	6,375			638	7,013
Totals	**£23,114**	**£5,389**	**£230,561**	**£25,908**	**£284,972**

Appendix C

ANALYSIS OF FINAL ACCOUNT

		Basic value	Addition for overheads and profit
		£	£
1.	Contract Sum	1,559,064	129,833
2.	Variations – net value		
	Drains	1,310	131
	Block A frame	5,507	551
	Kitchen decoration	(2,728)*	(273)
3.	Adjustment of PC and Provisional Sums – net value		
	Foundations	(1,827)	(183)
	Contingencies and daywork	(17,042)	–
	PC Sums	16,375	819
		1,560,659	130,878
			1,560,659
Total account excluding loss and/or expense and excluding fluctuations			£1,691,537

* Figures in brackets are net omissions.

Appendix D

INITIAL ASSESSMENT OF CONTRACTOR'S CLAIM SUBMISSION

Item	Amount claimed by Contractor	Approximate entitlement assessed by Quantity Surveyor	Comments upon the difference
	£	£	
Extended Preliminaries	56,994	14,250	Final extensions of time are likely to indicate that the Contractor was in culpable delay to some extent; he is not entitled to extra preliminaries for this period (say £28,500). Not all the remaining preliminaries will have been affected – allow say 50%.
Loss of output on labour	49,500	8,000	Again lost output due to the Contractor's delay is not reimbursable (say £29,500). Moreover, an extension of time is not pro rata to the loss and/or expense (para. 4.4.13). Many of the trades were not affected by the delays.
Loss and/or expense on concrete frame	780	–	This is not reimbursable (paras 3.3.17 to 3.3.19).
Inflation	4,643	1,250	The Employer is not responsible for the increase in inflation, nor for that occurring during the Contractor's delay. He is only responsible to the extent that his act or default has exacerbated the burden of inflation borne by the Contractor (section 4.7).
Overheads and Profit	33,778	7,500	Even if this formula is an acceptable method of calculating the amount due, no allowance has been made for changed market conditions nor for any recovery under the final account (para. 4.8.38).
Interest charges	2,385	3,000	This is payable from the time the loss was incurred until the time at which the amount is paid.
Cost of preparing the claim	7,000	–	This is not reimbursable (para. 4.10.5).
	£155,080	£34,000	

INDEX

In this index all references are to paragraph numbers; where a number of sequential paragraphs are to be referred to, only the first in that series is listed. The first digit in each reference is to the Chapter Number as set out below.

Direct loss and/or expense *(cont.)*
  global claims for 4.2.9, W.6.1
  grounds for claiming, in contrast to
    claiming extension of time 2.1.9, 2.5,
    3.5.3
  inadmissible items of, 4.2.4, 4.10, W.6.3
  initial assessment of, 5.3.3, W.Appx D
  items not constituting, 11.3.42
  JCT 98 PWQ provisions relating to,
    contrasted with
    GC/Works 98 11.1.8
    JCT 98 MC 10.3.9, 10.4
    JCT 98 MW 7.3.10, 7.4.1
    JCT 98 WCD 8.3.8, 8.4
    JCT IFC 9.3.8, 9.4
    NSC/C 6.3.5, 6.4
  keeping records for 3.5.25, 4.4.3
  List of Matters comprising 3.5.3, 6.3.8,
    7.3.14, 8.3.13, 9.3.12, 10.3.13,
    11.2.13, 11.3.35
  procedures for determining 6.3.5,
    7.3.10, 8.3.8, 9.3.8, 10.3.9, 11.3.24,
    11.4
  role of Quantity Surveyor and Contractor
    in assessing, contrasted 5.3.8
  rule of thumb for approximate calculation
    of 4.4.16, 5.3.3
  safeguards in calculating, abandoned 3.1.5,
    3.3.6, 3.4.9
  time limit for determining 11.3.29
Discrepancies 2.8.15, 3.5.5, 7.3.17, 8.1.8,
    8.2.10, 8.3.14, 11.3.3, W.6.2.6,
    W.6.2.11
  Contractor's obligation to search for
    2.8.16, W.6.2.11
Dissatisfaction by the Contractor 12.2.22

Early warning 5.2.11, 12.2.39, 12.4.18,
    12.4.22, 12.7.8, 12.7.12, 12.7.23,
    12.7.40
Egan, Sir John 12.2.32
Employer and Contractor to co operate
    11.2.15
Employer's Agent 8.1.4
Employer's Requirements 8.1.6
  in conflict with Contractor's Proposals
    8.1.8

Engineering and Construction Subcontract
    12.1.15, 12.7.25
Expense (see also direct loss and/or expense)
  definition of 11.3.24
  list of matters giving rise to 11.3.35
  provision of details by Contractor 11.3.33
  recovery of, due to:
    advice from Planning Supervisor 11.3.42
    delay in decisions by Project
      Manager 11.3.39
    delay in direct works 11.3.40
    delay in possession of site 11.3.38
    other works 11.3.37
    timetable for determining 11.3.29
Extension of time
  absence of express provisions for 2.1.4
  advantages of early decisions on 2.3.5,
    2.7.42
  after Practical Completion 2.7.32
  application made after completion 11.2.10
  Architect's obligations 2.7.25, 2.9.10
  Architect's role in granting 10.2.7
  causes giving rise to 11.2.13
    acts of prevention 7.2.6, 9.2.6
    adverse weather 2.8.4, 7.2.11, 11.2.19
    antiquities 2.8.25
    civil commotion 2.8.6
    compliance with instructions 2.8.8,
      10.2.11
    failure to provide labour or materials
      2.8.58
    suspension of work 2.8.71
    default of Management Contractor
      10.2.5
    deferment of Site Possession 1.2.10,
      2.8.24, 2.8.64, 2.11.2, 3.5.4, 7.2.12
    delay on part of other Works Contractors
      10.2.11
    discrepancies 2.8.15
    Employer's failure to grant access to site
      2.8.62
    Force Majeure 2.8.2, 11.2.18
    Governmental action 2.8.56
    insured risks 11.2.23
    late instructions 2.8.29, 2.8.33
    Nominations 2.8.26, 2.8.45

opening up and testing 2.8.28, 3.5.3,
    10.2.11
Project Manager's instructions 11.2.25
Specified Perils 2.8.5
Statutory undertaker's work 2.8.61
supply of materials by Employer 2.8.55
variations 2.8.18
work by Employer or his agents 2.8.51
data to support an application for 2.7.6
delay in granting 2.3
delay in submitting application for 2.7.11
detailed components of procedure for
    granting 2.7
effect of accepting a clause 13A quotation
    on 2.7.2
effect of float on 2.7.15
effect of fluctuation clauses on 2.12, 6.2.12,
    7.2.15, 8.2.11, 9.2.12, 10.2.15
effect of omissions of work on 2.7.36
Employer's benefit in granting 2.2.2
failure by Contractor to make application
    for 2.7.35
failure properly to implement the provisions
    of 2.2.2
giving notice by Contractor of 2.7.6, 11.2.4
grounds for claiming, in contrast to
claiming loss 2.1.9, 2.5, 3.5.2
inadequate data from Contractor in support
    of 2.7.23
JCT 98 PWQ contrasted with:
    GC/Works 98 11.1.8, 11.2.2
    JCT 98 MC 10.2.2, 10.2.10
    JCT 98 MW 7.2.2, 7.2.9
    JCT 98 WCD 8.2.1, 8.2.9
    JCT IFC 9.2.1, 9.2.10
    WC/2 10.22, 10.2.10
    NSC/C 6.2.2, 6.2.10
lack of endeavour not valid grounds for
    11.2.12
irreversible 7.2.7
procedures for granting 6.2.2, 7.2.1, 8.2.1,
    9.2.1, 10.2.2, 11.2.1
proceeding with diligence in support of
    11.2.12
relationship between Management
    Contractor and Works Contractor and
    Architect in granting 10.2.9

Relevant Events 2.8, 6.2.10, 7.2.9, 8.2.9,
    9.2.10, 10.2.10
retrospectively granted 2.3.2
time limit for 2.7.22, 11.2.8
timing of application for 2.7.7
timing of Architect's decision in granting
    2.7.11

Float 2.7.15, 2.10.8, 4.6.5, 12.2.36, 12.7.36,
    W.5.2, W.6.2.2
Finance charges 4.8.22, 11.2.8, 11.3.25,
    11.4.4 W.8.19
Fitness for purpose 11.1.32, 12.2.12
Flow charts 12.1.15, 12.4.4, 12.4.10
Fluctuations in prices 6.2.12, 7.2.15, 8.2.11,
    9.2.12, 10.2.15, W.8.13
freezing of 2.12.1, 12.7.35
Force majeure 2.8.2, 2.8.49, 4.4.17, 11.2.18
Forecast of actual cost 12.2.30, 12.3.2,
    12.4.54, 12.7.25, 12.7.29, 12.7.41

GC/Works 98
    acceleration under 11.1.35
    designer's role under 11.1.15
    insurances and bonds under 11.1.31
    parties to 11.1.14
    payments under 11.1.24
    programme and progress under 11.1.16
    role of Quantity Surveyor and Project
        Manager under 11.1.14
    structure of 11.1.9
GC/Works/1 Edition 3, aims of 11.1.4
Global claims 4.2.9, W.6.1.2
Government Agencies' use of GC/Works 98
    11.1.2
Guidance Notes 12.1.15

Head office overheads and profit 4.8, W.8.14
Heads of claim 4.2.3, 11.4.1
House of Lords 1.2.4

Institution of Civil Engineers 12.2.32
    Director General of 12.3.32
    form of contract 5th edition 12.4.36
    form of contract 6th edition 12.4.37, 12
        4.41
Identified Terms 12.2.4

Inconsistencies 12.4.2, 12.4.34, 12.5.11, 12.7.43

Inflation 4.7, 12.1.19, 12.7.33, W.6.3

Influences beyond the control of the parties 12.4.47

Information Release Schedule 2.8.29, 2.10.10, 4.3.6, 5.1.9

Instructions
  definition of 11.3.1
  failure of Contractor to comply with 11.3.4
  list of items constituting 11.3.1

Insurances 3.3.12, 4.6.2, 4.8.16, 11.1.31

Interest 3.5.31, 4.8.10, 4.8.24, 4.9, 11.3.24, 11.4.2, 11.4.6, 12.6.1, W.8.19

Joint Review of Procurement and Contractual Arrangements in the United Kingdom 12.1.6

Judgment in calculating loss and/or expense 1.2.2, 3.3.26, 3.5.24, 4.2.7, 4.4.11, 4.5.2, 7.3.12, 11.3.9, 11.3.19, W.6.3.6, W.8.13

Labour and materials, failure to procure, 2.8.58

Latham Report 11.1.6, 12.1.6, 11.1.44, 12.1.21, 12.2.32

Latham, Sir Michael 11.1.6, 11.2.16, 12.1.6, 12.1.11, 12.1.20, 12.2.28

Law, changes in 12.1.19

Legal response to NEC/2 12.1.23

Levene, Sir Peter 11.1.13, 12.1.21

Levene Report 11.1.13, 12.1.21

Liquidated and ascertained damages 2.1.3, 10.1.34
  as a measure of 'best endeavours' 2.7.45, 4.2.11, 11.2.12
  claimed by Employer from Nominated Sub-Contractor 2.8.49, 6.1.6, W.8.23
  conditions precedent for claiming 2.4.1
  effect on fluctuations payments on 2.14
  featuring in claim settlement calculations 5.4.9
  limitation in claiming 10.1.26, 10.2.11
  loss of entitlement to 2.2.2, 11.2.3
  method of calculation of 2.13, 2.14
  procedure for imposing 2.11.3, 2.11.6

repaying 2.7.30
  separate rates for, where sectional completion exists 2.9.4

List of Matters 3.5.3, 6.3.8, 7.3.14, 8.3.13, 9.3.12, 10.3.13, 11.2.13, 11.3.35

Loss and/or expense (see direct loss and/or expense

Loss, definition of 11.4.3

Low performance 12.1.15

Main options 12.1.14

Management Contractor
  obligation to Employer by 10.1.3
  services to be provided by 10.1.6
  to co operate with professional team 10.1.7

Management Contracts 10.1.1, 12.1.15

Management fee 10.1.10, 10.1.20, 10.2.17, 10.4.10

Management philosophy in NEC/2 12.2.28

Master programme (see also Programme) 2.10.1, 5.2.7, 11.1.16, W.3

Materials, facilities and samples, Employer's failure to supply 12.4.53

Matters (see List of Matters)

Milestone 11.1.26, 11.1.37

Milestone Payment Chart 11.1.26

Mobilisation payments 11.1.27, 11.1.33

Monthly advances 11.1.25

Multiple currencies 12.1.19

Named persons
  contrasted with sub- contractors 9.1.5
  determination of appointment of 9.1.15
  objections to appointment of 9.1.12
  procedure for naming 9.1.6

NEC, background to 12.1.3

NEC/1 12.1.2, 12.1.13, 12.1.20, 12.2.19, 12.2.23, 12.2.32, 12.4.38

NEC/2
  consideration of 12.1.20
  legal response to 12.1.23
  structure and content of 12.1.14
  terminology of 12.2.1

NEC Engineering and construction subcontract 12.1.12, 12.1.15, 12.2.13, 12.7.25

New Engineering Contract Working Group
    12.1.4
Nominated Sub-Contractor
    delay on part of 2.8.45
    relationship with Contractor of 6.1.4
    rights against Contractor 6.1.12
Notifications 12.2.27

Opening up for inspection 2.8.28, 3.5.8,
    6.2.11, 7.3.16, 10.2.11, 12.4.31
Oral instructions 7.3.2, 9.3.2, 10.3.5, 11.3.3
Options
    Main 12.1.14
    Secondary 12.1.18
Out of pocket expenses definition of 11.3.24
Outstanding information 2.8.29, 2.8.38
Overheads and profit 4.8
Overtime 4.2.4, 4.10.4, 11.3.1

PACE (Property Advisors to the Civil Estate)
    11.1.2, 11.1.7, 11.1.13, 11.1.26,
    11.1.33
Parent Company Guarantee 11.1.33, 12.1.19
Performance bond 11.1.33, 12.1.19
Period for reply 12.2.25, 12.2.35, 12.4.12
Phased completion 2.9.3
Physical conditions 12.4.35
Planning Supervisor 2.8.70, 11.1.14, 11.2.13,
    11.3.42
Possession date 12.2.6
Possession of Site
    extension of time due to delay in 2.8.64,
        7.2.12, 10.2.12, 11.3.38, 12.4.4
    to be granted as stated 2.11.2
Possession, Date of 2.11.2
Postponement 2.8.24, 3.5.32, 4.7.13
Pre-construction period 10.1.5
Preliminaries 3.1.8, 3.3.10, 3.3.23, 4.6, 5.2.6,
    W.4, W.8.2, W.8.11
Prices reduced in response to a Compensation
    Event 12.7.30
Prices in Contract Bills
    consistency required in, 3.3.8, 3.3.10
    correction of errors in, 3.3.14
Prime cost 3.3.37, 6.1.1, 10.1.9
Project Manager 11.1.4, 11.1.14, 12.2.15,
    12.7.45

Professional Services Contract 12.1.12,
    12.2.13
Professional Team, definition of 10.1.7
Programme (see also Accepted programme
    and Master programme) 2.10, 5.2.7,
    12.2.34
    amendments to 2.10.1, 2.10.7, 11.1.20,
        11.1.37, 11.1.46, 11.2.4
    as a Contract Document 2.10.1, 11.1.16,
        11.1.18
    as an aid to considering extension of time
        2.7.14
    Completion Date in, shown in advance of
        that in Contract 2.8.43, 2.10.9
    float allowed in 2.7.15, 12.2.37, 12.7.36
    format of 2.10.8, 11.1.17
    items to be allowed for in 2.7.17
    Named person's refusal to comply with
        9.1.13
    Possession of site granted later than shown
        on 12.4.4
    requirements of 12.2.36
    status of 2.10.1
    to be co-ordinated with Nominated Sub-
        Contractor's programme 2.8.50
Progress meetings 11.1.21, 11.2.1, 11.2.5
Project changes 10.3.6
Project programme 10.1.6

Quantification 3.5.15, 4.2.1, 4.10.5, 5.3.3,
    W.6.1, W.6.3
Quantity Surveyor
    instructed by Architect to ascertain loss
        and/or expense 3.1.4, 3.5.26, 6.3.6
    role contrasted with that of Architect 3.1.4,
        3.4.9, 3.7
    role in claims under Management Contracts
        10.4.6
    role under clause 13A 3.4.4
    to agree Contract Cost Plan 10.1.8
    to determine work to be valued by daywork
        3.3.42
    to give Contractor opportunity to be
        present 3.3.23
    to inspect rates 3.3.8, 3.3.15
    to use judgment in determining similar
        conditions 3.3.26

INDEX

Quantity Surveyor *(cont.)*
  to value work where quotation under clause
    13A or 13.4.1.2A not accepted 3.1.5
  where no formal role 7.3.6, 8.1.3

Rainfall 12.4.37, 12.4.44
Reasonable skill and care 6.1.6, 11.1.32,
    12.1.19, 12.2.12
Relevant Events (see also List of Matters)
    2.7.34, 2.8, 3.5.3, 6.2.10, 7.2.9, 8.2.9,
    9.2.10, 10.2.10
Retention 12.1.19
Retention payment bond 11.1.27
Risk
  apportionment of 1.2.15
  shouldered by Contractor 1.2.14
  shouldered by Employer 1.2.14
RICS Survey of Contracts in Use 7.1.4,
    12.1.25

Schedule of Time Limits 11.1.12
Scotland, laws of do not apply 1.2.5
Searches instructed by Supervisor 12.4.31
Secondary options 12.1.18
Sectional completion 2.9.3, 12.1.19
Set-off 6.1.7, 10.1.35, 11.1.29
Settlement of claims
  approaches to 5.4
  negotiations for 5.4.15
Site information 12.2.5, 12.4.39
Site meetings 5.2.24, W.6.2
Specified perils 2.8.5
Starting date 12.1.15, 12.2.6
Statutory undertaker 2.8.61, W.6.2
Supervision 4.6
Supervisor 12.1.15, 12.2.17. 12.3.3, 12.4.30,
    12.4.33, 12.4.53

Taking over the Works 12.2.8, 12.2.11,
    12.4.51
Target contracts 12.1.15
Temperature 12.4.44
Termination of the Works 12.4.49
Terminology 12.2.1
Terms of contract

deletion of, 1.2.6
effect of bespoke conditions in 1.2.12
Test or inspection, delay caused by 12.4.33
Time, at large 2.1.4, 2.3.1, 12.4.55
Time risk allowances 12.2.36, 12.7.37
Trust fund 12.1.19

Unnecessary delay 12.4.34

Variations
  alternative means of valuing 11.3.11
  as grounds for extension of time 2.8.18
  Contractor to provide data for valuing
    11.3.21
  Contractor's quotation for valuing 3.4
  data required in valuing 11.3.21
  definition of 3.2.2, 11.3.6
  detailed rules for valuing 3.3.21
  dispute by Contractor of valuation of
    11.3.22
  effect of omissions upon valuation of 3.3.43
  lump sum quotations for 3.4, 11.3.14
  Quantity Surveyor's role in valuing 11.3.17
  ranges of options for valuing 3.1
  reduction in rate for valuing 3.3.44
  respective role of Architect and Quantity
    Surveyor in valuing, 3.1.4
  resulting in changed character of work
    3.3.26
  resulting in changed conditions of working
    3.3.26
  rules for valuing, 3.1, 3.3, 11.3.18
  timetable for valuing, 11.3.23
  valuation to include cost of disruption and
    prolongation 11.3.19
  valuation at contract rates or at cost 3.1.3,
    3.1.11, 3.5.28, 3.7, 4.2.2, 11.3.18
  valuation by Alternatives A or B 3.3.2
  valuation of 6.3.2, 7.3.1, 8.3.1, 9.3.1,
    10.3.1, 11.3.11
  valuation under clause 13 or 26, 3.1.4, 3.7,
    8.3.7
  valuation on a daywork basis 3.3.35
Weather conditions exceptionally adverse
    2.8.4, 2.9.2, 7.2.11, 11.2.19, 12.4.43
Weather data 12.4.45